Plasmonic Sensors: A New Frontier in Nanotechnology

Plasmonic Sensors: A New Frontier in Nanotechnology

Editors

Samir Kumar
Sungkyu Seo

MDPI • Basel • Beijing • Wuhan • Barcelona • Belgrade • Manchester • Tokyo • Cluj • Tianjin

Editors

Samir Kumar
Korea University
Sejong
Republic of Korea

Sungkyu Seo
Korea University
Sejong
Republic of Korea

Editorial Office

MDPI
St. Alban-Anlage 66
4052 Basel, Switzerland

This is a reprint of articles from the Special Issue published online in the open access journal *Biosensors* (ISSN 2079-6374) (available at: https://www.mdpi.com/journal/biosensors/special_issues/plasm_sensors).

For citation purposes, cite each article independently as indicated on the article page online and as indicated below:

LastName, A.A.; LastName, B.B.; LastName, C.C. Article Title. *Journal Name* **Year**, *Volume Number*, Page Range.

ISBN 978-3-0365-7284-0 (Hbk)
ISBN 978-3-0365-7285-7 (PDF)

Contents

About the Editors

Samir Kumar

Samir Kumar, Ph.D., is a research professor (since 2021) at the Nano Biophotonics Lab, Korea University (Sejong, Korea). His previous academic positions were at Kyoto University (Japan, 2018–2021). He received his M.Sc. in physics from the University of Kalyani, India, in 2009, and his M.Tech. from the National Institute of Technology Hamirpur, India, in 2012. He received his Ph.D. in physics from the Indian Institute of Technology Delhi, India, in 2017. His research interests include designing, fabricating, and characterizing nano- and micro-electromechanical systems for sensor and actuator applications. He has authored more than 50 publications in scientific journals and proceedings.

Sungkyu Seo

Sungkyu Seo, Ph.D., is a professor (since 2009) in the Department of Electronics and Information Engineering at Korea University (Sejong, Korea). His previous academic positions were at UCLA (C.A., USA, 2008-2009). He is also an Adjunct Professor (since 2017) at the Department of Laboratory Medicine, Korea University (Guro Hospital, Seoul, Korea), and BK21 Plus Team Director (since 2014) at ICT Convergence Technology for Health & Safety, Korea University (Sejong, Korea). He received his B.S. degree in 1998 from Korea University (Seoul, Korea) and his Ph.D. degree in 2007 from Texas A&M University (TX, USA), all in electrical engineering. His previous academic positions were at UCLA (C.A., USA, 2008-2009). His research interests include bio photonics, nano optics, and MEMS/NEMS. He has published about 100 papers in scientific journals and proceedings.

biosensors

MDPI

Editorial

Plasmonic Sensors: A New Frontier in Nanotechnology

Samir Kumar * and Sungkyu Seo *

Department of Electronics and Information Engineering, Korea University, Sejong 30019, Republic of Korea
* Correspondence: skumar@korea.ac.kr (S.K.); sseo@korea.ac.kr (S.S.)

Plasmonics is the study of surface plasmons formed by the interaction of incident light with electrons to form a surface-bound electromagnetic wave [1]. Plasmons can configure light to nanoscale volumes, making them attractive for developing new technologies with a wide range of applications [2]. The intense fields created by plasmons can dramatically enhance various light–matter interactions, such as transmission [3], Raman scattering [4], and fluorescence [5], and can also lead to significant localized heating in metallic nanostructures, known as thermoplasmonics [6,7].

Surface plasmons are formed when the oscillation of free electrons on the surface of a metallic nanoparticle resonates with the incident light [8]. The resonance frequency of surface plasmons depends on the size, shape, and composition of the metallic nanoparticle, as well as on the dielectric properties of the surrounding medium [9]. Controlling the plasmon resonance frequency and tailoring the plasmon properties for specific applications is made possible by tuning these parameters. Applications of plasmonics can be found in nanotechnology, biophotonics, sensing, biochemistry, and medicine [10]. However, plasmonics has proven particularly useful in the development of biosensors [11]. The Scopus database contains 37,586 articles related to biosensors that have been published since 2018 (as of February 2023). Twenty-five percent of these studies contain plasmonics-related keywords.

Biosensors are devices that use a variety of physical, chemical, optical, electrochemical, and thermal processes to convert biological interactions into valuable information [12,13]. A biosensor has two components: a transducer and a bioreceptor. Bioreceptors are biomolecules that recognize the target analyte, whereas transducers convert this recognition into measurable signals [14]. Electrochemical and electrical transducers are at the forefront of revolutionizing biosensor technology. However, these systems are limited by their dependence on oxidizing/reducing agents [15]. Optical biosensors based on optical transducers play a prominent role in developing advanced biosensors due to their high detection accuracy and cable-free design [16]. Optical biosensors are capable of multiplex detection and remote sensing. Sensors based on absorption, surface plasmon resonance (SPR), and photoluminescence provide a next-generation optical biosensor platform [17–19]. There are several optical biosensing techniques which utilize plasmons, including surface-enhanced Raman spectroscopy (SERS), surface-enhanced fluorescence spectroscopy (SEFS), and surface-enhanced infrared absorption spectroscopy (SEIRAS) [20–22].

This *Biosensors* Special Issue focuses on the theory and fabrication of plasmonic nanostructures; patterned surfaces; and devices for SPR, SEFS, and SERS-based biosensors. This Issue contains five research articles, five review articles, and one perspective article covering various topics related to fabrication and recent developments in plasmonic biosensors.

Gaur et al. investigated the interplay between SPR and lossy mode resonance (LMR) in this Special Issue. They compared these two types of sensing techniques in terms of sensitivity, detection accuracy, and figure of merit [23]. The SPR effect is a complicated physical phenomenon that occurs when light strikes a conductive layer of noble metal. This phenomenon occurs at the boundary between a medium with a low refractive index (buffer) and a medium with a high refractive index (sensor glass surface) [24]. In most

Citation: Kumar, S.; Seo, S. Plasmonic Sensors: A New Frontier in Nanotechnology. *Biosensors* **2023**, *13*, 385. https://doi.org/10.3390/bios13030385

Received: 10 March 2023
Accepted: 13 March 2023
Published: 15 March 2023

plasmonic biosensors, the SPR effect detects the binding of biomolecules, such as proteins and DNA, to metallic nanoparticle surfaces. An interaction between a biomolecule and a nanoparticle changes the plasmon resonance frequency, which can be measured with a spectrophotometer. Unlike conventional biosensors, SPR-based biosensors offer several advantages, including label-free detection, high sensitivity, and real-time monitoring. In addition to SPR, LMR can also be used for optical sensing. LMR is a phenomenon that occurs when light is coupled into an optical waveguide and interacts with a lossy resonant mode of that waveguide [25]. The interaction between the light and the resonant mode results in a sharp dip in the transmission or reflection spectrum of the waveguide at the resonant wavelength. This effect is commonly observed in optical fibers, planar waveguides, and other structures supporting guided modes. LMR can be used in various applications, including sensing, filtering, and lasing. By detecting changes in the resonant wavelength, LMR sensors can measure changes in the refractive index, temperature, or other physical parameters in the waveguide environment [26]. The geometric configuration and the material supporting the SPR and LMR are key factors determining its performance. The authors investigated the properties of the bilayer (ITO + Ag structures) and trilayer (ITO + Ag + ITO structures) fiber-optic probes based on ITO, which enables simultaneous excitation of SPR and LMR. They found that LMR showed better sensitivity than SPR in the bilayer configuration. Moreover, as the thickness of the Ag layer increases, SPR depression becomes insensitive, and only LMR depression can be used for detection. Taking advantage of this property, a self-referencing sensor using the SPR depression as a reference point was proposed. In the three-layer configuration, they found that the first LMR depression in the visible region was less sensitive than the SPR depression when the thickness of the outermost ITO was varied. Conversely, when the thickness of Ag was varied, the resonance wavelength of the SPR depression shifted to the shorter wavelength side; however, the resonance wavelength of the LMR depression shifted to the longer wavelength side. The proposed sensor had a high sensitivity of 14 μm/RIU, good detection accuracy, and a Q-factor.

Label-free detection of biomolecules by localized surface plasmon resonance (LSPR) has excellent potential for point-of-care (POC) testing. SPR occurs at the metal–dielectric interface and is sensitive to changes in the refractive index of the dielectric layer. In contrast, LSPR occurs at the surface of metallic nanoparticles and is sensitive to changes in the local environment around the nanoparticle. LSPR results from the confinement of surface plasmons in nanoparticles (with a size comparable to the wavelength of the light used to excite the plasmons) [27]. In developing LSPR-based POC devices, a key challenge is to produce large-scale LSPR substrates that are reproducible and have high throughput. In a study included in this Special Issue, Kim et al. fabricated wafer-scale LSPR substrates using reproducible high-throughput techniques such as nanoimprint lithography, wet etching, and the glancing angle deposition (GLAD) technique [28]. First, they fabricated hard masks of SiO_2 nanodots on a transparent sapphire wafer using nanoimprint lithography, which was anisotropically etched with a solution of H_2SO_4 and H_3PO_4, resulting in a patterned sapphire substrate (PSS). Finally, they fabricated an LSPR substrate with Au on the PSS by GLAD and used it to detect biomolecule binding processes without labeling. The GLAD technique allowed for the formation of Au nanostructures on the PSS due to the PSS's three-dimensional structure (triangular–pyramidal). The Au nanostructures formed by the GLAD technique showed a red-shifted LSPR peak compared to those formed by vertical deposition. This suggests that the LSPR properties can be controlled by the shape of the PSS and deposition conditions, such as the PSS's angle and rotation.

GLAD is a physical vapor deposition technique used to fabricate complex nanostructures and thin films with controlled porosity and morphology [29,30]. In this technique, the material is deposited obliquely onto a substrate at a low angle of incidence, typically between 70° and 87°, resulting in highly anisotropic structures with directional growth [31]. Nanostructures fabricated by the GLAD technique exhibit high sensitivity, enhanced optical and catalytic properties, periodicity, and controlled morphology, which makes them attractive for sensory applications [32,33]. In another study in this Special Issue, Yadav et al. provided a detailed overview of recent advances in various nanostructures fabricated via GLAD and their applications in the biomedical field [34]. In addition to discussing the various ways to fabricate nanostructures using the GLAD technique, they also considered the assembly configuration, the effects of different growth parameters, and the advantages of GLAD-based nanostructures over conventional nanoparticles and substrates. In addition, the authors highlighted several advantages of GLAD compared to other chemical deposition processes. For example, GLAD does not require precursor materials, which provides safety from toxic precursors and by-products. The chemical composition and thickness in this deposition can be controlled at the atomic level. Higher temperatures are not required, so this technique can produce heat-sensitive substrates.

Of all the metals reported for plasmonic sensors, Ag and Au are considered the most useful plasmonic materials. They show the naturally lowest ohmic losses at optical frequencies due to their strong plasmon resonances in the visible range, their biocompatibility, and their stability [35,36]. In this Special Issue, Gahlaut et al. presented highlights of recent developments in the field of nanostructured Ag substrates for plasmonic sensing, with a focus on SPR and SERS, over the past decade [37]. They mainly focused on the chemical methods of solution phase synthesis and physical methods such as vapor phase deposition, GLAD, and lithographic techniques for synthesizing Ag nanostructures ranging from nanoparticles to nanocubes, nanotriangles, nanorods, and nanowires. In addition, the authors also discuss the latest spectroscopic techniques, focusing on plasmonic enhancement of biosensing methods such as SPR/LSPR, SERS, SEF, and SEIRAS using Ag as plasmonic material.

Metallic nanoparticles, especially Au, have been widely used in plasmonics due to their excellent plasmonic properties. However, the discovery of two-dimensional (2D) nanomaterials, such as MXenes, has revealed new possibilities for plasmonic applications [38,39]. MXenes are a class of 2D transition metal carbides, nitrides, and carbonitrides that have recently shown promise as materials for plasmonics [40]. They have excellent electrical conductivity, mechanical strength, and chemical stability, which makes them attractive for various applications. One of the advantages of MXenes over metallic nanoparticles is their tunability. The plasmonic properties of MXenes can be tuned in the mid-IR to THz regions of the spectrum, and their optical constants (such as the dielectric constant and refractive index) can be tuned by controlling their chemical composition, surface functionalization, number of layers, and morphology [41]. For example, the wavelength of the plasmon resonance of MXenes can be tuned by changing the thickness or the type of functional group on the surface [42]. MXenes have shown great potential in various plasmonic applications, including biosensing, imaging, and photovoltaics. They can also be integrated with other materials, such as polymers or nanoparticles, to create hybrid plasmonic systems with enhanced optical properties. Some researchers have coined the term "smart MXene" to describe MXene-based hybrid materials that exhibit unique properties related to their applications [43,44]. Smart MXene quantum dots (SMQDs), a new and rapidly emerging class of nanomaterials, are tantalizing candidates for SPR biosensor development given their intriguing optical properties, which include light absorption, photoluminescence, and electrochemiluminescence [43]. In their study, contained in this Special Issue, Mousavi et al. discussed recent advances in ultra-sensitive SPR nano biosensors based on SMQDs [45]. They began with an introduction to SPR and SPR biosensors. Then, they explained and discussed the different types of SMQDs and SPR biosensors based on SMQDs. Finally, they presented the useful characteristics of SMQDs for developing SPR biosensors and their biomedical applications, and discussed the current limitations of these biosensors. Overall, MXene plasmonics is a rapidly growing field with promising prospects for various applications, and it is expected to impact future plasmonic sensors' development significantly.

In this Special Issue, Park et al. discussed innovative advances in biosensors led by DNA nanotechnology [46]. Nanotechnology has made considerable progress in the last decade, allowing us to overcome the limitations of using DNA as the sole form of genetic material and thus develop a new method for constructing biosensors [47]. DNA aptamers can bind to a specific target with a strong affinity. DNA aptamers and enzymes can be produced via the systematic evolution of ligands by exponential enrichment (SELEX), and can bind to specific targets [48]. Aptamers are short, single-stranded DNA or RNA molecules that can bind to specific targets with high affinity and selectivity, similar to antibodies. Various research groups have presented plasmonic biosensing strategies based on aptamers [49]. In aptamer-based plasmonic sensors, the aptamer is immobilized on a metal surface, such as gold or silver. The binding event between the aptamer and the target molecule causes a change in the optical properties of the metal surface, which can be detected by various optical methods. The optical signal can be measured by various methods, e.g., fluorescence, absorbance, and SPR [49]. This review highlighted recent advances in the development of nucleic acids for the purpose of constructing plasmonic biosensors through studies published over the past five years. The authors discussed applying simple aptamers and origami-shaped structures in SPR/LSPR, SERS, and SEF sensing. They discussed several aspects that should be considered to improve the sensing process, including the nonspecific adsorption of unwanted molecules and the fabrication of homogeneous plasmon surfaces. The combination of aptamers and plasmonics has great potential to improve the sensitivity and selectivity of biosensors and enable the development of new diagnostic tools for various applications.

Plasmonics has made significant advances in the field of SERS, a highly effective analytical technique for identifying and detecting molecules [33,50]. SERS significantly amplifies Raman scattering signals when molecules are near plasmonic surfaces [4]. This enhancement is due to the excitation of LSP in metallic nanostructures, which generate strong electromagnetic fields that interact with molecules. SERS is a surface-sensitive technique that can detect analytes even at the level of single molecules [51]. In this Special Issue, Beeram et al. provide an excellent overview of recent trends in SERS-based plasmonic sensors for disease diagnostics, biomolecule detection, and machine learning techniques [52]. The authors review the work of the past decade and use simplified language to address the needs of an interdisciplinary audience. In the first section, the authors discuss the need for plasmonic sensors in biology and the advantages of SERS over existing technologies. Next, the authors discuss the use of SERS-based biosensors for disease diagnosis, focusing on cancer and respiratory disease detection, such as the recent detection of severe acute respiratory syndrome coronavirus 2 (SARS-CoV-2). They then discuss advances in the detection of microorganisms, such as bacteria, with particular emphasis on plasmonic sensors for the detection of biohazardous materials. Finally, the authors discuss machine learning techniques for identifying, classifying, and quantifying biological signals using SERS. They discuss the various machine learning models that have been developed for the trace detection, signal variation, quantification, and identification of SERS signals, including principal component analysis, support vector machines, partial least squares, decision trees, and convolutional neural networks. Considering that SERS is a complex system with many variables, machine learning techniques can help to identify patterns that no experts can identify.

The development of plasmonic biosensors has revolutionized not only the field of biomedical research and diagnostics, but also food safety and environmental research [53]. This Special Issue also contains research on biosensors for food and environmental contaminants such as mercury [54], tetrabromobisphenol A (TBBPA) [55], and thiamethoxam [56]. Kim et al. developed an Au nanoparticle (AuNP)-based immunochromatographic lateral flow assay (ICA) using the light scattering phenomenon of nanoparticles to rapidly detect mercury in rice, with a detection limit of 20 ng g^{-1} and a cut-off value of 500 ng g^{-1} [54]. The proposed ICA strip can be used to qualitatively determine mercury in rice, and the results agree with conventional instrumental methods such as inductively coupled plasma mass spectrometry. Based on aptamer technology, Yue et al. developed a biosensor system to detect thiamethoxam residues [56]. They developed a colorimetric sensor based on AuNPs to detect thiamethoxam in tea leaves. To screen ssDNA aptamers that bind specifically to thiamethoxam, they used graphene oxide SELEX (GO-SELEX) technology. They increased the amount of GO in the screening process to increase the screening pressure. The researchers applied the aptasensor system to the actual detection of samples. They achieved recoveries of 96.94% to 105.86% and RSD values of 0.41% to 3.76%, indicating that the aptasensor can be used for the rapid and sensitive detection of thiamethoxam residues. In another work, Dai et al. developed a hydrophobic Cu–Ag chip for SERS detection of TBBPA [57]. TBBPA is one of the world's most widely used brominated flame retardants, but its production and use can affect the environment and human health. Recently, there has been much interest in hydrophobic materials due to their wide range of applications in metal corrosion protection, self-cleaning, oil–water separation, and SERS [58,59]. According to Dai et al., the Cu–Ag chip's hydrophobicity increases the substrate's affinity for TBBPA, allowing TBBPA to approach the surface of the SERS substrate and subsequently combine with it via Ag–Br interactions. First, the researchers fabricated the hydrophobic copper-coated fabric with an ordered micro-nanostructure. Then, they constructed a hydrophobic Cu–Ag chip by introducing Ag onto the Cu surface through an exchange reaction. Combined with ultrasound-assisted extraction (UAE), they succeeded in quantitatively detecting TBBPA in electronics. After UAE, they extracted and determined the TBBPA content in the electronics with a detection limit of 2.0 mg kg^{-1} (0.01 mg L^{-1} for the TBBPA solution). SERS offers several advantages over conventional methods such as high-performance liquid chromatography (HPLC), including speed, convenience, and sensitivity.

The COVID-19 pandemic, caused by the novel coronavirus SARS-CoV-2, has resulted in a global public health crisis, affecting millions of people worldwide. Rapid, accurate, and sensitive diagnostic methods are critical for optimal patient care. Recently, SERS has emerged as a promising point-of-care testing technique for detecting various analytes, including viruses [60]. Unlike conventional methods such as polymerase chain reaction, which is limited to the analysis of genetic material, SERS offers high specificity and can detect a wide range of analytes. Furthermore, flexible SERS substrates are becoming increasingly important in practical application research due to their robustness and versatility [61]. In this Special Issue, Mousavi et al. evaluated the importance of the flexible SERS substrate for detecting SARS-CoV-2 [62]. They provided an overview of the flexible SERS substrates used to detect different subtypes of the novel SARS-CoV-2 coronavirus and discussed how this method has the potential to become a point-of-care diagnostic tool. In addition to discussing recent advances in SERS-based COVID-19 detection, they discussed the principles of SERS and their amplification mechanisms, the detection of analytes, and the use of multiplex analysis to detect coronaviruses.

In summary, plasmonic biosensors represent a promising technology for the label-free and sensitive detection of biomolecules. The sensitivity, specificity, and label-free detection offered by plasmonic biosensors make them an attractive alternative to conventional detection methods. This Special Issue on "Plasmonic Biosensors", published in *Biosensors*, provides a comprehensive overview of recent developments and applications in the field of plasmonic biosensors. The research articles present their original work in developing plasmonic biosensors for detecting various biomolecules, including DNA. In addition, this Special Issue highlights the significant advances in the development of plasmonic biosensors and their potential applications in various fields. This Special Issue also highlights the importance of interdisciplinary research and collaboration in developing plasmonic biosensors. Researchers from diverse backgrounds, including physics, chemistry, and biology, have contributed to progress in this field.

In conclusion, this Special Issue on "Plasmonic Biosensors" provides an excellent platform for researchers to present their latest work and exchange ideas regarding the future directions of plasmonic biosensors. The articles in this Issue demonstrate the enormous potential of plasmonic biosensors for various applications, and we can expect to see further advances in this field. As researchers continue to explore the fundamental principles of plasmonics and SERS, we can expect more exciting developments in this field in the coming years.

Funding: This study was supported by the Basic Science Research Program (Grant#: 2020R1A2C1012109) through the National Research Foundation (NRF) of Korea. This work was also supported by the Korea Medical Device Development Fund grant funded by the Korean Government (the Ministry of Science and ICT, the Ministry of Trade, Industry, and Energy, the Ministry of Health & Welfare, the Ministry of Food and Drug Safety) (Grant#: RS-2020-KD000142) and the Korea Institute of Marine Science & Technology Promotion (KIMST) funded by the Ministry of Oceans and Fisheries, Korea (Grant#: 20210660).

Acknowledgments: The authors are grateful for the opportunity to serve as the guest editors of the Special Issue "Plasmonic Sensors: A New Frontier in Nanotechnology", as well as for the contribution of all the authors to this Special Issue. The dedicated work of the Special Issue Editor of *Biosensors* and the editorial and publishing staff of *Biosensors* is greatly appreciated.

Conflicts of Interest: The authors declare no conflict of interest.

References

1. Maier, S.A. *Plasmonics;* Springer: New York, NY, USA, 2008; ISBN 9780387512938.
2. Lee, B.; Lee, I.-M.; Kim, S.; Oh, D.-H.; Hesselink, L. Review on subwavelength confinement of light with plasmonics. *J. Mod. Opt.* **2010**, *57*, 1479–1497. [CrossRef]
3. Ebbesen, T.W.; Lezec, H.J.; Ghaemi, H.F.; Thio, T.; Wolff, P.A. Extraordinary Optical Transmission through Sub-Wavelength Hole Arrays. *Nature* **1998**, *391*, 667–669. [CrossRef]
4. Moskovits, M. Surface-enhanced Raman spectroscopy: A brief retrospective. *J. Raman Spectrosc.* **2005**, *36*, 485–496. [CrossRef]
5. Geddes, C.D. (Ed.) *Surface Plasmon Enhanced, Coupled and Controlled Fluorescence*, 1st ed.; John Wiley & Sons: Nashville, TN, USA, 2017; ISBN 9781118027936.
6. Baffou, G.; Cichos, F.; Quidant, R. Applications and challenges of thermoplasmonics. *Nat. Mater.* **2020**, *19*, 946–958. [CrossRef]
7. Namura, K.; Okai, S.; Kumar, S.; Nakajima, K.; Suzuki, M. Self-oscillation of Locally Heated Water Vapor Microbubbles in Degassed Water. *Adv. Mater. Interfaces* **2020**, *7*, 2000483. [CrossRef]
8. Noguez, C. Surface Plasmons on Metal Nanoparticles: The Influence of Shape and Physical Environment. *J. Phys. Chem. C* **2007**, *111*, 3806–3819. [CrossRef]
9. Liz-Marzán, L.M. Tailoring Surface Plasmons through the Morphology and Assembly of Metal Nanoparticles. In *Colloidal Synthesis of Plasmonic Nanometals;* Jenny Stanford Publishing: Singapore, 2020; pp. 131–158, ISBN 9780429295188.
10. Wang, X.; Huang, S.-C.; Hu, S.; Yan, S.; Ren, B. Fundamental Understanding and Applications of Plasmon-Enhanced Raman Spectroscopy. *Nat. Rev. Phys.* **2020**, *2*, 253–271. [CrossRef]
11. Soler, M.; Lechuga, L.M. Principles, Technologies, and Applications of Plasmonic Biosensors. *J. Appl. Phys.* **2021**, *129*, 111102. [CrossRef]
12. Suvarnaphaet, P.; Pechprasarn, S. Graphene-Based Materials for Biosensors: A Review. *Sensors* **2017**, *17*, 2161. [CrossRef]
13. Seo, D.; Han, E.; Kumar, S.; Jeon, E.; Nam, M.-H.; Jun, H.S.; Seo, S. Field-Portable Leukocyte Classification Device Based on Lens-Free Shadow Imaging Technique. *Biosensors* **2022**, *12*, 47. [CrossRef] [PubMed]
14. Fatoyinbo, H.O.; Hughes, M.P. Biosensors. In *Encyclopedia of Nanotechnology;* Bhushan, B., Ed.; Springer: Dordrecht, The Netherlands, 2012; pp. 329–345, ISBN 9789048197514.
15. Tereshchenko, A.; Bechelany, M.; Viter, R.; Khranovskyy, V.; Smyntyna, V.; Starodub, N.; Yakimova, R. Optical Biosensors Based on ZnO Nanostructures: Advantages and Perspectives. A Review. *Sens. Actuators B Chem.* **2016**, *229*, 664–677. [CrossRef]
16. Coulet, P.R.; Blum, L.J. *Biosensor Principles and Applications;* CRC Press: Boca Raton, FL, USA, 2019; ISBN 9781000715675.
17. Kumar, S.; Namura, K.; Suzuki, M. Proposal for a gel-based SERS sensor. *Proc. SPIE* **2019**, *10894*, 1089414. [CrossRef]
18. Kumar, S.; Fukuoka, T.; Takahashi, R.; Yoshida, M.; Utsumi, Y.; Yamaguchi, A.; Namura, K.; Suzuki, M. Highly stable and reproducible Au nanorod arrays for near-infrared optofluidic SERS sensor. *Mater. Lett.* **2021**, *286*, 129106. [CrossRef]
19. Park, J.-H.; Cho, Y.-W.; Kim, T.-H. Recent Advances in Surface Plasmon Resonance Sensors for Sensitive Optical Detection of Pathogens. *Biosensors* **2022**, *12*, 180. [CrossRef] [PubMed]
20. Kumar, S.; Kanagawa, M.; Namura, K.; Fukuoka, T.; Suzuki, M. Multilayer thin-film flake dispersion gel for surface-enhanced Raman spectroscopy. *Appl. Nanosci.* **2020**, *13*, 155–163. [CrossRef]
21. Kumar, S.; Kumar, P.; Das, A. Surface-Enhanced Raman Scattering: Introduction and Applications. In *Recent Advances in Nanophotonics-Fundamentals and Applications;* IntechOpen: London, UK, 2020; pp. 137–160.
22. Soler, M.; Huertas, C.S.; Lechuga, L.M. Label-Free Plasmonic Biosensors for Point-of-Care Diagnostics: A Review. *Expert Rev. Mol. Diagn.* **2019**, *19*, 71–81. [CrossRef]
23. Gaur, D.S.; Purohit, A.; Mishra, S.K.; Mishra, A.K. An Interplay between Lossy Mode Resonance and Surface Plasmon Resonance and Their Sensing Applications. *Biosensors* **2022**, *12*, 721. [CrossRef]

24. Gosu, R.; Zaheer, S.M. Introduction to Surface Plasmon Resonance. In *Methods for Fragments Screening Using Surface Plasmon Resonance*; Zaheer, S.M., Gosu, R., Eds.; Springer: Singapore, 2021; pp. 1–4, ISBN 9789811615368.
25. Chiavaioli, F.; Janner, D. Fiber Optic Sensing with Lossy Mode Resonances: Applications and Perspectives. *J. Light. Technol.* **2021**, *39*, 3855–3870. [CrossRef]
26. Śmietana, M.; Koba, M.; Sezemsky, P.; Szot-Karpińska, K.; Burnat, D.; Stranak, V.; Niedziółka-Jönsson, J.; Bogdanowicz, R. Simultaneous Optical and Electrochemical Label-Free Biosensing with ITO-Coated Lossy-Mode Resonance Sensor. *Biosens. Bioelectron.* **2020**, *154*, 112050. [CrossRef]
27. Prabowo, B.A.; Purwidyantri, A.; Liu, K.-C. Surface Plasmon Resonance Optical Sensor: A Review on Light Source Technology. *Biosensors* **2018**, *8*, 80. [CrossRef] [PubMed]
28. Kim, K.; Lee, K.J.; Jo, N.R.; Jo, E.-J.; Shin, Y.-B.; Kim, M.-G. Wafer-Scale LSPR Substrate: Oblique Deposition of Gold on a Patterned Sapphire Substrate. *Biosensors* **2022**, *12*, 158. [CrossRef]
29. Hawkeye, M.M.; Taschuk, M.T.; Brett, M.J. *Glancing Angle Deposition of Thin Films: Engineering the Nanoscale*; John Wiley & Sons: Hoboken, NJ, USA, 2014; ISBN 9781118847336.
30. Kumar, S.; Gahlaut, S.K.; Singh, J.P. Sculptured Thin Films: Overcoming the Limitations of Surface-Enhanced Raman Scattering Substrates. *Appl. Surf. Sci. Adv.* **2022**, *12*, 100322. [CrossRef]
31. Namura, K.; Hanai, S.; Kondo, S.; Kumar, S.; Suzuki, M. Gold Micropetals Self-assembled by Shadow-sphere Lithography for Optofluidic Control. *Adv. Mater. Interfaces* **2022**, *9*, 2200200. [CrossRef]
32. Ai, B.; Zhao, Y. Glancing angle deposition meets colloidal lithography: A new evolution in the design of nanostructures. *Nanophotonics* **2018**, *8*, 1–26. [CrossRef]
33. Kumar, S.; Doi, Y.; Namura, K.; Suzuki, M. Plasmonic nanoslit arrays fabricated by serial bideposition: Optical and surface-enhanced Raman scattering study. *ACS Appl. Bio Mater.* **2020**, *3*, 3226–3235. [CrossRef]
34. Yadav, S.; Senapati, S.; Kumar, S.; Gahlaut, S.K.; Singh, J.P. GLAD Based Advanced Nanostructures for Diversified Biosensing Applications: Recent Progress. *Biosensors* **2022**, *12*, 1115. [CrossRef]
35. Baburin, A.S.; Merzlikin, A.M.; Baryshev, A.V.; Ryzhikov, I.A.; Panfilov, Y.V.; Rodionov, I.A. Silver-Based Plasmonics: Golden Material Platform and Application Challenges [Invited]. *Opt. Mater. Express* **2019**, *9*, 611. [CrossRef]
36. Kumar, S.; Namura, K.; Kumaki, D.; Tokito, S.; Suzuki, M. Highly reproducible, large scale inkjet-printed Ag nanoparticles-ink SERS substrate. *Results Mater.* **2020**, *8*, 100139. [CrossRef]
37. Gahlaut, S.K.; Pathak, A.; Gupta, B.D. Recent Advances in Silver Nanostructured Substrates for Plasmonic Sensors. *Biosensors* **2022**, *12*, 713. [CrossRef]
38. Huang, S.; Song, C.; Zhang, G.; Yan, H. Graphene Plasmonics: Physics and Potential Applications. *Nanophotonics* **2016**, *6*, 1191–1204. [CrossRef]
39. Elbanna, A.; Jiang, H.; Fu, Q.; Zhu, J.-F.; Liu, Y.; Zhao, M.; Liu, D.; Lai, S.; Chua, X.W.; Pan, J.; et al. 2D Material Infrared Photonics and Plasmonics. *ACS Nano* **2023**, *17*, 4134–4179. [CrossRef] [PubMed]
40. VahidMohammadi, A.; Rosen, J.; Gogotsi, Y. The World of Two-Dimensional Carbides and Nitrides (MXenes). *Science* **2021**, *372*, eabf1581. [CrossRef]
41. Ma, Q.; Ren, G.; Xu, K.; Ou, J.Z. Tunable optical properties of 2D materials and their applications. *Adv. Opt. Mater.* **2021**, *9*, 2001313. [CrossRef]
42. Iqbal, M.A.; Malik, M.; Shahid, W.; Ahmad, W.; Min-Dianey, K.A.A.; Pham, P.V. Plasmonic 2D Materials: Overview, Advancements, Future Prospects and Functional Applications. In *21st Century Nanostructured Materials: Physics, Chemistry, Classification, and Emerging Applications in Industry, Biomedicine, and Agriculture*; IntechOpen: London, UK, 2022; pp. 47–68.
43. Iravani, S.; Varma, R.S. Smart MXene Quantum Dot-Based Nanosystems for Biomedical Applications. *Nanomaterials* **2022**, *12*, 1200. [CrossRef] [PubMed]
44. Luo, J.; Gao, S.; Luo, H.; Wang, L.; Huang, X.; Guo, Z.; Lai, X.; Lin, L.; Li, R.K.Y.; Gao, J. Superhydrophobic and Breathable Smart MXene-Based Textile for Multifunctional Wearable Sensing Electronics. *Chem. Eng. J.* **2021**, *406*, 126898. [CrossRef]
45. Mousavi, S.M.; Hashemi, S.A.; Kalashgrani, M.Y.; Rahmanian, V.; Gholami, A.; Chiang, W.-H.; Lai, C.W. Biomedical Applications of an Ultra-Sensitive Surface Plasmon Resonance Biosensor Based on Smart MXene Quantum Dots (SMQDs). *Biosensors* **2022**, *12*, 743. [CrossRef]
46. Park, J.A.; Amri, C.; Kwon, Y.; Lee, J.-H.; Lee, T. Recent Advances in DNA Nanotechnology for Plasmonic Biosensor Construction. *Biosensors* **2022**, *12*, 418. [CrossRef] [PubMed]
47. Raveendran, M.; Lee, A.J.; Sharma, R.; Wälti, C.; Actis, P. Rational Design of DNA Nanostructures for Single Molecule Biosensing. *Nat. Commun.* **2020**, *11*, 4384. [CrossRef]
48. Komarova, N.; Kuznetsov, A. Inside the Black Box: What Makes SELEX Better? *Molecules* **2019**, *24*, 3598. [CrossRef]
49. Chang, C.-C. Recent Advancements in Aptamer-Based Surface Plasmon Resonance Biosensing Strategies. *Biosensors* **2021**, *11*, 233. [CrossRef]
50. Lai, H.; Xu, F.; Wang, L. A Review of the Preparation and Application of Magnetic Nanoparticles for Surface-Enhanced Raman Scattering. *J. Mater. Sci.* **2018**, *53*, 8677–8698. [CrossRef]
51. Xu, H.; Aizpurua, J.; Kall, M.; Apell, P. Electromagnetic contributions to single-molecule sensitivity in surface-enhanced raman scattering. *Phys. Rev. E Stat. Phys. Plasmas Fluids Relat. Interdiscip. Top.* **2000**, *62*, 4318–4324. [CrossRef] [PubMed]

52. Beeram, R.; Vepa, K.R.; Soma, V.R. Recent Trends in SERS-Based Plasmonic Sensors for Disease Diagnostics, Biomolecules Detection, and Machine Learning Techniques. *Biosensors* **2023**, *13*, 328. [CrossRef]
53. King, M.E.; Wang, C.; Fonseca Guzman, M.V.; Ross, M.B. Plasmonics for Environmental Remediation and Pollutant Degradation. *Chem Catal.* **2022**, *2*, 1880–1892. [CrossRef]
54. Lv, S.; Xu, X.; Song, S.; Xu, L.; Liu, L.; Xu, C.; Kuang, H. An Immunochromatographic Assay for the Rapid and Qualitative Detection of Mercury in Rice. *Biosensors* **2022**, *12*, 694. [CrossRef]
55. Dai, P.; Huang, X.; Cui, Y.; Zhu, L. Quantitative SERS Detection of TBBPA in Electronic Plastic Based on Hydrophobic Cu-Ag Chips. *Biosensors* **2022**, *12*, 881. [CrossRef] [PubMed]
56. Yue, Y.; Zhang, D.; Tian, K.; Ni, D.; Guo, F.; Yu, Z.; Wang, P.; Liang, P. Screening and Evaluation of Thiamethoxam Aptamer Based on Pressurized GO-SELEX and Its Sensor Application. *Biosensors* **2023**, *13*, 155. [CrossRef]
57. Sunday, O.E.; Bin, H.; Guanghua, M.; Yao, C.; Zhengjia, Z.; Xian, Q.; Xiangyang, W.; Weiwei, F. Review of the Environmental Occurrence, Analytical Techniques, Degradation and Toxicity of TBBPA and Its Derivatives. *Environ. Res.* **2022**, *206*, 112594. [CrossRef]
58. Kumar, S.; Goel, P.; Singh, D.P.; Singh, J.P. Fabrication of superhydrophobic silver nanorods array substrate using glancing angle deposition. *AIP Conf. Proc.* **2014**, *1591*, 872. [CrossRef]
59. Ahmad, D.; van den Boogaert, I.; Miller, J.; Presswell, R.; Jouhara, H. Hydrophilic and Hydrophobic Materials and Their Applications. *Energy Sources Recovery Util. Environ. Eff.* **2018**, *40*, 2686–2725. [CrossRef]
60. Yadav, S.; Senapati, S.; Desai, D.; Gahlaut, S.; Kulkarni, S.; Singh, J.P. Portable and sensitive Ag nanorods based SERS platform for rapid HIV-1 detection and tropism determination. *Colloids Surf. B Biointerfaces* **2021**, *198*, 111477. [CrossRef] [PubMed]
61. Kumar, S.; Goel, P.; Singh, J.P. Flexible and robust SERS active substrates for conformal rapid detection of pesticide residues from fruits. *Sens. Actuators B Chem.* **2017**, *241*, 577–583. [CrossRef]
62. Mousavi, S.M.; Hashemi, S.A.; Rahmanian, V.; Kalashgrani, M.Y.; Gholami, A.; Omidifar, N.; Chiang, W.-H. Highly Sensitive Flexible SERS-Based Sensing Platform for Detection of COVID-19. *Biosensors* **2022**, *12*, 466. [CrossRef] [PubMed]

biosensors

MDPI

Article

An Interplay between Lossy Mode Resonance and Surface Plasmon Resonance and Their Sensing Applications

Deependra Singh Gaur [1,†], Ankit Purohit [1,†], Satyendra Kumar Mishra [2,*] and Akhilesh Kumar Mishra [1]

1 Department of Physics, Indian Institute of Technology Roorkee, Roorkee 247667, India
2 Department of Electrical and Computer Engineering, Laval University, Quebec City, QC G1V 0A6, Canada
* Correspondence: satyendramishraiitd@gmail.com
† These authors contributed equally to this work.

Abstract: Conducting metal oxide (CMO) supports lossy mode resonance (LMR) at the CMO-dielectric interface, whereas surface plasmon resonance (SPR) occurs at the typical plasmonic metal-dielectric interface. The present study investigates these resonances in the bi-layer (ITO + Ag) and tri-layer (ITO + Ag + ITO) geometries in the Kretschmann configuration of excitation. It has been found that depending upon the layer thicknesses one resonance dominates the other. In particular, in the tri-layer configuration of ITO + Ag + ITO, the effect of the thickness variation of the sandwiched Ag layer is explored and a resonance, insensitive to the change in the sensing medium refractive index (RI), has been reported. Further, the two kinds of RI sensing probes and the supported resonances have been characterized and compared in terms of sensitivity, detection accuracy and figure of merit. These studies will not only be helpful in gaining a better understanding of underlying physics but may also lead to the realization of biochemical sensing devices with a wider spectral range.

Keywords: optical fiber sensor; surface plasmon resonance; lossy mode resonance; indium-tin oxide; silver; sensitivity; detection accuracy

Citation: Gaur, D.S.; Purohit, A.; Mishra, S.K.; Mishra, A.K. An Interplay between Lossy Mode Resonance and Surface Plasmon Resonance and Their Sensing Applications. *Biosensors* **2022**, *12*, 721. https://doi.org/10.3390/bios12090721

Received: 30 June 2022
Accepted: 2 September 2022
Published: 4 September 2022

Publisher's Note: MDPI stays neutral with regard to jurisdictional claims in published maps and institutional affiliations.

1. Introduction

Surface plasmon resonance (SPR) is generated at metal and dielectric interface by transverse magnetic fields (TM) or p-polarized light [1]. It is impossible to excite the SPR mode by direct light due to the momentum mismatch between the SPR mode and incident light [2]. In order to efficiently excite these modes, we need a momentum matching scheme. Several such schemes have been proposed, e.g., passing light through a high RI prism, using a grating, etc. [3]. Further, there exist two coupling configurations for SPR excitation—Otto and Kretschmann [4,5]. Owing to its ease of implementation, the Kretschmann configuration is preferred often.

The growing field of SPR has attracted significant research attention over the years due to its wide range of applications, which includes nano-antennas [6], imaging [7], biosensing, and so on and so forth [8]. Several extensive theoretical and experimental studies have been conducted on SPR-based sensors in the past [9–13]. A sensor's novelty is determined by the particulars of the plasmonic material used and the design implemented. Waveguide-based sensors have attracted a lot of attention due to their industrial applications. These sensors use a plasmonic material deposited as a thin film around the waveguide (e.g., an optical fiber). These materials can be classified into three categories based on the resonances they support. The first class of materials is plasmonic materials, which support SPR and have a real permittivity that is negative and larger in magnitude than both the imaginary permittivity and the permittivity of the surrounding medium. In the second category of materials, the real part of the material permittivity is positive and greater than both its imaginary part and the permittivity of the surrounding medium. The LMR phenomenon is observed in this category of materials. The third class of materials also exists for which the real part of the permittivity is close to zero and the imaginary part is large. Such a material supports long-range surface exciton-polariton [14]. The present work only focuses on the first and second classes of materials.

LMR results out of the coupling between lossy mode and evanescent wave at a particular thickness of the thin film [15]. Only a few studies are reported on the application of LMR to sensing because the selection of the appropriate material for the thin film is critical [16–19].

Different types of waveguide structures have been utilized to realize SPR and LMR-based sensors. In particular, plasmonic fiber grating based, U-shaped and D-shaped optical fiber-based biosensor are explored extensively [19–21]. The simultaneous generation of LMR and SPR on the same planer platform has also been reported [22].

Indium tin oxide (ITO) is one of the CMO materials that supports LMR. It is a transparent material with an optical band gap of 3.6 eV, which restricts band-to-band transitions. The electronic and optical properties of ITO can be tuned during fabrication, resulting in a significant variation in its characteristics [23]. This property can be used to shift the resonance wavelength of LMR. Unlike SPR, the excitation of LMR has the advantage of not requiring specific polarization for incident light. Additionally, it is possible to generate multiple dips in the transmission spectrum. LMR dips are usually found in the IR and UV regions but, with the proper optimization of thin-film and with the use of other materials, they can also be observed in the visible region [14]. The IR dip is observed due to oscillations in charge density along the metal-dielectric interface. In contrast, a charge density oscillation along the thickness of the metal film is responsible for UV dip [16].

LMR is also found suitable for sensing applications. The performance and novelty of the sensor are determined by the material and sensing probe design used. Because of the excellent characteristics of optical fibers, these are being used as substrates for depositing ITO thin films to constitute the sensor [24].

Our present study examines the characteristics of bi- and tri-layer fiber optic sensing probes based on ITO that enable simultaneous excitations of SPR and LMR both. In bi- and tri-layer geometries, we investigate ITO + Ag and ITO + Ag + ITO structures, respectively. Applied biosensing, chemical analysis, quality assurance of food, and wavelength filtering are some of the potential applications for the proposed sensing probe.

2. The Model

To generate LMR, the lossy mode must be coupled with the evanescent wave. At a particular angle or wavelength, the effective index of the evanescent wave matches with the effective index of the lossy mode. The effective RI of the evanescent wave is given by

$$n_{eff} = n_p \, sin(\theta_i) \tag{1}$$

where n_p is the RI of the substrate, and θ_i is the incident angle of the light. This relationship shows that the RI of evanescent waves can be controlled by the incident angle of incident light and/or corresponding wavelength.

Figure 1 schematically shows the proposed sensing probe in the Kretschmann configuration. The probe consists of a multimode fiber with a core diameter of 400 µm and a numerical aperture of 0.22. The 1 cm fiber cladding has been removed from the fiber probe. On top of the unclad (exposed) core, layers of ITO and Ag have been considered.

At one end of the fiber, light from a polychromatic source is launched, and the spectrometer records the corresponding transmission spectrum at the other end. At a certain wavelength, called the resonance wavelength, the spectrum exhibits a minimum transmitted power. A change in the sensing medium (i.e., RI of the analyte) will alter the resonance wavelength. The sensitivity of the sensor is defined as the shift in resonance wavelength corresponding to the change in the RI of the analyte. Another important characterization parameter, figure of merit (FOM) is defined as the ratio of sensitivity to the full width at half maximum (FWHM) of the transmission dip. Additionally, detection accuracy (DA) is another important parameter that measures the sharpness of the resonance. The following expression relates these parameters (sensitivity, FOM, and DA) with each other [25].

$$\text{FOM} = \frac{\text{Sensitivity}}{\text{FWHM}} = \text{Sensitivity} \times \text{DA} \tag{2}$$

where DA $\propto \frac{1}{\text{FWHM}}$.

Some materials are highly sensitive but have low FOM. In contrast, others have poor FOM. Hence, materials need to be carefully selected.

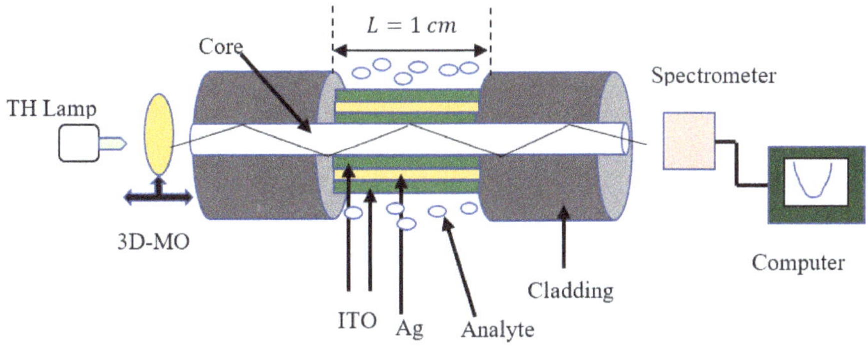

Figure 1. Schematic of the presented SPR setup (TH: tungsten halogen, 3D-MO: microscopic objective with 3D movement).

For multilayer structures, the transfer matrix method is used to calculate the transmission spectrum. Consider a kth layer of thickness d_k, having complex RI n_k, and dielectric coefficient ε_k. The transfer matrix for N layer system is expressed as

$$M = \prod_{k=2}^{N} M_k = \begin{bmatrix} M_{11} & M_{12} \\ M_{21} & M_{22} \end{bmatrix} = \begin{bmatrix} cos(\beta_k) & -i\, sin(\beta_k/q_k) \\ -i\, q_k sin(\beta_k) & cos(\beta_k) \end{bmatrix} \tag{3}$$

where β_k and q_k are defined as $(2\pi d_k/\lambda)\left(\varepsilon_k - n_1{}^2 sin^2\theta_1\right)^{1/2}$ and $\left(\varepsilon_k - n_1{}^2 sin^2\theta_1\right)^{1/2}/\varepsilon_k$, respectively and θ_1 is the incident angle of the ray, while λ is the wavelength of the incident light. The reflection coefficient r_p of p-polarized (TM polarized) incident wave through the film is expressed as:

$$r_p = \frac{(M_{11} + M_{12}q_N)q_1 - (M_{21} + M_{22}q_N)}{(M_{11} + M_{12}q_N)q_1 + (M_{21} + M_{22}q_N)} \tag{4}$$

The reflectance, R, for TM polarized light is given as

$$R = \left|r_p\right|^2 \tag{5}$$

A detailed description of this matrix method is given elsewhere [6,13]. The rays launched within the well-defined range of angle would be guided and the range is given by $\theta_1 = sin^{-1}(n_{cl}/n_1)$ to $\theta_2 = \pi/2$.

The transmitted power at the output end of the fiber is given by

$$P_{trans} = \frac{\int_{\theta_1}^{\theta_2} R_p^{N_{ref}(\theta)} n_1^2 \left(sin\theta cos\theta/\left(1 - n_1^2 cos^2\theta\right)^2\right) d\theta}{\int_{\theta_1}^{\theta_2} n_1^2 \left(sin\theta cos\theta/\left(1 - n_1^2 cos^2\theta\right)^2\right) d\theta} \tag{6}$$

where

$$N_{ref}(\theta) = \frac{L}{D\ tan\theta} \tag{7}$$

The number of reflections occurring in the sensing region is denoted by the Equation (7), where L is the length of the unclad region and D is the diameter of the fiber. The dielectric constant of the Ag and ITO layer is calculated by the Drude dispersive model expressed as

$$\epsilon(\lambda) = \epsilon_r + i\epsilon_i = 1 - \frac{\lambda^2\lambda_c}{\lambda_p^2(\lambda_c + i\lambda)} \tag{8}$$

and

$$\epsilon(\lambda) = \epsilon_r + i\epsilon_i = 3.8 - \frac{\lambda^2\lambda_c}{\lambda_p^2(\lambda_c + i\lambda)} \tag{9}$$

respectively, where λ_p and λ_c is the wavelength corresponding to bulk plasma frequency and collision wavelength. In the case of Ag, $\lambda_p = 0.14541$ µm and $\lambda_c = 17.6140$ µm, whereas $\lambda_p = 0.56497$ µm and $\lambda_c = 11.21076$ µm for ITO. The Sellmeier equation has been used to determine the RI of the fiber core [13]. We have assumed that above dispersion relations are valid in the whole wavelength range of investigation.

In order to fabricate the sensing probe, we use multimoded plastic clad fibers. The cladding can be removed by a few centimeters (a length of 1 cm of cladding is suitable for sensing applications) and then cleaned in a vacuum chamber using ion plasma bombardment and acetone. The unclad port of the fiber can be coated with metal or ITO after cleaning. Depending on the deposition techniques we have used, the uniformity of the films will vary. In order for the sensing probe to work correctly, the film uniformity must be good. A high-quality film can be achieved using sputtering and e-beam evaporation. The probe can be characterized by injecting light through one of the fiber faces and analyzing its sensing performance using a spectrometer at the other end of the fiber [26–30].

3. Results

The following two cases have been discussed in this section- in the first case, the ITO layer is deposited directly on the fiber core followed by the Ag layer (bi-layer sensing probe), and in the second case, an additional layer of the ITO is deposited over the Ag (tri-layer sensing probe).

3.1. Bi-Layer Configuration (ITO + Ag)

In this section, we numerically investigate a bilayer configuration of ITO + Ag coated fiber probe. In the first round of the simulation, the thickness of the ITO layer is fixed at 80 nm, whereas the thickness of the Ag layer is varied from 10 nm to 60 nm.

In the transmission spectrum, we observe two resonance dips for probe configuration with 80 nm ITO and 10 nm Ag for various values of analyte RIs, as shown in Figure 2a. Since the real part of the dielectric constant of the ITO is positive and larger than its imaginary part, at the lower wavelength region the condition of LMR generation is supported. Experimental evidence supports the LMR generation at short wavelengths and SPR excitation at long wavelengths [18]. The resonance dip in the visible region is caused by the LMR phenomena, while the second dip is the result of SPR. We would like to note here that the dielectric constant of ITO remains positive only for shorter wavelengths while at longer wavelengths it becomes negative (see Equation (9)). Therefore, for longer wavelengths, we see SPR resonance while in shorter wavelengths the probe supports LMR. These observations are well documented in the literature [14]. We would also like to note that for SPR excitation we require an interface of materials with opposite permittivity (one positive and other negative). An SPR dip can be observed even for a thin layer of Ag (i.e., 10 nm). Figure 2b illustrates that the SPR dip is more pronounced at a larger thickness of Ag. The thickness of the Ag layer, therefore, plays a critical role in the development of the SPR dip. In addition, as Ag thickness increases, the SPR dip becomes less sensitive to the RI variations in the analyte, while the LMR dip's sensitivity increases.

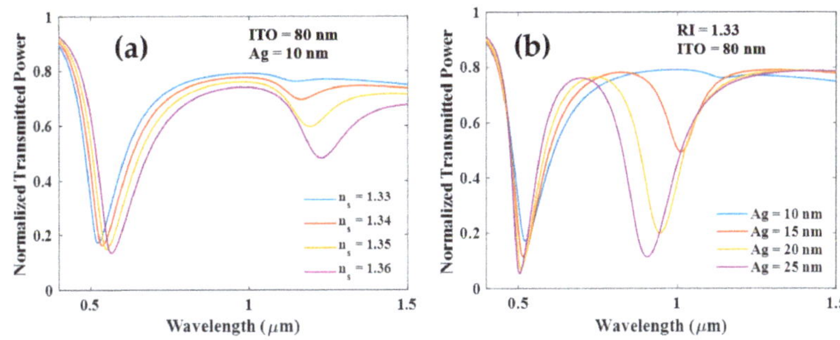

Figure 2. Normalized transmission spectra of sensing probe with (**a**) ITO (80 nm) + Ag (10 nm) and (**b**) for various thicknesses of Ag for 80 nm ITO and analyte RI $n_s = 1.33$.

The thickness of the ITO layer, however, significantly impacts the development of SPR and LMR resonances. As shown in Figure 3a, both resonances emerge with increasing thickness of the Ag layer for the 50 nm thick ITO layer. If the thickness of the ITO layer is less than 50 nm, SPR and LMR resonance dips still grow with the thickness of the Ag layer as shown in Figure 3b. Both of these cases show that SPR dip is insensitive to the RI of the sensing medium (transmitted power plot not shown).

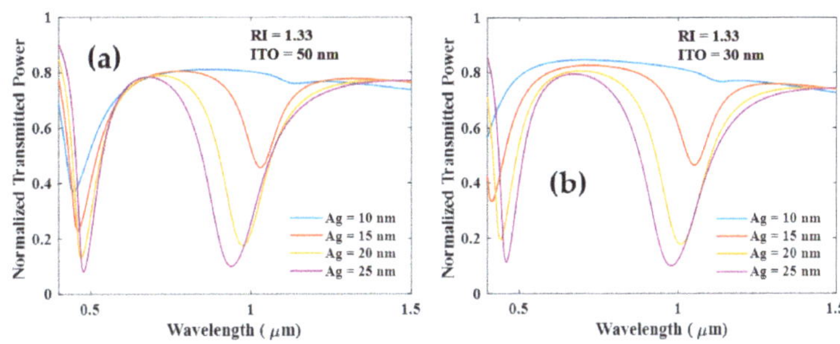

Figure 3. Normalized transmission spectra of sensing probe for various thicknesses of the Ag and (**a**) ITO (50 nm) (**b**) ITO (30 nm) and RI of the surrounding medium is 1.33.

In Figure 4a, we have plotted the LMR and SPR dip sensitivity against the thickness of the Ag layer for 80 nm thick ITO film. Both resonances are initially sensitive to changes in RI, but their sensitivities are drastically influenced by the thickness of the Ag layer. Nevertheless, the LMR sensitivity is improved as a result of the thicker Ag layer, while the SPR sensitivity is reduced as depicted in Figure 4a. At a very large thickness of the Ag layer, the SPR dip becomes insensitive to any RI variation of the analyte. This insensitive dip can be used as a reference point for characterizing the sensor's performance. The variation in the resonance wavelength of the LMR and SPR dip is also shown in Figure 4b. As Ag thickness is increased, the SPR resonance wavelength shifts slowly towards the smaller wavelength side. Also, shown is the transmitted power at the resonance wavelengths in Figure 4c. From the figure, we observe that the wavelength that corresponds to the LMR transmitted power minimum decreases to a minimum at a particular thickness of the Ag layer, and then increases. In contrast, the transmitted power for SPR dip is shifted toward the lower wavelength side with an increasing layer thickness of Ag. Previously a similar study is reported in [13], where resonance dip, observed in the visible region, was found useful for sensing applications and the second dip that appeared in the NIR region was insensitive to the surrounding RI, but we will not focus on this insensitive SPR dip here, since it is already detailed nicely in the literature [13].

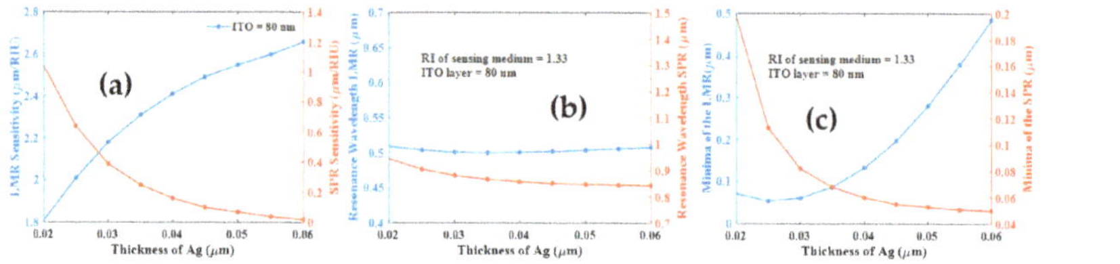

Figure 4. (**a**) Sensitivity (**b**) resonance wavelength and (**c**) minima of the transmission spectra at resonance wavelength of the LMR and SPR modes with the variation of Ag layer when ITO (80 nm) and RIs of the analyte are 1.33 and 1.34.

Further, we have studied the sensitivity and DA of both the resonances as a function of the analyte RI for different values of the Ag layer thickness as shown in Figures 5 and 6. The sensitivities of both the modes (LMR and SPR) increase with an increasing RI of the sensing medium as shown in Figure 5a,b. These plots also suggest that LMR dip is far more sensitive as compared to SPR dip. Figure 6 shows the corresponding DA variation as a function of the thickness of the Ag layer. DA for LMR dip (Figure 6a) is also relatively large as compared to that for SPR dip (Figure 6b). Also, note the opposite trends in Figures 5 and 6 with variations in the Ag layer thickness. Hence there is a trade-off between optimum values of sensitivity and DA for designing the bi-layer sensing probe.

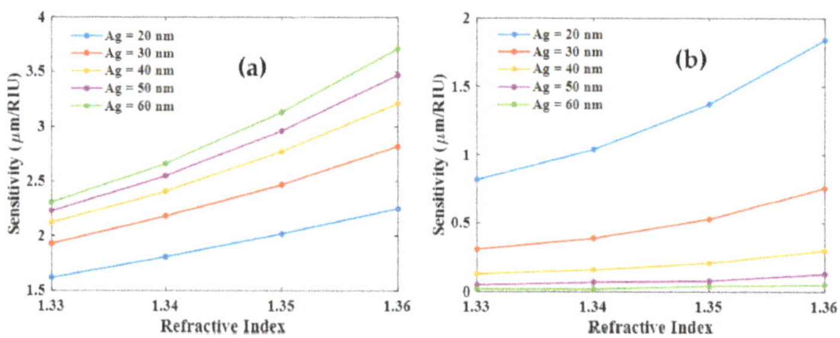

Figure 5. Sensitivity of the (**a**) LMR and (**b**) SPR dip with the variation of RI of the sensing medium for different thicknesses of Ag layer and 80 nm thickness of ITO layer.

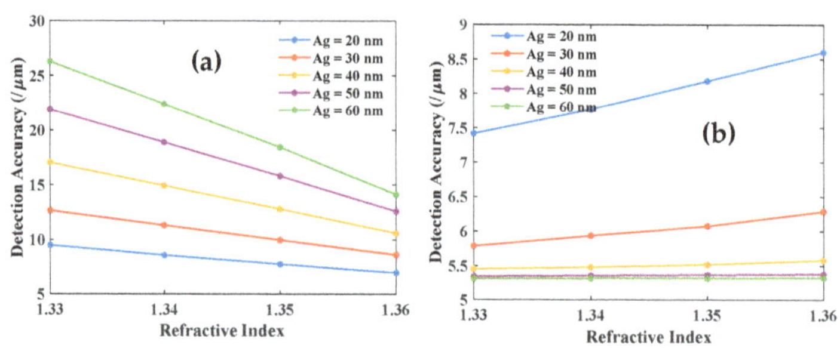

Figure 6. Detection accuracy of the (**a**) LMR and (**b**) SPR dip with the variation of RI of the sensing medium for different thicknesses of the Ag and 80 nm thickness of the ITO layer.

3.2. Tri-Layer Configuration

In this section, we investigate the tri-layer configuration (ITO + Ag + ITO). In the following tri-layer configuration, SPR is found to be more sensitive than the LMR dip. The following sub-sections analyze two important cases.

3.2.1. ITO (10 nm) + Ag (10 nm) + ITO (X nm)

This configuration examines the resonance characteristics of a tri-layer ITO + Ag + ITO coated fiber sensor, where a 10 nm layer of ITO is considered on the fiber core that is followed by a 10 nm thick layer of Ag, and then a third layer of ITO with varying thickness. Two resonance dips appear in the transmission spectra when the third layer of ITO is 10 nm thick, as shown in Figure 7a. The resonance dip in the visible range corresponds to LMR and the second dip to SPR. Increasing the thickness of the third ITO layer (40 nm) causes a new resonance dip to appear in the near-infrared region, known as LMR. The newly developed LMR dip (middle dip) is not affected by the change in RI of the surrounding medium, as shown in Figure 7b. Compared to the first LMR dip, the SPR dip (third dip) shows much better sensitivity.

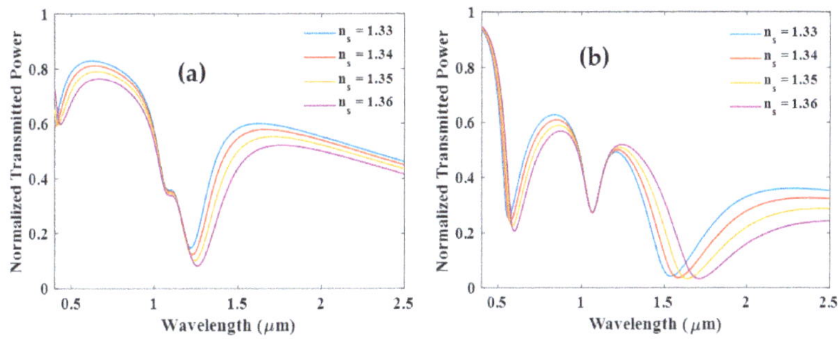

Figure 7. Normalized transmission spectra of sensing probe with (**a**) ITO (10 nm) + Ag (10 nm) + ITO (10 nm) and (**b**) ITO (10 nm) + Ag (10 nm) + ITO (40 nm).

The occurrence of the new LMR as a function of the thickness of the third ITO layer is shown in Figure 8a,b, where the sandwiched Ag layers are kept 10 nm and 20 nm thick, respectively. Figure 8a,b illustrates that increase in the thickness of the Ag layer from 10 nm to 20 nm shifts the resonance wavelength of the first LMR and SPR dips toward the longer wavelength side. However, the resonance wavelength of the insensitive LMR dip (central dip) remains nearly unchanged. Also, it seems that with the ITO layer thickness variation, the central dip merges with SPR dip. Alternatively, it also suggests a switch-over behavior between the two dips. This observation requires further exploration.

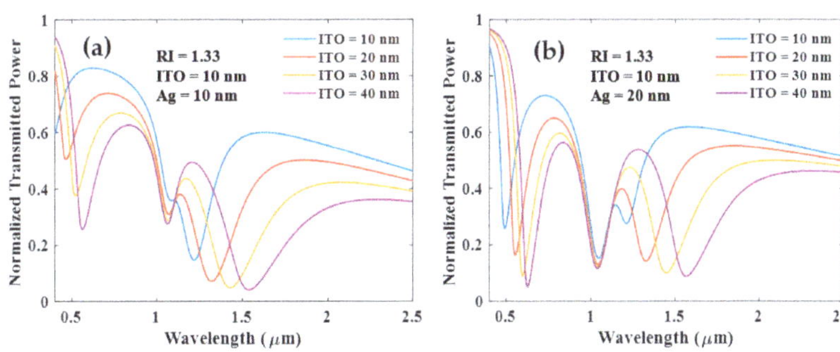

Figure 8. Normalized transmission spectra of sensing probe with (**a**) ITO (10 nm) + Ag (10 nm) + ITO (X nm) and (**b**) ITO (10 nm) + Ag (20 nm) + ITO (X nm).

As the thickness of the third ITO layer increases, the LMR dip's sensitivity decreases, whereas SPR's sensitivity increases as shown in Figure 9. The sensitivity variations of SPR and LMR dips in the tri-layer case are opposite to those in the bi-layer case (see Figure 4a).

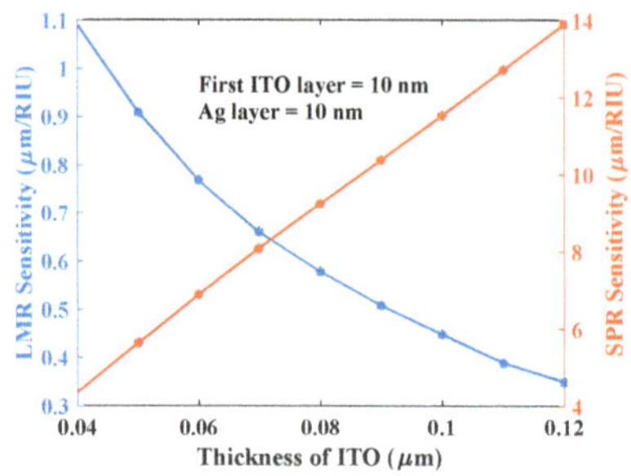

Figure 9. Sensitivity of the First LMR and SPR dip with the variation of the third ITO layer when the first layer is ITO (10 nm) and the second layer is Ag (10 nm) and RI of the sensing medium varies from 1.33 to 1.36 RIU.

Additionally, the DAs for the first LMR and SPR dip are shown in Figure 10a,b, respectively, for varying thicknesses of the third ITO layer. The DA for LMR decreases with sensing medium RI while the opposite trend is observed for SPR dip in Figure 10b.

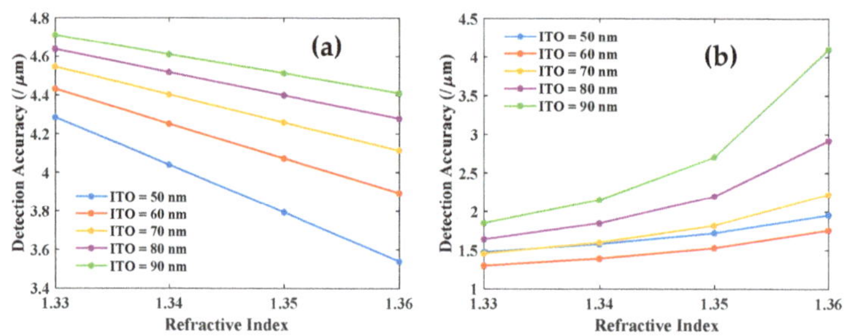

Figure 10. Detection accuracy of the (**a**) first LMR and (**b**) SPR dip with the variation of third ITO layer for 10 nm thick first layer ITO and 10 nm thick Ag layer.

Furthermore, the variation of the FOM with sensing medium RI for two dips is depicted in Figure 11a,b. The figure clearly shows the better performance of the SPR dip.

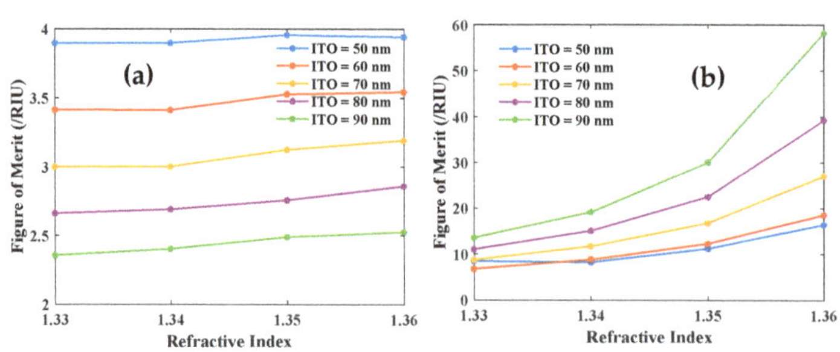

Figure 11. FOM of the (**a**) first LMR and (**b**) SPR dip with the variation of the third ITO layer thickness where the first layer is ITO (10 nm) and the second layer is Ag (10 nm).

3.2.2. (ITO (50) + Ag (X) + (ITO (50))

This section presents the normalized transmission spectrum for a tri-layer configuration with an Ag layer sandwiched between two ITO layers of thickness of 50 nm each. In this configuration too, two dips of LMR and one of SPR are observed, and the characteristics of these resonances are dependent on the thickness of the Ag layer, as shown in Figure 12a,b. Figure 12a shows transmittance variation with change in the sensing medium RI, while in Figure 12b, the RI of the sensing medium is kept fixed at 1.33 RIU and the thickness of the Ag layer is varied. Figure 12a depicts that SPR dip is relatively more sensitive and central LMR dip is completely insensitive. Figure 13 indicates that on further increase in the thickness of the Ag layer, the resonance wavelength of SPR dips shifts toward the shorter wavelengths; however, the resonance wavelength of the first LMR dip shifts toward the longer wavelength. The LMR dip observed in the NIR region is insensitive to the analyte RI variations. Moreover, the increased thickness of the Ag layer tends to annihilate the SPR dip as shown in Figure 13.

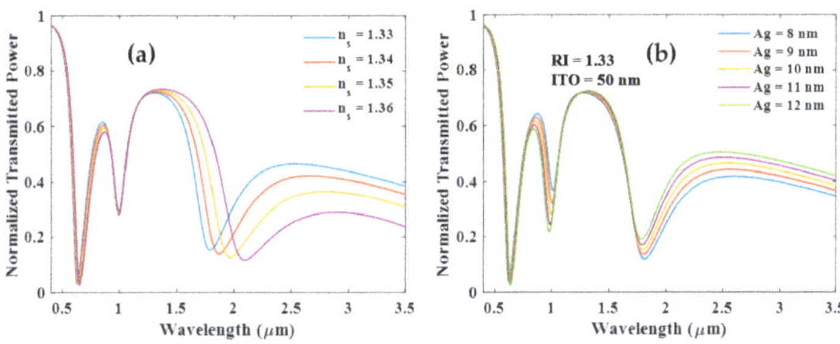

Figure 12. Normalized transmission spectra of sensing probe with (**a**) ITO (50 nm) + Ag (10 nm) + ITO (50 nm) and (**b**) for varying thickness of the Ag layer when both ITO layers are 50 nm thick.

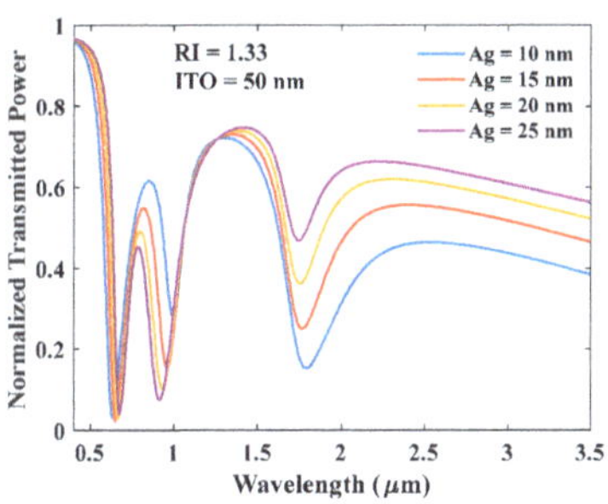

Figure 13. Normalized transmission spectra of sensing for varying thickness of the Ag layer when both ITO layers are 50 nm thick.

Further, we study the DA, sensitivity, and FOM of each resonance dip. We demonstrate that the DA, sensitivity, and FOM of the first LMR are improved as the Ag layer thickness is increased, as illustrated in Figure 14a,b,c, respectively.

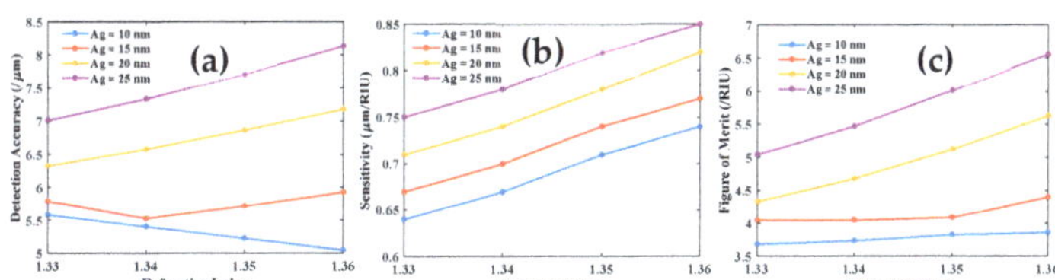

Figure 14. (**a**) DA, (**b**) sensitivity, and (**c**) FOM of the first LMR dip as a function of RI of the analyte.

This investigation is extended to the third resonance dip, as depicted in Figure 15. With the increase in the thickness of the Ag layer, DA, sensitivity, and FOM of SPR dip are improved as shown in Figure 15a,b,c respectively.

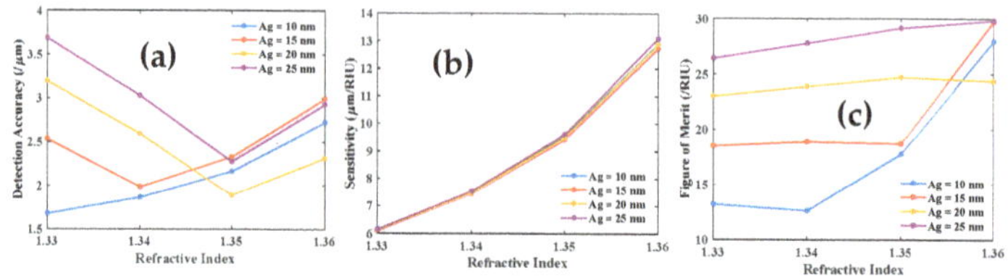

Figure 15. (**a**) DA, (**b**) sensitivity and (**c**) FOM of the SPR dip as a function of RI of the analyte.

Furthermore, we have plotted the absolute square of the electric field component along the interface across all the thicknesses of probes in two configurations- ITO (10 nm) +Ag (10 nm) +ITO (40 nm) and ITO (50 nm) +Ag (10 nm) +ITO (50 nm). The corresponding wavelength values are given in figure captions. These figures clearly show a large enhancement in the field at SPR resonance (see Figures 16c and 17c). These also corroborate the observed high sensitivity for SPR resonance.

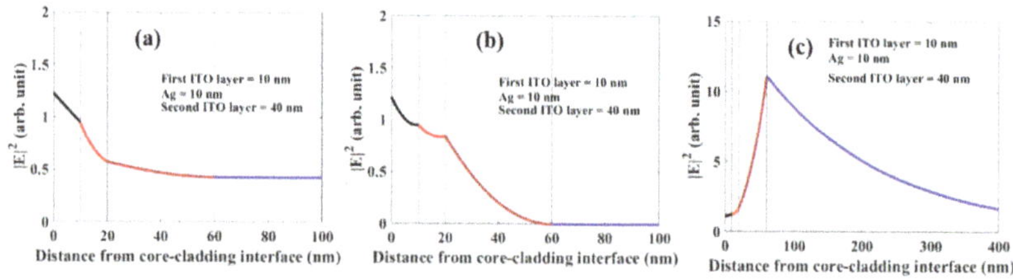

Figure 16. Electric field distribution corresponding to (**a**) first LMR (**b**) second LMR (**c**) SPR at their respective resonance wavelength at 561, 1065, and 1532 nm respectively for probe configuration ITO (10 nm) +Ag (10 nm) +ITO (40 nm).

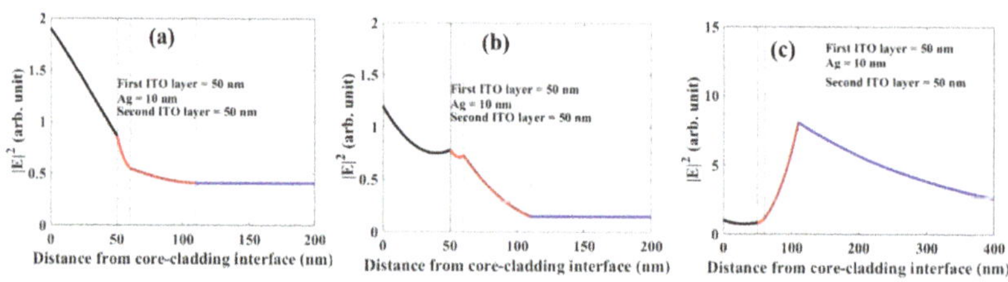

Figure 17. Electric field distribution corresponding to (**a**) first LMR (**b**) second LMR (**c**) SPR at their respective resonance wavelength at 631, 995, and 1798 nm respectively for probe configuration ITO (50 nm) +Ag (10 nm) +ITO (50 nm).

We would like to note that since in the present study a plastic-clad highly multimode fiber is considered, the proposed probe is only good for room temperature applications. Although slight variations in temperature do not influence the sensor performance, at high temperatures the fiber cladding will melt down.

4. Discussion and Conclusions

In conclusion, ITO + Ag-based bi-layer and tri-layer fiber-optic sensors have been studied. In the case of bi-layer geometry, two modes of resonances are possible. These are called LMR and SPR, and these resonances can be used for sensing purposes. In this bi-layer configuration, the LMR dip shows better sensitivity compared to the SPR dip. DA of the LMR dip is also far better than that of the SPR dip. As the thickness of the Ag layer increases further, the SPR dip becomes insensitive and only the LMR dip can be used for sensing. We suggest that at this thickness, the SPR dip can work as a reference, and this turns the sensor into a self-referenced sensor. By choosing the appropriate thickness of Ag, this configuration can be used in chemical, and bio-sensing, whereas the same configuration can also be utilized in wavelength filtering.

Further, two configurations of tri-layer geometry are explored, wherein one SPR and two LMR dips have been observed. In the first tri-layer configuration, the thickness of the outmost ITO layer was varied, the first LMR dip that arises in the visible region is found less sensitive as compared to the SPR dip. The other LMR dip (middle dip) that appears in the NIR region is found insensitive to any change in analyte RI. This insensitive LMR dip appears if the thickness of the third layer of ITO is increased. In the second configuration, the thickness of the Ag layer was varied. Particularly, it has been shown that the resonance wavelength of SPR dip shifts toward the shorter wavelength side; however, the resonance wavelength of the LMR dip shifts toward the longer wavelength side. Furthermore, we have plotted the electric field component along the interface across all the thicknesses of probes in two tri-layer configurations to demonstrate field enhancement. The observation of the insensitive second LMR dip and its manipulation with ITO layer thickness variation are the main contribution of this work as this suggests a switching of resonance type between LMR and SPR. Also, this work provides design rules of ITO-based bi- and tri-layer structures which support the excitations of LMR and SPR. The results of the study are summarized in Table 1. We see from the table that SPR dip exhibit a very high sensitivity of 14 μm/RIU and good DA and FOM.

Table 1. Summary of the results obtained in bi layer and tri layer configurations.

Configuration	Wavelength of Operation μm	Refractive Index Range	Sensitivity (μm/RIU)	DA (μm^{-1})	Figure of Merit (RIU^{-1})
ITO(10)/Ag(X)	0.4–0.8 (LMR) 0.8–1.5 (SPR)	1.33–1.36 1.33–1.36	~ 2.6 ($X = 60$ nm) ~ 1 ($X = 20$ nm)	~ 26 ($X = 60$ nm) ~ 7.5($X = 20$ nm)	
ITO(10)/Ag(10)/ ITO(X)	0.4–0.8 (LMR) 1.2–2.5 (SPR)	1.33–1.36 1.33–1.36	~ 1.1 ($X = 40$ nm) ~ 14 ($X = 120$ nm)	~ 4.7($X = 90$ nm) ~ 4($X = 90$ nm)	~ 4 ($X = 50$ nm) ~ 60 ($X = 90$ nm)
ITO(50)/Ag(X)/ ITO(50)	0.4–0.7 (LMR) 1.2–3 (SPR)	1.33–1.36 1.33–1.36	~ 0.7 ($X = 25$ nm) ~ 14 ($X = 25$ nm)	~ 7($X = 25$ nm) ~ 3.5($X = 25$ nm)	~ 5 ($X = 25$ nm) ~ 26 ($X = 25$ nm)

Author Contributions: A.K.M. and S.K.M. discussed the plan and agreed to it. The calculations were done by D.S.G. and A.P. The data was analyzed by all authors. The original manuscript was written by D.S.G. and A.P., S.K.M. and A.K.M. revised and finalized it. All authors have read and agreed to the published version of the manuscript.

Funding: This research received no external funding.

Informed Consent Statement: Not applicable.

Data Availability Statement: The data that support the findings of this study are available from the corresponding author upon request.

Conflicts of Interest: The authors declare no conflict of interest.

References

1. Homola, J.; Yee, S.S.; Gauglitzb, G. Surface plasmon resonance sensors: Review. *Sens. Actuators B Chem.* **1999**, *54*, 3–8. [CrossRef]
2. Verma, R.K.; Gupta, B.D. Surface plasmon resonance based fiber optic sensor for the IR region using a conducting metal oxide film. *JOSA A* **2010**, *27*, 846–851. [CrossRef] [PubMed]
3. Raether, H. *Surface Plasmons on Smooth and Rough Surfaces and on Gratings*; Springer: Berlin/Heidelberg, Germany, 1988.
4. Maier, S.A. *Plasmonics: Fundamentals and Applications*; Springer: Berlin/Heidelberg, Germany, 2007.
5. Mishra, A.K.; Mishra, S.K.; Verma, R.K. Graphene and beyond graphene MoS_2: A new window in surface plasmon resonance based fiber optic sensing. *J. Phys. Chem. C* **2016**, *120*, 2893–2900. [CrossRef]
6. Sun, S.; Zhang, T.; Liu, Q.; Ma, L.; Du, Q.; Duan, H. Enhanced directional fluorescence emission of randomly oriented emitter via a metal-dielectric hybrid antenna. *J. Phys. Chem. C* **2019**, *123*, 21150–21160. [CrossRef]
7. Puiu, M.; Bala, C. SPR and SPR Imaging: Recent Trends in Developing Nanodevices for Detection and Real-Time Monitoring of Biomolecular Events. *Sensors* **2016**, *16*, 870. [CrossRef]

8. Mishra, S.K.; Malviya, K.D.; Mishra, A.K. Highly sensitive bimetallic plasmonic sensing probe for aqueous samples. *Opt. Quantum Electron.* **2020**, *52*, 284. [CrossRef]
9. Mishra, S.K.; Verma, R.K.; Mishra, A.K. Versatile sensing structure: GaP/Au/Graphene/Silicon. *Photonics* **2021**, *8*, 547. [CrossRef]
10. Shalabney, A.; Abdulhalim, I. Electromagnatic fields distribution in multilayer thin film structures and the origin of sensitivity enhancement in surface plasmon resonance sensors. *Sens. Actuators A Phys.* **2010**, *159*, 24–32. [CrossRef]
11. Shukla, S.; Sharma, N.K.; Sajal, V. Sensitivity enhancement of a surface plasmon resonance based fiber optic sensor using ZnO Thin film: A theoretical study. *Sens. Actuators B Chem.* **2015**, *206*, 463–470. [CrossRef]
12. Hansen, W.N. Electric fields produced by the propagation of plane coherent electromagnetic radiation in stratified medium. *JOSA* **1968**, *58*, 380–390. [CrossRef]
13. Mishra, A.K.; Mishra, S.K.; Gupta, B.D. SPR based fiber optic sensor for refractive index sensing with enhanced detection accuracy and figure of merit in visible region. *Opt. Comm.* **2015**, *344*, 86–91. [CrossRef]
14. Villar, I.D.; Zamarreño, C.R.; Hernaez, M.; Arregu, F.J. Lossy Mode Resonance Generation with Indium-Tin-Oxide-Coated Optical Fibers for Sensing Applications. *J. Lightwave Technol.* **2010**, *28*, 111–117. [CrossRef]
15. Paliwal, N.; Joseph, J. Lossy Mode Resonance (LMR) Based Fiber Optic Sensors: A Review. *J. Lightwave Technol.* **2015**, *15*, 5361–5371. [CrossRef]
16. Franzen, S.; Rhodes, C.; Cerruti, M.; Gerber, R.W.; Losego, M.; Maria, J.; Aspnes, D.E. Plasmonic phenomena in indium tin oxide and ITO–Au hybrid films. *Opt. Lett.* **2009**, *34*, 2867–2869. [CrossRef]
17. Arregui, F.J.; Villar, I.D.; Zamarreno, C.R.; Zubiatea, P.; Matias, I.R. Giant sensitivity of optical fiber sensors by means of lossy mode resonance. *Sens. Actuators B* **2016**, *232*, 660–665. [CrossRef]
18. Villar, I.D.; Torres, V.; Beruete, M. Experimental demonstration of lossy mode and surface plasmon resonance generation with Kretschmann configuration. *Opt. Lett.* **2015**, *40*, 4739–4742. [CrossRef]
19. Zubiate, P.; Zamarreño, C.R.; Villar, I.D.; Matias, I.R.; Arregui, F.J. High sensitive refractometers based on lossy mode resonances (LMRs) supported by ITO coated D-shaped optical fibers. *Opt. Express* **2015**, *23*, 8045–8050. [CrossRef] [PubMed]
20. Caucheteur, C.; Loyez, M.; Gonzalez-vila, A.; Wattiez, R. Evaluation of gold layer configuration for plasmonic fiber grating biosensors. *Opt. Express* **2018**, *26*, 24154–24163. [CrossRef] [PubMed]
21. Wen, H.; Huang, C.H.; Li, Y.; Chen, J.; Yeh, Y.; Chiang, C.H. A lamping U-shaped fiber biosensor detector for microRNA. *Sensors* **2020**, *20*, 1509. [CrossRef]
22. Fuentes, O.; Villar, I.D.; Dominguez, I.; Corres, J.M.; Matias, I.R. Simultaneous generation of surface plasmon and lossy mode resonances in the same planar platform. *Sensors* **2022**, *22*, 1505. [CrossRef]
23. Rhodes, C.; Franzen, S. Surface plasmon resonance in conducting oxides. *J. App. Phys.* **2006**, *100*, 054905. [CrossRef]
24. Villar, I.D.; Hernaez, M.; Zamarreño, C.R.; Sánchez, P.; Fernández-Valdivielso, C.; Arregui, F.J.; Matias, I.R. Design rules for lossy mode resonance based sensors. *Appl. Opt.* **2012**, *51*, 4298–4307. [CrossRef] [PubMed]
25. Mishra, S.K.; Mishra, A.K. ITO/Polymer matrix assisted surface plasmon resonance based fiber optic sensor. *Results Opt.* **2021**, *5*, 100173. [CrossRef]
26. Azad, S.; Khosravi, M.; Nikzad, A.; Mishra, S.K. A novel contemporary molecular imprinting technique for non-enzymatic selective glucose detection. *Opt. Laser Technol.* **2022**, *148*, 107786. [CrossRef]
27. Mishra, S.K.; Singh, S.; Gupta, B.D. Surface plasmon resonance based fiber optic hydrogen sulphide gas sensor utilizing nickel oxide doped ITO thin film. *Sens. Actuators B: Chem.* **2014**, *195*, 215–222. [CrossRef]
28. Mishra, S.K.; Gupta, B.D. Surface plasmon resonance based fiber optic sensor for the detection of CrO_4^{2-} using Ag/ITO/hydrogel layers. *Anal. Methods* **2014**, *6*, 5191–5197. [CrossRef]
29. Mishra, S.K.; Chiang, K.S. Phenolic-compounds sensor based on immobilization of tyrosinase in polyacrylamide gel on long-period fiber grating. *Opt. Laser Technol.* **2020**, *131*, 106464. [CrossRef]
30. Mishra, S.K.; Usha, S.P.; Gupta, B.D. A lossy mode resonance-based fiber optic hydrogen gas sensor for room temperature using coatings of ITO thin film and nanoparticles. *Meas. Sci. Technol.* **2016**, *27*, 045103. [CrossRef]

biosensors

MDPI

Communication

Wafer-Scale LSPR Substrate: Oblique Deposition of Gold on a Patterned Sapphire Substrate

Kihyeun Kim [1,†], Ki Joong Lee [2,†], Na Rae Jo [2], Eun-Jung Jo [3], Yong-Beom Shin [2,4,*] and Min-Gon Kim [3,*]

1 Center for Systems Biology, Massachusetts General Hospital, Boston, MA 02114, USA; kkim50@mgh.harvard.edu
2 Bionanotechnology Research Center, Korea Research Institute of Bioscience and Biotechnology (KRIBB), 125 Gwahak-ro, Yuseong-gu, Daejeon 34141, Korea; kijoong123@kribb.re.kr (K.J.L.); nalgamgi@kimm.re.kr (N.R.J.)
3 Department of Chemistry, Gwangju Institute of Science and Technology (GIST), Gwangju 61005, Korea; jej@gist.ac.kr
4 BioNano Health Guard Research Center, 125 Gwahak-ro, Yuseong-gu, Daejeon 34141, Korea
* Correspondence: ybshin@kribb.re.kr (Y.-B.S.); mkim@gist.ac.kr (M.-G.K.)
† These authors contributed equally to this work.

Abstract: Label-free detection of biomolecules using localized surface plasmon resonance (LSPR) substrates is a highly attractive method for point-of-care (POC) testing. One of the remaining challenges to developing LSPR-based POC devices is to fabricate the LSPR substrates with large-scale, reproducible, and high-throughput. Herein, a fabrication strategy for wafer-scale LSPR substrates is demonstrated using reproducible, high-throughput techniques, such as nanoimprint lithography, wet-etching, and thin film deposition. A transparent sapphire wafer, on which SiO_2-nanodot hard masks were formed via nanoimprint lithography, was anisotropically etched by a mixed solution of H_2SO_4 and H_3PO_4, resulting in a patterned sapphire substrate (PSS). An LSPR substrate was finally fabricated by oblique deposition of Au onto the PSS, which was then applied to label-free detection of the binding events of biomolecules. To the best of our knowledge, this paper is the first report on the application of the PSS used as an LSPR template by obliquely depositing a metal.

Keywords: patterned sapphire substrate; localized surface plasmon resonance; oblique deposition; nanoimprint lithography; wafer-scale

Citation: Kim, K.; Lee, K.J.; Jo, N.R.; Jo, E.-J.; Shin, Y.-B.; Kim, M.-G. Wafer-Scale LSPR Substrate: Oblique Deposition of Gold on a Patterned Sapphire Substrate. *Biosensors* **2022**, *12*, 158. https://doi.org/10.3390/bios12030158

Received: 17 February 2022
Accepted: 1 March 2022
Published: 3 March 2022

Publisher's Note: MDPI stays neutral with regard to jurisdictional claims in published maps and institutional affiliations.

1. Introduction

Localized surface plasmon resonance (LSPR), which occurs in metal nanostructures, has attracted considerable attention from various optical- and electrical-based research fields, such as optical biosensors, color engineering, spectroscopy, metamaterials, and plasmonic optoelectronics [1–5]. Thus, large-scale, reproducible, high-throughput fabrication methods of wavelength-tunable LSPR substrates are in high demand not only for research but also for commercialization [6–8]. To fabricate LSPR substrates, two primary methods have been used until now: (i) the synthesis of metal nanoparticles (NPs), and covalent attachment of the NPs on substrates through self-assembled monolayers (SAMs) [9,10]; and (ii) the patterning of metal nanostructures by nanolithography techniques [7,11,12].

Regarding the former (i), various shapes of gold (Au) NPs were synthesized in a solution, such as spheres, rods, triangles, and spiked structures, which could easily tune LSPR properties depending on the shape [13,14]. However, the formation of SAMs on a substrate was usually unstable, which lowered reproducibility for the fabrication of LSPR substrates [15,16]. Moreover, it was difficult to align metal NPs on a substrate [17]; instability issues when metal NPs were immobilized on a substrate were also noted [18,19]. As for the latter (ii), various nanolithography techniques, such as colloidal, e-beam, and nanoimprint have been applied to fabricate metal nanostructures on substrates [20–22]. The limitations of e-beam and colloidal lithography are low-throughput and non-elaborate patterning, respectively [23]. Although nanoimprint lithography enabled a high-throughput process and elaborate patterning, it also encountered difficulties in the fabrication of various shapes of master molds, leading to limitations in the tuning of LSPR properties [24].

Herein, we report on a fabrication method for wafer-scale, elaborate LSPR substrates using high-throughput methods, such as nanoimprint lithography, wet-etch processing, and thin film deposition. In detail, the LSPR substrates were accomplished by oblique deposition of Au onto a (triangular-pyramid shape) patterned sapphire substrate (PSS) which was obtained through a wet-etch process, resulting in the formation of Au nanostructures on one side of the PSS. Thus, unlike typical strategies for fabrication of LSPR substrates that directly fabricated patterned metal nanostructure using nanolithography [20–22], the strategy suggested in this paper for LSPR-substrate fabrication was to pattern a substrate first, followed by oblique deposition of Au on the patterned substrate. This approach can variously tune LSPR properties according to the shape of PSS controlled by the wet-etch condition, deposition angle and/or direction of metals, which would be highly advantageous for obtaining wavelength-tunable LSPR substrates. Therefore, such a fabrication method could be an inspiring approach for nanofabrication and/or LSPR-based research fields.

2. Materials and Methods

2.1. Formation of a SiO_2-Nanodot on the Sapphire Wafer

SiO_2-nanodot was formed on a sapphire wafer (2 inches in diameter) via nanoimprint lithography, as reported in our previous study (Figure S1) [23]. In brief, a nanodot-patterned master stamp was used to pattern a thermoplastic resist layer that had been coated on the sapphire wafer. Then, Cr was obliquely deposited on the resist pattern to form the Cr hard masks only on the top of the resist layer that was not stamped by the master stamp. Next, O_2 plasma was treated to etch the resist layer where it was not covered by the Cr hard mask. A SiO_2-nanodot pattern was formed by the deposition of SiO_2 using an e-beam evaporator. Finally, the resist layer was completely removed using acetone. Such nanoimprint lithography is a highly advantageous technique for fabricating nanosized patterns on substrates because of its high speed and high-throughput, as well as high reproducibility and fidelity [23]. The diameter, height, and pitch of the SiO_2 nanodots were 130, 50, and 300 nm, respectively.

2.2. Fabrication of PSS via Wet-Etch

The SiO_2 nanodot-deposited sapphire wafer was annealed using a furnace under air at 750 °C for 12 h to densify the SiO_2. Then, the sapphire wafer was soaked in a beaker containing a mixture of H_2SO_4 and H_3PO_4 (1:3 v/v), and heated for 1 h at 310 °C to etch the sapphire wafers [25,26]. After finishing the wet-etch process, the beaker was cooled to 20 °C, and the PSS was washed with deionized (DI) water.

2.3. Formation of Metal Nanostructures on the PSS

Au nanostructures were formed on the PSS by oblique deposition (45°) of Au using a thermal evaporator. The deposition thickness and rate were set to 20 nm and 0.5 Å/s, respectively.

2.4. LSPR Shift of the Au Nanostructure-Formed PSS Due to Biomolecule Attachment

An LSPR shift was observed by exposing it to streptavidin (STA), biotin-bovine serum albumin (BSA), and STA, in turn. To attach STA onto the Au surface, 100-μM biotin-HPDP was attached to the Au nanostructure on the PSS overnight, followed by washing of the Au/PSS with EtOH and deionized (DI) water. Then, STA (50 μg mL^{-1}) and biotin-BSA (50 μg mL^{-1}) were attached in turns; and at each step, absorbance was measured after washing.

3. Results and Discussion

To fabricate a patterned sapphire substrate (PSS), SiO_2 nanodots—which played a role as hard masks—were well-orderly formed on the entire surface of the sapphire wafer (Figure 1a–c). Figure 1a shows blue light scattering observed when the sapphire wafer was exposed to light due to the well-ordered SiO_2 nanodots on the wafer. In addition, SEM images proved that precisely ordered SiO_2 patterns were present on the sapphire wafer; the diameter, height, and pitch of the pattern were 130, 50, and 300 nm, respectively (Figure 1b,c).

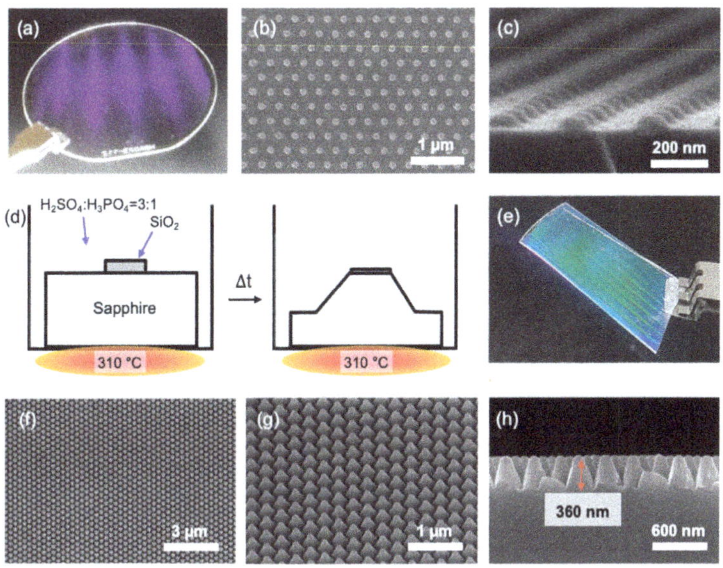

Figure 1. (**a**) A digital image and (**b**,**c**) SEM images of the SiO_2 nanodot array on the sapphire wafer (2 inches) ((**b**) top and (**c**) tiled view). (**d**) Schematic of the fabrication of the patterned sapphire substrate (PSS) via wet-etch. (**e**) A digital image of a piece of the PSS. (**f**–**h**) SEM images of the PSS for top, 45° tilted, and side views, respectively.

The SiO_2 nanodot pattern on the sapphire wafer was annealed using a furnace under air at 750 °C for 12 h to densify the SiO_2 nanodots before proceeding with the wet-etch process. This process was essential because as-deposited SiO_2 nanodots using the e-beam evaporator were not able to act as hard masks under high temperatures and acidic conditions (such as a mixture of H_2SO_4 and H_3PO_4), as shown in Figure S2. After the annealing process, the sapphire wafer was wet etched by a mixed solution composed of H_2SO_4 and H_3PO_4 at an elevated temperature of 310 °C for 1 h (Figure 1d). The composition of a wet-etching solution was 3:1 v/v.

As a result of the wet-etch, a PSS was fabricated, which was analyzed by digital (Figure 1e) and SEM images regarding top, 45° tilted, and side views (Figure 1f–h). Triangular-pyramid shape patterns in the PSS were observed across the entire wafer (Figure 1f,g). The triangular pyramid could have been formed because the SiO_2 hard mask prevented direct contact between the single crystal sapphire wafer and acidic etchant (H_2SO_4 and H_3PO_4). The height of the triangular pyramids on the PSS was approximately 360 nm (Figure 1h), which varied depending on the wet-etch temperature (Figure S3). This process occurred because although H_2SO_4 and H_3PO_4 can etch sapphire as follows (Equations (1) and (2)):

$$Al_2O_3 + 3H_2SO_4 \rightarrow Al_2(SO_4)_3 + 3H_2O, \quad (1)$$

$$Al_2O_3 + 2H_3PO_4 \rightarrow 2AlPO_4 + 3H_2O, \quad (2)$$

the roles of both etchants are different in terms of etching direction (H_2SO_4 and H_3PO_4 for perpendicular and lateral, respectively) [25]. In addition, these different behaviors also depended on the temperature. Thus, shape control was achieved by controlling the composition of the etchant. In addition, after the first etch of the sapphire wafer, the second etch of the wafer using a different composition of etchant resulted in sharp or broad triangular-pyramid shapes on the PSS (Figure S4).

Wafer-scale Au nanostructures were fabricated by oblique deposition of Au (20 nm) on the PSS using an evaporator. Oblique deposition (45°) of Au on the PSS resulted in Au nanostructures being only formed on only half of the available surface area of the triangular pyramids on the PSS, described as red lines in Figure 2a; moreover, a shadowing effect was observed due to adjacent triangular pyramids (red arrow in Figure 2b). For better visualization, a schematic illustration was demonstrated for obliquely deposited Au (45°)/PSS (Figure 3c). Whereas, vertical deposition resulted in nearly full coverage of Au on the PSS (Figure S5). In the case of oblique deposition of Au, the shape of the Au nanostructures was determined by the shape of triangular pyramids on the PSS, which can be controlled by the composition of etchant and etching temperature (Figures S3 and S4). Consequently, the shape of Au nanostructure can vary not only depending on the shape of the triangular pyramids on the PSS but also the deposition angle and/or direction with respect to the PSS.

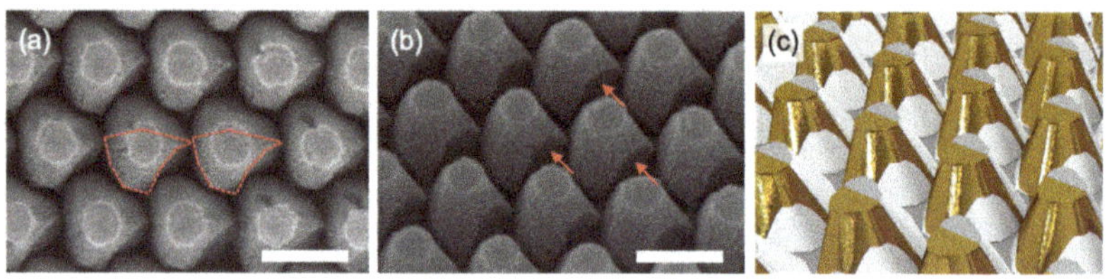

Figure 2. (**a**,**b**) SEM images after oblique (45°) deposition of Au onto a PSS ((**a**): top view, (**b**): 45° tilted view). Scale bars in the images are 300 nm. Red lines in (**a**) and arrows in (**b**) indicate Au nanostructure and the shadowing effect due to adjacent triangular pyramids, respectively. (**c**) Schematic illustration of obliquely deposited (45°) Au on PSS.

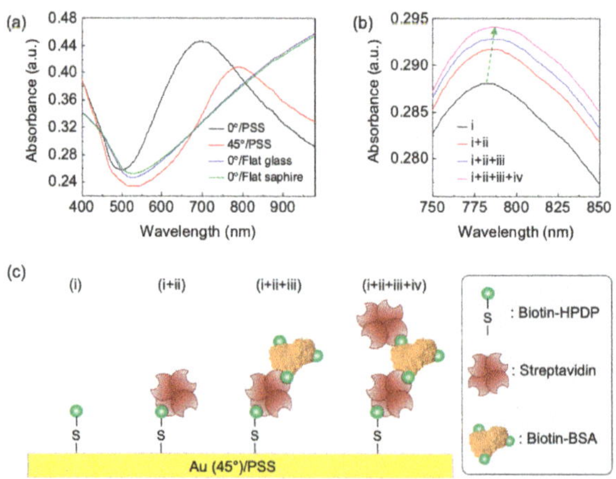

Figure 3. (**a**) Absorbance of obliquely and vertically deposited Au on PSSs, and vertically deposited Au on flat glass and sapphire substrates. (**b**) LSPR shift of the PSS with obliquely deposited Au: (i) biotin-HPDP, (ii) STA, (iii) biotin-BSA, and (iv) STA. (**c**) Schematic illustration for step-by-step biomolecule detection using an Au/PSS. The stages of (i), (i + ii), (i + ii + iii), and (i + ii + iii + iv) correspond to those in Figure 3b.

The phenomenon of a localized surface plasmon resonance (LSPR) of the Au nanostructures on the PSS was observed by measuring absorbance (Figure 3a). LSPR peaks were observed in the cases of Au deposition on a PSS, but not on flat substrates. The LSPR peak appeared in the vertically deposited case because the vertical deposition between the sample and Au sources in the thermal evaporator used in this work was not exactly vertical (Figure S6). Regarding the obliquely deposited case, the LSPR peak was red-shifted compared to that from the vertically deposited case; this was because the shapes of Au nanostructures in each case were different. Different shapes of metal nanostructure induce various LSPR properties [27]. Therefore, the location of the LSPR peak can be controlled by the shape of triangular pyramids on the PSS and deposition angle and/or direction of the PSS.

One of the most popular applications of the LSPR phenomenon is in biosensing fields because the LSPR shift occurs by perturbation in the presence of target molecules located near the metal nanostructures (~10 nm), which is a label-free tool that monitors molecular bindings in real-time, in small-volume samples [28]. Hence, Au nanostructures on PSS were exposed to biomolecules (STA, biotin-BSA, and STA, in turn), and LSPR shifts were observed (Figure 3b). A schematic illustration is depicted for the procedure for biomolecule detection (Figure 3c). In detail, to attach STA on the surface of the Au nanostructures, biotin-HPDP was, at first, covalently functionalized on the Au surface through the thiol-Au interaction [29]. As a result of the step-by-step attachment of STA/biotin-BSA/STA, the LSPR peak of the Au nanostructures was red-shifted step-by-step due to the attachment of the biomolecules (detailed values of LSPR are shown in Table S1). The redshift phenomenon occurred because the biomolecules attached to the Au nanostructures caused a change in the refractive index near the Au surface. Thus, our Au nanostructure on PSS could have an application as LSPR biosensors by manipulating the structures of the PSS and/or oblique deposition conditions. Furthermore, a thorough investigation of LSPR properties by controlling the PSS and deposition variables, such as the plasmonic substrate, could be further applicable to color engineering, spectroscopy, metamaterials, and plasmonic solar cells.

We also fabricated Au/PSSs where Au was deposited at different angles: 30°, 45°, and 60° (Figure S7). Among them, the LSPR peak obtained from Au (45°)/PSS showed the sharpest. This work is a preliminary study demonstrating the application of PSS for LSPR substrate by obliquely depositing Au on the PSS, and thus, LSPR properties of the Au nanostructure on PSS can be improved further, depending on Au deposition angle, the thickness of deposited Au, the direction of PSS under oblique deposition of Au, and the shape of PSS.

4. Conclusions

In this paper, we have suggested a fabrication method of wafer-scale LSPR substrates by engineering a single crystal substrate and obliquely depositing Au, using high-throughput and high-reproducibility techniques. As a single crystal substrate, sapphire wafers were chosen because of their transparency, which is highly advantageous for optical applications, such as optical biosensors, metamaterials, and solar cells. The PSS was fabricated using a nanoimprint lithography technique, thin film deposition, and wet-etching (using H_2SO_4 and H_3PO_4). The oblique deposition (45°) of Au on the PSS enabled the formation of Au nanostructures on the PSS due to the three-dimensional structure (triangular-pyramid shape) of the PSS. The Au nanostructures formed by the oblique deposition exhibited a red-shifted LSPR peak compared to those prepared by vertical deposition, implying that LSPR properties can be controlled not only by the shapes of the PSS but also by the deposition conditions, such as angle and/or rotation of the PSS. The Au nanostructure on PSS was applied to detect biomolecules by observing the LSPR shift; the LSPR peak was red-shifted by attaching STA, biotin-BSA, and STA, in turn.

PSS has been widely used as a substrate for light-emitting diode, so far due to its transparency, wafer-scale patterning, and high-throughput production. Here, we applied such a PSS to the fabrication of an LSPR substrate by obliquely depositing Au on the PSS. Further technical approaches could improve our fabrication strategy to expedite its application for commercially available products. As an example of the approaches, some experimental variables—Au deposition angle, the thickness of deposited Au, the direction of PSS under oblique deposition of Au, and the shape of PSS—could be addressed. In addition, due to the high expense of the sapphire wafer, the PSS might be used as a master mold to fabricate the reverse-shaped polydimethylsiloxane (PDMS) that could be used as patterned PDMS substrate for the angle deposition of Au.

Thus, such a fabrication method of wafer-scale plasmonic substrates could pave the way for use in LSPR biosensors by manipulating the structures of the PSS and/or oblique deposition conditions. This technology is also expected to be useful in other optical applications, such as color engineering, spectroscopy, metamaterials, and plasmonic optoelectronics.

Supplementary Materials: The following supporting information can be downloaded at: https://www.mdpi.com/article/10.3390/bios12030158/s1, Figure S1: Schematic of the formation of a SiO_2 nanodot array on a sapphire wafer, Figure S2: SEM image of the SiO_2 nanodot (not annealed)-patterned sapphire wafer after the wet-etch, Figure S3: SEM images of a PSS, Figure S4: SEM images after the second wet-etch of PSS, Figure S5: SEM images of vertically deposited Au on PSS. Figure S6: Schematic illustration of vertical and oblique depositions of Au on PSS, Figure S7: Analysis of Au/PSSs where Au was deposited at different angles, Table S1: LSPR-shift observation of obliquely deposited Au on PSS when exposed to biomolecules.

Author Contributions: Conceptualization, K.K., Y.-B.S. and M.-G.K.; methodology, K.K., K.J.L. and N.R.J.; software, N.R.J.; validation, K.K., K.J.L. and M.-G.K.; formal analysis, K.K. and K.J.L.; investigation, K.K. and K.J.L.; resources, K.K., Y.-B.S. and M.-G.K.; data curation, K.K. and K.J.L.; writing—original draft preparation, K.K., K.J.L. and E.-J.J.; writing—review and editing, K.K., K.J.L., E.-J.J., Y.-B.S. and M.-G.K.; visualization, K.K. and K.J.L.; supervision, Y.-B.S. and M.-G.K.; project administration, Y.-B.S. and M.-G.K.; funding acquisition, K.K., Y.-B.S. and M.-G.K. All authors have read and agreed to the published version of the manuscript.

Funding: This work was financially supported by a National Research Foundation grant funded by the Ministry of Science, ICT, and Future Planning as the MidCareer Researcher Program (NRF-2021R1A2B5B3001417); and Nano·Material Technology Development Program through the National Research Foundation of Korea (NRF) funded by Ministry of Science and ICT (Grant number NRF-2021M3H4A4079382). This research was also supported by a grant of the Korea Health Technology R&D Project through the Korea Health Industry Development Institute (KHIDI), funded by the Ministry of Health & Welfare, Republic of Korea (I II21C0957).

Institutional Review Board Statement: Not applicable.

Informed Consent Statement: Not applicable.

Data Availability Statement: The authors confirm that the data supporting the findings of this study are available within the article and its Supplementary Materials.

Conflicts of Interest: The authors declare no conflict of interest.

References

1. Kaur, B.; Kumar, S.; Kaushik, B.K. Recent Advancements in Optical Biosensors for Cancer Detection. *Biosens. Bioelectron.* **2022**, *197*, 113805. [CrossRef] [PubMed]
2. Wang, X.; Dai, C.; Yao, X.; Qiao, T.; Chen, M.; Li, S.; Shi, Z.; Wang, M.; Huang, Z.; Hu, X.; et al. Asymmetric Angular Dependence for Multicolor Display Based on Plasmonic Inclined-Nanopillar Array. *Nanoscale* **2021**, *13*, 7273–7278. [CrossRef] [PubMed]
3. Jaculbia, R.B.; Imada, H.; Miwa, K.; Iwasa, T.; Takenaka, M.; Yang, B.; Kazuma, E.; Hayazawa, N.; Taketsugu, T.; Kim, Y. Single-Molecule Resonance Raman Effect in a Plasmonic Nanocavity. *Nat. Nanotechnol.* **2020**, *15*, 105–110. [CrossRef] [PubMed]
4. Bai, J.; Yao, Y. Highly Efficient Anisotropic Chiral Plasmonic Metamaterials for Polarization Conversion and Detection. *ACS Nano* **2021**, *15*, 14263–14274. [CrossRef] [PubMed]
5. Rissi, N.C.; Comparetti, E.J.; Estevão, B.M.; Mastelaro, V.R.; Zucolotto, V. Doped Plasmonic Zinc Oxide Nanoparticles with Near-Infrared Absorption for Antitumor Activity. *ACS Appl. Nano Mater.* **2021**, *4*, 9779–9789. [CrossRef]
6. Fruncillo, S.; Su, X.; Liu, H.; Wong, L.S. Lithographic Processes for the Scalable Fabrication of Micro-and Nanostructures for Biochips and Biosensors. *ACS Sens.* **2021**, *6*, 2002–2024. [CrossRef]

7. Lu, X.; Yao, C.; Sun, L.; Li, Z. Plasmon-Enhanced Biosensors for MicroRNA Analysis and Cancer Diagnosis. *Biosens. Bioelectron.* **2022**, *203*, 114041. [CrossRef]
8. Chang, Y.-L.; Lai, I.-C.; Lu, L.-C.; Chang, S.-W.; Sun, A.Y.; Wan, D.; Chen, H.-L. Wafer-Scale Nanocracks Enable Single-Molecule Detection and on-Site Analysis. *Biosens. Bioelectron.* **2022**, *200*, 113920. [CrossRef]
9. Chi, J.; Xia, C.; Guo, Z.; Huang, G.; Lin, X. Gold Nanoparticle-Decorated Porous Silica for Surface-Enhanced Raman Scattering-Based Detection of Trace Molecules in Liquid Phase. *ACS Appl. Nano Mater.* **2022**, *5*, 527–536. [CrossRef]
10. Wang, Z.; Chen, J.; Khan, S.A.; Li, F.; Shen, J.; Duan, Q.; Liu, X.; Zhu, J. Plasmonic Metasurfaces for Medical Diagnosis Applications: A Review. *Sensors* **2021**, *22*, 133. [CrossRef]
11. Lee, S.-W.; Lee, K.-S.; Ahn, J.; Lee, J.-J.; Kim, M.-G.; Shin, Y.-B. Highly Sensitive Biosensing Using Arrays of Plasmonic Au Nanodisks Realized by Nanoimprint Lithography. *ACS Nano* **2011**, *5*, 897–904. [CrossRef]
12. Aćimović, S.S.; Šípová, H.; Emilsson, G.; Dahlin, A.B.; Antosiewicz, T.J.; Käll, M. Superior LSPR Substrates Based on Electromagnetic Decoupling for On-a-Chip High-Throughput Label-Free Biosensing. *Light Sci. Appl.* **2017**, *6*, e17042. [CrossRef] [PubMed]
13. Grzelczak, M.; Pérez-Juste, J.; Mulvaney, P.; Liz-Marzán, L.M. Shape Control in Gold Nanoparticle Synthesis. *Chem. Soc. Rev.* **2008**, *37*, 1783. [CrossRef] [PubMed]
14. Personick, M.L.; Mirkin, C.A. Making Sense of the Mayhem behind Shape Control in the Synthesis of Gold Nanoparticles. *J. Am. Chem. Soc.* **2013**, *135*, 18238–18247. [CrossRef] [PubMed]
15. Zhu, M.; Lerum, M.Z.; Chen, W. How to Prepare Reproducible, Homogeneous, and Hydrolytically Stable Aminosilane-Derived Layers on Silica. *Langmuir* **2012**, *28*, 416–423. [CrossRef]
16. Jones, J.A.; Qin, L.A.; Meyerson, H.; Kwon, I.K.; Matsuda, T.; Anderson, J.M. Instability of Self-Assembled Monolayers as a Model Material System for Macrophage/FBGC Cellular Behavior. *J. Biomed. Mater. Res.* **2008**, *86A*, 261–268. [CrossRef]
17. Pujari, S.P.; Scheres, L.; Marcelis, A.T.M.; Zuilhof, H. Covalent Surface Modification of Oxide Surfaces. *Angew. Chem. Int. Ed.* **2014**, *53*, 6322–6356. [CrossRef]
18. Karakouz, T.; Maoz, B.M.; Lando, G.; Vaskevich, A.; Rubinstein, I. Stabilization of Gold Nanoparticle Films on Glass by Thermal Embedding. *ACS Appl. Mater. Interfaces* **2011**, *3*, 978–987. [CrossRef]
19. Kyaw, H.H.; Al-Harthi, S.H.; Sellai, A.; Dutta, J. Self-Organization of Gold Nanoparticles on Silanated Surfaces. *Beilstein J. Nanotechnol.* **2015**, *6*, 2345–2353. [CrossRef]
20. Jensen, T.R.; Malinsky, M.D.; Haynes, C.L.; Van Duyne, R.P. Nanosphere Lithography: Tunable Localized Surface Plasmon Resonance Spectra of Silver Nanoparticles. *J. Phys. Chem. B* **2000**, *104*, 10549–10556. [CrossRef]
21. Near, R.; Tabor, C.; Duan, J.; Pachter, R.; El-Sayed, M. Pronounced Effects of Anisotropy on Plasmonic Properties of Nanorings Fabricated by Electron Beam Lithography. *Nano Lett.* **2012**, *12*, 2158–2164. [CrossRef]
22. Li, Z.; Zhang, X.; Ye, S.; Zhang, J.; Wang, T.; Fang, L.; Zhang, J.; Yang, B. The Fabrication of Long-Range Ordered Nanocrescent Structures Based on Colloidal Lithography and Parallel Imprinting. *Nanotechnology* **2013**, *24*, 105307. [CrossRef] [PubMed]
23. Jo, N.R.; Shin, Y.-B. Enhancing Biosensing Sensitivity of Metal Nanostructures through Site-Selective Binding. *Sci. Rep.* **2020**, *10*, 1024. [CrossRef] [PubMed]
24. Kwon, B.; Kim, J.H. Importance of Molds for Nanoimprint Lithography: Hard, Soft, and Hybrid Molds. *J. Nanosci.* **2016**, *2016*, 1–12. [CrossRef]
25. Aota, N.; Aida, H.; Kimura, Y.; Kawamata, Y.; Uneda, M. Fabrication Mechanism for Patterned Sapphire Substrates by Wet Etching. *ECS J. Solid State Sci. Technol.* **2014**, *3*, N69–N74. [CrossRef]
26. Shen, J.; Zhang, D.; Wang, Y.; Gan, Y. AFM and SEM Study on Crystallographic and Topographical Evolutions of Wet-Etched Patterned Sapphire Substrate (PSS): Part II. Cone-Shaped PSS Etched in H_2SO_4 and H_3PO_4 Mixture with Varying Volume Ratio at 230 °C. *ECS J. Solid State Sci. Technol.* **2017**, *6*, R122–R130. [CrossRef]
27. Xu, T.; Geng, Z. Strategies to Improve Performances of LSPR Biosensing: Structure, Materials, and Interface Modification. *Biosens. Bioelectron.* **2021**, *174*, 112850. [CrossRef]
28. Pellas, V.; Hu, D.; Mazouzi, Y.; Mimoun, Y.; Blanchard, J.; Guibert, C.; Salmain, M.; Boujday, S. Gold Nanorods for LSPR Biosensing: Synthesis, Coating by Silica, and Bioanalytical Applications. *Biosensors* **2020**, *10*, 146. [CrossRef]
29. Xue, Y.; Li, X.; Li, H.; Zhang, W. Quantifying Thiol–Gold Interactions towards the Efficient Strength Control. *Nat. Commun.* **2014**, *5*, 4348. [CrossRef]

biosensors

MDPI

Review

GLAD Based Advanced Nanostructures for Diversified Biosensing Applications: Recent Progress

Sarjana Yadav [1], Sneha Senapati [2], Samir Kumar [3], Shashank K. Gahlaut [1,†] and Jitendra P. Singh [1,*]

1 Department of Physics, Indian Institute of Technology Delhi, Hauz Khas, New Delhi 110016, India
2 School of Interdisciplinary Research, Indian Institute of Technology Delhi, Hauz Khas, New Delhi 110016, India
3 Department of Electronics and Information Engineering, Korea University, Sejong 30019, Republic of Korea
* Correspondence: jpsingh@physics.iitd.ac.in
† Current address: Institute of Chemistry, University of Potsdam, 14476 Potsdam, Germany.

Abstract: Glancing angle deposition (GLAD) is a technique for the fabrication of sculpted micro- and nanostructures under the conditions of oblique vapor flux incident and limited adatom diffusion. GLAD-based nanostructures are emerging platforms with broad sensing applications due to their high sensitivity, enhanced optical and catalytic properties, periodicity, and controlled morphology. GLAD-fabricated nanochips and substrates for chemical and biosensing applications are replacing conventionally used nanomaterials due to their broad scope, ease of fabrication, controlled growth parameters, and hence, sensing abilities. This review focuses on recent advances in the diverse nanostructures fabricated via GLAD and their applications in the biomedical field. The effects of morphology and deposition conditions on GLAD structures, their biosensing capability, and the use of these nanostructures for various biosensing applications such as surface plasmon resonance (SPR), fluorescence, surface-enhanced Raman spectroscopy (SERS), and colorimetric- and wettability-based bio-detection will be discussed in detail. GLAD has also found diverse applications in the case of molecular imaging techniques such as fluorescence, super-resolution, and photoacoustic imaging. In addition, some in vivo applications, such as drug delivery, have been discussed. Furthermore, we will also provide an overview of the status of GLAD technology as well as future challenges associated with GLAD-based nanostructures in the mentioned areas.

Keywords: glancing angle deposition (GLAD); biosensors; biomedical detection; GLAD substrates; plasmonics

Citation: Yadav, S.; Senapati, S.; Kumar, S.; Gahlaut, S.K.; Singh, J.P. GLAD Based Advanced Nanostructures for Diversified Biosensing Applications: Recent Progress. *Biosensors* **2022**, *12*, 1115. https://doi.org/10.3390/bios12121115

Received: 27 October 2022
Accepted: 23 November 2022
Published: 2 December 2022

1. Introduction

Biosensors have emerged in recent decades as a quintessential tool in the fields of healthcare, bioscience, the food industry, the chemical and biomedical industries, defense and security, and environmental monitoring [1]. Any system or chip consisting of a surface that is capable of binding to a specific biological substance that recognizes it, together with the ability to translate the biophysical or chemical reaction into a quantifiable signal, can be referred to as a biosensor [1]. The biosensor recognizing element may be nucleic acids, proteins, carbohydrates, lipids, antibodies, or an entire cell. These elements may be attached or immobilized on the sensor using various methods such as adsorption, bond formation, encapsulation, or entrapment [2]. Selectivity, specificity, reproducibility, stability, and sensitivity are various characteristics of a typical biosensor. Recently, the scientific community has witnessed vast development in optical, chemical, and physical transducers along with a specific affinity for biochemical interaction. This has led to vast progress in the development of biosensors in various fields. The most widely and extensively used biosensors are those that are based on nano-regimes due to their low detection limits, high stability, easy operation, low cost, short response time, and low power requirement [2]. Metallic-based nanostructures, therefore, possess specific unique features that ensure high sensitivity of the biosensors compared to other planar sensor configurations due to their nanoscale dimension and periodicity, thus ensuring strong binding of the target analyte and increased sensing properties [3].

The rapid advancement in the field of biosensing utilizing plasmonics has led to an extensive study of metallic nanostructures and their sensing applications [4–6]. One such emerging tool for fabricating metallic nanostructures is the glancing angle deposition (GLAD) technique [7]. It is a physical vapor deposition (PVD) process in which the gaseous flow of depositing atoms impinges on the substrate in a vacuum at an oblique angle such that the deposited film results in a columnar morphology [7,8] as a result of the shadowing effect. The tilting of the as-deposited column is controlled by the angle of incidence. The morphology of the columnar structure can be modified using a mobile substrate whose position can be managed during deposition [9,10]. The GLAD technique is considered better than the other chemical growth methods in terms of ease of fabrication of nanostructures due to controlled size, shape and spacing of columns in single step [11] Additionally, the nanostructures grown by this method possess high surface area to volume ratio [12], high porosity [13], crystallinity, and uniformity [11,14] making it the most suitable method for sensing applications. These structure has been widely investigated in designing different sensor devices [15], energy devices [16], and optical devices [17]. The GLAD technique has been extensively studied and developed over the past 15 years, but there is still much to explore.

The last few decades have witnessed a high increment in gold and silver-based nanostructures for catalytic [18] and biomedical applications such as immunotherapy, nanomedicine, and in vitro diagnostics [19–22]. Silver- and gold-based nanostructures are most widely used for sensing applications due to their high plasmonic properties [23–25], particularly their extinction band lying in the visible region [26] and low dispersion losses. Silver has a strong bactericidal effect due to the interaction of Ag ions with the thiol group and antimicrobial activity against microbes [27,28]. Gold has an excellent affinity for binding to antibodies [29], and hence, is most suited for antigen–antibody surface plasmon resonance [30,31] (SPR)-based detection. Silver- and gold-based nanostructures are shown to be the preferred transducers for LSPR-based sensing [32], surface-enhanced Raman scattering (SERS)-based sensing [33], as well as fluorescence resonance-based energy transfer [34] (FRET) sensing [35]. The highest SERS enhancement obtained using Ag nanorods prepared using the GLAD technique is 10^9 [36], so GLAD-based nanostructures combined with other techniques provides variations in metallic nanostructure shape, pattern, and period, extending the scope of their sensing abilities. The colorimetric- and wettability-based properties of these metallic nanostructures are found to be highly sensitive towards certain chemical and biological processes, revealing their use in biosensing detection [37–39]. The ongoing research in biosensor technology highlights an increase in the use of nanostructures and nanoparticles as a receptor or chip to enhance detection sensitivity and capability. Nanostructure-based biosensors or biochips fabricated using PVD methods exhibit advantages over conventionally prepared nanomaterials in terms of their high surface-to-volume ratio, fast response time, high versatility, multiple-fold increase in sensitivity, and better target-specific element recognition [40].

In this review, the biosensing and other imaging properties of GLAD-fabricated nanostructures are evaluated. We shall discuss the fabrication of these nanostructures with different types of substrate structures along with their implications in biosensing and other biomedical applications. The abilities of these nanostructures in various biosensing fields have been explored. A comparison of GLAD-based nanostructures with conventionally used nanoparticles and chips will be highlighted along with their advantages over the latter. There are various advantages of GLAD-based deposition over other chemical route depositions; for example, no precursor materials are needed in GLAD, which provides safety from toxic precursors and by-products [41]; atomic-level control of chemical composition and thickness can be afforded in this deposition [42]; and a higher temperature is not required and heat-sensitive substrates can be fabricated using this technique. GLAD is a green fabrication technique which produces very little pollution [43], promising high uniformity and repeatability in the structures [44–46].

2. Recent Strategies for Fabrication of Nanostructures Using the GLAD Technique

PVD methods such as sputtering, electron-beam vapor deposition, and thermal evaporation are most commonly used to fabricate thin films. When the vapor flux is normally incident over the substrate, thin-film formation takes place. However, if the angle of incidence is large with respect to the normal of the substrate surface (>70°), columnar structures are formed. This type of PVD deposition, carried out using either thermal evaporation or e-beam evaporation, is referred to as oblique-angle deposition (OAD) [3]. This extended OAD combined with azimuthal substrate rotation, giving various shapes and size of nano-columns, is referred as GLAD [14,47] or as dynamic oblique-angle deposition by some researchers [48]).

2.1. Setup Configuration of GLAD

Figure 1a shows the schematic of a GLAD setup where α is the angle subtended by the incident vapor with the substrate normal, called the deposition angle. Φ is the azimuthal angle that measures rotation about the substrate normal. The vapor flux falls on the substrate at a particular glancing angle (>700), initially leading to the formation of random nuclei, which is supported by the Volmer–Weber growth mechanism [49]. These further grow into isolated columns as a result of competition between shadowing and adatom diffusion due to larger nuclei and adatom surface diffusion. The columns lead to shadowing with size distribution and the screening of smaller neighboring nuclei from the coming flux. Therefore, the growth of the smaller nuclei is suppressed and ultimately ends due to complete shadowing [7]. This process of column extinction continues throughout the film's growth.

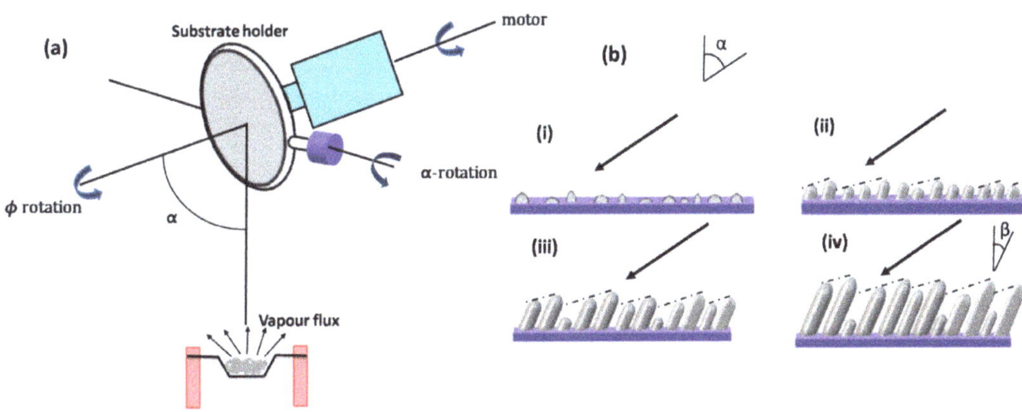

Figure 1. (**a**) Schematic of GLAD setup where α is the angle subtended by the incident vapor flux with respect to substrate normal and Φ is the azimuthal angle. (**b**) Mechanistic view of GLAD-based growth. (i) Incident vapor flux, shown by full arrows at an angle (α), resulting in random nucleation. (ii) Nucleus growth casting a nearby shadow, as shown by dotted lines. (iii) Column growth in progress, suppressing neighbors' growth via shadowing. (iv) Full column growth at a certain specific angle (β) with respect to the normal.

More vapor flux deposition takes place as the nucleus growth continues developing into isolated columns. Finally, the isolated columns are formed, and are tilted towards the source of the vapor flux. β is the column tilt angle. This process is shown in Figure 1b. It must be noted that β is different from deposition angle α, and $\beta = 0$ for the columns parallel to the substrate. The relationship between angles α and β is given by the tangent rule, given by Niewenhuizen and Haanstra [50], as

$$\tan\alpha = 2\tan\beta \tag{1}$$

when the deposition angle increases obliquely with respect to the normal, in which case, the tangent rule is not followed.

$$\beta = \arcsin\left(\frac{1-\cos\alpha}{2}\right) \tag{2}$$

2.2. *Effect of Various Growth Parameters*

2.2.1. Tilt Angle

As discussed earlier, the dependence of the tilt angle cannot be fully described in terms of angle of incidence by a single relationship. The general relationship is given by the tangent rule (Equation (1)), and Tait's rule [51] (Equation (2)) is satisfied only under certain conditions and must be cross-verified for a specific material under the given conditions such as deposition rate, pressure, substrate type, and temperature. All these factors influence the tilt angle β. Nakhodkin and Shaldervan investigated the nature of films of different materials (Ge, Si, GeTe) as a function of tilt angle [52]. They found that most of the films are formed of crystallites inclined to the substrate due to obliquely incident vapor flux. They classified and correlated the structures based on material, cross-section, and tilt angle up to a thickness of 300 nm. Salazar et al. used GLAD-fabricated Cu electrodes at a 65° tilt angle to study their sensing ability for the detection of glucose in blood samples [53]. Most of the recent reports showing GLAD-based biosensing applications have employed tilt angles ranging from 65° to 85° with high sensitivity and reproducibility [53,54].

2.2.2. Azimuthal Rotation (ϕ)

This represents the rotation of the substrate about the normal to the substrate. The azimuthal angle ϕ measures the rotation of the substrate, exploiting the growth direction steering the columnar structure along the growth path. The substrate position and the effect of azimuthal rotation on growth is depicted by α and ϕ and their variations. Control of both of these angles in a desirable manner during substrate rotation is the most crucial part of GLAD technology. If α and ϕ remain constant throughout the deposition, then it means that the substrate is stationary. α and ϕ may also change periodically at a given angle, but otherwise, may remain stationary [7]. There are six to nine basic GLAD structures that can be formed by different combinations of α and ϕ, which, along with their implications in biosensing, are discussed in detail in the next section. This variety of microstructures provides more enhanced properties that have a strong effect on the binding of biomolecules to the substrate and cause a change in the plasmonic properties of these substrates. Zhang et al. investigated the dependence of α and Φ on the biosensing ability of a Au nanostructured array fabricated over closely packed polystyrene balls and obtained the highest SPR and LSPR sensitivity for optimal values of α and Φ [55]. This regime as a function of tilt angle or azimuthal angle, due to the corresponding change in the substrate morphology and column growth, has not yet been fully explored and still has a long way to go.

2.3. *GLAD-Based Nanostructures*

The GLAD deposition technique can be used to create thin films of different morphology and sizes. As described in the previous section, the vapor flux falls on the substrate at a certain glancing angle ($\alpha > 70°$), resulting in the formation of microstructures because of competition between the shadowing effect and adatom diffusion. The formation and categorization of various GLAD structures depend on the variation in α and φ during the deposition process.

The formation of the plane metallic film takes place when the substrate is held perpendicular to the vapor flux ($\alpha = 0°$). However, this type of thin film is not very sensitive to biomolecule detection. Agrawal et al. has shown, using Rigorous coupled-wave analysis (RCWA) simulation, how nanohole array in a gold film can be employed for enhanced biosensing compared to planar gold film [56]. Some roughness over the surface of the nanostructure increases the sensitivity of biomolecules multiple-fold [57,58]. Slanted nanorods are the most commonly used GLAD structure for biosensing. These are formed when the substrate is held in the stationary position. α is taken to be greater than 70° (~86°). However, other desired micro- and nanostructures can be sculpted by varying the position of the substrate, and hence, the direction of the incoming vapor flux [59]. High sensitivity and plasmonic enhancement have been obtained using slanted nanorods with single arms and multiple arms [60]. Slanted nanorods have been utilized and studied widely in various biosensing applications using different materials (Ag, Si, Ni, TiO_2) [60–62] due to their ease of fabrication, lower material requirement, lower complexity, and high optimized sensitivity [36].

Chevron or zig-zag multi-armed nanorods are formed by a series of slanted nanorods. The substrate is initially held at a specific angle (α, φ) for the first arm, then, rotated by 180° (α, $\varphi + \pi$) for the second arm, and again by 180° (α, $\varphi + 2\pi$) for the third arm, etc.; that is, every time, the direction is rotated by 180° [7]. Square spirals can also be fabricated in the same way as chevrons with a rotation of 90°. The optimized slanted nanorods have shown enhanced biosensing and have been widely employed for SPR-, SERS-, colorimetric-, and wettability-based biosensing and detection. Sensitivity comparisons of various bioanalytes using various types of these GLAD-fabricated nanostructures have been performed by different groups [63]. Enhanced sensitivity, which can be employed for biological sensing using vertically standing nanoporous zig-zag or chevron nanorods providing hot spots with high SERS enhancement, has been shown by researchers [64,65]. Vertical nanorods are formed when the substrate is rotated at a constant rate during deposition. If the substrate is rotated at a constant slow rate, a helical columnar structure is formed with a constant pitch. Similarly, high and low morphology can be formed with a continuous change in φ while changing α in a discrete manner [66]. These structures form a Bragg stack and can be highly suitable for optical biosensing applications. Recently, Kumar et al. reviewed the fabrication and recent biochemical applications of various GLAD-fabricated substrates for SERS. They discussed, in depth, the advantages of these substrates and how these overcome the various limitations of conventional substrates [67], (Figure 2a–e). These GLAD-fabricated nanostructures have recently been employed in various biosensing detection settings. Recently, Kim et al. reported OAD-based LSPR substrates with a very high throughput by fabricating Au nanostructures (at 45°) over patterned sapphire substrates (PSS) [68] (Figure 2f). Very high LSPR sensitivity was obtained for various biomolecules and bioanalytes.

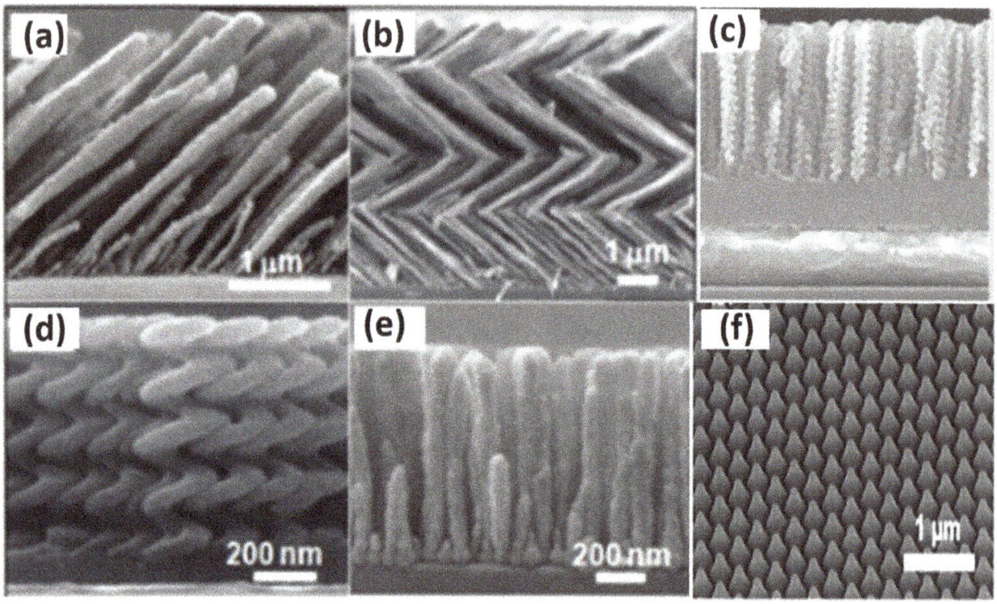

Figure 2. Different GLAD-structured thin films. (**a**) Slanted-post films; (**b**) zig-zag (chevron) films with 4 arms; (**c**) AgNR/SiO_2 SCL/SiO_2/PCL/Ag multilayer substrates; (**d**) helical film; (**e**) vertical post structure (reproduced with permission from [67]); (**f**) 45° tilted top view of Au mushrooms (reproduced with permission from [68]).

Hence, GLAD is a sophisticated technique used to design thin films with specific geometries. The geometries can be tailored in a controlled and desired manner to provide specific advantages for detection over conventional nanoparticles and nanofilms. The analyte binding sites, as well as the hotspots and scattering points (typically for optical sensing), can be increased in the desired way using this technique. The substrates can be fabricated in a controlled manner using multifunctional materials of different kinds by varying the substrate orientation and deposition conditions. The detailed bio-detection and sensitivity of various geometries and structures are yet to be explored in depth and could provide high scope in the future; they may provide certain specific binding sites for the proteins, amino acids, etc. present in biomolecules typically suited for SERS or fluorescence detection or antigen–antibody binding sites for SPR and LSPR.

3. Designing Substrates and Chips for Biosensing Applications

The sensitivity, binding, limit of detection, and other essential biosensing activities of a biosensor strongly depend on the substrate. However, there are only a few methods that can be used to develop uniform, reproducible, robust, stable, and cost-effective substrates. Recently, the GLAD technique has been employed to fabricate columnar thin-film substrates with high sensitivity [60,69]. Various studies have been conducted to study the sensing ability of the columnar Ag film [70,71]. Zhao et al. studied the s- and p-polarization absorbance spectra of Ag nanorod arrays with different heights and topologic shapes (cylindrical, needlelike, periodic, L, and Y shapes) in detail [72,73]. Additionally, the nanorods with complicated shapes, such as periodic, L, and Y shapes, had more hot spots for optical sensing applications than perfectly aligned nanorods. Thus, it is more suitable to observe the biosensing abilities of a small number of molecules on irregular nanorods than on a perfectly aligned nanorod. Recently, various types of GLAD-based nanostructures have been tested, optimized, and fabricated to increase the sensitivity of bioanalytes and expand their applications in the biomedical field. These substrates offer numerous advantages over the conventional GLAD-based substrates as they are more sensitive, durable, and flexible, and bind bioanalytes more effectively. The biosensing applications of the various newly fabricated substrates in the field of biomedicine and biosensing are discussed below.

3.1. Low-Temperature Biosensing Chips and Arrays

Aligned AgNR array films have been fabricated at low temperatures (LT; 140 K) and also at higher temperatures (HT; 373 K) [74,75]. The effects of temperature, from low to high, on the deposition of nanostructures and the morphology of films have already been discussed in the earlier section. Singh et al. deposited tilted AgNRs on glass slides at a substrate temperature of T = 140 K. The surface morphology changed drastically for the LT-deposited Ag nanostructured film. The aligned AgNR array SERS substrates deposited at a low-temperature oblique angle exhibited better SERS enhancement with good optical transmission compared to the standard room-temperature (RT)-deposited AgNR SERS substrates. The LT-deposited Ag nanostructured film was more porous, whereas the RT-deposited Ag nanostructured films appeared to be solid and rod-like. The nanorods with lengths of 1 μm, corresponding to a d = 2 μm QCM reading, deposited at room temperature showed the maximum spectral intensity, which is advantageous for optical (SERS, fluorescence, SPR) biosensing applications. However, for AgNR, the LT AgNR substrates, along with the maximum response, also provided substantial cost savings. A recent study by Yadav et al. demonstrated that magnetic surface-enhanced Raman spectroscopy (M-SERS) was effective in detecting hemozoin, a metabolic by-product of malarial parasites with paramagnetic properties. Using the GLAD technique, the SERS-active AgNRs were fabricated over neodymium substrates (0.3 T) kept at low temperature (120 K). Both hemozoin and hemoglobin were measured via a magnetic field-based SERS method on M-SERS substrates. In the absence of an external magnetic field, the intensity of the SERS signal was reportedly ten times greater than that of conventional SERS substrates. Figure 3a,b shows a SEM image of AgNR arrays fabricated over glass substrates at room temperature and at low temperature (120 K), respectively. Bunches of uniform nanorods were formed along the direction of vapor flux with slight variations in their lengths. The SEM images of AgNRs grown on magnetic substrates at a 120 K substrate temperature indicate the zone A type of morphology. A comparison in the XRD pattern between the crystalline phase of AgNRs at room temperature and at low temperature is shown in Figure 3c (JCPDS no. 04-0783). AgNR arrays exhibit a polycrystalline structure, as indicated by the XRD patterns. As a result of the differences in peak intensities and FWHM values between the two structures, a small difference in crystallinity can be observed.

3.2. Flexible Substrates

Compared to conventional hard and rigid substrates, flexible substrates and chips have several advantages. They are able to conform to the underlying object, allow efficient extraction of the target molecule from complex surfaces, have highly physicochemical properties, and can be integrated into wearable devices that provide bio-detection and health-monitoring services. Polydimethylsiloxane (PDMS) and polyethylene terephthalate (PET) sheets are the most common substrates used for the growth of AgNR on flexible substrates [76–78]. Singh et al. prepared a mechanical-strain-resistant AgNR on flexible PDMS and PET substrates [76]. The SERS response of these flexible SERS substrates was comparable to the AgNR on conventional glass slides. Flexible SERS substrates are capable of withstanding high tensile strain (ε) (30%) without losing their SERS performance. However, similar bending can reduce SERS performance by up to 13%. Kumar et al. fabricated flexible and robust SERS-active substrates by embedding the AgNR into the PDMS. Peipei and Young presented a flexible nanoprobe for high-performance real-time SPR biosensing. They constructed plasmonic optical fiber by transferring the metallic nanostructure from patterned templates to optical fiber using epoxy and reported a high figure of merit for SPR sensors.

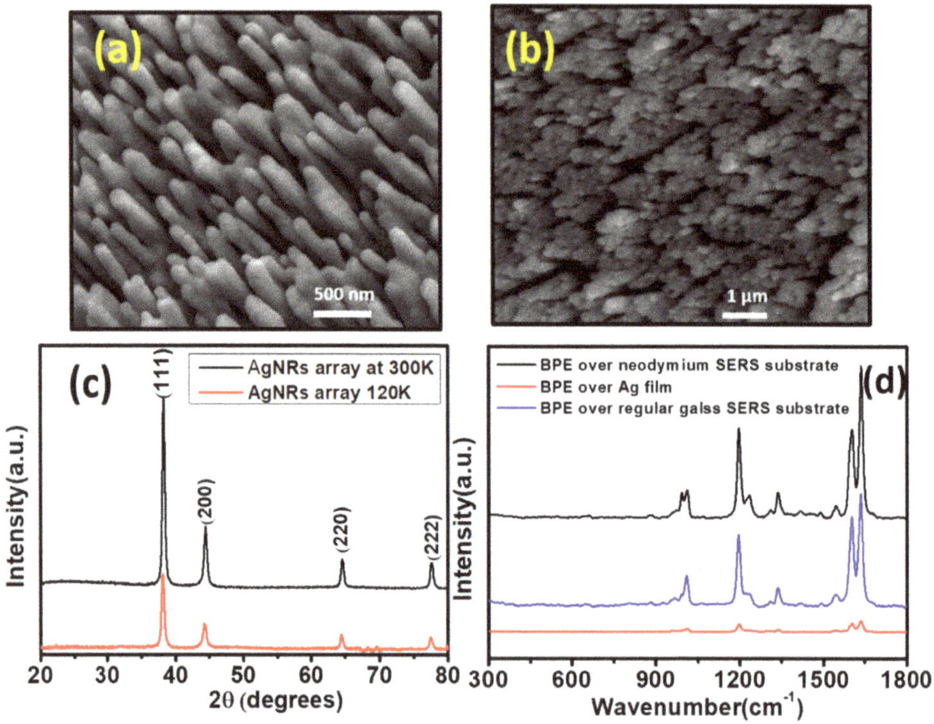

Figure 3. (**a**) SEM images of AgNRs over glass substrates at room temperature (300 K). (**b**) SEM images of AgNRs at 120 K deposited using GLAD technique. (**c**) XRD pattern for the two AgNR SERS substrates. (**d**) Comparison of the Raman and SERS spectra of BPE analyte over the three different substrates (c-SERS and M-SERS substrates, and 200 nm thin Ag film). Reproduced with permission from [63].

3.3. Multilayer Substrates

Increased sensitivity and strong field enhancement can be achieved in metal–dielectric–metal multilayer structures [79]. Suzuki et al. fabricated Ag and Au nanorods with quasi-parallel principal axes on a SiO_2 shape control layer (SCL) with a strongly anisotropic surface morphology [80,81]. They also demonstrated that the morphological and optical properties could be tuned by the deposition conditions without any pre- or post-treatment. Additionally, these structures supporting surface waves such as Bloch surface waves (BSW) could be used for advanced biosensing applications with enhanced sensitivity for protein interaction detection [82]. Multilayered structures have been used not only to tune the plasmon resonance but also to increase the number of hotspots [83,84]. Rajput et al. fabricated four-arm zig-zag Ag nanostructures of Ag and Al, with Al deposited on the edges of the zig-zag structure during growth [64]. The researchers etched less stable Al using a 2.5 wt% HCl solution from the Ag–Al zig-zag structure to generate intra-particle gaps, as depicted in Figure 4. These gaps at the elbows of the nanorods acted as additional hotspots while maintaining the zig-zag morphology. In their study, the SERS performance increased with the number of the increasing arms, which was attributed to the increased hot spots along the edges of the elbows of the zig-zag structure. The SERS performance further improved in the generation of nanoporous zig-zag Ag nanostructures obtained via de-alloying, and a high enhancement factor of $\sim 10^6$ was observed. These Ag structures are further under study for their applications in enhanced biosensing and early disease detection. Sun et al. constructed a sensitive biosensor based on self-assembled multilayer film for organophosphate pesticide using CdTe as a fluorescence probe. This multilayer film showed 100 times the sensitivity of CdTe in aqueous solution. Bhalla et al. created a new fabrication approach by exploiting the reactive plasma environment of SF_6 to assemble gold nanoplasmonic structures [85]. They developed mushroom-like gold-topped nanostructures on SiO_2 substrate. They used it as an LSPR chip for generic bioassay characterization and applications. A uniform protein coating was applied over the LSPR chip, achieving an LOD of 66 zM for biomolecules. These recent multilayered fabricated chips and nanostructures, therefore, are reported to show good stability and easy regeneration, along with increased bio-sensitivity. These can be employed as a promising platform for point-of-care diagnostics.

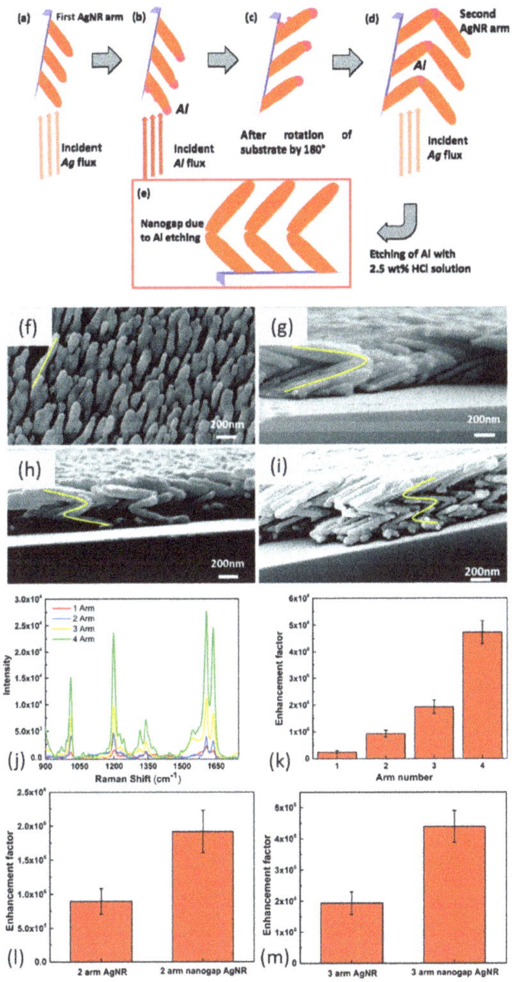

Figure 4. (**a**–**e**) An illustration of the fabrication of nano-porous zig-zag AgNR arrays; SEM images of zig-zag AgNR arrays on silicon substrate with: (**f**) 1 arm; (**g**) 2 arms; (**h**) 3 arms; and (**i**) 4 arms, respectively. Yellow lines show the approximate arm positions. (**j**) BPE SERS spectra (0.1 mM) acquired over zig-zag AgNR arrays with different numbers of arms. (**k**) Enhancement factor for the BPE SERS peak at 1200 cm^{-1} as a function of arm number. Calculated enhancement factor values for (**l**) 2-arm zig-zag AgNRs and 3-arm nanogap AgNR arrays and (**m**) 3-arm zig-zag AgNRs and 3-arm nanogap AgNR arrays. Reproduced with permission from [64].

3.4. Patterned Substrates

Patterned substrates have been demonstrated by various groups; periodically arranged nanostructures might result in enhancement in the localized electric field of plasmonic metals compared to that of randomly arranged nanostructures employed for optical sensing applications [86]. Studies have been performed to couple the periodicity of the grating with the metal nanostructure [87], localize the deposition within a micro-well, [88,89] formation of a hexagonal lattice of nanorods, [90,91] and couple discrete metal nanoparticles to photonic-crystal surface-resonant modes [92]. Researchers have demonstrated the feasibility of patterned substrate formation by combining GLAD and electron beam lithography; thus, they have designed numerous patterned structures in which structure period and structure diameter can be adjusted in a controlled manner, hence controlling the sensitivity [86]. Therefore, this type of substrate can be useful for numerous biosensing applications where patterned structure dimensions play an important role. Anisotropic organometallic halide perovskite nanowalls fabricated using the GLAD technique have recently been reported [93]. Their high plasmonic biosensing ability can be extracted and employed for early and low-cost efficient detection and analysis. Lee et al. [94] developed a unique 3-dimensional patterned substrate employed in ultra-thin-layer chromatography (UTLC) and SERS-based sensing consisting of Si nanowire arrays fabricated using a facile photolithographic and chemical etching technique, and further decorated it with Ag nano-dendrites. These patterned substrates can be further employed for SERS-based study for the distinction and detection of diseases and their co-infections. These multifunctional substrates could simultaneously be used for target localization, mixture separation and label free detection, and therefore, could be readily employed for biochemical assays and food safety. Kumar et al. proposed a flexible nanoslit-based sensor using patterns on periodic optical disks employing imprint lithography for periodic pattern printing on polydimethylsiloxane (PDMS) [95]. Patterned PDMS was obtained using a DVD-R disk and Ag was deposited over the patterned substrates using dynamic oblique-angle deposition at an angle 76° The sample was rotated by 180o after 5 nm deposition to form a nanoslit in the grating valley as a result of the shadowing effect. Films of thickness 50–400 nm were finally obtained. Figure 5 shows the SEM images (top view) of Ag nanoslit samples of varying thickness. They further investigated the SERS-based biosensing capability of these substrates using a bilirubin solution. Therefore, patterned substrates can be used as highly sensitive multipurpose substrates in biochemical sensing.

3.5. Coated and Co-Deposited Nanostructured Thin-Film Substrates

The practical application of Ag nanostructures has been limited by their chemical instability due to the oxidation, sulfuration, and etching of Ag. Alloyed or coated substrates made via the co-deposition of two or more materials have been studied for increased performance and shelf life [95]. There are various reports on increasing shelf life without compromising the sensing performance of a substrate either by coating the Ag with an ultra-thin layer of graphene, SiO_2, [96] Al_2O_3, [97] or by fabricating bi-metallic nanostructures [95,98–100]. The use of low-temperature atomic layer deposition (ALD) to wrap AgNR arrays with a very thin but dense Al_2O_3 layer can improve their robustness in morphology up to 400 °C, and can provide surface coverage that stabilizes SERS activity in air without reducing their sensitivity to SERS.

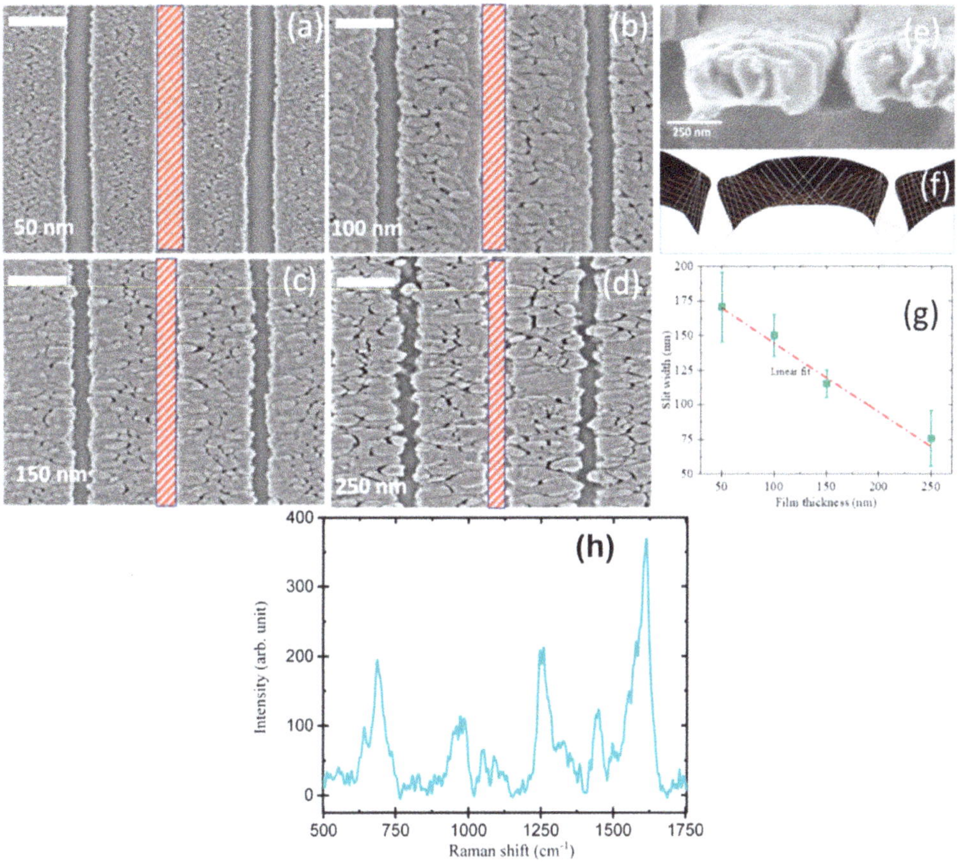

Figure 5. SEM images of Ag nanoslit substrates (top view) with thicknesses of (**a**) 50 nm, (**b**) 100 nm, (**c**) 150 nm, and (**d**) 250 nm. (**e**) Cross-sectional view of 100 nm thickness. (**f**) Net thickness of the film deposited at different substrate points (black region) with orange dotted lines showing incident flux direction and gray arrows showing columnar growth direction. (**g**) Plot of slit width vs. film thickness with a scale bar of 250 nm. (**h**) SERS spectra of 0.06 mg/mL bilirubin over Ag nanoslit substrate corresponding to 100 nm thickness. Reproduced with permission from [87].

3.6. Recyclable Chip and Substrates

Most of the chips and substrates fabricated for sensing applications are for one-time use only, and given the value of noble metals, these cannot be fully explored as a routine analytical technique. Much study has, therefore, been focused on developing reusable substrates [101]. Kumar et al. fabricated a reproducible and reusable Ag nanoparticle-decorated TiO_2 SERS substrate since TiO_2 is renowned for its photocatalytic properties. The intensity of the Raman signal decreases rapidly in the presence of UV light because the dye concentration decreases rapidly. Approximately 150 min after illumination with UV light, these Ag-TiO_2 nanorod SERS substrates recover completely. Hu et al. [102] reported a highly recyclable and sensitive biosensing platform comprising carbon-doped TiO_2 nanotube arrays. These can be refreshed and regenerated photo-catalytically, maintaining high selectivity and sensitivity when irradiated with UV or infrared light. Hunt and Armani [103] developed a new idea to generate recyclable optical microcavity via a dry chemistry method using oxygen plasma treatment. This recycling treatment, therefore, helped in reducing cost and preventing harsh and wet chemistry treatment for recycling.

3.7. *Microfluidics-Based Substrates*

There is tremendous interest in developing nanofluidic channels with a large surface-to-volume ratio that exhibit localized surface plasmon resonance due to the strong interaction between the fluids and the channel walls [104]. The advantages of the nanofluidic and microfluidic platforms are: reduced unwanted and nonspecific binding, and detection time minimization by studying multiple analytes or ligand interactions using a single device [105]. By using these nanofluidic channels with plasmonic nanostructures, SERS signals from molecules trapped near electromagnetic hotspots were enhanced. A monolayer of Ag film over silica nanospheres (AgFONs) was fabricated on a glass surface and covered with a polydimethylsiloxane (PDMS) slab. This amplifies SERS by a factor of ten or more, allowing label-free detection of important neurotransmitters in the nanomolar range, rather than using only plasmon resonance.

4. Biosensing Applications of GLAD-Fabricated Nanostructures

4.1. *SPR-Based Detection*

Surface plasmon resonance is a powerful analytical method for detecting the high sensitivity of biomolecules, and is based on plasmonic materials' optical properties. The SPR method has shown great promise with large sensitivity for biomolecular applications [106], especially for real-time biomolecule interaction, protein action, antigen, and nucleic acid detection [107–109]. SPR-based sensors are useful for label-free detection because of the very high sensitivity of the surface plasmon polaritons (SPP) to changes in the local environment in the measurement of both refractive indices and dielectric constants [110,111]. SPR-based biosensors possess the advantage of high versatility and can be tailored for the detection of a vast number of analytes since this type of detection does not demand any special characteristic of the bioanalyte, such as fluorescence properties, or absorption or scattering bands. Additionally, SPR-based bio-detection does not require analytes or the biomolecules to possess fluorescent or radioactive labels.

The primary sensing phenomenon in SPR is related to propagating surface plasmon polaritons. A surface plasmon polariton (SPP) is an electromagnetic wave propagating at the metal–dielectric boundary. The electromagnetic field of an SPP confined at the metal–dielectric interface decreases exponentially in both metals and dielectrics [112]. The dielectric constant of the propagating surface plasmon wave is highly sensitive to changes in the local refractive index of the dielectric. This property is the underlying principle of SPR biosensors. Metal-binding molecules or the recognition element attached to the surface of the metal capture the analyte, resulting in a change in the local refractive index at the metal surface. This change in the local refractive index creates a change in the propagation constant, which can be easily measured using various methods112. Figure 6 shows the underlying mechanism followed during SPR sensing using the Kretschmann configuration.

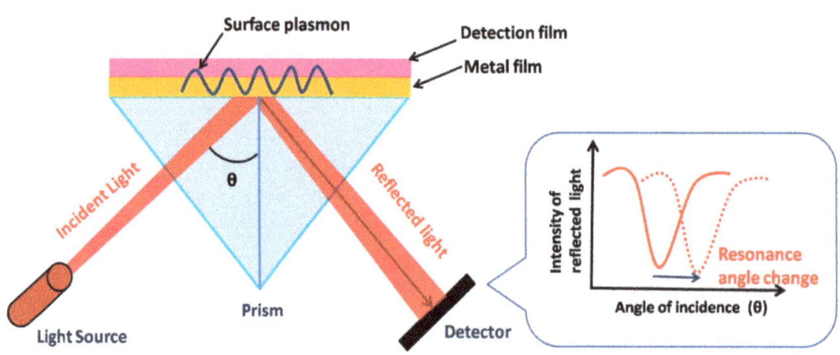

Figure 6. Schematic of SPR-based detection of biomolecules using Kretschmann configuration. The dip in the resonance peak shifts due to the change in the refractive index.

Recently, new strategies, such as the generation of frictional charges using triboelectric surfaces, have been employed for the detection and generation of localized surface plasmon resonance (LSPR) wavelength shifts in large-area Au nanostructured surfaces [113]. These LSPs with frictional charges, using recent methods such as spectroscopic and triboelectric measurements, have been found to show a complex interplay of bioconjugation and bio-complex detection. These budding strategies in combination with GLAD-fabricated nanostructures can be employed for enhanced biosensing detection and applications. GLAD-fabricated metallic thin films facilitate the development of SPR- and LSPR-based biosensors, possessing the advantage of a considerable increment in sensitivity. These nano-sculptured thin films have been demonstrated, both theoretically and experimentally, to possess high SPR sensitivity [114,115]. Additionally, their periodicity, gap, etc., can be tailored in a desired way, hence the binding of bioanalytes and sensitivity enhancement [115].

4.1.1. Bi-Material-Based SPR Enhancement

The properties of two different materials favorable for experimental sensing studies can be exploited together through the fabrication of biomaterial GLAD nanostructures. Additionally, using biomaterials also helps to eliminate the unfavorable properties of either of the two materials; for example, silver is highly plasmonic and provides a sharp SPR curve with high sensitivity but becomes easily oxidized. The oxidation of silver can be eliminated if a thin layer of some inert material is used over its surface. Graphene, possessing high impermeability [116] for gases and easy adsorption [117] ability for biomolecules, can be used to coat Silver to provide high SPR sensitivity. Wu et al. reported an SPR-based graphene biosensor in which the graphene layer was coated over a gold sheet. The sensitivity of the chip was shown to increase several times through calculations compared to the conventional gold film used for SPR biosensing. In another work, Choi et al. [118] presented a numerical model showing a layer of graphene coated over a 60 nm silver film. The SPR sensitivity was shown to increase several times due to the graphene coating and also to prevent the oxidation of silver. Over the graphene surface, a bimolecular DNA hybridization was modeled as a thin dielectric layer. The group showed improvement in the SPR bimolecular sensitivity of the graphene-coated substrate compared to conventional gold film chips employed for SPR biosensing applications. Graphene adds to the SPR sensitivity in comparison to the gold film due to its better adsorbing properties of biomolecules; this is due to the pi interactions between the biomolecules' carbon ring structure and graphene hexagonal cells [119,120]. Gold/zinc oxide (Au/ZnO), as well as gold/chromium (Au/Cr)-based nanocomposite thin films, have also been studied and were used effectively for enhancing SPR performance for the detection of a carbohydrate antigen, which is a tumor marker for breast cancer. The linear range for the SPR sensor was shown to increase from 1 to 40 U/mL [120]. Therefore, bimetal GLAD structures can be used to increase SPR-based bio-sensitivity along with increasing the inertness of the metallic structures.

4.1.2. Nanocolumnar Thin-Film-Based SPR Chips

As discussed in the earlier section, thin-film nanocolumns fabricated using the GLAD technique provide high sensitivity for the detection of bioanalytes. They are easy to fabricate, possess a high surface –volume ratio, require less material, and are broadly used as SPR chips. The use of these metallic nanocolumnar thin-film biosensors enhances both angular as well as spectral sensitivity due to increased roughness and porosity, and induces SPR broadening due to scattering losses inside the porous metallic film [120,121]. Kaur et al. [122] reported ZnO thin-film-based SPR detection for cholesterol biosensing applications. ZnO was coated over the gold film using the GLAD technique. ZnO/Au prism was used in the Kretschmann configuration. The cholesterol oxidase (ChOx) enzyme had been immobilized over the ZnO surface. PDMS microchannels fabricated over the ChOx/ZnO/Au/prism setup were used to pass different cholesterol concentrations, and SPR reflectance was obtained for both static and running mode, showing high SPR sensitivity. Figure 7a shows the SPR reflectance curves for various cholesterol concentrations over the ChOx/ZnO/Au prism, and Figure 7b shows the calibration curve for the variation in the resonance angle with cholesterol concentrations. However, in most of the work discussed above employing SPR detection using thin metallic films, the sensor chips were fabricated as planar thin films using GLAD-based techniques. The experimental use and applications of GLAD-based columnar thin films using different structural morphologies are still limited for biosensing applications. The effect of porosity on the SPR dip for different metallic nano-columnar thin films (Al, Ag, and Au) fabricated via GLAD using Kretschmann configuration has also been studied. The SPR dip widens with an increase in the porosity of the thin film, and finally, disappears near the TIR region. This group showed the vast possibility of SPR-based biosensors utilizing the porosity of the metals with increased sensitivity compared to nonporous films.

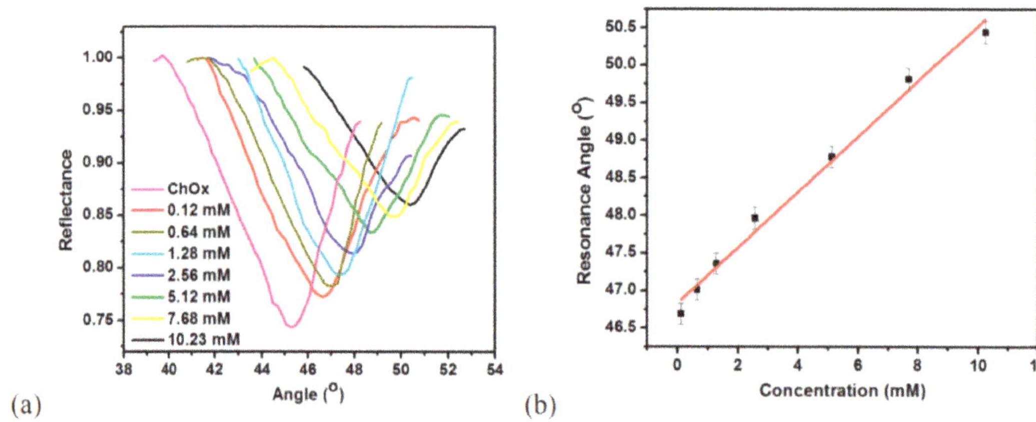

Figure 7. (**a**) SPR reflectance curve for ChOx/ZnO/Au prism corresponding to different cholesterol concentrations. (**b**) Calibration curve for variation in resonance angle with cholesterol concentrations. Reproduced with permission from [122].

Shalabney et al. [123] have explained this phenomenon very well both via theoretical calculations as well as experimentally. They carried out SPR measurements in Kretschmann configuration for GLAD-fabricated Ag nano-columnar thin films for different values of β and studied the SPR sensitivity in both angular as well as spectral modes. They indicated that for more porous films, the resonance wavelength shows a clear redshift of about 15 nm with respect to less porous films, and hence, higher sensitivity. Further, a theoretical calculation of the dielectric tensor of the material was carried out. There was an enhancement in the electromagnetic field intensity distribution as there was an increase in the volume interaction of the evanescent field in both the columnar void region and the sample area. These nanocolumnar thin films provide a high increment in sensitivity compared to planar metallic films. Agrawal et al. [114] presented a 2D nano-pillar array with nano-gaps for biosensing. They used rigorous coupled-wave analysis (RCWA) simulations to study the SPR sensitivity. The biosensor performance was studied in terms of sensitivity, characterized by spectral shift $\Delta\lambda$. The effect of polarization of the light incident was studied using simulations, which showed that nano-pillars' biosensing capability is independent of incoming light. The sensitivity of SPR chips can be most suitably defined in terms of differential reflectance amplitude (DRA) since the sensor performance also depends on reflectance contrast along with reflectance change. Agrawal et al. [56] also demonstrated, in another work, the increased sensitivity of the SPR detection of nanohole arrays of gold compared to planar gold film (Figure 8). The SPR sensitivity of the biosensor had been calculated in terms of DRA. They carried out RCWA simulations to study the effect of variation in the diameter and period of the nanoholes. Byun et al. [124] experimentally confirmed the high SPR bio-sensitivity of a gold nanowire array compared to conventional gold film. They used an angular interrogation scheme to perform biosensor characterization and showed that their results were in agreement with RCWA simulated results. They fabricated gold nanowires with two different periods—200 nm and 500 nm—with each showing increased sensitivity of 44% and 31%, respectively, when evaluated using ethanol of varying concentrations.

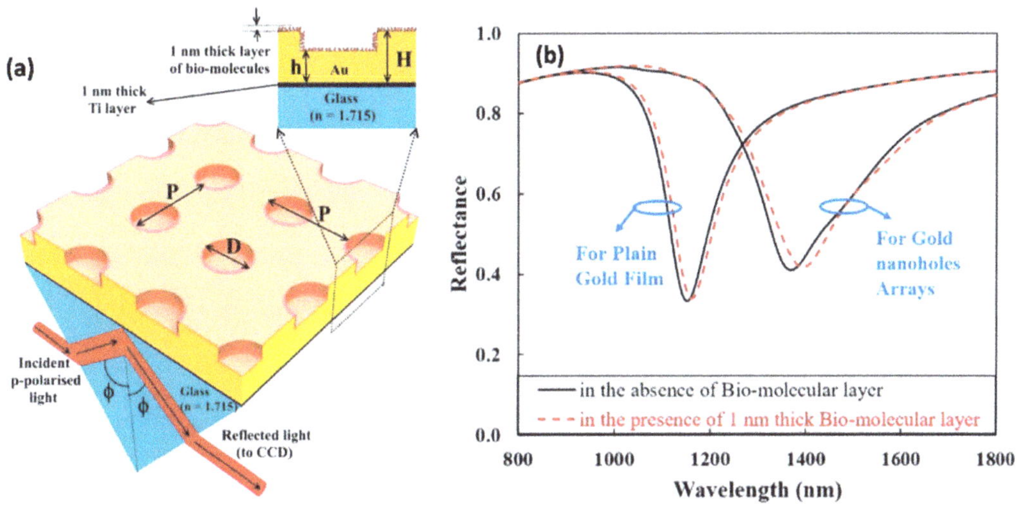

Figure 8. (**a**) Schematic diagram of circular nanohole array over the gold film. P, D, h, and H represent array period, array diameter, gold film thickness below the nanohole, and gold film thickness without the nanohole, respectively. This array is further coated with a 1 nm thick layer of biomolecules. (**b**) Reflectance spectra calculated for a planar gold film of thickness 60 nm and for gold film structure array with a period (P) and diameter (D) of nanohole array of 50 nm and 30 nm, respectively; gold film thickness without nanohole (H) of 60 nm; and gold film thickness below nanoholes of 40 nm. Reproduced with permission from [56].

Combining the two different ideas discussed above, one may suggest the fabrication of nanocolumnar thin films consisting of GLAD-engineered structures of different morphologies combined with other materials (ZnO, graphene, and gold) to increase bio-sensitivity. This may lead to us to ultra-high sophisticated SPR biosensors with multiple-fold sensitivity increments.

4.1.3. Microfluidics-Based SPR Chips

Microfluidic microchannels and systems are mainly employed for the SPR imaging approach. The advantages of the microfluidic platform are: reduced nonspecific binding and detection time minimization by studying multiple analytes or ligand interactions using a single device [105]. SPR-based microfluidic assays have been developed for various biomolecular interactions such as antibody-binding targets [125], multi ligand/analyte biosensing [126], and foodborne pathogen detection [127]. Malic et.al. [128] reported a multichannel droplet-based SPR detection platform comprising electrowetting on a dielectric microfluidic device for SPR imaging. The device was fabricated using Cr and Au layers as SPR supporters followed by the patterning of 24 um photoresist as a lift-off mask. A 50 nm Au film was etched to define a thin Au layer as ground. So, a spot-by-spot detection achievement was employed. Natarajan et al. [129] reported a 3D microfluidic system for protein sample deposition to discrete spots of a target. They showed exposure of each spot to a larger sample volume due to continuous flow and increased uniformity of the spots, and hence, expanded the protein array applications for protein printing. A compact and robust microfluidic flow cell for SiO_2 film followed by a PDMS microfluidic surface has been developed [130]. This biosensor flow has the advantage of operation up to 185 kPa in an aqueous environment. A unique droplet-based SPR biosensing platform coupled to a surface acoustic wave microfluidic system (SAW) has also been reported [131]. Surface acoustic wave excitation in continuous mode for different levels of power generation was studied. Droplet-based microfluidic SPR systems possess the advantage of limiting mass transport in comparison to conventional microfluidic flow.

Though most microfluidic systems focus on on-chip fabrication for SPR imaging, the major challenges of cost-effectiveness and the simplicity and compatibility of the detection and monitoring of SPR signals still require further effort and research. GLAD-fabricated arrays and chips can be very promising candidates for highly sensitive SPR imaging purposes.

4.2. *SERS-Based Biosensing*

Surface-enhanced Raman scattering (SERS) is the enormous amplification of Raman signals from molecules, by several orders of magnitude, when adsorbed on metal colloidal nanoparticles or a rough metal surface, discovered in 1974 by Fleischmann et al. [132,133]. The detection of analyte molecules even at a single-molecule level can be realized by employing the SERS technique, which expands its practical applications [134]. SERS has found applications in various fields, including the detection of trace chemicals [135] such as dye molecules, food additives, [136] and pesticides, [77] and in bioanalysis, medical diagnosis, [36] the detection of biomolecules, [137] cancer diagnosis, [138] in vivo molecular probing in live cells, [139] and explosives detection [140]. SERS substrates are the nanostructured platform that supports plasmon resonance and amplifies Raman signals [141–143] and are broadly classified as random morphology or ordered/periodic SERS substrates. Random morphology SERS substrates (which include roughened electrodes, metallic silver and gold colloids, and metal-island film on a planar substrate) are inhomogeneous and are not highly reproducible [144]. Periodic arrays of metallic nanostructures (using nanolithography and other physical vapor deposition techniques) can overcome this issue by providing uniform and controlled morphology of SERS substrates [145].

SERS has become a powerful technique for chemical and biological sensing applications due to its excellent sensitivity and specificity. Uniform and highly reproducible SERS substrates with batch-to-batch reproducibility that have a SERS signal variation of less than 15% can be fabricated using the GLAD technique [146]. Using SERS multivariate statistical analysis, various pathogen species or even strains can be differentiated. Therefore, extrinsic detection using Raman labels is more widely explored.

4.2.1. Detection of Pathogens

SERS substrates fabricated via GLAD have been utilized for the detection of human pathogens, particularly respiratory syncytial virus, human immunodeficiency virus, and rotavirus; the bacteria of mycoplasma pneumonia, tuberculosis, E. coli, and Pseudomonas aeruginosa; and the identification and classification of various virus strains. [54] Gahlaut et al. demonstrated the diagnosis of dengue (DENV) from clinical serum samples using a Ag nanorod array-based hand-held SERS platform [36]. Using the GLAD-deposited SERS substrate and principle component analysis (PCA), the DENV-positive, -negative, and healthy samples were analyzed and differentiated (Figure 9). They identified the SERS spectra of pure NS1 protein, both separately and in serum, and the corresponding characteristic peaks of NS1 protein. This portable, rapid, and cost-effective method can diagnose dengue early in the field, and with a high degree of sensitivity and reproducibility. Kumar et al. demonstrated a facile and straightforward method to increase the SERS enhancement of bacteria [78]. They deposited AgNRs on 30% pre-stretched PDMS using the GLAD technique and directly pipetted bacterial suspension (3 μL, 10^8 cells mL^{-1}) onto the AgNRs grown on the stretched PDMS substrate. They performed SERS measurements before and after releasing the strain of the stretched AgNR-PDMS SERS substrate. The release of the strain led to the formation of high-density hotspots in the AgNRs, allowing the bacteria and AgNRs to interact more effectively. Compared to a pre-stretched AgNR-PDMS film, these buckled AgNR-PDMS array substrates showed an approximately eleven-fold enhancement in SERS. Yadav et al. demonstrated a rapid SERS platform for rapid HIV-1 detection. The group performed a detailed study over five different HIV-1 subtypes. HIV-1 envelopes possessing glycoproteins bind to the AgNRs SERS substrate without any external reagent, thus providing specific signature peaks of HIV-1. Additionally, they showed a distinction in SERS spectra based on tropism. Their proposed technique could successfully distinguish between X-4 and R-5 tropic HIV viruses [54] (Figure 9d).

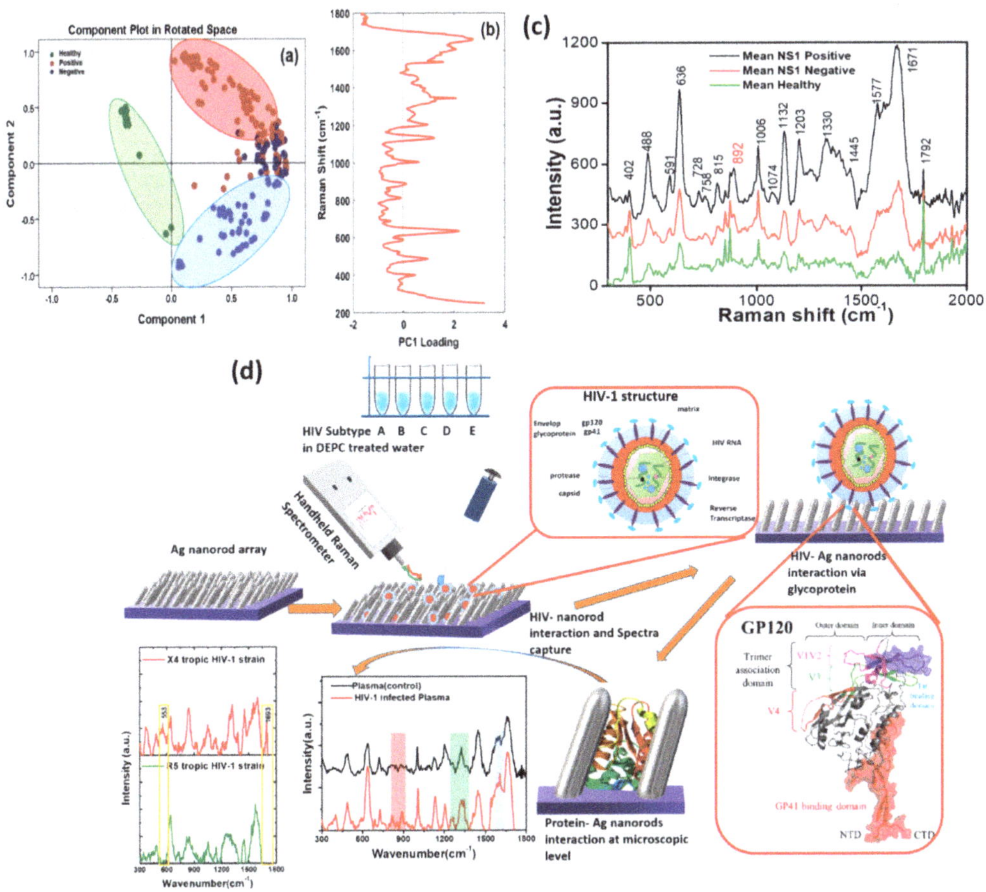

Figure 9. (**a**) Figure showing principal component 1 versus principal component 2 for three different groups of samples (antigen NS1-positive, -negative, and healthy). (**b**) Graph showing Raman shifts with the PC1 loading of plot (**a**). (**c**) SERS spectra obtained for the healthy, NS1-negative, and NS1-positive sera collected from the patients (each spectrum is an average of 20 spectra). Reproduced with permission from [36]. (**d**) Schematic diagram showing rapid handheld SERS platform for early HIV detection. Reproduced with permission from [54].

4.2.2. Detection of Biomarkers

A biomarker is any indicator of the severity or presence of a particular disease in an organism. Kumar et al. demonstrated the potential application of Ag nanoslit (AgNS) SERS substrates through bilirubin detection. Bilirubin is a toxic metabolite of heme reduction, and excessive levels of bilirubin result in well-known symptoms of liver jaundice. The group detected the SERS signal of bilirubin with a concentration much below the normal concentration in human blood using this AgNS substrate [87]. Yadav et al. used hemozoin and exploited its magnetic properties for the early and easy detection of malaria. They presented a unique magnetic field-based surface-enhanced Raman spectroscopy (M-SERS) technique for magnetic biomarkers. SERS-active Ag nanorods were grown over magnetic neodymium substrates (0.3 T) at low temperature using the GLAD technique. Figure 10 shows the complete mechanism of M-SERS and the obtained SERS spectra. The obtained limit of detection of hemozoin for M-SERS was as low as 10^{-11} M (< 10 parasites/μL), meaning it can be employed for early-stage malaria detection [63]. Srivastava et al. developed a biosensor chip for the detection of a protein biomarker of endocrine-disrupting compounds, vitellogenin (Vg), in an aquatic environment [147]. The researchers immobilized the anti-Vg antibody on 4-Aminothiophenol-coated Ag thin films for the specific detection of Vg. They also performed a control experiment to demonstrate the specificity of the sensor towards Vg using fetuin for the control experiment, which is a glycoprotein like Vg. The SERS substrate also exhibited high stability over a period of one year. SERS-based sensors for the detection of specific biomarkers is, therefore, a well-established sensing phenomenon and can be employed with advanced substrate growth techniques for enhancing sensitivity to the single-molecule level.

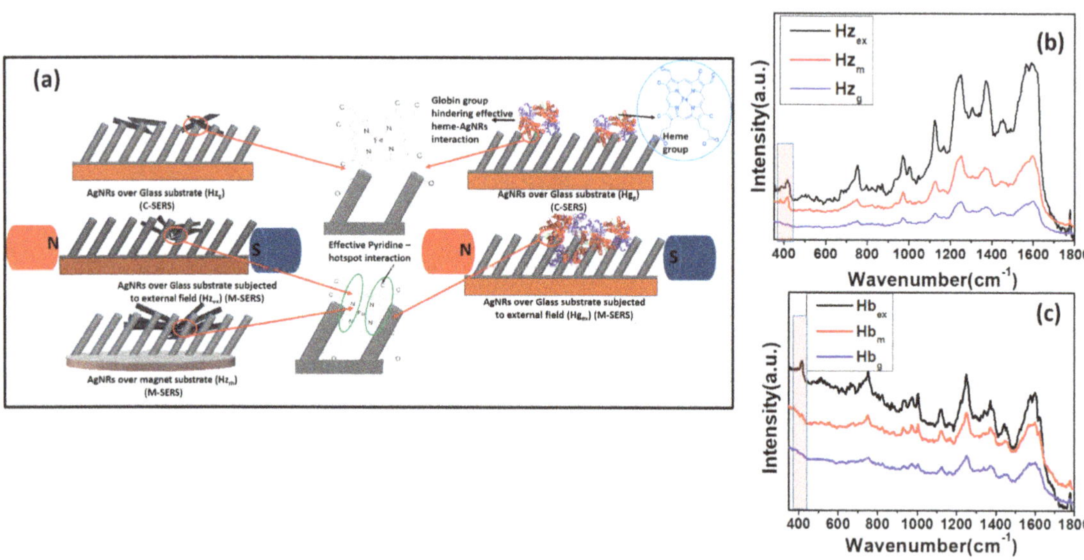

Figure 10. (**a**) Schematic diagram showing pyridine–AgNR interactions for the unmagnetized Hz and Hg over AgNS substrate (c-SERS), in the presence of a magnetic field (M-SERS). (**b**) Comparison of SERS spectra for Hz on AgNR arrays fabricated over glass (Hz_g) (neodymium magnet (Hz_m), Hz_g substrate kept at 0.3 T external magnetic field for 1 h (Hz_{ex})) and (**c**) Hg on AgNR arrays fabricated on glass (Hb_g) (neodymium magnet (Hb_m), Hb_g substrate kept at 0.3 T external magnetic field for 1 h (Hb_{ex})). Reproduced with permission from [63].

4.2.3. Detection of microRNA and Neurotransmitters

MicroRNAs (miRNAs) are a class of non-coding RNAs that play essential roles in regulating gene expression and also act as biomarkers for disease [148]. The evaluation of miRNA expression patterns is limited to cumbersome assays that often lack sensitivity and specificity to distinguish between different miRNA families and members. A SERS platform for the detection and classification of miRNA using GLAD was demonstrated by Driskell, and Tripp [149,150]. They detected and differentiated five unrelated miRNAs and eight members of the hsa-let-7 family using SERS. In their study, the SERS platform overcame the limitations of data normalization, empirical probe selection, and RNA labeling or amplification depending on the intrinsic stability and specificity of the reagents used in the PCR-based assays. By using SERS, microRNAs can be rapidly and specifically detected without amplification, providing a molecular fingerprint of each analyte and requiring only limited amounts of sample for analysis. Additionally, it improves comprehension of the spectrum properties of miRNAs, making it easier to interpret the results.

Oh et al. demonstrated the biosensing capability of microfluidic-based SERS substrates fabricated using GLAD by sensing two major neurotransmitters, named dopamine and gamma aminobutyric acids (GABA) [104]. Interstitial nanogaps between AgFONs and the PDMS channels were shown. Capillary force was used to inject a very small amount of analyte at a very low concentration into these nanofluidic channels. Due to the coexistence of nanofluidic stagnation points and hotspots with plasmon resonance, the SERS signals were greatly enhanced. Dopamine and GABA solutions at different concentrations were prepared in distilled water. SERS signals were detected at different concentrations in the nanofluidic channels and compared with those in the reservoir. It was found that the SERS signals increased with increasing concentrations of the neurotransmitter. While the peak intensity of GABA at 1370 cm^{-1} was 20 times higher in the nanofluidic channel than in the reservoir, the SERS peak intensity of dopamine at 1605 cm^{-1} was eight times higher in the nanofluidic channel.

4.2.4. Detection of Pesticides and Chemicals

Pesticide and trace pollutant detection are essential because of the hazard to the environment and public health caused by the accretion of these harmful chemicals. SERS substrates fabricated using the GLAD technique have been extensively utilized for the detection of these harmful chemicals [76].

Kumar et al. demonstrated a highly optimized SERS-active substrate for the easy and rapid detection and extraction of target molecules from complex surfaces [77]. The researchers directly collected traces of the pesticides thiram and chlorpyrifos from an apple surface using AgNRs embedded in PDMS substrates using a simple "paste and peel off" method (Figure 11). They were able to detect a Raman band at 1386 cm^{-1} even at a concentration of 10^{-6} M R6G, outperforming more established methods such as high-performance liquid chromatography (HPLC). They reported a limit of detection (LOD) of thiram of 2.4×10^{-9} g/cm^2, demonstrating the practical application of SERS substrates for the rapid trace detection of thiram, with an LOD much lower than the permissible limit for apple peels (~2×10^{-6} g/cm^2). Hou et al. developed SiO_2 nanorod (NR)@Au nanoparticles (NPs) for the detection of monochlorobiphenyl (CB) as low as 1×10^{-6} M with an enhancement factor of 10^8 [151]. Moreover, the authors also demonstrated the reusability of the SiO_2@Au SERS substrate in order to detect trace CB.

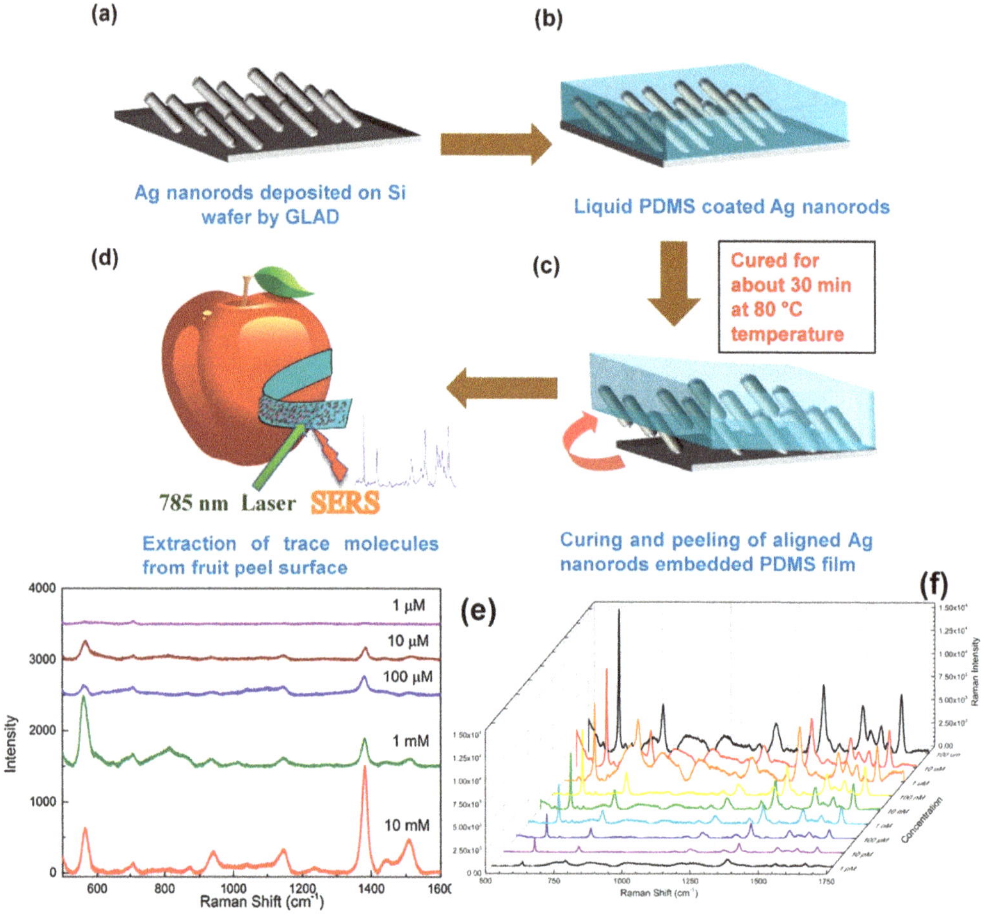

Figure 11. Schematic showing fabrication of AgNR-embedded PDMS SERS substrates (**a**–**d**). (**e**) SERS spectra of *thiram* pesticide. (**f**) SERS spectra of rhodamine 6G with different concentrations on AgNR-embedded substrate. Reproduced with permission from [77].

4.2.5. Detection of Drugs and Food Adulterants

The GLAD technique for SERS substrate fabrication has been used for the trace detection of methamphetamine and its primary metabolite, amphetamine, in human urine [152,153]. The rapid onsite screening of food contamination in quality control and market surveillance applications has also been reported [154]. Nuntawong et al. proposed an alternative method for the pre-treatment of urine-based samples to prevent urea from binding to the surface of SERS-active substrates [152]. Assuming that the acidic dilatation treatments would convert the urea to the much less soluble urea nitrate, allowing precipitation from the mixture, and that the remaining urea would be negatively inhibited from adhering to the SERS-active surface, they added dilute nitric acid to the urine samples. Consequently, they performed the SERS measurements from a series of mixtures of the urine-based samples and the diluted nitric acid via a compact SERS detection system using a portable Raman spectrometer.

Sodium saccharin (SS) is a commonly used synthetic sweetener in carbonated soft drinks and beverages. Han et al. developed a sensitive SERS method for the quantification of SS in soft drinks [155]. SERS peak intensity and SS concentration in the concentration range of 0.5–100 mg/L showed an excellent linear relationship, with a LOD of 0.3 mg/L. They also performed PLS-DA analysis to determine the LODs of SS in four soft drinks—Sprite (20 mg/L), Cola (5 mg/L), Fanta (10 mg/L), and Schweppes (20 mg/L)—which were far below the national standard.

Therefore, GLAD is a versatile nanofabrication technique for the design of nanostructured thin films for SERS sensing [67]. However, there are still a number of challenges that must be overcome before SERS can be used as a common analytical technique, like HPLC, for complex food matrices and pathogens in the real world. These include the high cost of the Raman spectrometer and the lack of commercially available, low-cost SERS substrates. When SERS measures trace amounts of any analyte in a food, background interference from food components is still a major problem. Therefore, further research is needed to successfully advance the SERS approach in practical applications of food analysis and disease identification.

4.3. Fluorescence-Based Biosensing

Fluorescence is defined as 'a short-lived kind of luminescence', created as a result of electromagnetic excitation caused when light energy is absorbed in a short wavelength, and then, emits the light at a longer wavelength [156]. Fluorophores or fluorochromes are the molecule that shows fluorescence. Fluorophores emit energy in the form of light radiation or sometimes dissipate in the form of heat. The basic concept of biosensing using a fluorophore involves the coupling of target recognition with some change in the fluorescence of the reporter molecule. Different types of fluorescence biosensors can be proposed based on the nature of the sensing and binding elements [157]. The change in the intensity of fluorophores attached to the recognition element is very sensitive to the local environment and is monitored directly. This change in the fluorescence intensity may be due to a change in fluorophore–biomolecule interactions, [158] or due to fluorophore–target molecule interactions, for example, the interaction of single-stranded DNA protein binding [159]. The basic advantages of fluorescence-based biosensors are its simplicity, its ability to detect smaller ligands, the availability of many donors and acceptors, its convenient transducing optical signals, and its suitability for short distances. Metallic nanostructures are known to enhance the fluorescence of fluorophores [160–162]. The fluorophores interact with the plasmonic field at the metal surface, therefore increasing the local field, resulting in enhanced fluorescence intensity [160,161,163]. The metallic structures are known to increase the local electric field as well as the radiation decay rate up to a factor of 1000 [161]. The sensitivity of fluorophore detection due to combined local field increments and increased photostability is enhanced by 10^5 [164]. Metallic nanostructures are employed for enhanced fluorescence detection of DNA [165], protein micro arrays [166], pathogens, cancer cells [167], and single species in tissue samples [168]. Enhanced fluorescence detection using chemical method-based metallic nanoparticles is, however, limited due to the inhomogeneity of the structure and non-proportion of fluorophore interaction with metals that provide enhancement. Plasmonic fluorescence enhancement by PVD-deposited metal nanostructures tends to provide increased enhancement due to the formation of periodic arrays over a large surface area. Controlled growth parameters and deposition rates may provide a large array of nano-columns with different shapes, sizes, periodicity, and interparticle separation, which can be regulated on the metal surface. Different applications of fabricated metal nanostructures in enhanced fluorescence-based bio-detection are discussed.

4.3.1. Protein- and DNA-Fluorescence-Enhanced Detection

The development of optically active metallic nanostructure substrates enhancing fluorescence can be used for detection using fluorescence-tagged analytes in DNA analysis and protein detection assays. Yang et al. [169] described, in their paper, a selective and sensitive carbohydrate microarray for glycan protein interaction detection. They deposited gold nanostructures using thermal evaporation, coating it further with a silicon carbon alloy layer, which was subjected to a surface-linked azide group for immobilization of glycan through the formation of covalent bonds. Investigation of the enhanced fluorescence-sensing ability of metal nanostructures via the characterization of carbohydrate lectin interactions has been conducted. The fluorescence-labeled lectins Concanavalin A (ConA) and peanut agglutinin (PNA) were chosen for specific interaction with lactose- and mannose-modified LSPR interfaces (Figure 12). In another work, Li et al. [166] presented highly sensitive fluorescence-enhanced detection in a silver metal microarray. Detection of IgE and streptavidin has been conducted using aptamers for tag-C preparation. The silver nanostructures formed using tags and fluorescence-enhanced solutions provided excellent enhancement. Fluorescence enhancement studies for DNA microarrays, with the help of GLAD-deposited silver nanorods of various lengths, have been performed; the highest enhancement of 200 folds is reported for Ag nanorods of 500 nm in length, which is multiple times higher than the commercially available amine slide [165]. Zang et al. [170] reported that fluorescence intensity enhancement increased 114 times using a GLAD-fabricated TiO_2 nanorod film applied over the surface of a label-free photonic crystal device. The fluorescence enhancement of the device was studied both experimentally and theoretically (RCWA simulations) for cy5-conjugated streptavidin application, and the enhancement of the signal was compared with the device without nanorods. These studies showed that GLAD-fabricated nanostructures can be an invaluable tool to study sensitive fluorescence enhancement for bio-detection and similar devices can be designed for sensing applications.

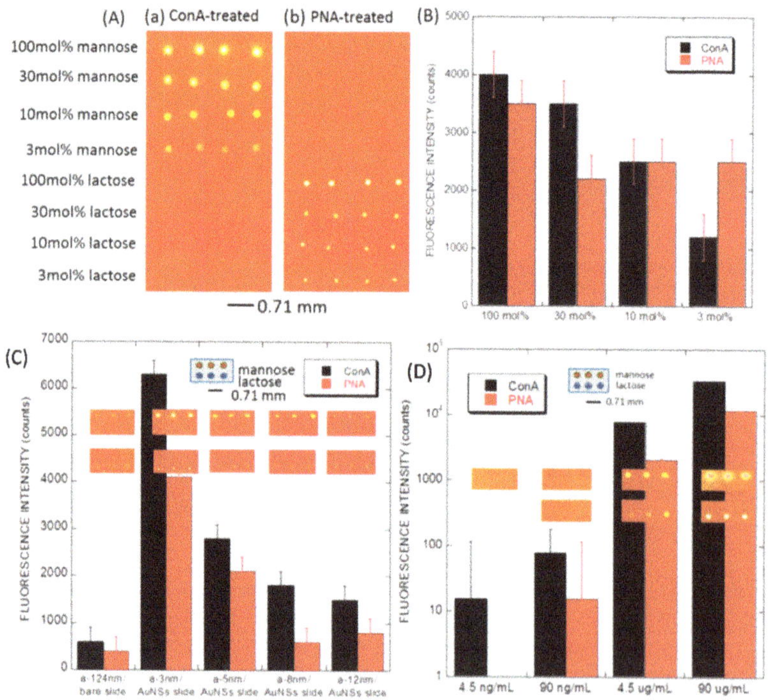

Figure 12. (**A**) Fluorescence images arrays spotted after interaction with (**a**) Con A(0.9 mg/mL) and (**b**) PNA (0.9 mg/mL). (**B**) Histogram of corresponding fluorescence intensity for diluted glycan spots. (**C**) Fluorescence histogram curve for spotted arrays with 100% lactose and 100% mannose after interaction with ConA and PNA (with a fluorescence exposure time of 2 s) followed by PBS rinse (**D**). Fluorescence histogram for spotted 3 nm thick Si 0.8 C0.2:H Au nanostructure array with 100% lactose and 100% mannose after ConA and PNA interaction (fluorescence exposure time of 5 s) followed by PBS rinse. Reproduced with permission from [169].

4.3.2. Detection of Pathogens

The rapid, cost-effective, selective, and sensitive detection of bacterial pathogens is very important for disease treatment, preventing bioterrorism, diagnosis, and foodborne disease identification. However, the majority of methods used are laborious, costly, and time-consuming. The fluorescence-based detection of these pathogens using nanostructures, however, can overcome these limitations. Geddes et al. [171] patented a method in which they provided a system whereby metal particles placed over a quartz substrate are attached to a biomolecular probe that possesses an affinity for the target pathogen. The binding target pathogen in contact with the biomolecular probe causes fluorophores to be at a sufficient distance from the metal structure, resulting in enhanced fluorescence emission when irradiated by a source. Therefore, they developed an entire platform for the rapid and sensitive identification of target pathogens via fluorescence emission. Park et al. [172] developed a novel biosensor for rapid pathogen detection using hetero-nanorods. The GLAD-fabricated silica nanorods were gold-sputtered and the dye immobilized via bond formation between the primary amine group and the dye ester. This was followed by antibody conjugation and bacteria detection. The entire detection process is shown in steps in Figure 13. The dye molecules attached to the nanorods resulted in an enhanced fluorescence signal and Salmonella detection using microscopic imaging. Hence, GLAD-fabricated nanostructures of different materials can be employed for fluorescence-based biosensing to detect foodborne pathogenic bacteria.

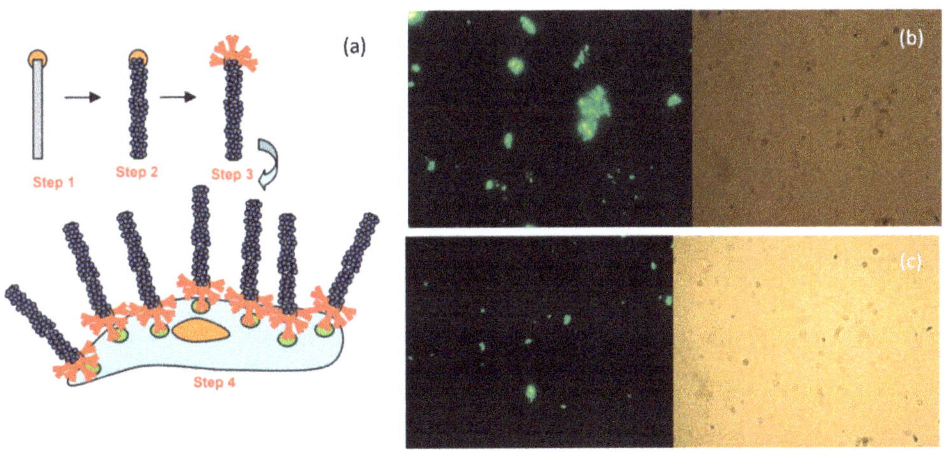

Figure 13. (**a**) Step-by-step process for Salmonella detection using Au/Si rods comprising nanorod fabrication, dye immobilization, antibody conjugation, and bacteria detection through antigen–antibody conjugation, respectively. Fluorescence microscopic images with BSA blocking under fluorescent light and white light for (**b**) anti-salmonella/Au/Si nanorods and fixed salmonella and (**c**) Au/Si nanorods and fixed salmonella. Reproduced with permission from [172].

4.3.3. Single-Molecule and Tissue Detection

Single-molecule detection techniques are, nowadays, vastly explored. These techniques provide a new powerful idea to explore biochemical and life science components at the elementary level. However, conventional optical methods face diffraction-based limitations [173]. Fluorescence-based detection enhancing the signals, even for very small analyte concentrations, opens up new possibilities for single-molecule detection.

Zhang et al. [174] reported target molecule imaging using fluorescent material probes. They synthesized silica metal nanoshells encapsulated by $Ru(by)_3^{2+}$ to obtain fluorescence properties. Avidin silver complexes were formed, and imaging measurements indicated that individual complexes could be distinctly isolated while the tissue remained stained by organic dyes. Hence, the group successfully demonstrated that metal-based nanoprobes could be employed for single-target-molecule tissue detection during fluorescence imaging. Metal nanoprobes have also been used to detect CCR5 receptors over the cell surface using a covalent bond between anti CCR5 monoclonal antibodies and silver nanoparticles [168]. The mAb-metal complex showed a single nanoparticle emission. Punj et al. [175] described that plasmonic nanoparticle dimer antennas can be extremely promising for single-molecule enhanced fluorescence detection at micromolar concentrations. They quantified the fluorescence enhancement and detection volume for various self-assembled nanoantennas using fluorescence spectroscopy. Hence, the combination of metallic nanostructures along with fluorescence enhancement can be used for an extended variety of single-molecule detection applications.

4.4. Colorimetric- and Wettability-Based Detection

Sensing based on the change in color and surface wettability has also been realized on GLAD substrates, especially plasmonic metallic arrays. Silver has the highest reflectance among all metals, at over 97% throughout most of the visible region and about 99% in the IR region. Pure silver looks shiny and whitish in color, as seen by the naked eye. However, silver nanoparticles exhibit a plasmonic effect with a very high extinction coefficient, and plasmonic absorption lies in the visible range and varies with particle size and shape, interparticle separation, and the refractive index of the surrounding medium [176]. A homogenous colloidal solution of nanoparticles of a particular size scatters a specific wavelength of light, which makes them promising for distinguishable colorimetric visual readout sensors. Such properties of metal nanoparticles enable them to be exploited in various analytical tools, e.g., the absorbance or fluorescence spectroscopy. Colorimetric-based assays have been developed by illustrating changes in the color associated with the aggregation of noble metal nanoparticles [177–179]. However, the employment of GLAD-fabricated silver nano-columnar thin film in such sensing applications is limited. The pristine aligned AgNR array fabricated by GLAD in a high vacuum looks bright due to the multiple scattering and multimode localized surface plasmon. In addition to optical-based sensing, these nanorod substrates have also been studied for gas and biosensing by exploiting their novel characteristic colorimetric properties. The chemistry between silver and sulfur has been studied by different groups [180–182]. Graedel et al. extensively explored the reaction between silver- and sulfur-containing gaseous molecules and found out the dependence of the reaction rate on relative humidity. Thereafter, Chen et al. demonstrated the use of AgNP films as H_2S gas sensors. According to their findings, the reaction between AgNPs and H_2S gas follows a first-order reaction rate law and is proportional to the 1.3 power of the H_2S gas concentration [182–184]. This relationship was used to determine the H_2S gas concentration under ambient conditions. The intensity of the LSPR peak of the AgNP's films decreases and exhibits a shift upon exposure to H_2S gas. Though they have several applications in art conservation, AgNPs are used to detect the emission of H_2S from degraded materials, e.g, aged wool fabrics, rubbers, etc. under ambient conditions. A GLAD-fabricated AgNR array was employed as an H_2S gas sensor by Gahlaut et al. [185]. The authors demonstrated a dual-mode, i.e., colorimetric- as well as wettability-based detection, using a mobile app which calculates the change in greyscale intensity of the AgNR array upon exposure to H_2S gas. The optical darkness ratio (ODR) was defined as the relative change in the average pixel intensity of pristine and exposed AgNR arrays. The variation in ODR values is shown in Figure 14b.

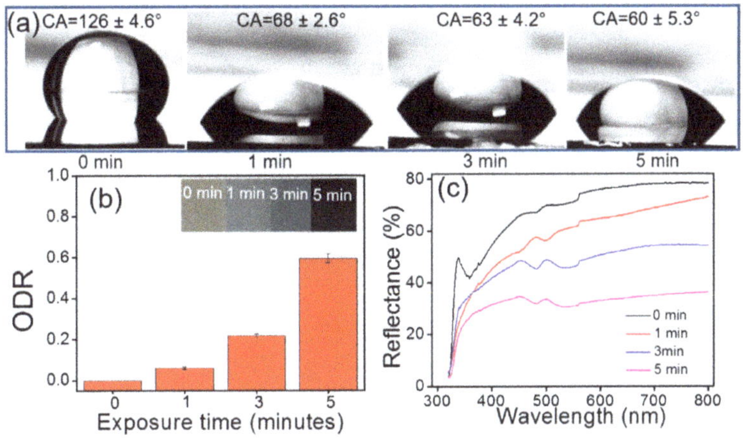

Figure 14. Variation in (a) water contact angle and (b) optical darkness ratio (ODR); inset shows respective photographs. (c) Surface reflectance of AgNR array samples after 5 ppm H_2S exposure times of 1, 3 and 5 min. Reproduced with permission from [185].

The limit of detection and response time of the detector were determined as 5 ppm and 30 sec, respectively. The method was also applied to the detection of H_2S emission from aged wool fabric placed in textile museums and art galleries to prevent artifacts made of silver. In another report [38], the same group also demonstrated the determination of viability and antibiotic resistance in bacterial cells based on the colorimetric change in an AgNR array, as shown in Figure 15. The nanorod array was found to be highly sensitive and selective toward H_2S gas in the presence of sub-ppm concentrations under ambient conditions.

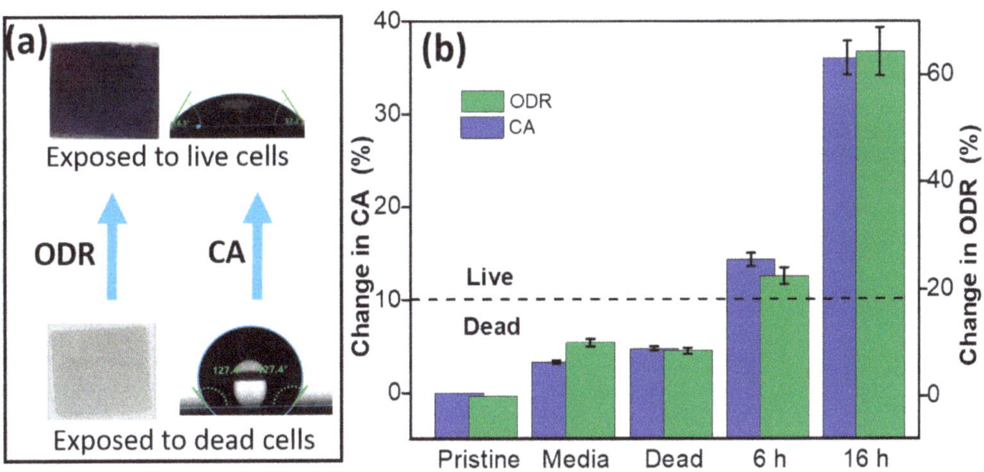

Figure 15. Determination of live and dead *E. coli* bacteria using a change in ODR and CA. (a) A digital photograph of the sensing array and water droplet on the array for the ODR and CA measurements. (b) Graph showing the percentage change in ODR and CA for pristine array and after 6 h and 16 h of culture. The media and heat-killed bacteria were considered as a control. A demarcation line is drawn representing the threshold values above which the live bacteria are distinguishable from dead bacteria. Reproduced with permission from [38].

Ahn et al. [186] introduced a highly inexpensive and facile colorimetric system for selective H_2S detection in living cells using a silver-embedded Nafion/polyvinylpyrrolidone (PVP) membrane. The membrane was applied to a polystyrene microplate cover. A known concentration of the $H2_S$-donor Na_2S solution was used to evaluate the performance of the in vitro colorimetric detection assay with the silver/Nafion/PVP membrane, which was performed both at room temperature and at 37 °C in an incubator containing 5% CO_2. Last but not least, they used the silver/Nafion/PVP membrane to determine the endogenous H_2S concentration in live C6 glioma cells stimulated with and without Cys and L-homocysteine. The rationale behind the detection of H_2S gas was that silver reacts with the gas to form Ag_2S, which has a brown color.

Hao et al. [187] developed a highly sensitive H_2S detection platform by using single gold and silver core-shell nanosphere for mapping the local distribution of H_2S in living cells. The concept of sensing was the same colorimetric variation of the Ag shell in the presence of sulfide. The formation of Ag_2S on the surface led to a large difference in the refractive index between Ag (~0.17) and Ag_2S (∼2.2), which resulted in a redshift in the plasmonic spectrum, allowing highly sensitive detection of H_2S.

Their biocompatibility and efficient thiol binding ability make gold nanoparticles (AuNPs) an ideal candidate for rapid biosensing based on the sensitivity of the surface plasmon resonance to the aggregation state, which produces a visible color change. Additionally, therefore, antibody-conjugated AuNPs have been used in the rapid colorimetric detection of pathogenic bacterial species in multiple systems, [188–191] Huan et al. [191] have developed a simple phage display for calorimetric detection of a variety of bacterial species, (Figure 16). This stage involeves Au nanoparticles for simple, efficient and sensitive detection of bacterial cells upto 100 cells with no cross reactivity between different species.

In addition to gold, silver nanostructures have been employed in plasmonic-based colorimetric sensors [192,193]. Dong et al. [194] demonstrated colorimetric detection of thiosulphate using AgNPs by measuring the variation in RGB value via a smartphone application. Another novel selective colorimetric method for the detection of iodide using citrate-stabilized silver triangular nanoplates was reported by Yang et al. [195]. Researchers have also demonstrated a colorimetric detector using helically structured Ni nanofilms on silicon substrates fabricated via GLAD and found that these films exhibited peroxidase-like activity to catalyze the oxidation of tetramethylbenzidine (TMB) [196]. This colorimetric detection of uric acid was able to achieve a low LOD (3.3 μM) and was found to be sensitive throughout the entire clinically relevant range (15−500 μM).

GLAD-fabricated sculptured thin films have been extensively used for chemical and biosensing, especially for exploiting their LSPR and SERS properties. However, for colorimetric- and wettability-based applications, there are only a few reports are available; therefore, there is a gap yet to be filled. In recent years, considering the huge possibility of tuning the porosity of GLAD-fabricated substrates, there is large scope for surface wetting modification in a wide hydrophilic to superhydrophobic range. Some studies have been carried out on anisotropic wetting and water droplet evaporation on nano-columnar thin films for self-cleaning and the Leidenfrost effect [197–200]. Wettability-based sensing could be a promising method for the detection of analytes. Gahlaut et al. demonstrated a GLAD nanorod array of silver for H_2S gas sensing by observing a rapid and drastic change in the water wetting property of the array. As-grown AgNR arrays were found to be hydrophobic with a contact angle of 128°. Exposure to H_2S gas led to the formation of Ag_2S on the surface of the AgNR array and resulted in the enhancement of wettability with a water contact angle of 60° [185]. They also presented a novel method of bacterial viability detection using the wetting behavior of the AgNR array and further discriminated antibiotic resistance in bacterial species [38]. The used AgNR array was reutilized in the form of Ag-Ag_2S nano-heterostructures for various energy and environmental applications, e.g., water purification, hydrogen evolution, and antibacterial activity [201]. All these novel applications of nano-sculptured thin films grown via oblique-angle deposition signify promising future scope in various domains.

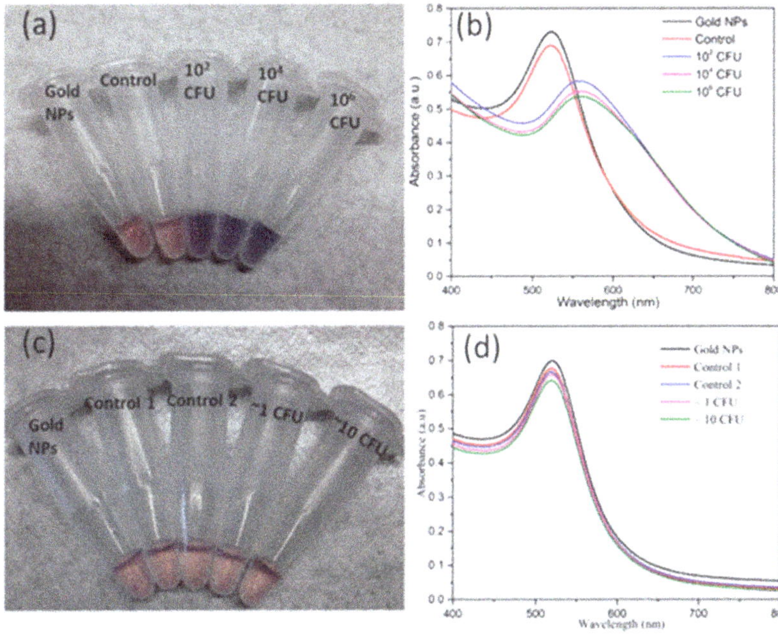

Figure 16. (**a**,**c**) Digital photographs showing detection of *E. coli* using thiolated M13KE and AuNPs. (**b**,**d**) UV–vis spectra. From left to right in (**a**), samples containing AuNPs and: no bacteria or phages (black line in (**b**)), unmodified M13KE with 10^6 CFU *E. coli* (red line in (**b**)), and thiolated M13KE with *E. coli* at different CFU's (10^2, 10^4, and 10^6) (blue, magenta, and green lines, respectively, in (**b**)). From left to right in (**c**), samples containing AuNPs and: no bacteria (black line in (**d**)), unmodified M13KE phage and ~10 CFU of *E. coli* (red and blue lines in (**d**), respectively), and thiolated M13KE phage and ~10 CFU of *E. coli* (magenta and green lines in (**d**), respectively). Reproduced with permission from [191].

4.5. Molecular Imaging

The complex biological processes that occur in the cell and its microenvironment can be visualized using a molecular technique. Using this technique, the internal mechanisms of living systems can be visualized. Additionally, certain specific molecules can be studied and the pathways related to the molecules can be unveiled [202]. Disease progression and drug intervention can be effectively studied using imaging. The resolution and noise-to- sound ratio (NSR) is a challenge in molecular imaging. To improve both of these features, the design and development of probes used in imaging in nanoscale regimes have an encouraging future [203]. Contrast materials show unique physical and chemical properties in the nanoscale zone, and this feature is exploited in this molecular imaging. Different chemical and biological methods are used for the fabrication of NPs used in the molecular imaging technique. However, the fabrication cost and the complexity of its preparation has limited its use in biomedical applications. In this current section, we will be discussing the GLAD-based fabrication of NPs and its application in different molecular imaging applications [204,205].

4.5.1. Fluorescence Imaging

Being a very versatile technique, fluorescence finds a wide variety of applications in biomedical fields; these include sensing, bioimaging, diagnosis, and detection [206]. Signal intensity is one of the challenges of fluorescence and it can be met by using metal-enhanced fluorescence (MEF) [207]. MEF is a very powerful and sensitive technique whereby the fluorophores interact with metallic nanoparticles, resulting in an enhancement in fluorescence. Different NP structures affect the plasmonic resonances, which, in turn, enhance fluorescence. The plasmonic effect, along with the enhanced electric field, increases the signal intensity of fluorophores [172,208,209].

Dhruv et al. used a GLAD-fabricated Ag nanorod array to study surface-enhanced fluorescence (SEF), which showed very good enhancement due to the morphology of nanostructures. The spectra of Rh6G were taken on both Ag nanorods and reference substrates. SEF with the length of the nanorods was also found by the group [162]. Jonghyun et al. fabricated Ag nanorods using oblique-angle deposition (OAD) for enhanced fluorescence substrates for improving the sensitivity of microarray analysis used in medical diagnostics. A maximum enhancement of 23 was obtained from the substrates [209].

M.A. Badshah et.al used GLAD-fabricated vertical nanorods for understanding their effect on the sensitivity of DNA microarrays. The group used a kallikrein-related peptidase 7 (KLK7) ssDNA duplex with cyanine 5 dye for checking the sensitivity of the Ag nanorods [165]. Similarly, X Ji et al. used zig-zag Ag nanorod arrays fabricated using GLAD and studied the improved protein and DNA detection. Biotin–neutravidin fluorospheres were attached to the zig-zag Ag nanorods. The enhancement was found to increase with the folding number of the zig-zag nanorods. Additionally, temperature was found to affect the florescence enhancement. Finally, the DNA detection limit was found to be 0.01 pM [210].

4.5.2. Super-Resolution Imaging

On the grounds of medicine, imaging plays a very significant role. Optical microscopy, due to non-invasiveness and simplicity in use, has always been a priority for biologists. Its spatial resolution is a major setback in the case of imaging [211]. Over the years, super-resolution microscopy (SRM) has taken place in imaging techniques to overcome the drawback of optical imaging. Super-resolution microscopy's higher resolution helps researchers study single molecules and their interactions [212]. Different SRM methods have been developed over the years, such as stochastic optical reconstruction microscopy (STORM), stimulated emitted depletion (STED), photoactivated localization microscopy (PALM), near-field scanning optical microscopy (NSOM), and fluorescence photoactivation localization microscopy (FPLM) [213–215]. Along with the high-frequency spatial information, contained in evanescent waves, this type of microscopy uses mathematical modeling and a series of diffraction-limited images, and also optical, electrical setups, to increase the diffraction limit to sublevels. Still, there exist some limitations in super-resolution microscopy; for example, NSOM requires lengthy scan times, whereas FPLM works only with a single-wavelength light source [216,217]. Overall, super-resolution microscopy uses sophisticated instrumentation and is expensive. So, in order to improve the signal-noise-ratio without using a complex setup, different nanostructures are used. Additionally, one of the important applications of super-resolution microscopy is single-molecule imaging; single-molecule imaging is feasible within a limited fluorophore concentration range of 1 pM to 10 nM, whereas protein interaction and enzyme activity require a fluorophore conc. of 1 uM or more. Hence, single-molecule imaging is difficult for biomolecules [218–220]. So, in order to tackle this challenge, the density of hotspots should be increased.

Chen et al. used GLAD to fabricate a patterned substrate. To improve the hotspots and enhancement, plasmonic gratings with periodic nanogaps were developed. Silver film growth on polymethylsilsesquioxane (PMSQ) polymer gratings were replicated from HD DVD gratings molds. The silver was deposited at an angle of $\alpha = 60°$ using GLAD. SM fluorescence images were taken on these GLAD-based gratings for a DNA/RNA duplex tagged with cyanine 3 and cyanine 5 labels. GLAD gratings with incorporated nanoprotrusions enabled SM imaging over a wide range of fluorescence from 50 pM to 10 uM. Additionally, the nanoprotrusions allowed the group to use multiple-emitter fitting analysis to solve the crowded field problem created by higher fluorophore concentration. Patchy microspheres fabricated using GLAD setup have also shown potential application in improving the diffraction limit of super-resolution microscopy [221]. Wood et al. performed similar work whereby they used DVD to produce gratings in polymethylsilsesquioxane

(PMSSQ) deposited on a glass substrate via microcontact lithography. Ag was deposited using GLAD fabrication. This formed nanoprotrusions and nanogaps that enhanced the LSPR activity and also increased the hotspots, exciting the fluorophores [222].

Shang et al. fabricated microsphere-nanospheres by coating 100 nm thick Ag films using GLAD on the BaTiO3 glass ($\alpha = 60°$), and the patchy microsphere was made from p-BTG particles. The imaging contrast of p-BTG was found to be improved by a factor of 6.5. Overall, this patchy microsphere improved the imaging contrast without the use of immersion liquid and led to greater advances in the field of super-resolution microscopy.

Liu et al. worked on improving super-resolution optical imaging by using core-shell microfibers. Etching of the optical microfiber was performed to a 6 um diameter followed by the deposition of different metallic thin films using GLAD (Figure 17). By controlling the angle of deposition and substrate orientation, a uniform metallic shell was deposited on the dielectric microfiber core surface. The core-shell microfiber was placed directly on the nanoscale grating surface and SRM was performed. Due to the surface plasmonic properties due to thin metal films, high-definition imaging was obtained compared to the microsphere and microcylinder counterparts. The gold shell produced the best resolution and magnification, with higher contrast [223]. Hence, it was shown by the group that with GLAD-based metallic coating, the intensity of a photonic nanojet can be adjusted and higher-definition imaging is possible.

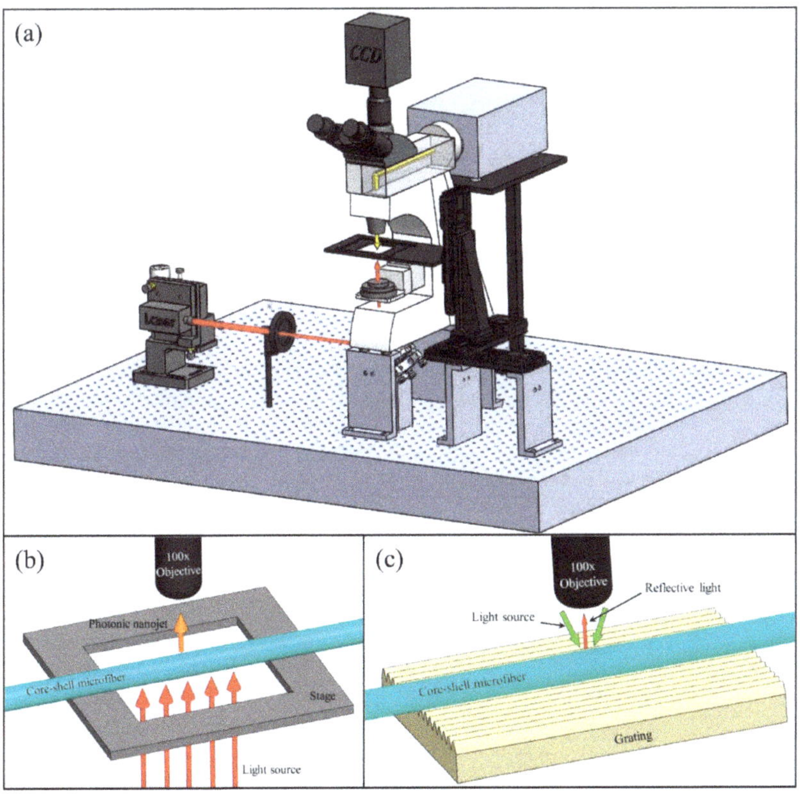

Figure 17. (**a**) Schematic diagram showing SRM microscopy performed. (**b**) Core-shell microfiber shown in the setup whereby the core shell consists of metallic films deposited by GLAD (**c**) Core shell microfiber placed on the top of the specimen. Reproduced with permission from [223].

4.5.3. Photoacoustic Imaging (PA)

A non-invasive biomedical imaging technique exixsts whereby ultrasonic waves are excited into the tissues via irradiation with pulsed or modulated light [224]. The main principle of this method is to convert electromagnetic radiation into acoustic signals through thermoelastic expansion. PA imaging is considered to be one of the prevalent techniques for in vivo studies. This technique plays a crucial role in cancer diagnosis and progression [225], blood oxygen monitoring [226], brain mapping [227], etc. The acoustic signals are detected using an ultrasound receiver which is mostly piezoelectric in nature. Currently, piezoelectric receivers are slowly being replaced by optical receivers to acquire ultrasonic signals, due to their advantage of reduced electrical interconnects [228]. Some of the optical detectors include the Fabry–Perot interferometer (FPI) and micro-ring resonators (MRR) [229]. Hajireza et.al. used an FPI-based detector that utilized nano-thin films fabricated using GLAD. GLAD allows low acoustic impedance of the FP device. These thin films are used for in vivo PA imaging. Multilayers of thin films were deposited using GLAD to act as a mirror in FPI fabrication. TiO_2 was used as the material due to its high refractive index. An alternate layer of GLAD mirror and parylene C was prepared and a final layer of SiO2 antireflection coating was deposited to improve the optical coupling. This multilayer GLAD optical-resolution PAM(OR-PAM) was used for the demonstration of in vivo imaging of the chorioallantoic membrane (CAM) of a five-day-old chicken embryonic model, showing its application in in vivo imaging [230]. This shows the potential application of GLAD in in vivo PA imaging. The same group also used GLAD-based FPI for measuring carbon fiber networks with a 7 um diameter of each fiber [231].

High resolution PA imaging is possible using multilayer mirrors fabricated using GLAD. This shows the potential of GLAD in improving the resolution of PA microscopy. Additionally, more intense work should be conducted in this field to establish FPI using GLAD and improve its applicability.

4.6. *In Vivo Application of GLAD*

Biomimicking and biomaterial synthesis are new necessary fields in biomedical research. They have an enormous and substantial effect on healthcare, as new materials provide novel properties that can be used in prosthetics and drug carriers [232]. Biomaterials with an anisotropic composition are now currently used in drug delivery for cancer treatment, as biomarkers, as bactericidal agents, for tissue engineering, and for vaccine development [233]. Due to their multi-applicability, anisotropic NPs are currently in demand. They have varied in composition, functionality, shape, and size on the different surfaces of single NPs [234,235]. With this huge diversity, various in vivo applications are possible. Janus particles (JP) and patchy particles, and are the types of NPs used for such applications [236].

JPs or patchy particles are asymmetric colloidal particles with more than one composition and chemical modification at different sites [237]. JPs, due to their multicomposition, can be used for binding with specific molecules and probes [238]. Various methods are used for fabricating JPs, including physical deposition, chemical routes [239], electrochemical methods [240], microfluidics [241], electrohydrodynamic methods, and lithography [242].

As the position of the substrate, as well as the angle, can be varied using the GLAD setup, this vapor flux deposition method is used for obtaining different geometries required in Janus particles [243]. Both the thickness and geometric structures can be controlled by adjusting the flux rate and rotation of the substrate [244]. Xuan et al.'s group designed self-propelling Janus micromotors whereby GLAD was used for depositing Pt at a specific angle over the Si microparticles (Figure 18). Further, the micromotors were modified with biotin so that the charged organic dyes could be transported while maintaining the fast speed of the micromotor. This showed the efficacy of Janus particles in drug delivery applications [245].

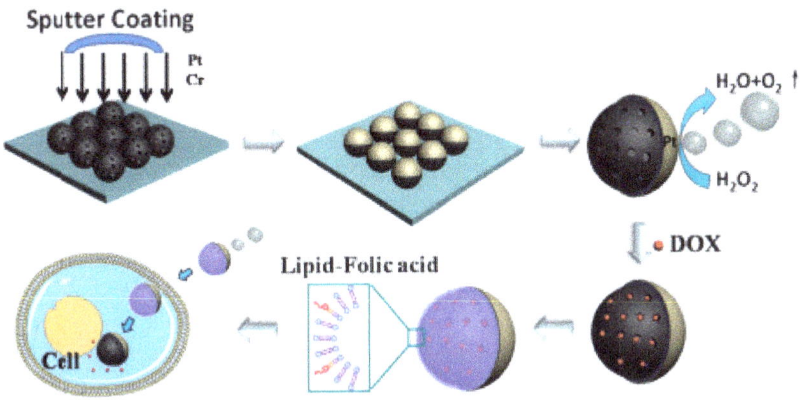

Figure 18. Janus micromotor with one hemisphere coated with Pt and the other hemisphere coated with biotin for specific drug delivery. Reproduced with permission from [245].

Peng et al. combined both top-down and bottom-up fabrication techniques to prepare nanomotors, which were polymeric vesicles deposited with Pt that showed very high drug-loading efficiency. The Janus polymersome nanomotor showed enhanced permeability and retention (EPR) enhancement and also released the encapsulated cargo in a controlled manner under external stimuli [246].

Tejeda-Rodriguez et al. and his group reported making a Janus nanomotor with the capsid from a plant virus on one hemisphere and Pt on the other hemisphere. The Janus viral nanomotor was found to carry and deliver the chemotherapeutic drug tamoxifen to breast tumor cells. The drug release was controlled by a pH-shift mechanism. As the capsid was a biomaterial, it showed an immense advantage over other materials; moreover, surface modification was easy to achieve in this case.

Zhiguang et al.'s group made an extraordinary development in this area by designing micropropellers that can penetrate the delicate vitreous humor of the eye and can perform drug delivery in the retina. The helical magnetic micropropellers were fabricated using GLAD. Ni was deposited onto the Si nanoparticles at an oblique angle. Then, Si was sequentially deposited upon rotation, forming a helical structure. Further, inspired by the sticky liquid layer found on the carnivorous Nepenthes pitcher plant, a non-toxic silicone oil and fluorocarbon coating was used as a slippery surface. Under the influence of a magnetic field, the coated micropropellers showed controlled movement and were able to reach the retina within half an hour [247].

GLAD is a versatile and cheap fabrication method, which has future potential in the fields of drug delivery and biomolecule transportation. Further research could be conducted in the area on its use as a cargo transporter and gene delivery system. GLAD, being a simple and cheap fabrication method, could be explored for various in vivo applications in biological fields. Very little exploration has been done in this field.

4.7. Optical and Electrochemical GLAD-Based Sensors

Though GLAD-based biosensors are vastly employed in various sensing systems such as gas sensors [185], optical sensors [6], and electrochemical sensors [248], most of the work that has been conducted and reported employs optical and electrochemical systems due to the high porosity, plasmonic nature, and greater diffusion properties of these metallic nanostructures. Plasmonic metals such as Ag, Au, Cu, and Pt exhibiting high plasmonic resonance in the optical region and the formation of dense electromagnetic hotspots among these nanostructures further increase their sensitivity as optical biosensors. A comparative study highlighting the advantages, LOD, and limitations of GLAD-based optical and electrochemical biosensors is shown below (Table 1).

Table 1. Comparative study of GLAD-based optical biosensors and electrochemical biosensors.

GLAD-Based Optical Biosensors	GLAD-Based Electrochemical Biosensors
1.The aspect ratio and morphology of the nanostructures are tuned and optimized to improve sensitivity to a variety of optical properties (fluorescence, absorption, etc.).	1. An electrode surface is coated with biological sensing material for potentiometric, amperometric, or conductimetric measurements [249].
2. Enhanced sensitivity is provided by the plasmonic nature of metals such as Ag, Au, Cu, and Pt.	2. Involves the modulation of electrical properties such as potential, current, or impedance associated with the interaction of biomolecules with the working electrodes [250].
3. By forming electromagnetic hotspots, GLAD-based nanostructures significantly improve optical spectroscopy (Raman, fluorescence, and infrared) and plasmon resonance sensing [24].	3. High porosity, large exposed areas, and excellent diffusion properties make GLAD-based metallic nanostructures excellent electrochemical sensors [251,252].
4. LOD ~ 1 fM [253]	4. LOD ~ 1 μM [251]
5. Nanostructures must be optimized to match optical measurements, must possess high sensitivity, and require trained personnel to operate [54]	5. Comparatively low sensitivity, costly instruments, trained personnel required for their operation [248].

5. Limitations of GLAD Technique

The glancing angle deposition technique possess certain limitations. The nucleation process that occurs is random and cannot be controlled. Due to the shadowing effect taking place, the substrate used for deposition cannot be rough or uneven; otherwise, the portion pointing inside will not be able receive sufficient nucleation due to the shadowing of the outward-pointing portion of the substrate. The technique also involves a lot of material wastage and is not very economical. Due to the high sensitivity and considerable expertise required, limited research has been conducted using this technique. In areas such as molecular imaging, multimodal imaging, and in vivo biosensing, limited research has been conducted using the GLAD technique due to most of these limitations. These gaps can be filled by precise usage and expertise, and the technique can be explored to its fullest potential.

6. Conclusions

In conclusion, GLAD is emerging as a convenient technique for the fabrication of nanostructures for different biosensing applications. The homogeneous structures fabricated by GLAD allow reproducible, enhanced, and predictable signals over a large volume. We discussed, in detail, recent advances and strategies in the fabrication techniques of GLAD, their geometrical aspects, and their applications in a number of emerging biochemical and biomedical sensing techniques. In addition, the superiority of the GLAD-based fabrication technique over other conventional techniques and overcoming the limitations of other techniques were discussed in detail. In addition, various types of GLAD -based substrates with improved sensitivity, longevity, and flexibility that have been used in biomedical detection in recent years were discussed. The versatile use of these nanostructures in optical-based biosensing applications such as SPR-, SERS-, and fluorescence-based sensing, as well as colorimetric- and wettability-based detection, has been extensively discussed and studied. The GLAD method in combination with certain new techniques, such as triboelectric methods, new spectroscopic methods, etc., has recently flourished because of their high reproducibility, durability, and sensitivity. The emerging use of GLAD in molecular techniques and in vivo detection was highlighted. A discussion of work conducted by different groups shows the high potential of these fabricated metal nanostructures for biological applications including effective target analyte immobilization. These studies open new horizons for improved bio-detection and applications, and form the basis for further development in the biomedical field, as well as for clinical trials and studies. We also pointed out the challenges of the technology that limit its use in certain applications and emerging fields. Certain limitations of the technique, such as high cost, the need for trained personnel, and sample loss, still need to be addressed. Our review shows the high potential of GLAD-fabricated nanostructures of different morphologies and materials for future biosensing applications. These nanostructures can be a promising tool for certain applications such as single-molecule detection, in vivo biosensing, multimodal imaging, as well as diagnosis, although the use of GLAD-based nanostructures in these fields is yet to be extensively explored.

Author Contributions: Writing—original draft preparation: S.Y., S.S., S.K. and S.K.G.; writing—review and editing: S.Y., S.S., S.K. and J.P.S. All authors have read and agreed to the published version of the manuscript.

Funding: This research received no external funding.

Institutional Review Board Statement: Not applicable.

Informed Consent Statement: Not applicable.

Data Availability Statement: Not applicable.

Conflicts of Interest: The authors declare no conflict of interest.

References

1. Jianrong, C.; Yuqing, M.; Nongyue, H.; Xiaohua, W. Nanotechnology and biosensors. *Biotechnol. Adv.* **2004**, *22*, 505–518. [CrossRef]
2. Biosensors Nanotechnology—Google Books. Available online: https://books.google.co.in/books?hl=en&lr=&id=TG_rAwAAQBAJ&oi=fnd&pg=PA1916&dq=Silver+and+gold+based+nanostructures+for+biosensing&ots=L91KN0-F6r&sig=QFNHttYRTC-r0lIn1_poN68Blik#v=onepage&q=Silver+and+gold+based+nanostructures+for+biosensing&f=false (accessed on 8 July 2020).
3. Psychoyios, V.N.; Nikoleli, G.-P.; Tzamtzis, N.; Nikolelis, D.P.; Psaroudakis, N.; Danielsson, B.; Israr, M.Q.; Willander, M. Potentiometric Cholesterol Biosensor Based on ZnO Nanowalls and Stabilized Polymerized Lipid Film. *Electroanalysis* **2013**, *25*, 367–372. [CrossRef]
4. Bauch, M.; Toma, K.; Toma, M.; Zhang, Q.; Dostalek, J. Plasmon-Enhanced Fluorescence Biosensors: A Review. *Plasmonics* **2014**, *9*, 781–799. [CrossRef] [PubMed]
5. Han, X.; Liu, K.; Sun, C. Plasmonics for Biosensing. *Materials* **2019**, *12*, 1411. [CrossRef]
6. Bochenkov, V.; Baumberg, J.; Noginov, M.; Benz, F.; Aldewachi, H.; Schmid, S.; Podolskiy, V.; Aizpurua, J.; Lin, K.; Ebbesen, T.; et al. Applications of Plasmonics: General Discussion. *Faraday Discuss.* **2015**, *178*, 435–466. [CrossRef]
7. Robbie, K.; Brett, M.J. Sculptured Thin Films and Glancing Angle Deposition: Growth Mechanics and Applications Sculptured Thin Films and Glancing Angle Deposition: Growth Mechanics and Applications. *J. Vac. Sci. Technol. A* **1997**, *15*, 1460. [CrossRef]
8. Jensen, M.O.; Brett, M.J. Periodically Structured Glancing Angle Deposition Thin Films. *IEEE Trans. Nanotechnol.* **2005**, *4*, 269–276. [CrossRef]
9. Hawkeye, M.M.; Taschuk, M.T.; Brett, M.J. *Glancing Angle Deposition of Thin Films: Engineering the Nanoscale*; John Wiley & Sons: Hoboken, NJ, USA, 2014; Available online: https://www.wiley.com/en-in/Glancing+Angle+Deposition+of+Thin+Films:+Engineering+the+Nanoscale-p-9781118847565 (accessed on 7 July 2020).
10. Hawkeye, M.M.; Brett, M.J. Glancing Angle Deposition: Fabrication, Properties, and Applications of Micro- and Nanostructured Thin Films. *J. Vac. Sci. Technol. A Vac. Surf. Film.* **2007**, *25*, 1317. [CrossRef]
11. Zhao, Y.-P.; Ye, D.-X.; Wang, G.-C.; Lu, T.-M. Designing Nanostructures by Glancing Angle Deposition. In *Nanotubes and Nanowires*; SPIE: Bellingham, WA, USA, 2003.
12. Van Dijken, J.G.; Fleischauer, M.D.; Brett, M.J. Advanced Nanostructuring of Metal Phthalocyanines for Organic Photovoltaic Devices. In Proceedings of the Conference Record of the IEEE Photovoltaic Specialists Conference, Seattle, WA, USA, 19–24 June 2011; pp. 3469–3471.
13. Jensen, M.O.; Brett, M.J. Porosity Engineering in Glancing Angle Deposition Thin Films. *Appl. Phys. A Mater. Sci. Process.* **2005**, *80*, 763–768. [CrossRef]
14. Robbie, K.; Sit, J.C.; Brett, M.J. Advanced Techniques for Glancing Angle Deposition. *J. Vac. Sci. Technol. B Microelectron. Nanometer Struct.* **1998**, *16*, 1115–1122. [CrossRef]
15. Taghavi, A.; Rahbarizadeh, F.; Abbasian, S.; Moshaii, A. Label-Free LSPR Prostate-Specific Antigen Immune-Sensor Based on GLAD-Fabricated Silver Nano-Columns. *Plasmonics* **2020**, *15*, 753–760. [CrossRef]
16. González-García, L.; González-Valls, I.; Lira-Cantu, M.; Barranco, A.; González-Elipe, A.R. Aligned TiO 2 Nanocolumnar Layers Prepared by PVD-GLAD for Transparent Dye Sensitized Solar Cells. *Energy Environ. Sci.* **2011**, *4*, 3426–3435. [CrossRef]
17. Sit, J.C.; Broer, D.J.; Brett, M.J. Optical Devices Fabricated from Porous Thin Films Embedded with Liquid Crystals. In Proceedings of the Technical Digest—International Electron Devices Meeting, Washington, DC, USA, 5–8 December 1999; IEEE: Piscataway, NJ, USA, 1999; pp. 123–126.

18. Zhao, P.; Feng, X.; Huang, D.; Yang, G.; Astruc, D. Basic Concepts and Recent Advances in Nitrophenol Reduction by Gold- and Other Transition Metal Nanoparticles. *Coord. Chem. Rev.* **2015**, *287*, 114–136. [CrossRef]
19. Alex, S.; Tiwari, A. Functionalized Gold Nanoparticles: Synthesis, Properties and Applications-A Review. *J. Nanosci. Nanotechnol.* **2015**, *15*, 1869–1894. [CrossRef] [PubMed]
20. Dykman, L.A.; Khlebtsov, N.G. Uptake of Engineered Gold Nanoparticles into Mammalian Cells. *Chem. Rev.* **2014**, *114*, 1258–1288. [CrossRef] [PubMed]
21. Goswami, N.; Yao, Q.; Chen, T.; Xie, J. Mechanistic Exploration and Controlled Synthesis of Precise Thiolate-Gold Nanoclusters. *Coord. Chem. Rev.* **2016**, *329*, 1–15. [CrossRef]
22. Dykman, L.; Khlebtsov, N. Gold Nanoparticles in Biomedical Applications: Recent Advances and Perspectives. *Chem. Soc. Rev.* **2012**, *41*, 2256–2282. [CrossRef]
23. Kolwas, K.; Derkachova, A. Plasmonic Abilities of Gold and Silver Spherical Nanoantennas in Terms of Size Dependent Multipolar Resonance Frequencies and Plasmon Damping Rates. *Opto-Electron. Rev.* **2010**, *18*, 429–437. [CrossRef]
24. Gahlaut, S.K.; Pathak, A.; Gupta, B.D. Recent Advances in Silver Nanostructured Substrates for Plasmonic Sensors. *Biosensors* **2022**, *12*, 713. [CrossRef]
25. Namura, K.; Hanai, S.; Kondo, S.; Kumar, S.; Suzuki, M. Gold Micropetals Self-Assembled by Shadow-Sphere Lithography for Optofluidic Control. *Adv. Mater. Interfaces* **2022**, *9*, 2200200. [CrossRef]
26. Moores, A.; Goettmann, F. The Plasmon Band in Noble Metal Nanoparticles: An Introduction to Theory and Applications. *New J. Chem.* **2006**, *30*, 1121–1132. [CrossRef]
27. Marambio-Jones, C.; Hoek, E.M.V. A Review of the Antibacterial Effects of Silver Nanomaterials and Potential Implications for Human Health and the Environment. *J. Nanoparticle Res.* **2010**, *12*, 1531–1551. [CrossRef]
28. Tran, Q.H.; Nguyen, V.Q.; Le, A.T. Silver nanoparticels: Synthesis, properites, toxicology, applications and perspective. *Adv. Nat. Sci. Nanosci. Nanotechnol.* **2013**, *4*, 03300. [CrossRef]
29. Wang, Y.; Qian, W.; Tan, Y.; Ding, S. A Label-Free Biosensor Based on Gold Nanoshell Monolayers for Monitoring Biomolecular Interactions in Diluted Whole Blood. *Biosens. Bioelectron.* **2008**, *23*, 1166–1170. [CrossRef] [PubMed]
30. Nath, N.; Chilkoti, A. Label-Free Biosensing by Surface Plasmon Resonance of Nanoparticles on Glass: Optimization of Nanoparticle Size. *Anal. Chem.* **2004**, *76*, 5370–5378. [CrossRef] [PubMed]
31. Gu, Y.; Song, J.; Li, M.X.; Zhang, T.T.; Zhao, W.; Xu, J.J.; Liu, M.; Chen, H.Y. Ultrasensitive MicroRNA Assay via Surface Plasmon Resonance Responses of Au@Ag Nanorods Etching. *Anal. Chem.* **2017**, *89*, 10585–10591. [CrossRef] [PubMed]
32. Gish, D.A.; Nsiah, F.; McDermott, M.T.; Brett, M.J. Localized Surface Plasmon Resonance Biosensor Using Silver Nanostructures Fabricated by Glancing Angle Deposition. *Anal. Chem.* **2007**, *79*, 4228–4232. [CrossRef]
33. Shanmukh, S.; Jones, L.; Driskell, J.; Zhao, Y.; Dluhy, R.; Tripp, R.A. Rapid and Sensitive Detection of Respiratory Virus Molecular Signatures Using a Silver Nanorod Array SERS Substrate. *Nano Lett.* **2006**, *6*, 2630–2636. [CrossRef]
34. Deng, J.; Jin, Y.; Wang, L.; Chen, G.; Zhang, C. Sensitive Detection of Endonuclease Activity and Inhibition Using Gold Nanorods. *Biosens. Bioelectron.* **2012**, *34*, 144–150. [CrossRef]
35. Qaddare, S.H.; Salimi, A. Amplified Fluorescent Sensing of DNA Using Luminescent Carbon Dots and AuNPs/GO as a Sensing Platform: A Novel Coupling of FRET and DNA Hybridization for Homogeneous HIV-1 Gene Detection at Femtomolar Level. *Biosens. Bioelectron.* **2017**, *89*, 773–780. [CrossRef]
36. Gahlaut, S.K.; Savargaonkar, D.; Sharan, C.; Yadav, S.; Mishra, P.; Singh, J.P. SERS Platform for Dengue Diagnosis from Clinical Samples Employing a Hand Held Raman Spectrometer. *Anal. Chem.* **2020**, *92*, 2527–2534. [CrossRef]
37. Goel, P.; Kumar, S.; Sarkar, J.; Singh, J.P. Mechanical Strain Induced Tunable Anisotropic Wetting on Buckled PDMS Silver Nanorods Arrays. *ACS Appl. Mater. Interfaces* **2015**, *7*, 8419–8426. [CrossRef]
38. Gahlaut, S.K.; Kalyani, N.; Sharan, C.; Mishra, P.; Singh, J.P. Smartphone Based Dual Mode in Situ Detection of Viability of Bacteria Using Ag Nanorods Array. *Biosens. Bioelectron.* **2019**, *126*, 478–484. [CrossRef]
39. Hawkeye, M.M.; Brett, M.J. Optimized Colorimetric Photonic-Crystal Humidity Sensor Fabricated Using Glancing Angle Deposition. *Adv. Funct. Mater.* **2011**, *21*, 3652–3658. [CrossRef]
40. Jeevanandam, J.; Barhoum, A.; Chan, Y.S.; Dufresne, A.; Danquah, M.K. Review on Nanoparticles and Nanostructured Materials: History, Sources, Toxicity and Regulations. *Beilstein J. Nanotechnol.* **2018**, *9*, 1050–1074. [CrossRef] [PubMed]
41. Pandey, P.A.; Bell, G.R.; Rourke, J.P.; Sanchez, A.M.; Elkin, M.D.; Hickey, B.J.; Wilson, N.R. Physical Vapor Deposition of Metal Nanoparticles on Chemically Modified Graphene: Observations on Metal-Graphene Interactions. *Small* **2011**, *7*, 3202–3210. [CrossRef]
42. Mattox, D.M. Physical Vapor Deposition (PVD) Processes. *Met. Finish.* **2000**, *98*, 410–423. [CrossRef]
43. Mahmoud, S.A.; Ibrahim, A.A.; Riad, A.S. Physical Properties of Thermal Coating CdS Thin Films Using a Modified Evaporation Source. *Thin Solid Film.* **2000**, *372*, 144–148. [CrossRef]
44. Messier, R.; Venugopal, V.C.; Sunal, P.D. Origin and Evolution of Sculptured Thin Films. *J. Vac. Sci. Technol. A Vac. Surf. Film.* **2000**, *18*, 1538–1545. [CrossRef]
45. Reina, A.; Jia, X.; Ho, J.; Nezich, D.; Son, H.; Bulovic, V.; Dresselhaus, M.S.; Jing, K. Large Area, Few-Layer Graphene Films on Arbitrary Substrates by Chemical Vapor Deposition. *Nano Lett.* **2009**, *9*, 30–35. [CrossRef]
46. Kumar, S.; Goel, P.; Singh, D.P.; Singh, J.P. Fabrication of Superhydrophobic Silver Nanorods Array Substrate Using Glancing Angle Deposition. *AIP Conf. Proc.* **2015**, *1591*, 872. [CrossRef]

47. Maithani, Y.; Choudhuri, B.; Mehta, B.R.; Singh, J.P. Self-Adhesive, Stretchable, and Dry Silver Nanorods Embedded Polydimethylsiloxane Biopotential Electrodes for Electrocardiography. *Sens. Actuators A Phys.* **2021**, *332*, 113068. [CrossRef]
48. Suzuki, M.; Taga, Y. Integrated Sculptured Thin Films. *Jpn. J. Appl. Phys.* **2001**, *40*, L358. [CrossRef]
49. Ratsch, C.; Venables, J.A. Nucleation Theory and the Early Stages of Thin Film Growth. *J. Vac. Sci. Technol. A Vac. Surf. Film.* **2003**, *21*, S96–S109. [CrossRef]
50. Dirks, A.G.; Leamy, H.J. Columnar Microstructure in Vapor-Deposited Thin Films. *Thin Solid Film.* **1977**, *47*, 219–233. [CrossRef]
51. Tait, R.N.; Smy, T.; Brett, M.J. Modelling and Characterization of Columnar Growth in Evaporated Films. *Thin Solid Film.* **1993**, *226*, 196–201. [CrossRef]
52. Nakhodkin, N.G.; Shaldervan, A.I. Effect of Vapour Incidence Angles on Profile and Properties of Condensed Films. *Thin Solid Film.* **1972**, *10*, 109–122. [CrossRef]
53. Salazar, P.; Rico, V.; Rodríguez-Amaro, R.; Espinós, J.P.; González-Elipe, A.R. New Copper Wide Range Nanosensor Electrode Prepared by Physical Vapor Deposition at Oblique Angles for the Non-Enzimatic Determination of Glucose. *Electrochim. Acta* **2015**, *169*, 195–201. [CrossRef]
54. Yadav, S.; Senapati, S.; Desai, D.; Gahlaut, S.; Kulkarni, S.; Singh, J.P. Portable and Sensitive Ag Nanorods Based SERS Platform for Rapid HIV-1 Detection and Tropism Determination. *Colloids Surf. B Biointerfaces* **2021**, *198*, 111477. [CrossRef]
55. Zhang, N.; Su, X.; Free, P.; Zhou, X.; Neoh, K.G.; Teng, J.; Knoll, W. Plasmonic Metal Nanostructure Array by Glancing Angle Deposition for Biosensing Application. *Sens. Actuators B Chem.* **2013**, *183*, 310–318. [CrossRef]
56. Agrawal, A.K.; Member, S.; Yadav, S.; Dhawan, A. Enhanced Sensitivity of SPR Biological Sensor Based on Nanohole Arrays in Gold Films. In Proceedings of the 2018 IEEE 18th International Conference on Nanotechnology, Cork, Ireland, 23–26 July 2018; pp. 1–4.
57. Almaguer-Flores, A.; Ximénez-Fyvie, L.A.; Rodil, S.E. Oral Bacterial Adhesion on Amorphous Carbon and Titanium Films: Effect of Surface Roughness and Culture Media. *J. Biomed. Mater. Res. Part B Appl. Biomater.* **2010**, *92B*, 196–204. [CrossRef] [PubMed]
58. Aschauer, E.; Fasching, R.; Varahram, M.; Jobst, G.; Urban, G.; Nicolussi, G.; Husinsky, W.; Friedbacher, G.; Grasserbauer, M. Surface Modification of Platinum Thin Film Electrodes towards a Defined Roughness and Microporosity. *J. Electroanal. Chem.* **1997**, *426*, 157–165. [CrossRef]
59. Steele, J.J.; Brett, M.J. Nanostructure Engineering in Porous Columnar Thin Films: Recent Advances. *J. Mater. Sci. Mater. Electron.* **2007**, *18*, 367–379. [CrossRef]
60. Abdulhalim, I. Plasmonic Sensing Using Metallic Nano-Sculptured Thin Films. *Small* **2014**, *10*, 3499–3514. [CrossRef] [PubMed]
61. Fu, J.; Park, B.; Siragusa, G.; Jones, L.; Tripp, R.; Zhao, Y.; Cho, Y.J. An Au/Si Hetero-Nanorod-Based Biosensor for Salmonella Detection. *Nanotechnology* **2008**, *19*, 155502. [CrossRef] [PubMed]
62. Tyagi, M.; Tomar, M.; Gupta, V. Glad Assisted Synthesis of NiO Nanorods for Realization of Enzymatic Reagentless Urea Biosensor. *Biosens. Bioelectron.* **2014**, *52*, 196–201. [CrossRef]
63. Yadav, S.; Khanam, R.; Singh, J.P. A Purview into Highly Sensitive Magnetic SERS Detection of Hemozoin Biomarker for Rapid Malaria Diagnosis. *Sens. Actuators B Chem.* **2022**, *355*, 131303. [CrossRef]
64. Rajput, A.; Kumar, S.; Singh, J.P. Vertically Standing Nanoporous Al-Ag Zig-Zag Silver Nanorod Arrays for Highly Active SERS Substrates. *Analyst* **2017**, *142*, 3959–3966. [CrossRef]
65. Yarbakht, M.; Nikkhah, M.; Moshaii, A.; Abbasian, S.; Dellith, A. Fabrication of Silver Chevron Arrays as an Efficient and Stable SERS Substrate: Implications in Biological Sensing. *Plasmonics* **2018**, *13*, 715–726. [CrossRef]
66. Dick, B.; Brett, M.J.; Smy, T. Investigation of Substrate Rotation at Glancing Incidence on Thin-Film Morphology. *J. Vac. Sci. Technol. B Microelectron. Nanometer Struct.* **2003**, *21*, 2569–2575. [CrossRef]
67. Kumar, S.; Gahlaut, S.K.; Singh, J.P. Sculptured Thin Films: Overcoming the Limitations of Surface-Enhanced Raman Scattering Substrates. *Appl. Surf. Sci. Adv.* **2022**, *12*, 100322. [CrossRef]
68. Kim, K.; Lee, K.J.; Jo, N.R.; Jo, E.J.; Shin, Y.B.; Kim, M.G. Wafer-Scale LSPR Substrate: Oblique Deposition of Gold on a Patterned Sapphire Substrate. *Biosensors* **2022**, *12*, 158. [CrossRef] [PubMed]
69. Jen, Y.-J.; Huang, J.-W.; Liu, W.-C.; Chan, S.; Tseng, C.-H. Glancing Angle Deposited Gold Nanohelix Arrays on Smooth Glass as Three-Dimensional SERS Substrates. *Opt. Mater. Express* **2016**, *6*, 697. [CrossRef]
70. Martínez, J.L.; Gao, Y.; Lopez-Ríos, T.; Wirgin, A. Anisotropic Surface-Enhanced Raman Scattering at Obliquely Evaporated Ag Films. *Phys. Rev. B* **1987**, *35*, 9481–9488. [CrossRef]
71. Martínez, J.L.; Gao, Y.; Lãpez-Ríos, T. Surface-Enhanced Raman Scattering of Obliquely Evaporated Ag Films. *Phys. Rev. B* **1986**, *33*, 5917–5919. [CrossRef]
72. Zhao, Y.P.; Chaney, S.B.; Zhang, Z.Y. Absorbance Spectra of Aligned Ag Nanorod Arrays Prepared by Oblique Angle Deposition. *J. Appl. Phys.* **2006**, *100*, 063527. [CrossRef]
73. Zhang, Z.Y.; Zhao, Y.P. Extinction Spectra and Electrical Field Enhancement of Ag Nanorods with Different Topologic Shapes. *J. Appl. Phys.* **2007**, *102*, 113308. [CrossRef]
74. Oh, M.K.; Shin, Y.S.; Lee, C.L.; De, R.; Kang, H.; Yu, N.E.; Kim, B.H.; Kim, J.H.; Yang, J.K. Morphological and SERS Properties of Silver Nanorod Array Films Fabricated by Oblique Thermal Evaporation at Various Substrate Temperatures. *Nanoscale Res. Lett.* **2015**, *10*, 259. [CrossRef]
75. Singh, J.P.; Lanier, T.E.; Zhu, H.; Dennis, W.M.; Tripp, R.A.; Zhao, Y. Highly Sensitive and Transparent Surface Enhanced Raman Scattering Substrates Made by Active Coldly Condensed Ag Nanorod Arrays. *J. Phys. Chem. C* **2012**, *116*, 20550–20557. [CrossRef]

76. Singh, J.P.; Chu, H.; Abell, J.; Tripp, R.A.; Zhao, Y. Flexible and Mechanical Strain Resistant Large Area SERS Active Substrates. *Nanoscale* **2012**, *4*, 3410–3414. [CrossRef]
77. Kumar, S.; Goel, P.; Singh, J.P. Flexible and Robust SERS Active Substrates for Conformal Rapid Detection of Pesticide Residues from Fruits. *Sens. Actuators B Chem.* **2017**, *241*, 577–583. [CrossRef]
78. Kumar, S.; Lodhi, D.K.; Goel, P.; Neeti, N.; Mishra, P.; Singh, J.P. A Facile Method for Fabrication of Buckled PDMS Silver Nanorod Arrays as Active 3D SERS Cages for Bacterial Sensing. *Chem. Commun.* **2015**, *51*, 12411–12414. [CrossRef] [PubMed]
79. Yang, Y.; Hu, Z.; Wang, Y.; Wang, B.; Zhan, Q.; Zhang, Y.; Ao, X. Broadband SERS Substrates by Oblique Angle Deposition Method. *Opt. Mater. Express* **2016**, *6*, 2644. [CrossRef]
80. Suzuki, M.; Maekita, W.; Wada, Y.; Nakajima, K.; Kimura, K.; Fukuoka, T.; Mori, Y. Surface Enhanced Raman Scattering on Physically Self-Assembled Ag Nanorod Arrays. *MRS Online Proc. Libr.* **2005**, *900*, 217–222. [CrossRef]
81. Suzuki, M.; Nakajima, K.; Kimura, K.; Fukuoka, T.; Mori, Y. Au Nanorod Arrays Tailored for Surface-Enhanced Raman Spectroscopy. *Anal. Sci.* **2007**, *23*, 829–833. [CrossRef]
82. Bužavaitė-Vertelienė, E.; Maciulis, V.; Anulytė, J.; Tolenis, T.; Baskys, A.; Plikusiene, I.; Balevičius, Z. Total Internal Reflection Ellipsometry Approach for Bloch Surface Waves Biosensing Applications. *Biosensors* **2022**, *12*, 584. [CrossRef]
83. Kumar, S.; Kanagawa, M.; Namura, K.; Fukuoka, T.; Suzuki, M. Multilayer Thin-Film Flake Dispersion Gel for Surface-Enhanced Raman Spectroscopy. *Appl. Nanosci.* **2020**, *1*, 1–9. [CrossRef]
84. Sharma, Y.; Dhawan, A. Plasmonic "Nano-Fingers on Nanowires" as SERS Substrates. *Opt. Lett.* **2016**, *41*, 2085. [CrossRef] [PubMed]
85. Bhalla, N.; Sathish, S.; Galvin, C.J.; Campbell, R.A.; Sinha, A.; Shen, A.Q. Plasma-Assisted Large-Scale Nanoassembly of Metal-Insulator Bioplasmonic Mushrooms. *ACS Appl. Mater. Interfaces* **2018**, *10*, 219–226. [CrossRef]
86. Patzig, C.; Zajadacz, J.; Zimmer, K.; Fechner, R.; Khare, C.; Rauschenbach, B. Patterning Concept for Sculptured Nanostructures with Arbitrary Periods. *Appl. Phys. Lett.* **2009**, *95*, 103107. [CrossRef]
87. Kumar, S.; Doi, Y.; Namura, K.; Suzuki, M. Plasmonic Nanoslit Arrays Fabricated by Serial Bideposition: Optical and Surface-Enhanced Raman Scattering Study. *ACS Appl. Bio Mater.* **2020**, *3*, 3226–3235. [CrossRef] [PubMed]
88. Xiao, C.; Cao, Z.; Huang, Z.; Xu, Z.; Fu, J.; Yobas, L. Microfluidic Integration of Plasmonic Applications for Highly Sensitive Bioanalysis. In Proceedings of the 17th International Conference on Miniaturized Systems for Chemistry and Life Sciences (MicroTAS 2013), Freiburg, Germany, 27–31 October 2013; Zengerle, R., Ed.; Curran: Freiburg, Germany, 2014; Volume 2.
89. Hu, X.; Meng, G.; Huang, Q.; Zhu, C.; Chen, B.; Huang, Z.; Li, F.; Wang, Z. Nano-Petri-Dish Array Assisted Glancing Angle Sputtering for Ag-NP Assembled Bi-Nanoring Arrays as Effective SERS Substrates. *ACS Appl. Mater. Interfaces* **2014**, *6*, 7991–7995. [CrossRef] [PubMed]
90. Zhang, X.; Zhou, Q.; Wang, W.; Shen, L.; Li, Z.; Zhang, Z. Latticing Vertically Aligned Ag Nanorods to Enhance Its SERS Sensitivity. *Mater. Res. Bull.* **2012**, *47*, 921–924. [CrossRef]
91. Giordano, M.C.; Foti, A.; Messina, E.; Gucciardi, P.G.; Comoretto, D.; Buatier De Mongeot, F. SERS Amplification from Self-Organized Arrays of Plasmonic Nanocrescents. *ACS Appl. Mater. Interfaces* **2016**, *8*, 6629–6638. [CrossRef] [PubMed]
92. Kim, S.; Zhang, W.; Cunningham, B.T. Coupling Discrete Metal Nanoparticles to Photonic Crystal Surface Resonant Modes and Application to Raman Spectroscopy. *Opt. Express* **2010**, *18*, 4300. [CrossRef]
93. Castillo-Seoane, J.; Contreras-Bernal, L.; Obrero-Perez, J.M.; García-Casas, X.; Lorenzo-Lázaro, F.; Aparicio, F.J.; Lopez-Santos, C.; Rojas, T.C.; Anta, J.A.; Borrás, A.; et al. Highly Anisotropic Organometal Halide Perovskite Nanowalls Grown by Glancing-Angle Deposition (Adv. Mater. 18/2022). *Adv. Mater.* **2022**, *34*, 2270137. [CrossRef]
94. Lee, B.S.; Lin, P.C.; Lin, D.Z.; Yen, T.J. Rapid Biochemical Mixture Screening by Three-Dimensional Patterned Multifunctional Substrate with Ultra-Thin Layer Chromatography (UTLC) and Surface Enhanced Raman Scattering (SERS). *Sci. Rep.* **2018**, *8*, 516. [CrossRef] [PubMed]
95. Eiamchai, P.; Chananonnawathorn, C.; Horprathum, M.; Patthanasettakul, V.; Limwichean, S.; Nuntawong, N. Spatial Elemental Investigations in Nanostructured Alloyed Ag/Au SERS Substrates by Magnetron Sputtering Oblique-Angle Co-Deposition towards Increased Performance and Shelf Life. *Appl. Surf. Sci.* **2020**, *513*, 145748. [CrossRef]
96. Kim, S.M.; Zhang, W.; Cunningham, B.T. Photonic Crystals with SiO_2-Ag "Post-Cap" Nanostructure Coatings for Surface Enhanced Raman Spectroscopy. *Appl. Phys. Lett.* **2008**, *93*, 143112. [CrossRef]
97. Ma, L.; Huang, Y.; Hou, M.; Xie, Z.; Zhang, Z. Silver Nanorods Wrapped with Ultrathin Al2O3layers Exhibiting Excellent Sers Sensitivity and Outstanding Sers Stability. *Sci. Rep.* **2015**, *5*, 12890. [CrossRef]
98. Savaloni, H.; Babaei, R. Surface Enhanced Raman Spectroscopy and Structural Characterization of Ag/Cu Chiral Nano-Flower Sculptured Thin Films. *Appl. Surf. Sci.* **2013**, *280*, 439–445. [CrossRef]
99. Liu, Y.; Wu, H.; Ma, L.; Zou, S.; Ling, Y.; Zhang, Z. Highly Stable and Active SERS Substrates with Ag-Ti Alloy Nanorods. *Nanoscale* **2018**, *10*, 19863–19870. [CrossRef] [PubMed]
100. Savaloni, H.; Babaei, R.; Goli-Haghighi, S. Application of Mn–Cu Helical Star-Shaped (Pine-Tree-Like) Sculpted Thin Films with Different Symmetries Using Surface-Enhanced Raman Spectroscopy (SERS). *Appl. Spectrosc.* **2019**, *73*, 879–892. [PubMed]
101. Kumar, S.; Lodhi, D.K.; Singh, J.P. Highly Sensitive Multifunctional Recyclable Ag-TiO_2 Nanorod SERS Substrates for Photocatalytic Degradation and Detection of Dye Molecules. *RSC Adv.* **2016**, *6*, 45120–45126. [CrossRef]
102. Hu, L.; Huo, K.; Chen, R.; Gao, B.; Fu, J.; Chu, P.K. Recyclable and High-Sensitivity Electrochemical Biosensing Platform Composed of Carbon-Doped TiO_2 Nanotube Arrays. *Anal. Chem.* **2011**, *83*, 8138–8144. [CrossRef]

103. Hunt, H.K.; Armani, A.M. Recycling Microcavity Optical Biosensors. *Opt. Lett.* **2011**, *36*, 1092. [CrossRef]
104. Oh, Y.-J.; Park, S.-G.; Kang, M.-H.; Choi, J.-H.; Nam, Y.; Jeong, K.-H. Beyond the SERS: Raman Enhancement of Small Molecules Using Nanofluidic Channels with Localized Surface Plasmon Resonance. *Small* **2011**, *7*, 184–188. [CrossRef]
105. Abbas, A.; Linman, M.J.; Cheng, Q. New Trends in Instrumental Design for Surface Plasmon Resonance-Based Biosensors. *Biosens. Bioelectron.* **2011**, *26*, 1815–1824. [CrossRef]
106. Homola, J. Present and Future of Surface Plasmon Resonance Biosensors. *Anal. Bioanal. Chem.* **2003**, *377*, 528–539. [CrossRef]
107. Liao, H.; Nehl, C.L.; Hafner, J.H. Biomedical Applications of Plasmon Resonant Metal Nanoparticles. *Nanomedicine* **2006**, *1*, 201–208. [CrossRef]
108. Willets, K.A.; Van Duyne, R.P. Localized Surface Plasmon Resonance Spectroscopy and Sensing. *Annu. Rev. Phys. Chem.* **2007**, *58*, 267–297. [CrossRef] [PubMed]
109. Schuck, P. Use of Surface Plasmon Resonance to Probe The Equilibrium and Dynamic Aspects of Interactions between Biological Macromolecules. *Annu. Rev. Biophys. Biomol. Struct.* **1997**, *26*, 541–566. [CrossRef] [PubMed]
110. Guedon, P.; Livache, T.; Martin, F.; Lesbre, F.; Roget, A.; Bidan, G.; Levy, Y. Characterization and Optimization of a Real-Time, Parallel, Label-Free, Polypyrrole-Based DNA Sensor by Surface Plasmon Resonance Imaging. *Anal. Chem.* **2000**, *72*, 6003–6009. [CrossRef] [PubMed]
111. Jordan, C.E.; Frutos, A.G.; Thiel, A.J.; Corn, R.M.; Peterlinz, K.A.; Georgiadis, R.M.; He, L.; Musick, M.D.; Nicewarner, S.R.; Salinas, F.G.; et al. Surface Plasmon Resonance Imaging Measurements of DNA and RNA Hybridization Adsorption onto DNA Microarrays. *Gen. Anal. Biomol. Eng* **1997**, *69*, 1–7. [CrossRef]
112. Homola, J.; Yee, S.S.; Gauglitz, G. Surface Plasmon Resonance Sensors: Review. *Sens. Actuators B Chem.* **1999**, *54*, 3–15. [CrossRef]
113. Bhalla, N.; Yu, Z.; Pauly, S.; Kumar, A.; Maddi, C.; Mariotti, D.; Zhao, P.; Payam, A.F.; Soin, N. Total Electrification of Large-Scale Nanophotonic Arrays by Frictional Charges. *Nanoscale Horiz.* **2022**, *7*, 1513–1522. [CrossRef]
114. Kumar Agrawal, A.; Das, A.; Dhawan, A. Enhanced Sensitivity of SPR Sensing and Imaging Using Plasmonic Nanopillar Arrays. In Proceedings of the IEEE Conference on Nanotechnology, Cork, Ireland, 23–26 July 2018; IEEE Computer Society: Washington, DC, USA, 2019; Volume 2018, pp. 1–4.
115. Agrawal, A.; Gupta, N.; Das, A.; Ahmed, K.; Dhawan, A. Nanostructured Plasmonic Gold Films for Enhanced Sensitivity of SPR Biological Sensing and Imaging. In Proceedings of the Plasmonics in Biology and Medicine XVI, San Francisco, CA, USA, 7 March 2019; Vo-Dinh, T., Ho, H.-P.A., Ray, K., Eds.; SPIE: Bellingham, WA, USA, 2019; Volume 10894, p. 21.
116. Bunch, J.S.; Verbridge, S.S.; Alden, J.S.; Van Der Zande, A.M.; Parpia, J.M.; Craighead, H.G.; McEuen, P.L. Impermeable Atomic Membranes from Graphene Sheets. *Nano Lett.* **2008**, *8*, 2458–2462. [CrossRef]
117. Wu, L.; Chu, H.S.; Koh, W.S.; Li, E.P. Highly Sensitive Graphene Biosensors Based on Surface Plasmon Resonance. *Opt. Express* **2010**, *18*, 14395. [CrossRef]
118. Choi, S.H.; Kim, Y.L.; Byun, K.M. Graphene-on-Silver Substrates for Sensitive Surface Plasmon Resonance Imaging Biosensors. *Opt. Express* **2011**, *19*, 458. [CrossRef]
119. McGaughey, G.B.; Gagné, M.; Rappé, A.K. π-Stacking Interactions. Alive and Well in Proteins. *J. Biol. Chem.* **1998**, *273*, 15458–15463. [CrossRef]
120. Tang, Z.; Wu, H.; Cort, J.R.; Buchko, G.W.; Zhang, Y.; Shao, Y.; Aksay, I.A.; Liu, J.; Lin, Y. Constraint of DNA on Functionalized Graphene Improves Its Biostability and Specificity. *Small* **2010**, *6*, 1205–1209. [CrossRef]
121. Leong, H.S.; Guo, J.; Lindquist, R.G.; Liu, Q.H. Surface Plasmon Resonance in Nanostructured Metal Films under the Kretschmann Configuration. *J. Appl. Phys.* **2009**, *106*, 124314. [CrossRef]
122. Kaur, G.; Tomar, M.; Gupta, V. Nanostructured Zinc Oxide Thin Film for Application to Surface Plasmon Resonance Based Cholesterol Biosensor. In Proceedings of the International Workshop on Thin Films for Electronics, Electro-Optics, Energy, and Sensors, Dayton, OH, USA, 4–6 July 2015; Subramanyam, G., Ed.; SPIE: Bellingham, WA, USA, 2015; Volume 9667, p. 966706.
123. Shalabney, A.; Khare, C.; Rauschenbach, B.; Abdulhalim, I. Sensitivity of Surface Plasmon Resonance Sensors Based on Metallic Columnar Thin Films in the Spectral and Angular Interrogations. *Sens. Actuators B Chem.* **2011**, *159*, 201–212. [CrossRef]
124. Byun, K.M.; Yoon, S.J.; Kim, D.; Kim, S.J. Experimental Study of Sensitivity Enhancement in Surface Plasmon Resonance Biosensors by Use of Periodic Metallic Nanowires. *Opt. Lett.* **2007**, *32*, 1902. [CrossRef] [PubMed]
125. Ouellet, E.; Lausted, C.; Lin, T.; Yang, C.W.T.; Hood, L.; Lagally, E.T. Parallel Microfluidic Surface Plasmon Resonance Imaging Arrays. *Lab Chip* **2010**, *10*, 581–588. [CrossRef] [PubMed]
126. Krishnamoorthy, G.; Carlen, E.T.; Deboer, H.L.; Van Den Berg, A.; Schasfoort, R.B.M. Electrokinetic Lab-on-a-BioChip for Multi-Ligand/Multi-Analyte Biosensing. *Anal. Chem.* **2010**, *82*, 4145–4150. [CrossRef]
127. Zordan, M.D.; Grafton, M.M.G.; Park, K.; Leary, J.F. The Design of a Microfluidic Biochip for the Rapid, Multiplexed Detection of Foodborne Pathogens by Surface Plasmon Resonance Imaging. In Proceedings of the Frontiers in Pathogen Detection: From Nanosensors to Systems, San Francisco, CA, USA, 23–25 January 2010; Fauchet, P.M., Ed.; SPIE: Bellingham, WA, USA, 2010; Volume 7553, p. 755307.
128. Malic, L.; Veres, T.; Tabrizian, M. Two-Dimensional Droplet-Based Surface Plasmon Resonance Imaging Using Electrowetting-on-Dielectric Microfluidics. *Lab Chip* **2009**, *9*, 473–475. [CrossRef]
129. Natarajan, S.; Katsamba, P.S.; Miles, A.; Eckman, J.; Papalia, G.A.; Rich, R.L.; Gale, B.K.; Myszka, D.G. Continuous-Flow Microfluidic Printing of Proteins for Array-Based Applications Including Surface Plasmon Resonance Imaging. *Anal. Biochem.* **2008**, *373*, 141–146. [CrossRef]

130. Liu, C.; Cui, D.; Li, H. A Hard-Soft Microfluidic-Based Biosensor Flow Cell for SPR Imaging Application. *Biosens. Bioelectron.* **2010**, *26*, 255–261. [CrossRef]
131. Galopin, E.; Beaugeois, M.; Pinchemel, B.; Camart, J.-C.; Bouazaoui, M.; Thomy, V. SPR Biosensing Coupled to a Digital Microfluidic Microstreaming System. *Biosens. Bioelectron.* **2007**, *23*, 746–750. [CrossRef]
132. Moskovits, M. Surface-Enhanced Spectroscopy. *Rev. Mod. Phys.* **1985**, *57*, 783–826. [CrossRef]
133. Fleischmann, M.; Hendra, P.J.; McQuillan, A.J. Raman Spectra of Pyridine Adsorbed at a Silver Electrode. *Chem. Phys. Lett.* **1974**, *26*, 163–166. [CrossRef]
134. Kumar, S.; Taneichi, T.; Fukuoka, T.; Namura, K.; Suzuki, M. Study on Transport of Molecules in Gel by Surface-Enhanced Raman Spectroscopy. *Cellulose* **2021**, *28*, 10803–10813. [CrossRef]
135. Craig, A.P.; Franca, A.S.; Irudayaraj, J. Surface-Enhanced Raman Spectroscopy Applied to Food Safety. *Annu. Rev. Food Sci. Technol.* **2013**, *4*, 369–380. [CrossRef] [PubMed]
136. Granger, J.H.; Schlotter, N.E.; Crawford, A.C.; Porter, M.D. Prospects for Point-of-Care Pathogen Diagnostics Using Surface-Enhanced Raman Scattering (SERS). *Chem. Soc. Rev.* **2016**, *45*, 3865–3882. [CrossRef]
137. Hughes, J.; Izake, E.L.; Lott, W.B.; Ayoko, G.A.; Sillence, M. Ultra Sensitive Label Free Surface Enhanced Raman Spectroscopy Method for the Detection of Biomolecules. *Talanta* **2014**, *130*, 20–25. [CrossRef]
138. Kong, K.V.; Leong, W.K.; Lam, Z.; Gong, T.; Goh, D.; Lau, W.K.O.; Olivo, M. A Rapid and Label-Free SERS Detection Method for Biomarkers in Clinical Biofluids. *Small* **2014**, *10*, 5030–5034. [CrossRef]
139. Radziuk, D.; Moehwald, H. Prospects for Plasmonic Hot Spots in Single Molecule SERS towards the Chemical Imaging of Live Cells. *Phys. Chem. Chem. Phys.* **2015**, *17*, 21072–21093. [CrossRef]
140. Hakonen, A.; Andersson, P.O.; Stenbæk Schmidt, M.; Rindzevicius, T.; Käll, M. Explosive and Chemical Threat Detection by Surface-Enhanced Raman Scattering: A Review. *Anal. Chim. Acta* **2015**, *893*, 1–13. [CrossRef]
141. Kumar, S.; Kumar, P.; Das, A.; Pathak, C.S.; Kumar, S.; Kumar, P.; Das, A.; Pathak, C.S. Surface-Enhanced Raman Scattering: Introduction and Applications. *Recent Adv. Nanophotonics-Fundam. Appl.* **2020**. [CrossRef]
142. Sharma, B.; Fernanda Cardinal, M.; Kleinman, S.L.; Greeneltch, N.G.; Frontiera, R.R.; Blaber, M.G.; Schatz, G.C.; Van Duyne, R.P. High-Performance SERS Substrates: Advances and Challenges. *MRS Bull.* **2013**, *38*, 615–624. [CrossRef]
143. Kumar, S.; Tokunaga, K.; Namura, K.; Fukuoka, T.; Suzuki, M. Experimental Evidence of a Twofold Electromagnetic Enhancement Mechanism of Surface-Enhanced Raman Scattering. *J. Phys. Chem. C* **2020**, *124*, 21215–21222. [CrossRef]
144. Ko, H.; Singamaneni, S.; Tsukruk, V.V. Nanostructured Surfaces and Assemblies as SERS Media. *Small* **2008**, *4*, 1576–1599. [CrossRef] [PubMed]
145. Kumar, S.; Fukuoka, T.; Takahashi, R.; Yoshida, M.; Utsumi, Y.; Yamaguchi, A.; Namura, K.; Suzuki, M. Highly Stable and Reproducible Au Nanorod Arrays for Near-Infrared Optofluidic SERS Sensor. *Mater. Lett.* **2021**, *286*, 129106. [CrossRef]
146. Song, C.; Chen, J.; Zhao, Y.; Wang, L. Gold-Modified Silver Nanorod Arrays for SERS-Based Immunoassays with Improved Sensitivity. *J. Mater. Chem. B* **2014**, *2*, 7488–7494. [CrossRef]
147. Srivastava, S.K.; Shalabney, A.; Khalaila, I.; Grüner, C.; Rauschenbach, B.; Abdulhalim, I. SERS Biosensor Using Metallic Nano-Sculptured Thin Films for the Detection of Endocrine Disrupting Compound Biomarker Vitellogenin. *Small* **2014**, *10*, 3579–3587. [CrossRef]
148. O'Brien, J.; Hayder, H.; Zayed, Y.; Peng, C. Overview of MicroRNA Biogenesis, Mechanisms of Actions, and Circulation. *Front. Endocrinol.* **2018**, *9*, 402. [CrossRef]
149. Driskell, J.D.; Seto, A.G.; Jones, L.P.; Jokela, S.; Dluhy, R.A.; Zhao, Y.P.; Tripp, R.A. Rapid MicroRNA (MiRNA) Detection and Classification via Surface-Enhanced Raman Spectroscopy (SERS). *Biosens. Bioelectron.* **2008**, *24*, 917–922. [CrossRef]
150. Driskell, J.D.; Tripp, R.A. Label-Free SERS Detection of MicroRNA Based on Affinity for an Unmodified Silver Nanorod Array Substrate. *Chem. Commun.* **2010**, *46*, 3298–3300. [CrossRef]
151. Hou, M.; Huang, Y.; Ma, L.; Zhang, Z. Sensitivity and Reusability of SiO2 NRs@ Au NPs SERS Substrate in Trace Monochlorobiphenyl Detection. *Nanoscale Res. Lett.* **2015**, *10*, 444. [CrossRef]
152. Nuntawong, N.; Eiamchai, P.; Somrang, W.; Denchitcharoen, S.; Limwichean, S.; Horprathum, M.; Patthanasettakul, V.; Chaiya, S.; Leelapojanaporn, A.; Saiseng, S.; et al. Detection of Methamphetamine/Amphetamine in Human Urine Based on Surface-Enhanced Raman Spectroscopy and Acidulation Treatments. *Sens. Actuators B Chem.* **2017**, *239*, 139–146. [CrossRef]
153. Li, B.; Wang, T.; Su, Q.; Wu, X.; Dong, P. Fabrication of Au Nanorods by the Oblique Angle Deposition Process for Trace Detection of Methamphetamine with Surface-Enhanced Raman Scattering. *Sensors* **2019**, *19*, 3742. [CrossRef] [PubMed]
154. Hou, M.; Huang, Y.; Ma, L.; Zhang, Z. Quantitative Analysis of Single and Mix Food Antiseptics Basing on SERS Spectra with PLSR Method. *Nanoscale Res. Lett.* **2016**, *11*, 296. [CrossRef] [PubMed]
155. Han, C.; Yao, Y.; Wang, W.; Qu, L.; Bradley, L.; Sun, S.; Zhao, Y. Rapid and Sensitive Detection of Sodium Saccharin in Soft Drinks by Silver Nanorod Array SERS Substrates. *Sens. Actuators B Chem.* **2017**, *251*, 272–279. [CrossRef]
156. Devor, E.J.; Behlke, M.A.; Huang, L.; Bogh, L.; Rose, S. *Fluorescence and Fluorescence Applications*; Integrated DNA Technologies: Coralville, IA, USA, 2005.
157. Nanobiosensors and Fluorescence Based Biosensors: An Overview. Available online: http://www.ijnd.ir/article_660965.html (accessed on 19 July 2020).

158. Hirshberg, M.; Henrick, K.; Lloyd Haire, L.; Vasisht, N.; Brune, M.; Corrie, J.E.T.; Webb, M.R. Crystal Structure of Phosphate Binding Protein Labeled with a Coumarin Fluorophore, a Probe for Inorganic Phosphate. *Biochemistry* **1998**, *37*, 10381–10385. [CrossRef] [PubMed]
159. Dillingham, M.S.; Tibbles, K.L.; Hunter, J.L.; Bell, J.C.; Kowalczykowski, S.C.; Webb, M.R. Fluorescent Single-Stranded DNA Binding Protein as a Probe for Sensitive, Real-Time Assays of Helicase Activity. *Biophys. J.* **2008**, *95*, 3330–3339. [CrossRef] [PubMed]
160. Aslan, K.; Gryczynski, I.; Malicka, J.; Matveeva, E.; Lakowicz, J.R.; Geddes, C.D. Metal-Enhanced Fluorescence: An Emerging Tool in Biotechnology. *Curr. Opin. Biotechnol.* **2005**, *16*, 55–62. [CrossRef]
161. Matveeva, E.G.; Gryczynski, Z.; Lakowicz, J.R. Myoglobin Immunoassay Based on Metal Particle-Enhanced Fluorescence. *J. Immunol. Methods* **2005**, *302*, 26–35. [CrossRef]
162. Singh, D.P.; Kumar, S.; Singh, J.P. Morphology Dependent Surface Enhanced Fluorescence Study on Silver Nanorod Arrays Fabricated by Glancing Angle Deposition. *RSC Adv.* **2015**, *5*, 31341–31346. [CrossRef]
163. Zhong, W. Nanomaterials in Fluorescence-Based Biosensing. *Anal. Bioanal. Chem.* **2009**, *394*, 47–59. [CrossRef]
164. Aslan, K.; Lakowicz, J.R.; Geddes, C.D. Metal-Enhanced Fluorescence Using Anisotropic Silver Nanostructures: Critical Progress to Date. *Anal. Bioanal. Chem.* **2005**, *382*, 926–933. [CrossRef] [PubMed]
165. Badshah, M.A.; Ju, J.; Lu, X.; Abbas, N.; Kim, S. min Enhancing the Sensitivity of DNA Microarrays by Metal-Enhanced Fluorescence Using Vertical Nanorod Structures. *Sens. Actuators B Chem.* **2018**, *274*, 451–457. [CrossRef]
166. Li, H.; Chen, C.Y.; Wei, X.; Qiang, W.; Li, Z.; Cheng, Q.; Xu, D. Highly Sensitive Detection of Proteins Based on Metal-Enhanced Fluorescence with Novel Silver Nanostructures. *Anal. Chem.* **2012**, *84*, 8656–8662. [CrossRef] [PubMed]
167. Xing, H.; Bu, W.; Zhang, S.; Zheng, X.; Li, M.; Chen, F.; He, Q.; Zhou, L.; Peng, W.; Hua, Y.; et al. Multifunctional Nanoprobes for Upconversion Fluorescence, MR and CT Trimodal Imaging. *Biomaterials* **2012**, *33*, 1079–1089. [CrossRef] [PubMed]
168. Zhang, J.; Fu, Y.; Li, G.; Zhao, R.Y.; Lakowicz, J.R. Direct Observation of Chemokine Receptors 5 on T-Lymphocyte Cell Surfaces Using Fluorescent Metal Nanoprobes 2: Approximation of CCR5 Populations. *Biochem. Biophys. Res. Commun.* **2011**, *407*, 63–67. [CrossRef]
169. Yang, J.; Moraillon, A.; Siriwardena, A.; Boukherroub, R.; Ozanam, F.; Gouget-Laemmel, A.C.; Szunerits, S. Carbohydrate Microarray for the Detection of Glycan-Protein Interactions Using Metal-Enhanced Fluorescence. *Anal. Chem.* **2015**, *87*, 3721–3728. [CrossRef]
170. Barranco, A.; Borras, A.; Gonzalez-Elipe, A.R.; Palmero, A. Perspectives on Oblique Angle Deposition of Thin Films: From Fundamentals to Devices. *Prog. Mater. Sci.* **2016**, *76*, 59–153. [CrossRef]
171. Geddes, C.D.; Gryczynski, I.; Malicka, J.; Gryczynski, Z.; Lakowicz, J.R. Metal enhanced fluorescence: Potential applications in HTS. *Comb. Chem. High Throughput Screen.* **2003**, *6*, 109–117. [CrossRef]
172. Park, B.; Fu, J.; Zhao, Y.; Siragusa, G.R.; Cho, Y.-J.; Lawrence, K.C.; Windham, W.R. Bio-Functional Au/Si Nanorods for Pathogen Detection. In Proceedings of the Nanosensing: Materials, Devices, and Systems III, Boston, MA, USA, 11–12 September 2007; Islam, M.S., Dutta, A.K., Eds.; SPIE: Bellingham, WA, USA, 2007; Volume 6769, pp. 159–167.
173. Haustein, E.; Schwille, P. Single-Molecule Spectroscopic Methods. *Curr. Opin. Struct. Biol.* **2004**, *14*, 531–540. [CrossRef]
174. Zhang, J. Target Molecule Imaging on Tissue Specimens by Fluorescent Metal Nanoprobes. *J. Biomed. Opt.* **2011**, *16*, 116004. [CrossRef]
175. Punj, D.; Regmi, R.; Devilez, A.; Plauchu, R.; Moparthi, S.B.; Stout, B.; Bonod, N.; Rigneault, H.; Wenger, J. Self-Assembled Nanoparticle Dimer Antennas for Plasmonic-Enhanced Single-Molecule Fluorescence Detection at Micromolar Concentrations. *ACS Photonics* **2015**, *2*, 1099–1107. [CrossRef]
176. Mayer, K.M.; Hafner, J.H. Localized Surface Plasmon Resonance Sensors. *Chem. Rev.* **2011**, *111*, 3828–3857. [CrossRef] [PubMed]
177. Fu, J.X.; Collins, A.; Zhao, Y.P. Optical Properties and Biosensor Application of Ultrathin Silver Films Prepared by Oblique Angle Deposition. *J. Phys. Chem. C* **2008**, *112*, 16784–16791. [CrossRef]
178. Zhang, X.; Zhou, W.; Yuan, Z.; Lu, C. Colorimetric Detection of Biological Hydrogen Sulfide Using Fluorosurfactant Functionalized Gold Nanorods. *Analyst* **2015**, *140*, 7443–7450. [CrossRef] [PubMed]
179. Jarosz, A.P.; Yep, T.; Mutus, B. Microplate-Based Colorimetric Detection of Free Hydrogen Sulfide. *Anal. Chem.* **2013**, *85*, 3638–3643. [CrossRef] [PubMed]
180. Lilienfeld, S.; White, C.E. A Study of the Reaction between Hydrogen Sulfide and Silver. *J. Am. Chem. Soc.* **1930**, *52*, 885–892. [CrossRef]
181. Graedel, T.E.; Franey, J.P.; Gualtieri, G.J.; Kammloty, G.W.; Malm, D.L. On the mechanism of silver and copper sulfidation by atmospheric h2s and ocs. *Corros. Sci.* **1985**, *25*, 1163–1180. [CrossRef]
182. Graedel, T.E. Corrosion Mechanisms for Silver Exposed to the Atmosphere. *J. Electrochem. Soc.* **1992**, *139*, 1963. [CrossRef]
183. Kim, H. Corrosion Process of Silver in Environments Containing 0.1 Ppm H2S and 1.2 Ppm NO2. *Mater. Corros.* **2003**, *54*, 243–250. [CrossRef]
184. Chen, R.; Morris, H.R.; Whitmore, P.M. Fast Detection of Hydrogen Sulfide Gas in the Ppmv Range with Silver Nanoparticle Films at Ambient Conditions. *Sens. Actuators B Chem.* **2013**, *186*, 431–438. [CrossRef]
185. Gahlaut, S.K.; Yadav, K.; Sharan, C.; Singh, J.P. Quick and Selective Dual Mode Detection of H2S Gas by Mobile App Employing Silver Nanorods Array. *Anal. Chem.* **2017**, *89*, 13582–13588. [CrossRef] [PubMed]

186. Ahn, Y.J.; Lee, Y.J.; Lee, J.; Lee, D.; Park, H.K.; Lee, G.J. Colorimetric Detection of Endogenous Hydrogen Sulfide Production in Living Cells. *Spectrochim. Acta-Part A Mol. Biomol. Spectrosc.* **2017**, *177*, 118–124. [CrossRef] [PubMed]
187. Hao, J.; Xiong, B.; Cheng, X.; He, Y.; Yeung, E.S. High-Throughput Sulfide Sensing with Colorimetric Analysis of Single Au-Ag Core-Shell Nanoparticles. *Anal. Chem.* **2014**, *86*, 4663–4667. [CrossRef] [PubMed]
188. Shafiei, M.; Ghasemian, A.; Mostafavi, S.K.S.; Teimouri, M.; Vardanjani, H.R.; Mirforughi, S.A. Gold Nanoparticle-Based Colorimetric Platform Technology as Rapid and Efficient Bacterial Pathogens Detection Method from Various Sources. *Rev. Med. Microbiol.* **2019**, *30*, 128–132. [CrossRef]
189. Saha, K.; Agasti, S.S.; Kim, C.; Li, X.; Rotello, V.M. Gold Nanoparticles in Chemical and Biological Sensing. *Chem. Rev.* **2012**, *112*, 2739–2779. [CrossRef]
190. Aldewachi, H.; Chalati, T.; Woodroofe, M.N.; Bricklebank, N.; Sharrack, B.; Gardiner, P. Gold Nanoparticle-Based Colorimetric Biosensors. *Nanoscale* **2018**, *10*, 18–33. [CrossRef]
191. Peng, H.; Chen, I.A. Rapid Colorimetric Detection of Bacterial Species through the Capture of Gold Nanoparticles by Chimeric Phages. *ACS Nano* **2019**, *13*, 1244–1252. [CrossRef]
192. Wang, G.; Chen, C.; Chen, Y.; Kang, X.; Yang, C.; Wang, F.; Liu, Y.; Xiong, X. Self-Stabilized and Strongly Adhesive Supramolecular Polymer Protective Layer Enables Ultrahigh-Rate and Large-Capacity Lithium-Metal Anode. *Angew. Chemie Int. Ed.* **2020**, *59*, 2055–2060. [CrossRef]
193. Pathak, A.; Gupta, B.D. Fiber-Optic Plasmonic Sensor Utilizing CTAB-Functionalized ZnO Nanoparticle-Decorated Carbon Nanotubes on Silver Films for the Detection of Catechol in Wastewater. *ACS Appl. Nano Mater.* **2020**, *3*, 2582–2593. [CrossRef]
194. Dong, C.; Wang, Z.; Zhang, Y.; Ma, X.; Iqbal, M.Z.; Miao, L.; Zhou, Z.; Shen, Z.; Wu, A. High-Performance Colorimetric Detection of Thiosulfate by Using Silver Nanoparticles for Smartphone-Based Analysis. *ACS Sensors* **2017**, *2*, 1152–1159. [CrossRef]
195. Yang, X.H.; Ling, J.; Peng, J.; Cao, Q.E.; Ding, Z.T.; Bian, L.C. A Colorimetric Method for Highly Sensitive and Accurate Detection of Iodide by Finding the Critical Color in a Color Change Process Using Silver Triangular Nanoplates. *Anal. Chim. Acta* **2013**, *798*, 74–81. [CrossRef] [PubMed]
196. Tripathi, A.; Harris, K.D.; Elias, A.L. Peroxidase-Like Behavior of Ni Thin Films Deposited by Glancing Angle Deposition for Enzyme-Free Uric Acid Sensing. *ACS Omega* **2020**, *5*, 9123–9130. [CrossRef]
197. Kruse, C.; Anderson, T.; Wilson, C.; Zuhlke, C.; Alexander, D.; Gogos, G.; Ndao, S. Extraordinary Shifts of the Leidenfrost Temperature from Multiscale Micro/Nanostructured Surfaces. *Langmuir* **2013**, *29*, 9798–9806. [CrossRef] [PubMed]
198. Singh, D.P.; Singh, J.P. Delayed Freezing of Water Droplet on Silver Nanocolumnar Thin Film. *Appl. Phys. Lett.* **2013**, *102*, 243112. [CrossRef]
199. Tyrrell, J.W.G.; Attard, P. Images of Nanobubbles on Hydrophobic Surfaces and Their Interactions. *Phys. Rev. Lett.* **2001**, *87*, 176104. [CrossRef]
200. Li, C.; Wang, Z.; Wang, P.I.; Peles, Y.; Koratkar, N.; Peterson, G.P. Nanostructured Copper Interfaces for Enhanced Boiling. *Small* **2008**, *4*, 1084–1088. [CrossRef]
201. Gahlaut, S.K.; Devi, P.; Singh, J.P. Self-Sustainable and Recyclable Ag Nanorods for Developing Ag-Ag2S Nano Heterostructures Using Sewage Gas: Applications in Photocatalytic Water Purification, Hydrogen Evolution, SERS and Antibacterial Activity. *Appl. Surf. Sci.* **2020**, *528*, 147037. [CrossRef]
202. Cassidy, P.J.; Radda, G.K. Molecular Imaging Perspectives. *J. R. Soc. Interface* **2005**, *2*, 133–144. [CrossRef]
203. Minchin, R.F.; Martin, D.J. Minireview: Nanoparticles for Molecular Imaging—An Overview. *Endocrinology* **2010**, *151*, 474–481. [CrossRef]
204. Sheng, Y.; De Liao, L.; Thakor, N.V.; Tan, M.C. Nanoparticles for Molecular Imaging. *J. Biomed. Nanotechnol.* **2014**, *10*, 2641–2676. [CrossRef]
205. Padmanabhan, P.; Kumar, A.; Kumar, S.; Chaudhary, R.K.; Gulyás, B. Nanoparticles in Practice for Molecular-Imaging Applications: An Overview. *Acta Biomater.* **2016**, *41*, 1–16. [CrossRef] [PubMed]
206. Jin, S.; Ma, X.; Ma, H.; Zheng, K.; Liu, J.; Hou, S.; Meng, J.; Wang, P.C.; Wu, X.; Liang, X.J. Surface Chemistry-Mediated Penetration and Gold Nanorod Thermotherapy in Multicellular Tumor Spheroids. *Nanoscale* **2013**, *5*, 143–146. [CrossRef] [PubMed]
207. Hayakawa, T.; Selvan, S.T.; Nogami, M. Field Enhancement Effect of Small Ag Particles on the Fluorescence from Eu3+-Doped SiO2 Glass. *Appl. Phys. Lett.* **1999**, *74*, 1513–1515. [CrossRef]
208. Fulmore, J.S.; Geiger, B.F.; Werner, K.A.; Talbott, L.L.; Jones, D.C. Feature Article. *Child. Educ.* **2009**, *85*, 293–299. [CrossRef]
209. Ju, J.; Byeon, E.; Han, Y.A.; Kim, S.M. Fabrication of a Substrate for Ag-nanorod Metal-enhanced Fluorescence Using the Oblique. *Micro Nano Lett.* **2013**, *8*, 370–373. [CrossRef]
210. Ji, X.; Xiao, C.; Lau, W.F.; Li, J.; Fu, J. Metal Enhanced Fluorescence Improved Protein and DNA Detection by Zigzag Ag Nanorod Arrays. *Biosens. Bioelectron.* **2016**, *82*, 240–247. [CrossRef] [PubMed]
211. Novotny, L.; Hecht, B. *Principles of Nano-Optics*; Cambridge University Press: Cambridge, UK, 2006; ISBN 9780511813535.
212. Schermelleh, L.; Heintzmann, R.; Leonhardt, H. A Guide to Super-Resolution Fluorescence Microscopy. *J. Cell Biol.* **2010**, *190*, 165–175. [CrossRef]
213. Neuman, T.; Alonso-González, P.; Garcia-Etxarri, A.; Schnell, M.; Hillenbrand, R.; Aizpurua, J. Mapping the near Fields of Plasmonic Nanoantennas by Scattering-Type Scanning near-Field Optical Microscopy. *Laser Photonics Rev.* **2015**, *9*, 637–649. [CrossRef]

214. Villarrubia, J.S.; Vladár, A.E.; Ming, B.; Kline, R.J.; Sunday, D.F.; Chawla, J.S.; List, S. Scanning Electron Microscope Measurement of Width and Shape of 10nm Patterned Lines Using a JMONSEL-Modeled Library. *Ultramicroscopy* **2015**, *154*, 15–28. [CrossRef]
215. Hess, S.T.; Girirajan, T.P.K.; Mason, M.D. Ultra-High Resolution Imaging by Fluorescence Photoactivation Localization Microscopy. *Biophys. J.* **2006**, *91*, 4258–4272. [CrossRef]
216. Kawata, S.; Inouye, Y.; Verma, P. Plasmonics for Near-Field Nano-Imaging and Superlensing. *Nat. Photonics* **2009**, *3*, 388–394. [CrossRef]
217. Wang, Y.; Srituravanich, W.; Sun, C.; Zhang, X. Plasmonic Nearfield Scanning Probe with High Transmission. *Nano Lett.* **2008**, *8*, 3041–3045. [CrossRef] [PubMed]
218. Holzmeister, P.; Acuna, G.P.; Grohmann, D.; Tinnefeld, P. Breaking the Concentration Limit of Optical Single-Molecule Detection. *Chem. Soc. Rev.* **2014**, *43*, 1014–1028. [CrossRef]
219. Stranahan, S.M.; Willets, K.A. Super-Resolution Optical Imaging of Single-Molecule SERS Hot Spots. *Nano Lett.* **2010**, *10*, 3777–3784. [CrossRef]
220. Willets, K.A. Super-Resolution Imaging of Interactions between Molecules and Plasmonic Nanostructures. *Phys. Chem. Chem. Phys.* **2013**, *15*, 5345–5354. [CrossRef] [PubMed]
221. Chen, B.; Wood, A.; Pathak, A.; Mathai, J.; Bok, S.; Zheng, H.; Hamm, S.; Basuray, S.; Grant, S.; Gangopadhyay, K.; et al. Plasmonic Gratings with Nano-Protrusions Made by Glancing Angle Deposition for Single-Molecule Super-Resolution Imaging. *Nanoscale* **2016**, *8*, 12189–12201. [CrossRef]
222. Qi, H.; Xiao, S.; Shi, R.; Ward, M.O.; Chen, Y.; Tu, W.; Su, Q.; Wang, W.; Wang, X.; Zhang, Z. Enhanced Reader.Pdf. *Nature* **2018**, *388*, 539–547.
223. Liu, C.Y.; Lo, W.C. Large-Area Super-Resolution Optical Imaging by Using Core-Shell Microfibers. *Opt. Commun.* **2017**, *399*, 104–111. [CrossRef]
224. Xu, M.; Wang, L.V. Photoacoustic Imaging in Biomedicine. *Rev. Sci. Instrum.* **2006**, *77*, 041101. [CrossRef]
225. Su, J.L.; Wang, B.; Wilson, K.E.; Bayer, C.L.; Chen, Y.S.; Kim, S.; Homan, K.A.; Emelianov, S.Y. Advances in Clinical and Biomedical Applications of Photoacoustic Imaging. *Expert Opin. Med. Diagn.* **2010**, *4*, 497–510. [CrossRef]
226. Laufer, J.; Delpy, D.; Elwell, C.; Beard, P. Quantitative Spatially Resolved Measurement of Tissue Chromophore Concentrations Using Photoacoustic Spectroscopy: Application to the Measurement of Blood Oxygenation and Haemoglobin Concentration. *Phys. Med. Biol.* **2007**, *52*, 141–168. [CrossRef] [PubMed]
227. Wang, X.; Pang, Y.; Ku, G.; Xie, X.; Stoica, G.; Wang, L.V. Noninvasive Laser-Induced Photoacoustic Tomography for Structural and Functional in Vivo Imaging of the Brain. *Nat. Biotechnol.* **2003**, *21*, 803–806. [CrossRef] [PubMed]
228. Zhang, E.; Laufer, J.; Beard, P. Backward-Mode Multiwavelength Photoacoustic Scanner Using a Planar Fabry-Perot Polymer Film Ultrasound Sensor for High-Resolution Three-Dimensional Imaging of Biological Tissues. *Appl. Opt.* **2008**, *47*, 561–577. [CrossRef] [PubMed]
229. Xie, Z.; Roberts, W.; Carson, P.; Liu, X.; Tao, C.; Wang, X. Evaluation of Bladder Microvasculature with High-Resolution Photoacoustic Imaging. *Opt. Lett.* **2011**, *36*, 4815. [CrossRef]
230. Hajireza, P.; Sorge, J.; Brett, M.; Zemp, R. In Vivo Optical Resolution Photoacoustic Microscopy Using Glancing Angle-Deposited Nanostructured Fabry–Perot Etalons. *Opt. Lett.* **2015**, *40*, 1350. [CrossRef] [PubMed]
231. Hajireza, P.; Krause, K.; Brett, M.; Zemp, R. Glancing Angle Deposited Nanostructured Film Fabry-Perot Etalons for Optical Detection of Ultrasound. *Opt. Express* **2013**, *21*, 6391. [CrossRef]
232. Champion, J.A.; Katare, Y.K.; Mitragotri, S. Making Polymeric Micro- and Nanoparticles of Complex Shapes. *Proc. Natl. Acad. Sci. USA* **2007**, *104*, 11901–11904. [CrossRef]
233. Tirrell, D.A.; Langer, R. Materials for Biology and Medicine. *Nature* **2012**, *428*, 25.
234. Love, J.C.; Gates, B.D.; Wolfe, D.B.; Paul, K.E.; Whitesides, G.M. Fabrication and Wetting Properties of Metallic Half-Shells with Submicron Diameters. *Nano Lett.* **2002**, *2*, 891–894. [CrossRef]
235. Sun, Y.; Xia, Y. Shape-Controlled Synthesis of Gold and Silver Nanoparticles. *Science* **2002**, *298*, 2176–2179. [CrossRef]
236. Pawar, A.B.; Kretzschmar, I. Fabrication, Assembly, and Application of Patchy Particles. *Macromol. Rapid Commun.* **2010**, *31*, 150–168. [CrossRef] [PubMed]
237. Zhang, J.; Grzybowski, B.A.; Granick, S. Janus Particle Synthesis, Assembly, and Application. *Langmuir* **2017**, *33*, 6964–6977. [CrossRef] [PubMed]
238. Yi, Y.; Sanchez, L.; Gao, Y.; Yu, Y. Janus Particles for Biological Imaging and Sensing. *Analyst* **2016**, *141*, 3526–3539. [CrossRef] [PubMed]
239. Wu, Z.; Li, L.; Liao, T.; Chen, X.; Jiang, W.; Luo, W.; Yang, J.; Sun, Z. Janus Nanoarchitectures: From Structural Design to Catalytic Applications. *Nano Today* **2018**, *22*, 62–82. [CrossRef]
240. Loget, G.; Roche, J.; Gianessi, E.; Bouffier, L.; Kuhn, A. Indirect Bipolar Electrodeposition. *J. Am. Chem. Soc.* **2012**, *134*, 20033–20036. [CrossRef]
241. Nie, Z.; Li, W.; Seo, M.; Xu, S.; Kumacheva, E. Janus and Ternary Particles Generated by Microfluidic Synthesis: Design, Synthesis, and Self-Assembly. *J. Am. Chem. Soc.* **2006**, *128*, 9408–9412. [CrossRef]
242. Roh, K.H.; Martin, D.C.; Lahann, J. Biphasic Janus Particles with Nanoscale Anisotropy. *Nat. Mater.* **2005**, *4*, 759–763. [CrossRef]
243. Pawar, A.B.; Kretzschmar, I. Multifunctional Patchy Particles by Glancing Angle Deposition Technique. In Proceedings of the 2008 Annual Meeting, Washington, DC, USA, 13 October 2008; pp. 355–358.

244. Imura, Y.; Kato, M.; Kondo, T.; Kawai, T. Strings of Metal Half-Shells Fabricated Using Colloidal Particle Monolayer as a Template. *Langmuir* **2010**, *26*, 11314–11318. [CrossRef]
245. Xuan, M.; Shao, J.; Lin, X.; Dai, L.; He, Q. Self-Propelled Janus Mesoporous Silica Nanomotors with Sub-100 Nm Diameters for Drug Encapsulation and Delivery. *ChemPhysChem* **2014**, *15*, 2255–2260. [CrossRef]
246. Peng, F.; Men, Y.; Tu, Y.; Chen, Y.; Wilson, D.A. Nanomotor-Based Strategy for Enhanced Penetration across Vasculature Model. *Adv. Funct. Mater.* **2018**, *28*, 1706117. [CrossRef]
247. Wu, Z.; Troll, J.; Jeong, H.H.; Wei, Q.; Stang, M.; Ziemssen, F.; Wang, Z.; Dong, M.; Schnichels, S.; Qiu, T.; et al. A Swarm of Slippery Micropropellers Penetrates the Vitreous Body of the Eye. *Sci. Adv.* **2018**, *4*, eaat4388. [CrossRef] [PubMed]
248. Martín, M.; Salazar, P.; Álvarez, R.; Palmero, A.; López-Santos, C.; González-Mora, J.L.; González-Elipe, A.R. Cholesterol Biosensing with a Polydopamine-Modified Nanostructured Platinum Electrode Prepared by Oblique Angle Physical Vacuum Deposition. *Sens. Actuators B Chem.* **2017**, *240*, 37–45. [CrossRef]
249. El Kazzy, M.; Weerakkody, J.S.; Hurot, C.; Mathey, R.; Buhot, A.; Scaramozzino, N.; Hou, Y. An Overview of Artificial Olfaction Systems with a Focus on Surface Plasmon Resonance for the Analysis of Volatile Organic Compounds. *Biosensors* **2021**, *11*, 244. [CrossRef] [PubMed]
250. Wilson, A.D.; Baietto, M. Applications and Advances in Electronic-Nose Technologies. *Sensors* **2009**, *9*, 5099–5148. [CrossRef]
251. Salazar, P.; Rico, V.; González-Elipe, A.R. Nickel–Copper Bilayer Nanoporous Electrode Prepared by Physical Vapor Deposition at Oblique Angles for the Non-Enzymatic Determination of Glucose. *Sens. Actuators B Chem.* **2016**, *226*, 436–443. [CrossRef]
252. Yadav, J.; Raturi, P.; Yadav, S.; Singh, J.P. Zig-Zag Ag2S Nanostructures for Superior Optical Absorption and Photoelectrochemical Water Splitting Performance. *Renew. Energy* **2021**, *179*, 2256–2266. [CrossRef]
253. Norrod, K.L.; Sudnik, L.M.; Rousell, D.; Rowlen, K.L. Quantitative Comparison of Five SERS Substrates: Sensitivity and Limit of Detection. *Appl. Spectrosc.* **2016**, *51*, 994–1001. [CrossRef]

biosensors MDPI

Review

Recent Advances in Silver Nanostructured Substrates for Plasmonic Sensors

Shashank K. Gahlaut [1], Anisha Pathak [1] and Banshi D. Gupta [2,*]

[1] Institute of Chemistry, University of Potsdam, 14476 Potsdam, Germany
[2] Department of Physics, Indian Institute of Technology Delhi, New Delhi 110016, India
* Correspondence: bdgupta@physics.iitd.ac.in

Abstract: Noble metal nanostructures are known to confine photon energies to their dimensions with resonant oscillations of their conduction electrons, leading to the ultrahigh enhancement of electromagnetic fields in numerous spectroscopic methods. Of all the possible plasmonic nanomaterials, silver offers the most intriguing properties, such as best field enhancements and tunable resonances in visible-to-near infrared regions. This review highlights the recent developments in silver nanostructured substrates for plasmonic sensing with the main emphasis on surface plasmon resonance (SPR) and surface-enhanced Raman spectroscopy (SERS) over the past decade. The main focus is on the synthesis of silver nanostructured substrates via physical vapor deposition and chemical synthesis routes and their applications in each sensing regime. A comprehensive review of recent literature on various possible silver nanostructures prepared through these methodologies is discussed and critically reviewed for various planar and optical fiber-based substrates.

Keywords: silver; plasmonics; sensors; surface-enhanced Raman spectroscopy; surface-enhanced fluorescence; surface-enhanced infrared spectroscopy

Citation: Gahlaut, S.K.; Pathak, A.; Gupta, B.D. Recent Advances in Silver Nanostructured Substrates for Plasmonic Sensors. *Biosensors* **2022**, *12*, 713. https://doi.org/10.3390/bios12090713

Received: 31 July 2022
Accepted: 22 August 2022
Published: 2 September 2022

1. Introduction

The study of light–matter interaction in coinage metals has been of great interest since ancient times, and their ability to reflect light was admired for ages [1]. The other historical uses of these metals have also been widely found in artifacts such as medieval stained glasses, the Lycurgus Cup and the global antibacterial use of silver by the Greeks and Romans [2]. However, in addition to more sophisticated scientific developments and understandings in the field of the atomic world, the recent use of these metals has gained intensive interest, touching almost every possible area of science and technology. With precise control over the nano dimensions of these metals, a very strong light–matter interaction arises through the free electrons of the metals, popularly known as the field of plasmonics. Initially this light confinement was only possible with dielectrics, which is diffraction limited to areas smaller than the wavelength of light [3]. Plasmonic modes, however, can localize light in the dimensions of the supporting metallic structure, i.e., nanostructured metals can tightly concentrate and manipulate light with unrivaled accuracy in the nanometer regime. The two major roles served by metal nanostructures in plasmonics are to guide light to desired locations and to serve as nano-antennas for localized strong electric fields, thus bridging the gap between the micro and nano worlds. As a proof-of-concept, plasmons have been known for over 150 years since the documentation by Michael Faraday in 1857 [4]. However, the recent applications of plasmonics have made significant developments since the discovery of nanoscience and have opened doors for processes that were considered impossible earlier. Thus, plasmonics is considered a burgeoning field of nanotechnology with an unraveled level of control over light manipulation, with applications including catalysis [5], photovoltaics [6], superlenses [7], plasmonic circuitry [8], waveguides [9], SPASER [10], single-photon transistors [11] and sensors [3,12–14].

The confinement and enhancement of light by metal nanostructures have led to point-of-care plasmonic sensors achieving new competencies in their sensitivities. Plasmonic sensors are one of the first, most widely used and successful applications that have gained intensive research interest due to their ultrahigh sensitivity now reaching the single-molecule level [15]. These sensors gain an advantage from the highly enhanced electromagnetic field around the metal nanoparticles, which is extraordinarily sensitive to the surroundings. These highly sensitive plasmonic sensors can provide a sound interpretation of biological processes in a simple and noninvasive manner for an improved healthcare system and, hence, have been surveyed by recent interesting review articles [16]. Since the discovery of plasmonic biosensors about four decades ago, technological advances over time have improved the fundamental and advanced understanding of this field, while many challenges are still being actively looked into within the research community. Almost every spectroscopic technique, such as Raman [17], fluorescence [18], infrared and UV–vis [19], have gained advantages from the field of plasmonics in the context of enhanced light–matter interaction and, hence, higher sensitivity. The most important factor in the growth of this field is the explicit control of nanostructured components in terms of their ease of fabrication, cost-effectiveness and efficacy in general environments, along with unprecedented sensitivity.

Among all the metals reported for plasmonic sensors, silver and gold are considered the most useful owing to their strong plasmon resonances in the visible range, biocompatibility and stability [13]. Silver has been long known for its antibacterial applications even in historical times. It was considered an efficacious weapon against pathogens due to the interaction of silver ions with the thiol group in bacteria and proteins leading to cell death [2]. However, the efficiency of silver at nano dimensions was discovered recently due to the growing field of plasmonics. In plasmonics, the interaction of light with a metal nanoparticle is best described by Mie scattering, which provides an extinction (absorption + scattering) cross-section of a spherical nanoparticle as

$$C_{ext} = \frac{24\,\pi^2 R^2 \varepsilon_m^{3/2}}{\lambda}\left[\frac{\varepsilon_i^2}{(\varepsilon_r + 2\varepsilon_m)^2 + \varepsilon_i^2}\right] \quad (1)$$

where R is the radius of the nanoparticle, λ is the excitation wavelength, ε_r and ε_i represent the real and imaginary parts of the complex dielectric function of the metal and ε_m is the dielectric constant of the medium surrounding the metal nanoparticle [3,20]. The dielectric properties of a metal are governed by ε_r and ε_i, which depend highly on the excitation wavelength. Thus, the interaction between a metal nanoparticle and light depends on its dielectric properties, as stated in Equation (1). Among all other factors important to the engineering of plasmonic sensors, the optical properties of the metal are key. Examining Equation (1) tells us that C_{ext}, which signifies the strength of this electromagnetic interaction, increases to reach infinity when the denominator in the bracket approaches zero, indicating resonance in the excitation light and electronic oscillations in the metal. This will occur when ε_r is nearly equal to $-2\varepsilon_m$ and ε_i is near zero, which is not possible for dielectrics, which typically have ε_r values greater than one. These conditions can be satisfied only by some of the metals, and their plasmon strength depends on the quality factor (QF) broadly described as the ratio of ε_r to ε_i. The high value of QF is associated with strongly confined plasmons, whereas a low QF is associated with lossy plasmons with low C_{ext}. Silver has the highest QF across most of the visible spectrum, from 300 to 1200 nm, although aluminum (Al) provides a higher value in the UV region [3,21]. Considering the interband transitions from the conduction band to higher energy levels, which weaken SPs modes, Au and copper (Cu) are limited by their localized surface plasmon resonance (LSPR) excitation, above 500 and 600 nm [22]. However, for Ag, these transitions occur far above the LSPR frequencies. The other factors to be considered for the best plasmonic material for biosensors are their toxicity and stability. Although these factors make Au the choice over silver, a passivating layer may overcome these challenges for silver very easily. The controlled synthesis techniques and overall cost of the metal will also greatly determine its feasibility for large-scale applications. Keeping in mind all the above factors, silver offers a unique choice for plasmonic biosensors owing to the strongest plasmonic resonances, cost-effectiveness and the ease of synthesis of Ag structures with controlled shape/size [3,23].

This review focuses on the various synthesis methodologies for diverse silver nanostructures reported for plasmonic sensors. The aim of the article is to compile the literature specifically on silver as a plasmonic material for various sensing applications, as compared to the more generalized articles reported so far, to emphasize the peculiarity of silver over other plasmonic materials in terms of cost and efficacy [16]. It mainly focuses on the chemical methods involving solution-phase synthesis and physical methods such as vapor deposition; glancing angle deposition (GLAD); and lithographic techniques for silver nanostructures, ranging from nanoparticles, nanocubes, nanotriangles, nanorods and nanowires. Further, the review elaborates on recent spectroscopic techniques, focusing on plasmonic enhancement for biosensing methods such as SPR/LSPR, SERS, surface-enhanced fluorescence (SEF) and surface-enhanced infrared spectroscopy (SEIRS) using Ag as the plasmonic material. In addition, the focus will be on the advancements made in these fields underlying the fundamental mechanisms and applications specific to silver nanostructures for environmental and food monitoring, defense applications and biological detection. The scope of the article is schematically presented in Figure 1.

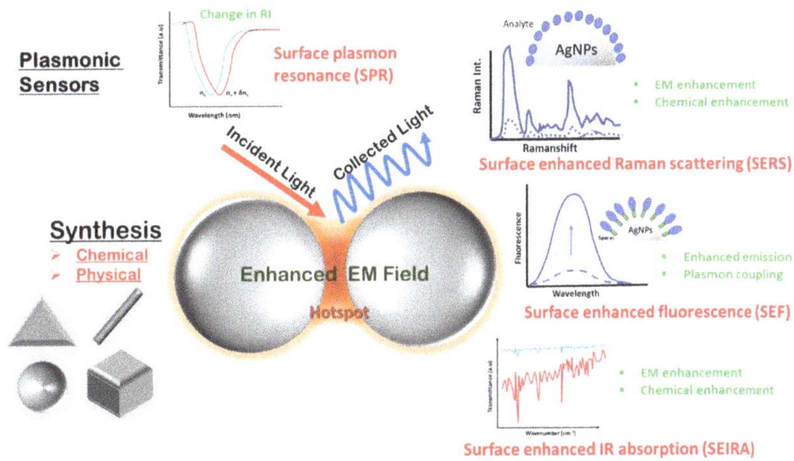

Figure 1. Schematic showing the scope of the review article.

2. Synthesis of Silver Nanostructures

A wealth of synthesis methods has been reported and modified over time to determine a precise control over the shape and size of AgNPs. These methods may be broadly classified into two categories as physical methods (lithography, vapor deposition, microwave-assisted synthesis, laser ablation) and chemical methods (involving reduction, precipitation, photoreduction, biological/green synthesis, hydrolysis, etc.). The focus will be on the major methods reported for the synthesis of silver nanostructures for plasmonic sensors, i.e., chemical reduction, green synthesis, photoreduction, physical vapor deposition and lithography.

2.1. Chemical Synthesis

A fine tuning of the shape, size and composition of silver nanoparticles can greatly affect their optical, thermal, electronic and catalytic properties for a wide range of plasmonic applications. Solution-phase synthesis is the most widely accepted method for maneuvering the shape and composition of silver nanostructures with great uniformity [3].

As compared to isotropic nanoparticles such as spheres, tuning the shape of nanoparticles to obtain anisotropic structures has been a focus of researchers due to the very high enhancement of plasmonic electromagnetic fields and the possibility of tuning the resonances over a wide wavelength range, spanning from the visible to IR spectrum. Thus, the engineering of anisotropic metal nanoparticles is very important for plasmonic sensors, especially for SERS-based sensing, where tuning the size of a hotspot may lead to the detection sensitivities up to single-molecule level [24,25]. Chemical methods provide good control of the morphology of nanoparticles to obtain various shapes. Here, a metal salt (generally silver nitrate; $AgNO_3$) is reduced in the presence of a stabilizer to provide innumerable possibilities of morphological control for the synthesis of silver nanostructures [14]. The reduction of Ag^+ ions in the precursor causes elemental Ag atoms to grow into clusters and, finally, into nanostructures. The major factors controlling the growth and stabilization of nanostructures are the choice of reducing agent, stabilizer, temperature, relative concentrations and time. The reduction of metal salt/precursor may be performed in several ways, i.e., chemical reduction (using reductants such as sodium borohydride, hydroxylamine, citrate, hydrogen, hydrazine, etc.), photoreduction, electrochemical or sonochemical reduction. This reduction process is then stabilized using surfactants, ligands or organic molecules, which inhibit columbic repulsion/steric hindrance by adsorbing on the nanoparticle surface [26].

2.1.1. Polyol Reduction

Polyol-based chemical reduction is the most widely accepted method for the synthesis of silver nanoparticles of various shapes and sizes. Basically, a polyol, typically ethylene glycol, is used as a solvent and reducing agent in this synthesis procedure. At an elevated temperature, silver precursor (mainly $AgNO_3$) and a stabilizer or capping agent such as PVP (polyvinylpyrrolidone) are added to polyols for the reduction of Ag^+ ions to silver nanostructures of various shapes. Several parameters, such as temperature, pH and trace ions, have been reported for the shape engineering of silver nanostructures over time to tune the SPR properties of AgNPs in polyol synthesis [27–30]. Several reports have provided detailed mechanisms for the reduction process in polyol synthesis where ethylene glycol (EG) is used both as a reducing agent and a solvent for the silver precursor. It was later proposed that EG forms glycolaldehyde in the presence of oxygen at high temperatures, and then it is the major reductant in EG-based polyol processes [31]. In these processes, the final shape of the nanostructure is controlled by the twin planes present in the seed formed at the initial stage. When silver nitrate is reduced by EG, the initial seeds may take single-twin, multiple-twin or single-crystal shapes depending on the more thermodynamically favored system, as shown in Figure 2a [27]. The nucleation and growth of one of these structures may be controlled by selecting the reaction conditions to obtain the desired shape of the final nanostructure. Single-twin seeds can be modified into right bipyramids and beams, multiple-twinned seeds can be grown into 1D nanowires and nanorods and single-crystal seeds can produce nanocubes [27]. Various morphologies have been reported by the introduction of the NaCl- and NaBr-etching of these seeds, as shown in Figure 2b. The capping agent also plays a major role in controlling the growth of selective crystal planes (100) or (111) and has been extensively used to see the effect on the morphology of the final structure [28]. On the other hand, it is believed that anisotropic structures, such as silver nanoplates, where the lateral dimension is much larger than their thickness, have the highest electromagnetic field enhancement and tunability of the LSPR peak. A simple preparation of silver nanoplates with spherical colloids of 3.5 nm was reported after refluxing in ambient light conditions. Light and the driving force provided by refluxing is assumed to transform spherical seeds into nanoplate-like structures, as shown in Figure 2c [32]. A photochemical synthesis route has also been proposed to produce silver nanoplates from citrate-capped spherical nanoparticles [33]. Here, the anisotropic growth of these structures is favored in the presence of excess Ag^+ ions and sodium citrate. It is believed that the presence of twin planes in the seeds and the preferential capping of (111) facets by citrate are responsible for the growth of the plate structure. Later, a DFT theoretical model of this preferential binding was provided [34]. It was noticed that three-fold symmetry in citric acid matches with Ag (111). This leads to stabilization of this particular facet and growth in the lateral dimension. Zhang et al. [35] provided a detailed understanding of the formation of these kinds of structures and the most important factors involving the role of citrates and other similar carboxyl compounds. They carefully analyzed the role of the most extensively used PVP ligands and explained their trivial role in a much-simplified process of large-scale synthesis. The critical role of hydrogen peroxide (H_2O_2) in the formation of planar-twinned seeds to produce silver nanoplates was carefully examined and emphasized in their process. Their versatile method was later utilized by Song et al. [36] for a sensitive fiberoptic LSPR sensor.

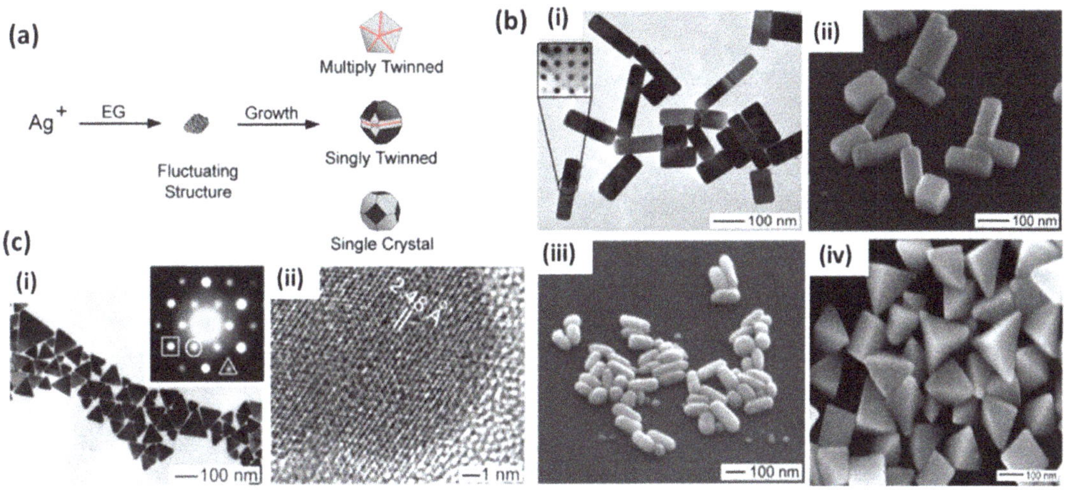

Figure 2. (**a**) The growth of silver nanostructures in polyol synthesis caused by the formation of silver nuclei with low surface energy twin boundary defects in order to multiply twinned, singly twinned or single-crystal seeds. The final morphology may be controlled by engineering the growth of these seeds with various parameters [27]. (**b**) (**i**,**ii**) TEM and SEM images of Ag nanobars produced with NaBr; (**iii**) single-crystal nanorice formed with the storage of nanobars in PVP; (**iv**) the formation of right bipyramids just by reducing the NaBr concentration to half, producing single-twinned seeds by causing the degree of etching to be moderated [27]. Reproduced with permission from ref. [27]. Copyright 2007, American Chemical Society. (**c**) (**i**) Triangular silver nanoplate self-assembled monolayer formed by a simple reflux of silver seeds in ambient laboratory conditions, and (**ii**) a high-resolution TEM showing single-crystallinity in the (111) direction [32]. Reproduced with permission from ref. [32]. Copyright 2003, American Chemical Society.

Sun and Xia [26] reported the synthesis of silver nanocubes through such a polyol process in a very nice, shape-controlled manner. Many modifications with the involvement of Cl^- [37], SH^-, HCl [38], Fe^{2+}/Fe^{3+} [39], and Br^- [40] have been reported for monodispersed cubic particles. Most of the studies conclude that PVP chain length and the molecular weight on the final structure are important. The main hypothesis is the selective binding of PVP to (100) facets as compared to (111) facets, resulting in shapes mainly terminated by (100) facets [26].

A very well-studied polyol synthesis was reported recently with an emphasis on the role of Cl^- ions in the formation of well-oriented uniform silver nanocubes [37]. Experimental and theoretical considerations were taken into account to elucidate the role of Cl^- in the formation of Ag nanocubes. It was found that adding HCl, on the one hand, controls the rate of $AgNO_3$ reduction due to the formation of HNO_3 and, on the other hand, controls the formation of silver cubes due to the release of Cl^- ions, which preferentially stabilize Ag (100). Thus, by increasing the concentration of Cl^-, the shapes change from truncated octahedra to truncated cubes and, finally, to cubes, as shown in Figure 3.

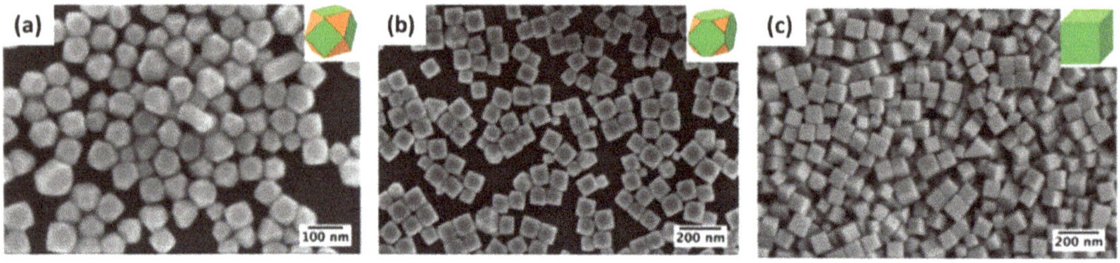

Figure 3. SEM images of AgNPs synthesized with varying concentrations of Cl^-: (**a**) 3 mM HNO_3 + 0.03 mM NaCl, (**b**) 3 mM HNO_3 + 0.3 mM NaCl and (**c**) 3 mM HNO_3 + 3 mM NaCl. Insets show the validation of experimental observation with abinitio thermodynamic calculations. Reproduced with permission from ref. [37]. Copyright 2019, American Chemical Society.

Zhu et al. [41] reported the synthesis of silver nanowires with a polyol process using PVP as the capping agent. The growth mechanism is nicely explained in terms of increasing molecular weight and the chain length of PVP due to the chemical adsorption of Ag^+ ions on PVP chains. Tsuji et al. [29] also reported a microwave-assisted polyol process for the fast synthesis of silver nanostructures from 2D nanosheets and nanoplates at short PVP chain length (10 K) into 1D nanorods and nanowires with an increasing PVP chain length (40, 360 K). Xia's group extensively studied the growth mechanism and many factors related to the growth of silver nanowires in the polyol process [30]. They observed pentagonal cross-section NWs in a PVP-controlled polyol process. Here, the initial molar ratio of PVP and $AgNO_3$ at a particular temperature results in multiply twinned nanoparticles (MTPs) through Ostwald ripening, which further governs the growth of NWs. PVP acts as a capping and stabilizing agent that selectively passivates the {100} facets rather than {111} and facilitates the uniaxial growth of silver in 1D leading into uniform nanowires, as shown in Figure 4a [30]. Later, they also studied the role of Fe (II), Fe (III) [39] and Cu (I) and Cu (II) [42] ions in the polyol reduction process. The iron ion concentration greatly affects the nanostructure and controls the formation of nanocubes (Figure 4b(i)) or nanowires (Figure 4b(ii)) [39]. It was observed that the lower iron ion concentration results in the selective etching of MTPs, as they are unable to completely remove atomic oxygen, which is adsorbed on the surface of nanoparticles and leads to the formation of nanocubes. In a similar way, the higher concentration prevents etching and allows for the growth of twinned seeds in nanowires by removing the adsorbed oxygen. In the case of copper salt, the rapid synthesis of nanowires was attributed to the presence of both cation and anion, where Cl^- helps control the amount of Ag^+ in the initial stage, and Cu(I) helps prevent oxidative etching, as shown in Figure 4c [42].

The synthesis of branched nanowires was realized by Cong et al. [43] for SERS-sensing applications using polyethylene glycol (PEG) as a solvent and reducing agent for silver nitrate and PVP as the capping agent. The branches were observed to be grown anisotropically from the defects on silver nanowire surfaces.

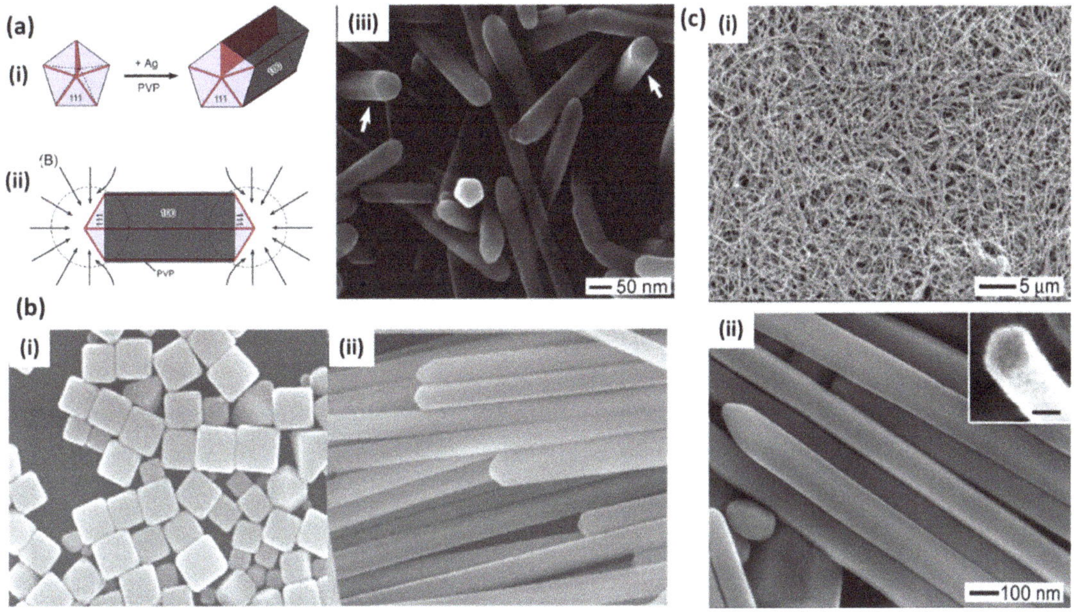

Figure 4. (**a**) Mechanism for the growth of pentagonal cross-section silver nanowires. (**i**) Formation of a nanorod from a multiply twinned nanoparticle (MTP) of silver due to the strong interaction between PVP and {100} facets and weak interaction with the {111} facets. (**ii**) Diffusion of silver atoms toward the two ends of a nanorod, with the side surfaces completely passivated by PVP. (**iii**) SEM image of pentagonal facet silver NWs [30]. Reproduced with permission from ref. [30]. Copyright 2003, American Chemical Society. (**b**) (**i**) Formation of nanocubes (<0.44 µM) at a low concentration of iron ions and (**ii**) nanowires at high concentrations (2.2 µM) of iron ions from silver MTPs [39]. Reproduced with permission from ref. [39]. Copyright 2005, American Chemical Society. (**c**) (**i**,**ii**) SEM images of Ag nanowires prepared with the polyol reduction of $AgNO_3$ in the presence of $CuCl_2$ and PVP at different magnifications [42]. Reproduced with permission from ref. [42]. Copyright 2008, Royal Society of Chemistry.

2.1.2. Citrate Reduction

The second most popular chemical synthesis technique for the production of silver nanoparticles is citrate reduction, which was proposed in 1982 [44]. Here, typically sodium citrate is used in a dual role, i.e., the reduction and stabilization of silver NPs. In a typical synthesis, an aqueous solution of sodium citrate is added to a boiling solution of silver nitrate to obtain diverse-shaped nanoparticles. A detailed study on the control of the shape and size of AgNPs in this method was conducted by Pillai and Kamat [45] using pulse radiolysis. Here, mostly large-sized silver NPs (50–100 nm) were obtained with well-defined facets. The concentration of citrate ions plays a critical role in controlling the kinetics of Ag^+ reduction and, hence, defines the final morphology. The increase in the concentration of sodium citrate decreases the growth of silver particles by forming a complex with Ag^{2+} dimers, thereby producing larger clusters. Citrate has also been reported to be used for the photoinitiated conversion of silver nanoparticles into nanoplates [46,47]. The role of pH in this process was also investigated due to the change in the activity of citrate with pH [48]. It was found that, at high pH, citrate shows a higher reduction rate for silver precursor, resulting in spherical and rod-like nanoparticles, whereas low pH results in the formation of triangles and polygons. The TEM images of various nanoparticles prepared using this method are shown in Figure 5.

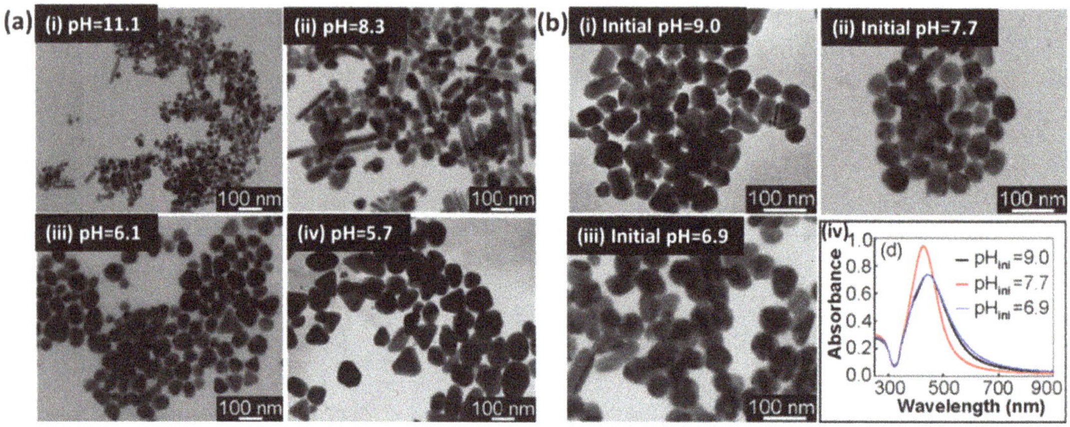

Figure 5. (**a**) TEM images of silver nanoparticles synthesized under pH values of (**i**) 11.1, (**ii**) 8.3, (**iii**) 6.1 and (**iv**) 5.7. (**b**) TEM images of silver nanoparticles synthesized using the stepwise method with initial pH values of (**i**) 9.0, (**ii**) 7.7 and (**iii**) 6.9, as well as (**iv**) their corresponding UV–visible absorption spectra. Reproduced with permission from ref. [48]. Copyright 2009, American Chemical Society.

A modified citrate synthesis method has also been reported for "clean surface" nanowires without any surfactant or seed processes [49]. Here, citrate serves as a reducing agent and hydroxyl ions in NaOH facilitate the nanowires' growth at elevated temperatures. It is believed that, although citrates provide isotropic structures at room temperature, at high temperatures, the equilibrium constant of this process may differ, and citrate binding is restricted at certain crystal faces, allowing for the growth of wire structure. The concentration of hydroxyl ions was found to be responsible for the growth of nanowires and their aspect ratios. The advantage of these methods is the clean surface, as compared to template-directed methods (using polymers and surfactants), where multiple washings are required to remove the template for sensing applications such as SERS. Molecular dynamics simulations were also provided by the same group to better understand the growth of these nanowires in citrate-mediated processes [50]. The citrate-capped AgNPs have been the choice for SERS-based studies for a long time. The LSPR properties of single and dimeric silver nanoparticles prepared with citrate reduction have been studied thoroughly with electron microscopy [51] and later applied to single-molecule SERS studies [52].

More recently, the field of the green synthesis of silver nanostructures has picked up pace due to its ample availability and environmentally friendly route [53–56]. A green synthesis approach has been proposed for spherical and crystalline silver NPs, self-assembled on NH_2-modified glass substrates using citrus peel extracts from orange fruit (AgNP-Ora), tangerine fruit (tangerine fruit, AgNP-Tan) and lemon fruit (AgNP-Lem) [57]. These substrates were utilized for SERS studies of 4-aminobenzenethiol, rhodamine 6G and methylene blue as Raman probe molecules in µM concentrations. However, a lack of shape-controlled structures still limits their use in real applications and provides a lot of room for the research community. Various other methodologies such as homogenous and heterogenous seed-mediated growth [58–60], template-directed growth [61–64] and light-mediated [65] chemical synthesis methods have also been reported for AgNPs.

Therefore, till now we discussed major routes of chemical synthesis of AgNPs. However, a major issue with the colloidal substrate is the tendency to aggregate after the addition of analyte solution, which makes the colloid unstable and often leads to the poor reproducibility of the signal. In addition, there is very low control over the tuning of analyte-NP surfaces in colloidal nanoparticles, leading to the comparatively low enhancement of spectroscopic signals. Moreover, the transfer or deposition of colloidal NPs on any surface has always been a concerning factor in many applications due to inherited inhomogeneity at the macro- and nano-level [66,67]. For example, deposition on curved substrates, e.g., optical fiber, is not an easy task. Binding chemistry to attach NPs needs to be meticulously designed to ensure uniform coating. Hence, although the wet synthesis of NPs is found to be a very easy and cost-effective synthesis method, at the same time, the response and enhancement are compromised. This leads to very few commercially available plasmonic sensors fabricated with colloidal nanoparticles due to difficulty in reusability, homogeneity and disposal after every use. In contrast, physical vapor deposition techniques, such as thermal and e-beam evaporation, sputtering, etc., ensure very high reproducibility, purity, uniformity and the high enhancement of plasmonic signals. These substrates may prove to be very reliable plasmonic nanosensors with great stability for commercial and large-scale fabrication needs. Thus, the next section is devoted to discussing the physical deposition techniques for plasmonic sensors.

2.2. Lithography-Based Silver Substrates

Photolithography is the most used technique for the deposition of patterned arrays of metal and semiconductor materials on planar substrates, especially for device fabrication. Due to the diffraction limit of light, it does not fulfill the resolution requirement for smaller feature sizes. Electron beam lithography (EBL) and focused ion beam lithography (FIB) are top-down approaches utilizing polymeric resists to fabricate masks for the deposition or etching of metal with high resolutions. They have the ability to control the size, shape and periodicity of the nanostructured substrates [68]. In EBL, a focused electron beam is scanned over a substrate, mainly silicon (Si), through a programmed desired mask. The Si wafer is already spin-coated with a special polymer layer called a resist. This polymer is sensitive to exposure to e-beams. It could be a positive or a negative resist. In the case of a positive resist, for example, polymethyl methacrylate (PMMA), the exposed area breaks down after being illuminated by an e-beam, and then it dissolves into a solvent (developer). The substrate is coated with a silver layer of desired thickness using a metal evaporation process, and the rest of the metal, which is a resist, is lifted off, leaving a regular array of silver nanoislands. The resolution capability of EBL has repeatedly been reported to be up to 10 nm [68]. Similarly, in FIB, a gallium ion (Ga^+) beam is used in place of an electron beam for metal deposition and etching. However, FIB includes the possibility of maskless deposition. Very thin, rectangular nanoarrays and overhanging nanostructures with very high resolution are possible with FIB [69,70]. The major issue with these conventional lithographic techniques is that they are expensive, time-consuming and need sophisticated equipment, which limits their use for scalable fabrication. Consequently, a facile lithographic technique known as nanosphere lithography (NSL) was developed by Fischer and Zingsheim [71]. NSL is cost-effective and timesaving as compared to the aforementioned techniques. In parallel, Deckman and Dunsmuir also successfully illustrated this technique in 1982 by preparing a monolayer of spherical particles on a substrate to use as a mask for the fabrication of nanomaterials and called this technique "Natural Lithography" [72].

NSL, also known as natural lithography (as previously noted), colloidal lithography and shadow nanosphere lithography, is a promising technique for the fabrication of two-dimensionally arranged periodic, disconnected nanostructures on both rigid and flexible substrates. In this technique, a monolayer of colloidal nanospheres is deposited on the substrate, e.g., Si or glass, to use the pattern for the mask [72,73]. Over time, different methods have been invented, e.g., a self-assembled monolayer of polystyrene nanospheres on the air/water interface, spin-coating, rubbing, etc., in order to design a template for fabrication in a large area. Creating a closely packed hexagonal monolayer of spherical particles on a larger area of the substrate, with a high-quality, single-crystalline domain, is still a challenging task. Recently, Zhao's group extensively exploited NSL for patterning 2D arrangements of tunable plasmonic substrates, also shown in Figure 6a [74]. Zhu et al. demonstrated Ag nanorod bundles array vertically when grown on a Cu/Au substrate using colloidal lithography combined with the binary-template-assisted electrodeposition method, as shown in Figure 6b [75]. Lee et al. demonstrated a defect-free silver nanohole (diameter 300 nm) array over a millimeter-sized area. A colloidal, self-assembled monolayer of polystyrene nanospheres was deposited on a substrate at the air–water interface. Then, reactive ion etching (RIE) was employed to shrink the nanospheres before silver deposition, which led to the formation of a metallic nanohole array (shown in Figure 6c) after nanosphere removal with extraordinary optical transmission [76]. The fabricated array exhibited a high density of hotspots for the SERS-based sensing of phenolic pollutants. Ingram et al. demonstrated Ag–Cu mixed phase nanopatterns at different compositions of metals, combining shadow nanosphere lithography and glancing angle co-deposition (Figure 6d) [77]. Combining NSL with oblique angle deposition, they designed a transparent metallic nanohole array over a large surface area with improved electrical properties [78]. A thick layer of Ag (>100 nm) was made transparent with a periodic array of hollow nanocones using NSL. The optical transmission was measured with the height of the tips and the diameter of the holes. A SEM image of hollow nanocones, with a height/diameter of 500/350 nm, is shown in Figure 6e [79]. By extension, the same group demonstrated a surface plasmon sensor by fabricating disk-in-volcano array structures using NSL, as shown in Figure 6f [80].

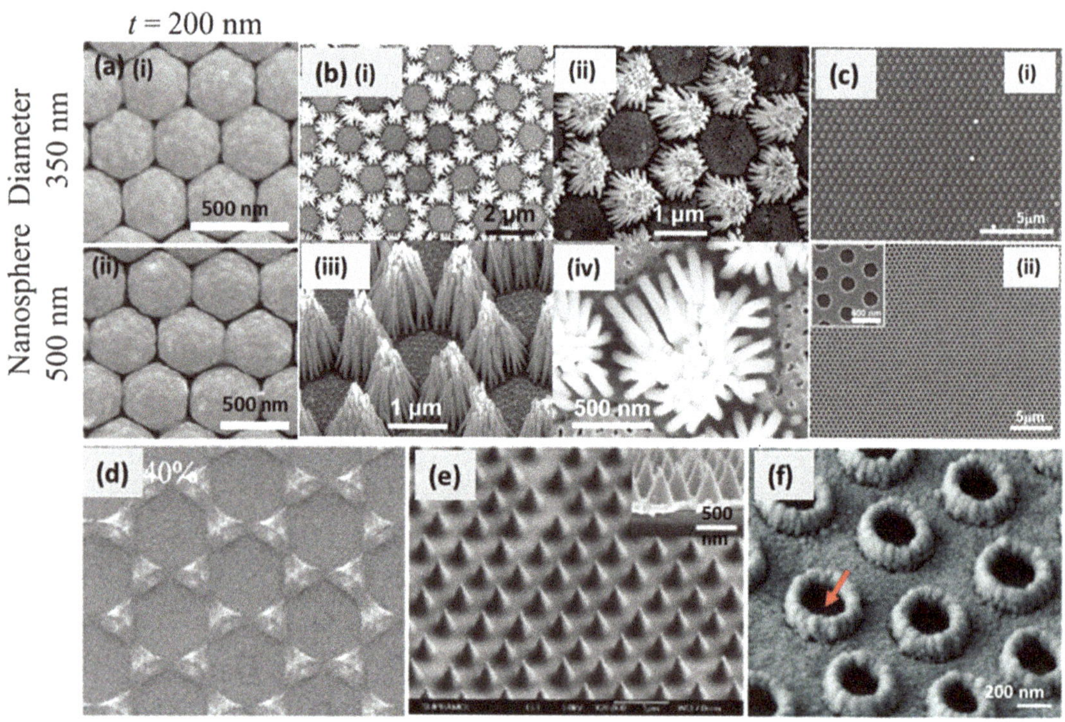

Figure 6. Ag nanostructured substrates fabricated by NSL. (**a**) Top-view SEM images of a 200 nm thick layer of Ag, deposited on monolayers of polystyrene nanospheres of (**i**) 350 nm (**ii**) 500 nm diameter at a tilt angle of θ = 55° [74]. Reproduced with permission from [74]. Copyright 2020, IOP. (**b**) SEM images of as-prepared Ag-nanorod bundle arrays: (**i**,**ii**) top views at different magnifications, (**iii**) side view of the bundle arrays and (**iv**) top view of a single bundle of Ag-nanorods [75]. Reproduced with permission from [75]. Copyright 2016, John Wiley & Sons. (**c**) (**i**) SEM showing an Ag nano-mesh film with a PS 600 nm nanosphere template after 50 nm thick Ag deposition and (**ii**) the resulting large area (~30 × 30 μm^2), a single-crystalline, hexagonal-aligned hole array [76]. Reproduced with permission from [76]. Copyright 2009, American Chemical Society. (**d**) Ag–Cu mixed phase nanopatterns at the calculated composition of Ag 40% using shadow nanosphere lithography and glancing angle co-deposition [77]. Reproduced with permission from [77]. Copyright 2017, IOP. (**e**) SEM image

(45° tilting views) of hollow nanocones with a height/diameter of 500/350 nm; the insets show the cross-sectional view of the corresponding samples [79]. Reproduced with permission from [79]. Copyright 2014, American Chemical Society. (**f**) SEM of disk-in-volcano arrays at a 30° tilting view with a bottom hole diameter of D = 340 nm; the top-hole diameter is d = 220 nm; the height of the volcano is H = 250 nm; the height of the disk is h = 100 nm; and the total thickness of the film = 150 nm. The red arrow points to the disks inside the volcanos [80]. Reproduced with permission from [80]. Copyright 2015, Royal Society of Chemistry.

In another report, a close-packed monolayered polystyrene (PS) sphere array was successfully prepared using the H_2O_2-assisted air–water interfacial floating method. The advantage of H_2O_2 over water is the decreasing number of second-layer defects and the promotion of the self-assembly of PS nanospheres to form a close-packed SAM [81]. For more details about NSL, readers are referred to a comprehensive review [82] and other useful research articles [83,84].

NSL, nonetheless, has been found to be a promising and easy-to-use technique to fabricate a regular pattern of plasmonic nanostructures in a 2D arrangement, but, like other techniques, it also has some limitations, for example, the moderate ability to tune the shape of nanostructures and fabrication on large surface areas. In addition, self-assembled metallic colloid is another method to produce regular-patterned nanostructures [85]. Yet again, the reduction of metallic salt takes place on the substrate, which can lead to random aggregation and low reproducibility. To overcome all these challenges, the entire thin film community came up with a solution by using physical vapor deposition (PVD) techniques.

2.3. Glancing Angle Deposition (GLAD)

In general, PVD comprising thermal evaporation, e-beam evaporation and sputtering has been extensively used for uniform thin film deposition on a large-area solid substrate. If the vapor flux is incident on the substrate at a normal incidence angle, a thin film or nano-island growth be achieved [86]. Conversely, if the substrate is placed at an angle (generally > 70°) from the incoming vapor flux, a new era of sculptured thin film deposition starts. This specialized PVD method is generally known as oblique angle deposition (OAD) or glancing angle deposition (GLAD) [87–90]. As this tool has much potential to fabricate a variety of nanostructured substrates with high purity, uniformity and reproducibility, it has been accepted by numerous researchers worldwide, specifically those working on plasmonic sensors [89,91]. GLAD is a specialized version of physical vapor deposition (PVD), offering a provision of manipulation of the substrate orientation during thin film deposition in a high-vacuum chamber. The substrate can be rotated in polar and azimuthal directions with respect to the direction of the incoming vapor flux. The ballistic shadowing effect plays a major role in the foundation of columnar thin films in GLAD, which is only possible when the incoming vapor flux is well collimated toward the substrate, which could be glass, Si or sapphire. A large angular spread in incoming vapor flux may result in poor shadowing [88]. There are two prominent approaches to obtaining collimated vapor flux: a large distance between the vapor source and substrate or physical obstacles that select a subset of the uncollimated vapor flux. A great distance from the source to the substrate not only improves the collimation of the incidental vapor flux, but also increases the number of collisions of atoms before reaching the substrate. Vapor's mean free path is another important parameter that should be larger than the source–substrate distance in GLAD configuration. One needs to optimize all the deciding parameters to attain high-quality nanostructured films using GLAD. To evaporate the metal powder or pellets, electron beam and thermal sources are usually employed in GLAD depositions. At a larger angle of incidence (>75°), the low surface diffusion of adatoms and the shadow cast by the larger nucleation collectively result in one-dimensional columnar growth. Initially, the incident atoms are randomly deposited on the substrate, having some defects or roughness. Following the Volmer–Weber growth model, these adatoms form small islands and then 3D nuclei on the substrate. As deposition proceeds, the initial nucleated islands start projecting ballistic shadows on the surrounding region. The larger islands receive more than the smaller ones or the shadowing regions, which converts them into a columnar structure tilted in the direction of the incident flux. The direct and spontaneous growth of nanostructured thin films of a wide range of materials on various substrates can be easily attained using GLAD [87–91]. M.J. Brett et al. [87,88] pioneered this technique, and then it was widely explored by many researchers for the fabrication of various metals and insulator-sculptured thin films.

In recent years, a large number of 3D nanostructures, e.g., columns, rods, helices, zigzags, springs, etc., have been obtained using GLAD, as shown in Figure 7. In the scope of this article, we focused on various silver-sculptured thin films manifested by GLAD. Ag-decorated SiO_2 helical films were deposited on large-area transparent substrates, and the plasmonic chiroptical properties of these arrays were studied using circular dichroism. The chiral nanohelices are shown in Figure 7a [92]. This could be an exceptionally facile method to fabricate metamaterials on any solid surface. In recent years, GLAD-fabricated nanocolumnar pure Ag nanorod arrays were used for gas-sensing applications for the first time (Figure 7b) [93]. A drastic change in the colorimetric and water wettability properties of AgNR arrays was observed in the presence of a low amount of H_2S gas. The same feature of these substrates was exploited for the detection of viability and antimicrobial resistance in bacteria [94]. Further, sulfurized AgNRs were utilized in the form of multifunctional material Ag-Ag_2S nanoheterostructures, on which photocatalysis, hydrogen evolution, SERS and antimicrobial properties were investigated [95]. In another report, the same group studied the effect of a number of Ag arms in a zigzag array (shown in Figure 7d,e) on the EM enhancement of the SERS signal [96]. Moreover, Jen et al. fabricated a chiral Ag nanohelix array, and the effect of GLAD parameters (angles, deposition rate and rotation speed) on their growth was studied in detail (Figure 7f–k) [97].

The fabrication of silver film on patterned or templated substrates can also be realized using GLAD. Aligned tilted AgNR arrays have been fabricated on commercial compact discs (CDs) and digital versatile discs (DVDs) by using this technique. Before depositing Ag, the thin polycarbonate protective coating found on these discs was removed by immersing them in a concentrated nitric acid solution for a certain amount of time. The disc was then washed and rinsed with deionized water and blow-dried with nitrogen gas. It was then loaded into a vacuum chamber for metal evaporation at an angle of 86° with respect to the vapor direction. Due to the shadowing effect caused by deposition, the resulting hybrid structure acts as an effective anisotropic grating, with a period for the disc. The structure exhibited an optical anisotropy that depends on the polarization of the incident light. The variations in the plasmonic resonant shift with respect to the aspect ratio of the AgNRs and film thickness were studied [98,99]. Ag nanostructured grating substrates show significant plasmonic enhancement and, therefore, have been exploited for surface plasmon-based refractive index sensors [100,101] and SERS-based sensors [102,103].

There are tremendous possibilities in designing and fabricating numerous nanostructures by controlling deposition parameters, such as incident angle θ, azimuthal angle φ, substrate rotation, deposition rate, substrate temperature, material to be deposited and substrate [104,105]. Sculptured thin films fabricated using GLAD have potential applications in numerous fields, and above all, plasmonics is the major one. The GLAD-fabricated substrates emerged to overcome the challenges of uniformity and reproducibility in SERS substrates, which were extensively reviewed recently [106]. The simplicity, flexibility, cost-effectiveness and versatility to deposit a variety of materials are the main features that make GLAD stand out as the most promising tool for micro- and nanofabrication [107].

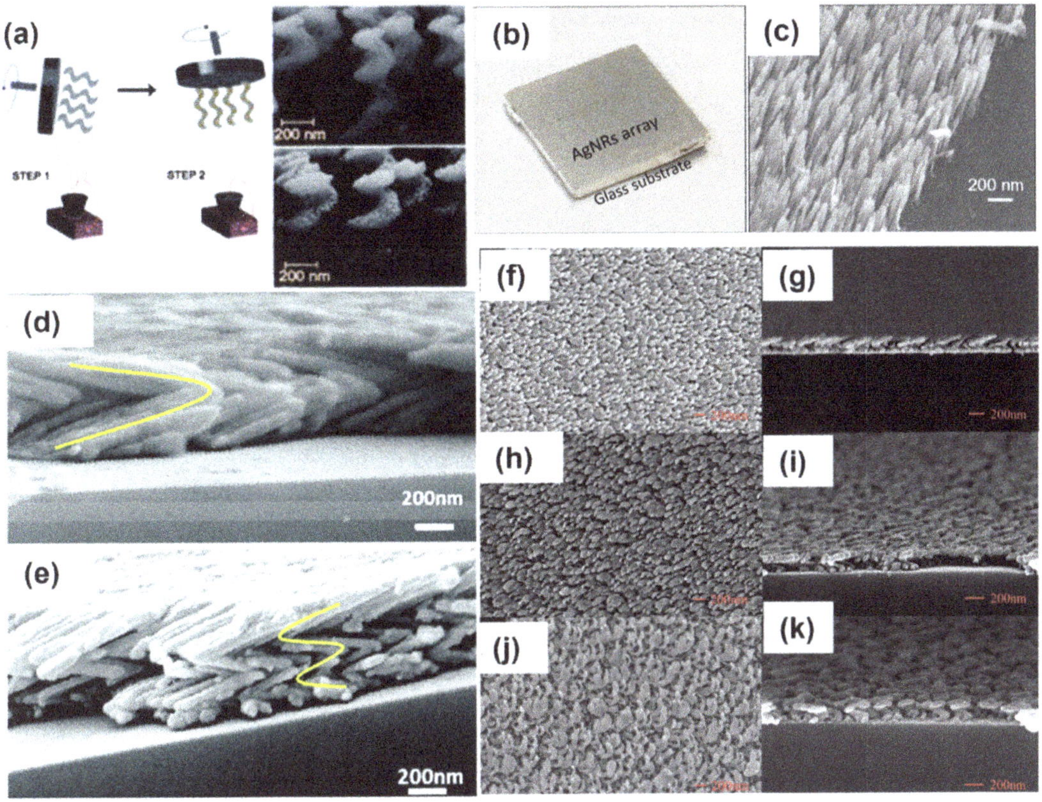

Figure 7. (**a**) Schematic of the GLAD technique used to fabricate a dielectric 3D template in the first step, followed by the evaporation of Ag metallic islands in the second step. SEM images of the nanostructured surfaces without (top) and with (bottom) the Ag metallic islands [92]. Reproduced with permission from [92]. Copyright 2013, Royal Society of Chemistry. (**b**) Photograph of the substrate, (**c**) SEM image of Ag nanorod array fabricated on glass substrate by using GLAD [93]. Reproduced with permission from [93]. Copyright 2017, American Chemical Society (**d**) SEM images of zigzag silver nanostructures on Si substrates with different bending numbers: two arms; (**e**) four arms. Yellow lines indicate the approximate arm positions [96]. Reproduced with permission from [96]. Copyright 2017, Royal Society of Chemistry. Top-view and cross-section SEM images of Ag nanohelix arrays deposited at different deposition angles: (**f**,**g**) 80°, (**h**,**i**) 83° and (**j**,**k**) 86° [97]. Reproduced with permission from [97]. Copyright 2017, MDPI.

3. Detection Methodologies for Silver Nanostructure-Based Plasmonic Sensors

The field of sensing has found its quintessential implications in next-generation devices for environmental monitoring, food safety, defense applications, medical diagnostics and the development of smart medicines, etc. Portable, fast and easy-to-use sensors are in demand with the fast pace of development in every area of human life. Plasmonic sensing has continuously emerged as a vital tool for fast and specific detection in the past decade. The enhanced and tightly confined electromagnetic (EM) field associated with plasmonic nanostructures influences almost all optical spectroscopic methods and is at the forefront of rapidly emerging surface-enhanced spectroscopies such as SERS, SEF and SEIS. In the next sections, the fundamentals of these technologies and their applications in sensing will be discussed in the context of silver nanostructures.

3.1. Surface Plasmon Resonance

Over the past two decades, the field of plasmonic sensors based on SPR and LSPR schemes has sufficiently matured, with a plethora of label-free and continuous monitoring applications. Since the initial reports studying interactions on metal surfaces and gas-sensing, there has been a rapid development in the fundamental mechanisms and applications of this field, making it a prime tool to monitor label-free surface interactions in real-time. Recent developments have focused on sensitivity enhancement and the quantum aspects of these sensors, as well as their applicability in the form of industrial devices that will clearly revolutionize the field of optical sensors [108,109]. The goal now is to focus the attention of the SPR community on advancing the technology from proof-of-concept to industrial devices, such as the first one commercialized in 1990 by Biacore, by taking note of thousands of articles focusing on the quantification of clinical and environmentally relevant analytes at desired concentration levels. The fundamentals of SPR sensors have been extensively reported in various reviews and books over the past decade, with various possible transducer geometries such as the prism-based Kretschmann configuration, planar waveguides, gratings, optical fibers and many more [110–117]. A brief overview of the concept is provided here for the purpose of completion.

Plasmons are the collective electron oscillations in metals that can be confined to metal–dielectric interfaces known as propagating surface plasmons (PSPs) or to the surface of a metallic nanostructure (of dimensions less than the wavelength of light) that are termed localized surface plasmons (LSPs). The coupling of these modes to incident light gives rise to resonances that strongly depend on the compositions, shapes and sizes of the metal nanostructure, as well as the dielectric properties of the surrounding medium, making it a critical tool to probe surface processes. The EM field associated with SPs and LSPs is bound to the surface and decays exponentially in the surrounding medium with decay lengths of ~30 nm and ~ 200 nm, respectively. Thus, the sensors based on these mechanisms are extremely sensitive to changes near the surface. SPR and LSPR sensors are based on modulation in the refractive index of the sensing layer around the metallic nanostructure due to physicochemical interaction with the analyte.

The sensors based on silver nanostructures are discussed here with widespread applications in environmental monitoring, biosensing, food safety and defense security. Recently, a fiberoptic SPR sensor based on silver thin film for the sensing of the phenolic compound catechol, a dangerous environmental pollutant, was reported [118]. The thermal deposition of 40 nm silver film was used as a plasmonic metal with the sensing layer comprising a ZnO/CNT nanocomposite for the interaction of catechol. The nanocomposite was functionalized with cetyltrimethylammonium bromide (CTAB), which governs the sensing performance of two pH regimes. The schematic of the sensing mechanism, the experimental setup and the SPR response of the two pH regimes are shown in Figure 8. The performance was analyzed over a concentration range of 0–100 μM, with the detection limit reaching 0.1 μM, making it feasible for real applications.

Moreover, Ag has also been incorporated with other metals and semiconductors to enhance their optical and electrical properties in low-dimensional hybrid nanomaterials. Nonlinear optical absorption and electrical conduction in multiwalled carbon nanotubes (MWCNTs) were observed to be enhanced when they are coated with AgNPs. The optical nonlinearities exhibited by Ag/MWCNTs can be attributed to the saturable absorption regarding the absorption band associated with the SPR of AgNPs [119]. In another report, MWCNTs decorated with AgNPs were demonstrated to improve acetone gas-sensing at room temperature using the resistive sensing method [120]. Some other SPR-based sensors utilizing carbon nanomaterials and silver-based nanocomposites were reviewed by Gupta et al. [121]. In addition, Ag-metal oxide nanohybrid systems have been widely explored in sensing applications [86,122].

Recently, a self-referenced optical fiber LSPR sensor for the detection of environmental pollutants, mercury and hydrogen peroxide (Hg^{2+} and H_2O_2) was reported, using the layer-by-layer nanoassembly of silver and gold NPs [123]. The plasmonic substrates were prepared using the electrostatic attraction between poly(acrylic acid) (PAA)-capped AuNPs and poly(allylamine hydrochloride) (PAH)-capped AgNPs. The AuNPs showed a higher affinity for Hg^{2+} as compared to other metal ions and resulted in a wavelength shift of LSPR spectra, whereas the AgNPs showed degradation due to the catalytic decomposition of H_2O_2. Thus, a self-referenced sensor was realized as SPR wavelengths of either Ag or Au shifted upon exposure to Hg^{2+} and H_2O_2. Similarly, numerous sensors have been realized based on plasmonic silver for the detection of environmental analytes such as nitrates, arsenic ions [124], phenolic compounds [125], hydrogen peroxide [126], methane gas [127] and ammonia gas [86].

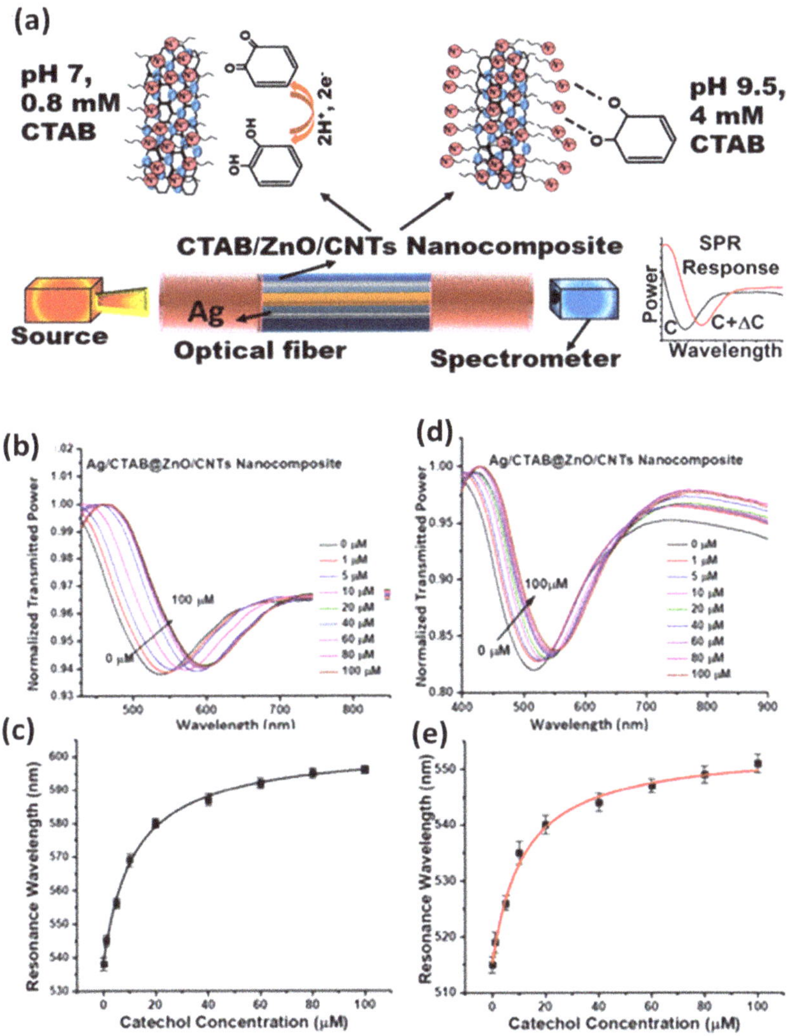

Figure 8. (**a**) The sensing mechanism and experimental setup for the sensing of catechol using silver thin film as the plasmonic material and CTAB-functionalized ZnO/CNT nanocomposite as the sensing layer. (**b**) SPR response and (**c**) calibration curve for the regime: pH of the sample, 9.5, and CTAB concentration = 4 mM. (**d**) SPR curves. (**e**) Calibration curve for the regime: pH of the sample, 7.0, and CTAB concentration = 0.8 mM. Reproduced with permission from [118]. Copyright 2020, American Chemical Society.

A great deal of progress has also been made in the field of SPR-based biosensors using silver nanostructures. A silver nanoparticle-based SPR/LSPR fiber probe was reported for the sensing of cholesterol using the enzyme cholesterol oxidase (ChOx) and graphene oxide (GO) sheets in the sensing layer [128]. The performance of three types of probes shown in Figure 9 was compared and optimized for a range of pH. The first probe involves a ChOx-entrapped hydrogel layer over Ag thin film; the second probe has a layer of GO nanosheets, along with ChOx, over Ag thin film; and the third probe contains ChOx immobilized on AgNPs embedded in PVA over GO nanosheets. It was noticed that the probe with silver nanoparticles and GO showed the best response due to the combined effect of SPR and LSPR, along with the role of hydrogen peroxide decomposition due to AgNPs. The third probe showed the best sensitivity and the lowest limit of detection (LOD) value for cholesterol, reaching 1.131 mM.

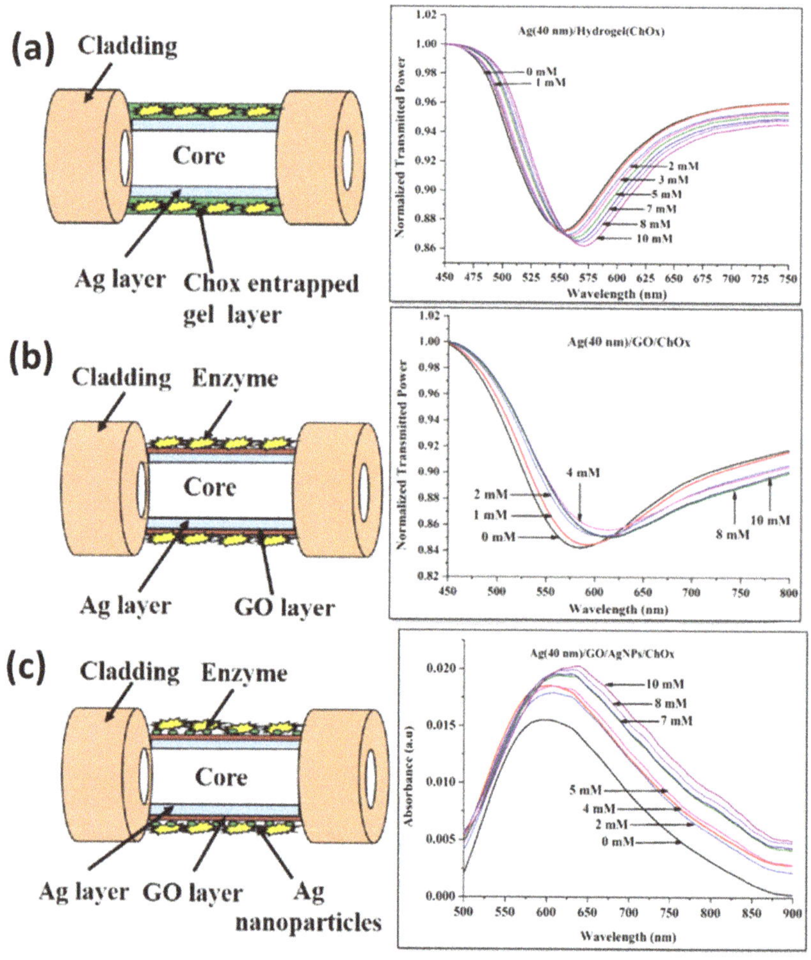

Figure 9. Schematic and corresponding SPR response for (**a**) ChOx-entrapped hydrogel layer over Ag thin film, (**b**) GO nanosheets along with ChOx over Ag thin film and (**c**) ChOx immobilized on AgNPs embedded in PVA over a GO nanosheet probe. Reproduced with permission from [128]. Copyright 2018, IEEE.

Similarly, very recently, a biosensor for detecting NS1 antigen related to dengue fever that uses silver thin film and an antibody–antigen interaction mechanism was reported [129]. The probe was optimized for various self-assembled monolayers of alkanethiols for effective antibody attachment, which finally governs the stable interaction with the NS1 antigen. The schematic for the probe design and interaction mechanism is shown in Figure 10a. The SPR response is depicted as a red shift in the resonance wavelength with an increasing antigen concentration (Figure 10b). The feasibility of the sensor was tested for dengue-positive blood serum samples, underlying its usage in direct clinical applications for early dengue diagnosis with a very fast response time of 20 min.

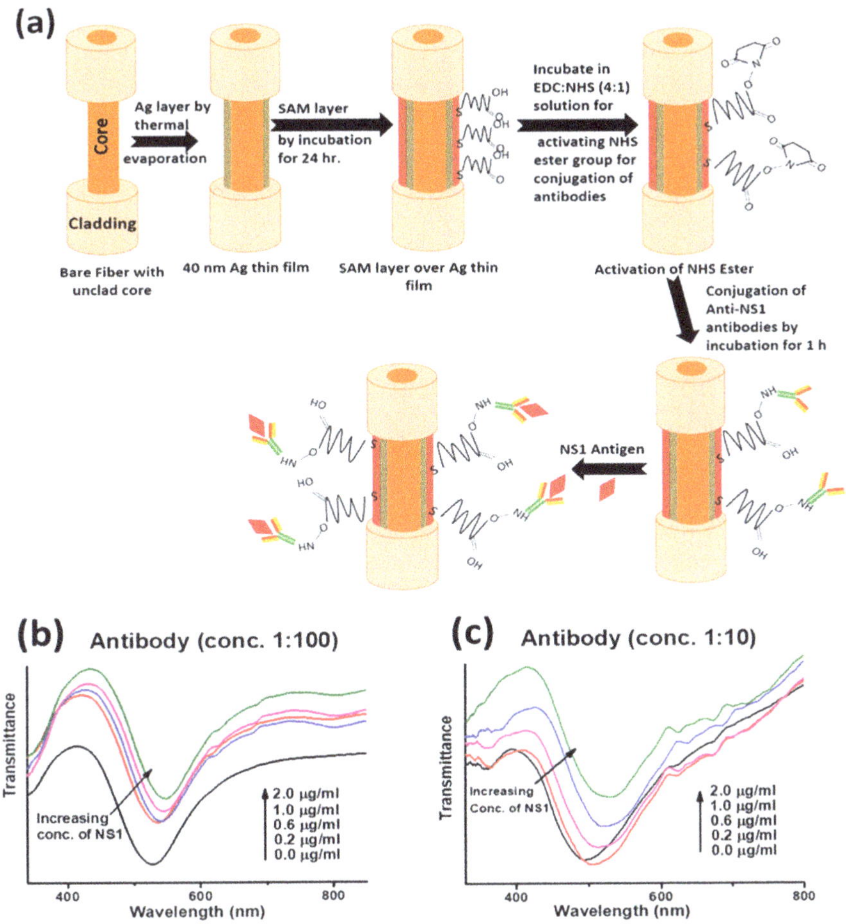

Figure 10. (a) Fabrication steps of the fiber probe for NS1 antigen detection, and (b,c) SPR response curves with changes in NS1 concentration at two antibody concentrations. Reproduced with permission from [129]. Copyright 2022, Elsevier.

A similar rapid biosensor was also reported for SARS-CoV-2 detection using a silver nanotriangle array as an LSPR sensing platform functionalized with human angiotensin-converting enzyme 2 protein (ACE2). The sensor selectively and rapidly detected the COVID virus, with LOD values of 0.83 pM, 391 PFU/mL and 625 PFU/mL in the spike RBD protein and CoV NL63 in the buffer, and untreated saliva providing a very effective alternative to rapid antigen tests [130]. A plethora of biosensor reports are available in the literature for the silver-based LSPR sensing of various biomarkers and bioanalytes for, e.g., glucose [131], ErbB2 breast cancer biomarker [132], triacylglycerides [133], cysteine [134], endotoxins [135] dopamine [136], ascorbic acid [137], urea and glucose [138].

Another important domain of SPR-based sensors is food monitoring and homeland security. A recent review summarizes plasmonic sensors for food security [139]. Shrivastav et al. proposed a molecular imprinting (MIP)-based fiberoptic SPR sensor for the detection of erythromycin (ERY) residue in milk and honey [140]. To overcome the limitation of the lower sensitivity of bulk MIP, MIP nanoparticles prepared with the two-phase mini-emulsion method were utilized. The sensing principle relies on a change in the refractive index of the MIP layer coated over the silver thin film on the optical fiber. ERY is imprinted in the MIP matrix during the synthesis step using certain monomers and a cross-linker. Thus, during the sensing step, ERY molecules bind noncovalently with these imprinted sites, bringing the change in the refractive index with a changing concentration. A schematic of the sensing mechanism, optical fiber probe and SPR response for ERY quantification are shown in Figure 11. ERY was spiked in milk and honey to test the feasibility of the probe in a real sample analysis. Another report also explored the sensing of the adulterated, nitrogen-rich compound melamine based on a similar MIP technique [141] for food safety.

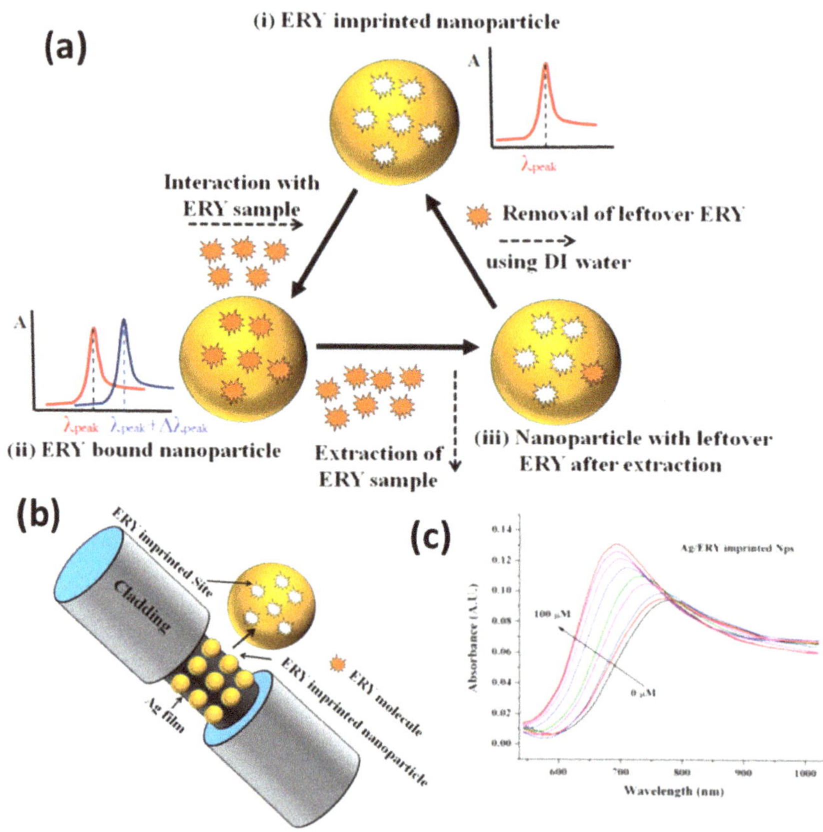

Figure 11. Application of silver-based SPR sensors for food safety. (**a**) Sensing mechanism in a MIP-based approach for the sensing of ERY in milk and honey. (**b**) Optical fiber probe. (**c**) SPR response with an increasing concentration of ERY. Reproduced with permission from [140]. Copyright 2017, Elsevier.

A very recent report on homeland security [142] utilized a molecularly imprinted (MIP) AgNPs composite for the LSPR sensing of an explosive taggant of trinitrotoluene (TNT), i.e., 3-nitrotoluene (3-NT). A one-step synthesis of MIP nanocomposite embedded with AgNPs was reported in the PEI matrix, which also serves as a reducing agent for Ag (I) to AgNPs. The synthesis steps are shown in Figure 12a, along with the sensing mechanism in Figure 12b. The sensor had a highly sensitive response due to the synergistic approach of LSPR and MIP, as shown in Figure 12c. LOD for 3-NT was reported to be 54.8 ng, with an extremely specific response in the presence of other NO_2-containing molecules, as shown in Figure 12d. The approach can easily be implemented for portable readout sensors in remote detection and bomb-disposal robots. Another group [143] also reported porous silica embedded with silver clusters for the sensing of nitroaromatic and nitroaliphatic explosives. Thus, silver-based SPR and LSPR sensors are extensively reported in the literature for all possible sensing applications.

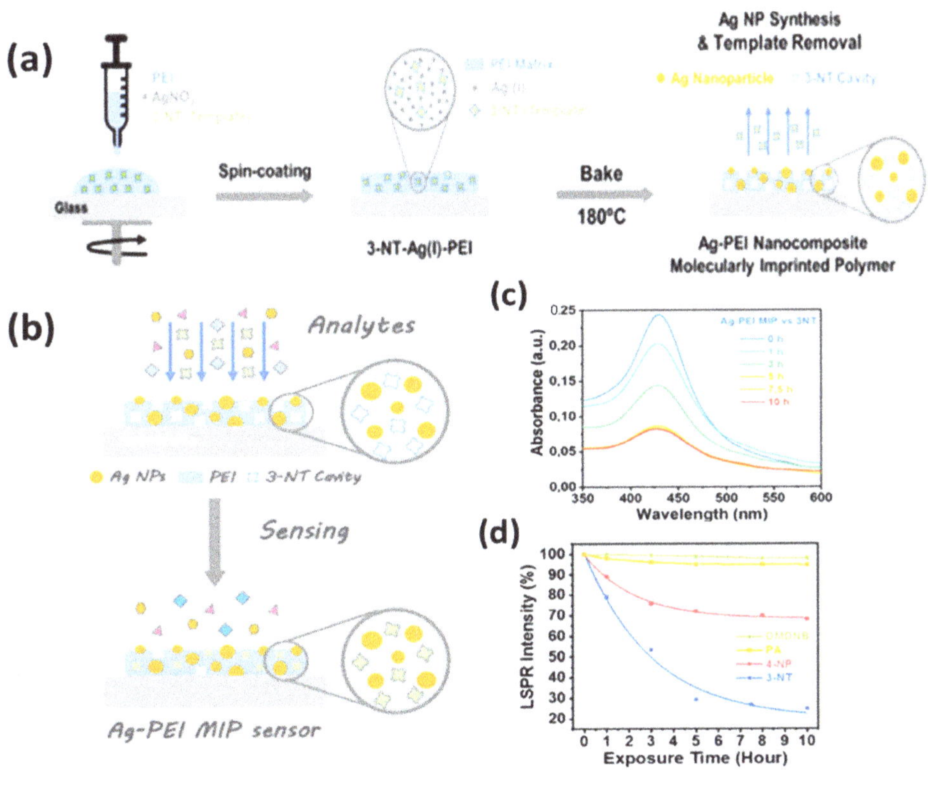

Figure 12. (**a**) Synthesis of a MIP layer for detecting explosive taggant 3-NT; (**b**) sensing mechanism; (**c**) SPR response; and (**d**) selectivity. Reproduced with permission from [142]. Copyright 2021, American Chemical Society.

3.2. *Surface-Enhanced Raman Spectroscopy (SERS)*

As the name suggests, SERS is a technique to enhance the originally weaker Raman signal on the surface of plasmonic nanostructures. Before the discovery of SERS in 1974, [144], Raman was only used as a characterization tool within a limited area of interest. However, after the discovery of SERS, Raman spectroscopy gained tremendous attention in many avenues of research. SERS has the potential to detect chemical and biomolecules at a single-molecule level [15,145]. Therefore, is employed in various fields, e.g., medical diagnostics [15,17,146], food safety and security [147,148], environmental monitoring [149,150] and defense and security [151–154]. An appropriate laser excitation induces a local intense electromagnetic field in metal nanoparticles, which enormously enhances the Raman (as well as the fluorescence) signal of the molecule adsorbed on the surface. The magnitude of the signal enhancement is referred to as the enhancement factor (EF) [155], which is provided as

$$EF = \frac{I(SERS) \times N(Bulk)}{N(SERS) \times I(Bulk)} \tag{2}$$

where *I(SERS)* and *I(Bulk)* are the signal intensities of the analyte molecule under SERS and the bulk (normal) Raman, respectively; *N(SERS)* and *N(Bulk)* are the number of analyte molecules being probed with SERS and the bulk Raman excitation, respectively. In general, the EF provided by equation 2 is in the order of 10^5–10^6; however, several reports claimed it to be up to 10^{14}–10^{15} [145]. The SERS enhancement mechanism is still debatable, and, globally, physicists and chemists are continuously working to find more insights into this phenomenon. However, most of them agree on the two popular mechanisms known as (i) electromagnetic (EM) enhancement (plasmonic) and (ii) chemical enhancement (charge transfer) [20,21,156]. The first one has a higher contribution of about four folds, whereas the latter amplifies the signal by two orders of magnitude. Figure 13a shows the schematic of the enhancement mechanism (EM), and Figure 13b shows the charge transfer mechanism.

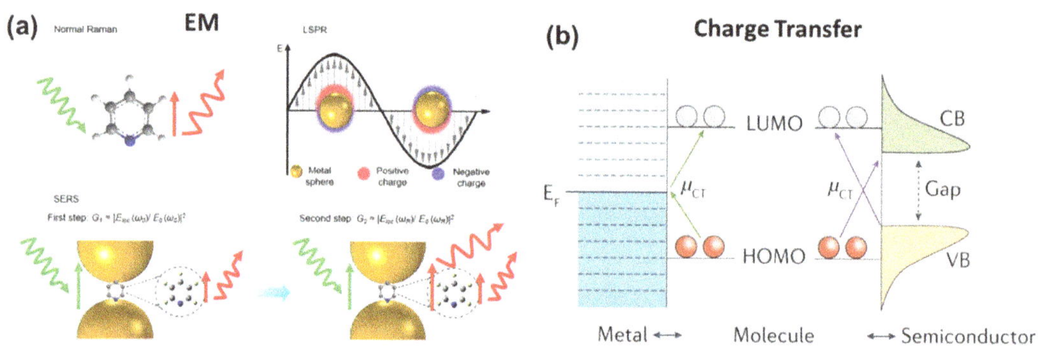

Figure 13. Schematic of enhancement mechanisms in SERS. (**a**) Electromagnetic (EM) enhancement mechanism in SERS, including the two-step enhancements illustrated in the nanogap of two metal nanoparticles. Reproduced with permission from [157]. Copyright 2018, American Chemical Society. Reproduced with permission from [158]. Copyright 2016, Springer Nature Limited. (**b**) Chemical enhancement through charge transfer (CT) between metal/semiconductor and the adsorbed molecule. The CT transitions (μCT) arrows show the CT directions. Red and white circles represent molecular orbitals. CB, conduction band; EF, Fermi level; HOMO, highest occupied molecular orbital; LUMO, lowest unoccupied molecular orbital; VB, valence band. Reproduced with permission from [159]. Copyright 2021, Springer Nature Limited.

The quality of the SERS substrate is a prominent factor in the enhancement of the signal, which opens a new domain of research focusing on the fabrication and engineering of SERS-active substrates. Generally, there are two important criteria in developing a useful SERS substrate. First, the material should demonstrate superior optical responses (plasmonic activity) in the visible or NIR regime. Ag and Au qualify this very well, as the real and imaginary parts of their dielectric function have a large negative and a small positive value, respectively, in this region of wavelength. Second, the surface should be roughened or nanostructured to generate hotspots (the sharp curvatures or nanogaps on the structures) for sufficient E-field enhancement. For the highest enhancement, the molecules should be present in these hotspots. Here, anisotropy in nanostructures plays a very important role. Anisotropic nanoparticles, e.g., rods, flowers, stars, etc., provide higher surface area and a large density of hotspots as compared to isotropic particles such as nanospheres. The array, clusters or aggregates of these nanoparticles exhibit the enormous enhancement of the Raman signal [25]. The signal collected from the hotspot (also called hotspot SERS) is always at least 10^3 times greater than the signal coming from other surfaces of the nanoparticles. For example, the dimer of two closely placed nanospheres only generates one hotspot region, whereas the dimer of nanostars or flowers generates many hotspots. Even if a single molecule is placed in the hotspot region, SERS will be observed. Therefore, single-molecule detection is possible and is attained using Ag hotspot SERS.

Not only plasmonic metals, such as Ag, Cu, Pt, Al and Cu, but many other nonmetallic materials (including metal oxides, sulfides, graphene, etc.) have also been investigated for SERS. However, the remarkable response of Ag is still unparalleled in this field. In addition, cost-effectiveness could be considered the second reason to select Ag over Au. In the following sections of this review, some important reports on silver-based SERS substrates and their applications will be discussed. Silver being the best choice amongst plasmonic materials, Ag-based SERS substrates have been fabricated by various methods and implemented in different domains of applications. The single-molecule detection limit by SERS has been attained by using Ag nanostructures [15]. AgNR arrays, vertically standing over a large-area substrate fabricated by GLAD, were optimized to provide a very high SERS enhancement. The effects of the aspect ratios of the NRs, as well as the gap between them and underneath reflecting silver thin film, were extensively studied experimentally and theoretically by Zhao's group [160,161]. Subsequently, similar AgNR arrays were substantially explored in biosensing applications on the planar and flexible substrates by our group [162,163]. SERS-assisted single-molecule detection using a AgNP uniform monolayer was demonstrated by Chen et al. [164], as shown in Figure 14a–d. We reported an enhancement of about 10^8 in the Raman signal of a probe molecule trans-1,2-bi-(4-pyridyl) ethylene (BPE) on a AgNR array fabricated over glass substrates, as illustrated in Figure 14e,f [165].

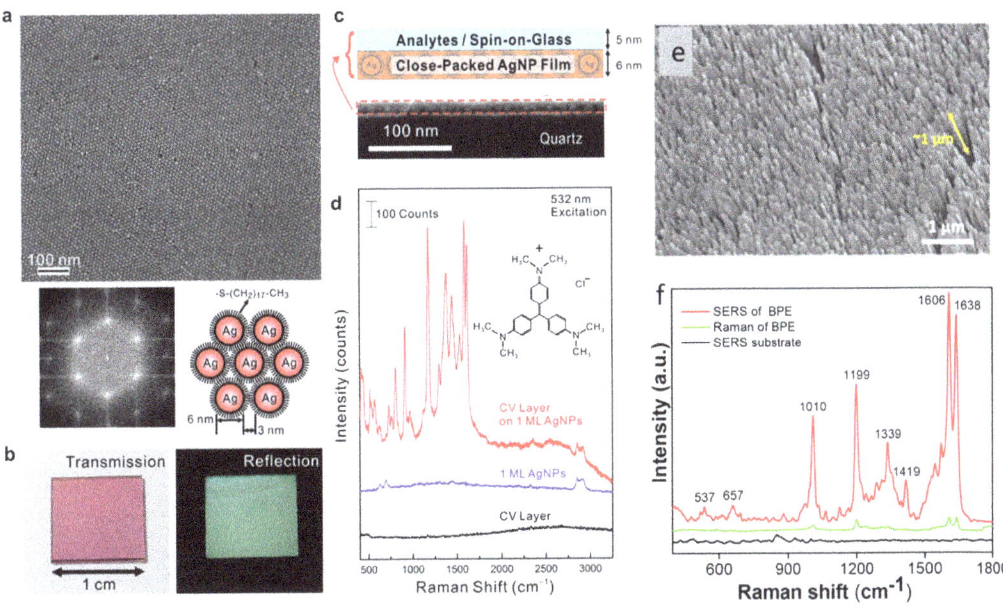

Figure 14. (**a**) FESEM image of a one-monolayer (1 ML), close-packed AgNP film on quartz, the diffraction pattern acquired from fast Fourier transform with an image area of 0.7×0.7 μm^2 and the schematic showing the interparticle gap regulated by the thiolate chain length. (**b**) Transmission and reflection photographs illustrating the effect of collective plasmonic resonance. (**c**) Structure of the

SERS substrate (close-packed AgNP film) and the analyte layer, which is a spin-coated layer of crystal violet (CV) molecules embedded in spin-on-glass (SOG); the cross-sectional SEM image shows the thicknesses of the CV/SOG layer and AgNP film are about 5 and 6 nm, respectively. (**d**) Raman spectra (vertically offset for clarity) obtained from a CV/SOG layer on AgNP film, a bare AgNP film and a bare CV layer. Reproduced with permission from [164]. Copyright 2015, American Chemical Society. (**e**) SEM of a GLAD-fabricated AgNR array. (**f**) Enhanced Raman signal of a BPE molecule deposited on the SERS-active array compared with normal the Raman of the molecule and substrate background signal. Reproduced with permission from [165]. Copyright 2020, American Chemical Society.

Further, to create a high density of hotspots (defined in the previous section), the substrates were modified into zigzag and multiple-armed geometry [96]. The effectiveness of the substrate is defined by the order of enhancement factor. Probe molecules with a high Raman cross-section are generally used, e.g., BPE, Rhodamine G (RhG), Methylene Blue (MB), Nitrothiophenol (NTP), Aminothiophenol (ATP), etc., for the determination of EF.

Raman spectroscopy is considered a vibrational fingerprint of a molecule, and its advanced version, SERS, has overcome the low-intensity limitation of normal Raman. Therefore, SERS has become an established technique and has attracted immense attention for all the sensing applications in chemical and biosensing. There is hardly any domain left untouched by SERS-based detection. Recently, to make the technique user-friendly and cost-effective, various academic and industrial collaborations have developed portable/handheld or field-deployable SERS-based platforms, which have been implemented in biomedicine, defense and security. In view of rapid health monitoring for mass communities, especially in developing and highly populous countries such as India, SERS-based portable devices have been developed and tested on real clinical samples of dengue- [165] and HIV-infected [166] patients in hospitals. The label-free detection of the NS1 antigen present in dengue-infected patients was illustrated on an AgNR array, as illustrated in Figure 15a,b. The study was carried out over 100 subjects and the collected SERS data were successfully classified with the help of the principal component analysis (PCA) statistical tool [165]. The same strategy was applied for the detection of human immunodeficiency virus (HIV1) in clinical samples and also determined the tropism. The SERS spectra of control and HIV-infected blood plasma are shown in Figure 15c, and their classification using PCA is plotted in Figure 15d [166].

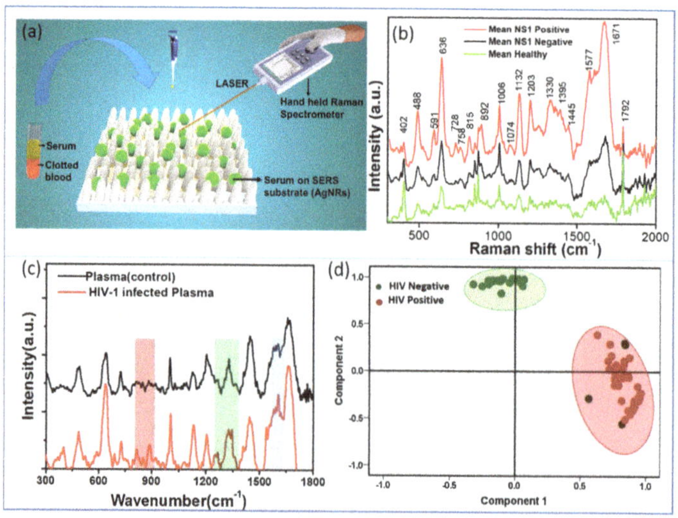

Figure 15. AgNRs as a portable SERS platform in disease diagnostics. (**a**) Schematic of the process of

handheld SERS using a AgNR array. (**b**) The average SERS spectra of dengue-infected (NS1 positive), non-infected (NS1 negative) and healthy subjects. Reproduced with permission from [165]. Copyright 2020, American Chemical Society. (**c**) SERS spectra of healthy plasma (control) and HIV subtype D virus spiked in plasma (conc. 10^5 copies/mL). (**d**) Principal component analysis (PCA) plot of HIV-1-positive and -negative plasma samples. Reproduced with permission from [166]. Copyright 2021, Elsevier.

Recently, the AgNP-based, label-free SERS detection of SARS-CoV-2, human adenovirus 3 and the H1N1 influenza virus were demonstrated. AgNPs were modified by utilizing calcium ions as aggregators, citrate ions were removed from the surface and acetonitrile was added to ensure the formation of high-quality hotspots. SERS measurements were performed on the samples obtained from 20 random groups of SARS-CoV-2 subjects. The schematic of the whole protocol is shown in Figure 16a [167].

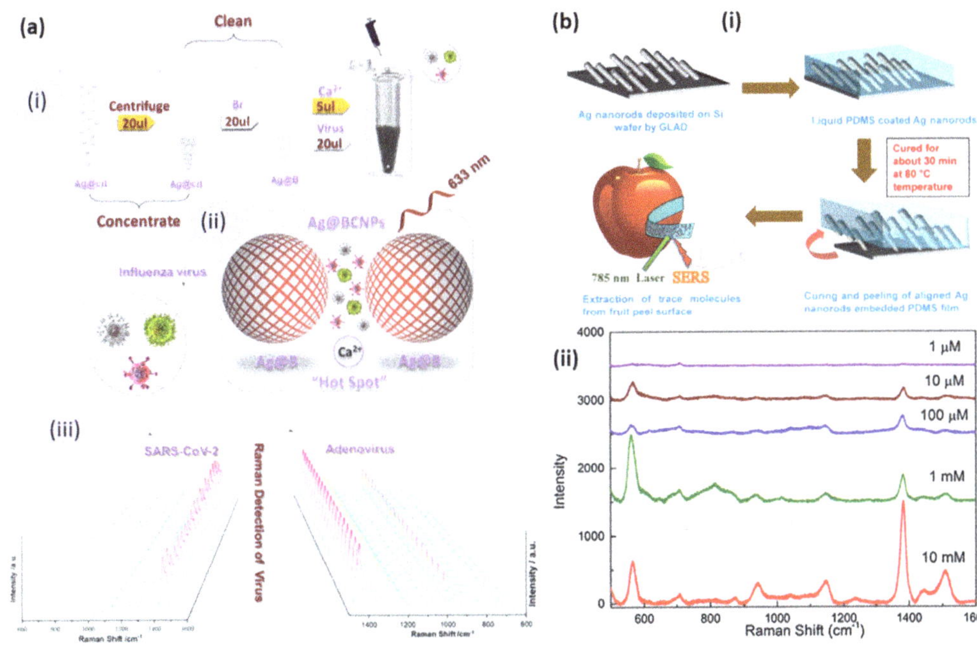

Figure 16. (**a**) (**i**) Schematic diagram of the preparation of silver-enhanced substrate and virus detection using SERS. Ag@cit: Silver nanoparticles obtained via the reduction of citrate; Ag@B: Silver nanoparticles modified by bromide ion; Ag@BCNPs: Ag@B with acetonitrile and calcium ions added. (**ii**) Conceptual schematic diagram of the relationship between the virus sample and the "hotspots" generated by the silver-enhanced substrate. (**iii**) SERS spectra obtained by 20 random groups of SARS-CoV-2 patients (10^4 PUF/test) and HAdV (10^5 copies/test) samples under the current method (Ag@BCNPs). Reproduced with permission from [167]. Copyright, 2022, Elsevier. (**b**) (**i**) Schematic showing the preparation of AgNR-embedded, PDMS-based flexible SERS substrate and (**ii**) SERS spectra of different concentrations of the pesticide thiram. Reproduced with permission from [148] Copyright, 2017, Elsevier.

Food safety is another domain in which SERS could be a promising tool. For detailed literature, readers are referred to the review papers [168,169]. As flexible substrates, AgNRs embedded in PDMS polymer were utilized as SERS tape for the detection of pesticide in fruits [148]. The flexibility and robustness of the substrate were examined under mechanical tensile strain conditions and using the scotch tape peeling test. A schematic of the substrate preparation is shown in Figure 16b(i) and the trace level detection of the pesticide thiram successfully achieved by the SERS spectra at different concentrations, as shown in Figure 16b(ii). Thus, SERS has proven to be a versatile tool for detecting and studying the kinetics of various chemical reactions. To obtain more insight into the present and future of this technique, readers are directed to several extensive reviews [158,170–173].

3.3. Surface Enhance Fluorescence Spectroscopy (SEFS)

Detection at the single-molecule level has gained considerable attention in the sensing and imaging community over the last few years. Among all the single-molecule optical spectroscopies, single-molecule fluorescence is the oldest and most widely applied spectroscopic technique due to inherited advantages such as noninvasive detection, fast and simple application and high contrast. Fluorescence is the property of certain organic molecules (fluorophores) to absorb and emit light through a transition in their electronic energy levels. However, for most fluorophores, fluorescence is limited by the low quantum yield, long relaxation time, poor photostability or photobleaching, which limit the single-molecule measurement due to a low signal-to-noise ratio. Purcell suggested in 1946 that the spontaneous emission properties of a molecule may be modified by controlling the external EM field in close vicinity [174]. Thus, the optical properties of a fluorophore may be effectively modified by keeping it in proximity to the nanostructured metal. The low fluorescence efficiency benefits from the interaction between fluorophores and the high nearfield enhancement caused by metal NPs, and it is referred to as metal-enhanced fluorescence (MEF) or surface-enhanced fluorescence (SEF) [67,175,176].

MEF emerged as the most effective and important technique in improving the low fluorescence of molecules after the first classical interpretation of environmental effects on the excited state electronic transition of fluorophores by Drexhage in the 1970s [177]. Both the excitation and emission properties of a fluorophore may be modulated by controlling the EM field around it. Metal NPs can majorly enhance fluorescence efficiency with two processes [175]. Firstly, at the plasmon resonance wavelength, the nearfield around the NP is strongly enhanced, resulting in an increased absorption cross-section of vicinal fluorophores, which, consequently, enhances excitation and emission efficiency. Fluorescence is strongly quenched in the directly adsorbed molecules on the metal surface, but at a few nanometer distances, fluorescence can be strongly enhanced. The second process is the increased radiative decay rates of molecules due to coupling with metal NPs. Here, the excited state fluorophores may transfer their energy to surface plasmons, resulting in an increased decay rate and emission intensity from the metal–fluorophore complex system. This improves both the fluorescence intensity and photostability (less time in an excited state) of the molecule or fluorophore. In addition, MEF is confined to a volume surrounding the plasmonic particle and, hence, greatly limits the background signal from freely diffusing molecules. Thus, the interaction between a fluorophore and metal NPs, in general, may lead to the quenching of fluorescence or its enhancement depending on various parameters such as fluorophore–metal distance and relative orientation, shape and size of metal NPs; spectral overlap between LSPR modes; and fluorophores [178–180]. Chen et al. studied DNA linkers for the attachment of dye molecules to silver nanoprisms. They analyzed the near-filed effects leading to the enhancement in fluorescence intensity as a function of the spectral overlap between the LSPR resonance of silver nanoprisms and the dye's emission and absorption spectra [181]. A detailed mechanism of MEF is available in a wealth of literature covering various possible mechanisms and factors responsible for the enhancement. A schematic depicting processes in MEF is shown in Figure 17 in terms of the spectral overlap of a fluorophore and a metal NP [16].

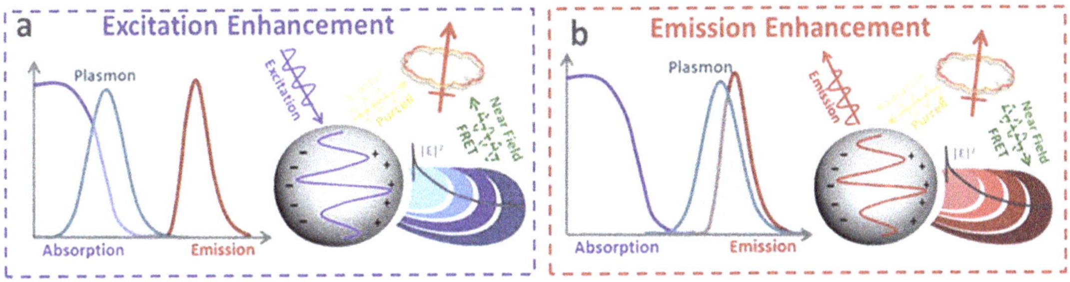

Figure 17. Depiction of MEF. (**a**) If the plasmon overlaps with the absorption of the fluorophore, an excitation enhancement is possible. (**b**) If the plasmon overlaps with the emission of the fluorophore, an emission enhancement is possible. Reproduced with permission from [16]. Copyright 2015, Royal Society of Chemistry.

Numerous AgNPs based fluorescence enhancement methods have been reported in the literature with experimental and theoretical validations [182–184]. Several reviews highlight the critical applications of these methods for sensing [18,185,186]. Lin et al. proposed a silver nanoprism-based MEF sensor for the detection of sulfides in an aqueous solution. Atto550 has been used as a fluorophore attached via polymer COOH-PEG-SH and a streptavidin–biotin bond to silver nanoprisms. A series of nanoprisms were tested with resonance wavelength tuning in the range 500–900 nm. The highest enhancement in the fluorescence intensity was 10-fold, and it was obtained for prisms with an LSPR wavelength of 570 nm [187]. Ray et al. also demonstrated a several-hundred-fold fluorescence enhancement in Cy5 dye assembled on a silver NPs-dielectric-mirror (PDM) substrate. They elucidate the importance of single-molecule spectroscopy through these kinds of ensembles with a several-fold enhancement in fluorescence intensity and an up-to-10-fold enhancement in the decay rates of Cy5 [184]. Similarly, a core-shell silver-poly(3-acrylamidephenylboronic acid-co-acrylic acid) (Ag@PAPBA-PAA) structure was reported for pH and glucose sensing, incorporating porphyrin molecules (Por4+) as a fluorophore. The schematic of the fluorescence response and fluorescence spectra are shown in Figure 18. The response clearly indicates the distance-dependent fluorescence enhancement mechanism due to the swelling and shrinking of the PAPBA-PAA shell in response to the increased glucose concentration or pH [188].

Similarly, aptamer-modified AgNPs were used for the fluorescence-based detection of single and multiplexed proteins. The sandwich assay of aptamer-coated AgNPs as capture probes, as well as fluorophore-labeled aptamers as detection probes, improved the detection limit of thrombin and platelet-derived growth factor BB (PDGF-BB) by 80 or 8 times, reaching 21 pM and 625 pM, respectively. A schematic of the sandwich-type assay for multiplex detection is shown in Figure 19a, and the corresponding fluorescence readout for varying concentrations of thrombin and PDGF-BB is shown in Figure 19b [189].

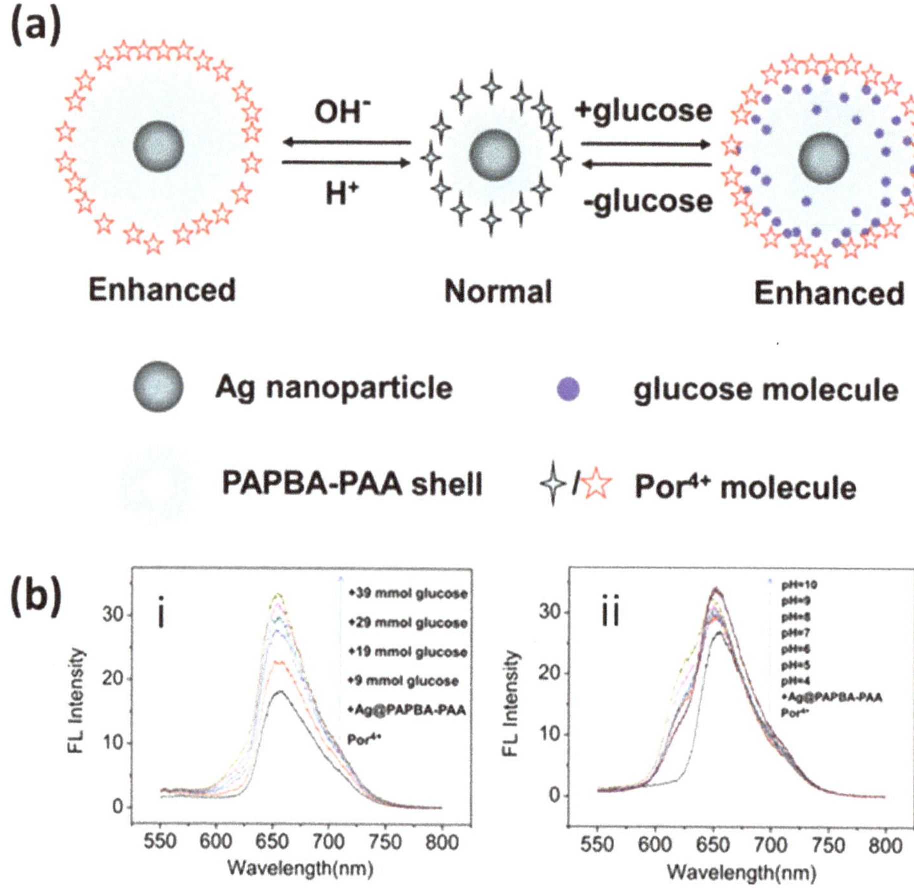

Figure 18. (**a**) Schematic illustration of the swelling–shrinking Ag@PAPBA-PAA hybrid nanoparticles with pH- and glucose concentration and the corresponding controllable MEF effects. (**b**) Fluorescence spectra of the Por4+/Ag@PAPBA-PAA nanocomposite as tuned by (**i**) various glucose concentrations and (**ii**) the pH of the environment. Reproduced with permission from [188]. Copyright 2012, American Chemical Society.

As ordered metallic arrays have been found to be better substrates as compared to aggregated or randomly prepared nanoparticle suspensions, Ag nanorod arrays prepared with GLAD have also been explored for effective MEF substrates [190,191]. Enhancement factors have been analyzed in detail with a number of branches in zigzag Ag nanorod arrays. The detection limits of 0.01 pM have been achieved with an enhancement factor of 28 considering the hybridization of two oligonucleotides containing 33 base pairs using Alexa448. Here, the plasmon resonance of the nanorods may be tuned by varying their size to have a good spectral overlap with the fluorophore. Thus, it was concluded that a maximum 14-fold enhancement can be achieved for 7-fold zigzag nanorods due to an increase in the scattering intensity of the emission wavelength of the fluorophore and with an increase in the nanorods' folding number. Several other sensors for the detection of insulin [192], heparin [193], proteins [194] and tetracycline [195] have been reported based on silver nanostructured MEF.

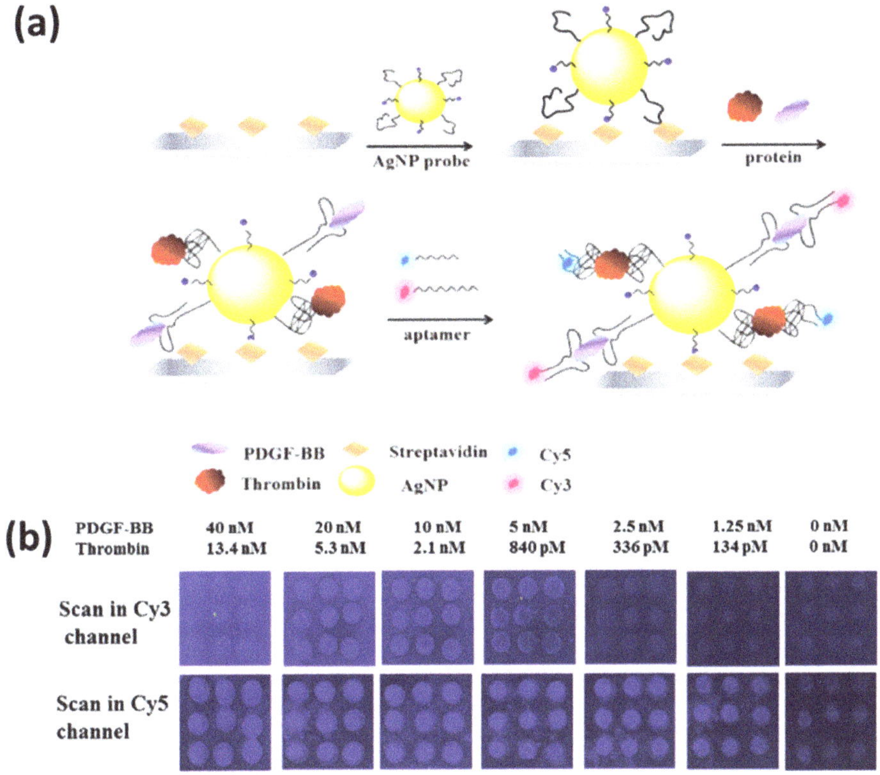

Figure 19. (**a**) Schematic for an aptamer-modified AgNP-based sandwich assay for multiplexed protein (thrombin and PDGF-BB) detection. (**b**) Fluorescence images for the multiplex analysis of the mixture of thrombin and PDGF-BB in different concentrations. Reproduced with permission from [189]. Copyright 2016, Elsevier.

3.4. Surface-Enhanced Infrared Absorption (SEIRA)

Infrared (IR) spectroscopy measures the vibration of molecular bonds by considering the absorption in the mid-IR spectral region (3000–600 cm^{-1}). IR spectra are intrinsically endowed with the chemical fingerprint of the molecule and, hence, may be leveraged for the direct measurement of molecular mechanisms. However, conventional IR spectroscopy has limitations such as low sensitivity and difficulty measuring aqueous solutions due to the IR activity of water. The enhancement of IR absorption in the molecules adsorbed on metallic nanoparticle surfaces is referred to as surface-enhanced infrared absorption spectroscopy (SEIRAS) and can significantly overcome the limitations of IR spectroscopy. The first study on SEIRA was reported after almost a decade of SERS by Hartstein et al. [196]. The mechanism of enhancement is quite similar to the SERS phenomenon, as Raman and IR are just two complementary vibrational spectroscopies. The detailed mechanism can be found in a number of review articles, chapters and books dedicated to this topic [19,197–199]. Of note, the localized EM field around metal NPs and the charge transfer between the molecule physiosorbed or chemisorbed on the surface of metal NPs give rise to a 10–1000-fold enhancement of IR absorption as compared to conventional techniques. Although the enhancement factors are much lower than SERS, the much higher infrared absorption cross-section compared to Raman scattering has maintained the interest of researchers in this field. Depending on the polarization and angle of incidence of the IR light, enhanced spectra may be collected in the attenuated total reflection mode (ATR), the transmission mode or the diffuse reflection mode.

The potential of SEIRA has been recently explored over citrate-stabilized AgNP substrates for the identification of microorganisms: *Candida albicans* (*C. albicans*), Escherichia coli (*E. coli*) and Staphylococcus aureus (*S. aureus*). It was concluded that SEIRA provides more explicit molecular information about these species as compared to conventional FTIR spectroscopy [200]. Similarly, a novel chalcogenide waveguide sensor using Ag-island film was reported exploiting SEIRA for the detection of the gaseous and liquid phases. Various thicknesses of Ag-island film were fabricated for the best sensing performance, and it was found that a 1.8 nm thickness provides the best results. The absorbance enhancement factors of >1.5 for ethanol (C_2H_6O) at 1654 nm and >2.3 for methane gas (CH_4) at 3291 nm were also obtained. Lift-off and GLAD were used for the fabrication of this SEIRA waveguide sensor [201]. The application of this sensor for shale gas measurement was reported, which is very useful for exploring natural resources. Similarly, an ATR SEIRA analysis of fatty acids was also successfully performed on silver nanoparticles [202]. Apart from this, many recent review articles have covered the field of biosensors using SEIRA [203,204]. The aforementioned plasmonic sensing techniques and their applications are summarized in Table 1.

Table 1. Ag-based plasmonic sensing techniques and their applications for sensing various analytes.

Sensing Technique	Ag Substrate Structure	Sensing Method	Sensing Layer	Analyte	Ref.
SPR	Thin film	Thermal evaporation	ZnO/CNT nanocomposite	Catechol	[118]
SPR	Thin film	Thermal evaporation	SnO_2 thin film	Ammonia gas	[86]
LSPR	Nanoparticles	Chemical reduction	PAH capped Ag NPs	Hydrogen peroxide	[123]
SPR/LSPR	Thin film and NPs	Thermal evaporation and chemical	Cholesterol oxidase in polyacrylamide gel and graphene oxide	Cholesterol	[128]
SPR	Thin film	Thermal evaporation	Anti-NS1 antibody over SAM layer	Dengue	[129]
LSPR	Nanotriangle array	NSL and e-beam evaporation	ACE2 protein	SARS-CoV-2	[130]
SPR	Thin film	Thermal evaporation	MIP	Erythromysin	[140]
LSPR	AgNPs	Chemical reduction	MIP	3-NT	[142]
SERS	AgNP film	Chemical	Closely packed AgNP film on quartz	CV	[164]
SERS	AgNPs	Chemical	AgNPs modified by bromide ions, acetonitrile and calcium ions	SARS-CoV-2, H1N1 influenza virus, Human adenovirus-3	[167]
SERS	AgNR array	GLAD	AgNR array embedded in PDMS	Thiram	[148]
SERS	AgNR array	GLAD	AgNR array on glass	Dengue	[165]
SERS	AgNR array	GLAD	AgNR array on glass	HIV-1	[166]
SERS	Zigzag Ag–Al array	GLAD	Ag–Al zigzag array on glass	BPE	[96]
SERS	AgNR bundles	NSL and electrodeposition	Porous AAO coated with AgNR bundles	Phenolic pollutants	[75]
SERS	Ag nanotriangles	Chemical reduction	Ag nanotriangles on silicon	Picric acid and ammonium nitrate	[153]
SEF	Ag nanoprisms	Chemical reduction	ATTO550 attached via COOH-PEG-SH and streptavidin–biotin	Sulfides	[187]
SEF	AgNPs	Sputtering	AgNPs on silica	Cy5	[184]
SEF	AgNPs	Chemical	Ag@PAPBA-PAA gel embedded with porphyrin as a fluorophore	pH and glucose	[188]
SEF	AgNPs	Chemical	Aptamer-modified AgNPs	Thrombin and PDGF-BB	[189]
SEF	AgNR array	GLAD	Biotin-coated AgNRs	Neutravidin and DNA	[190]
SEIRA	AgNP suspension	Chemical	AgNP direct detection	*C. albicans*, *E. coli*, and *S. aureus*	[200]
SEIRA	Ag nanoislands film	GLAD	Direct detection by Ag nanoislands	Ethanol and methane gas	[201]
SEIRA	AgNPs	Chemical	Direct detection	Fatty acids	[202]

4. Conclusions and Outlook

In the current scenario, most of the research on nanotechnology has somehow moved toward the involvement of plasmonics due to the attraction capability of plasmonic nanomaterials for light confinement and manipulation. In other words, plasmonic metal nanostructures act as optical antennae to covert light into localized electric fields and route it to a desired location with nanometric precision. This field is growing so profoundly that the entire fraternity of nanoscience is engaged in understanding more insights into the phenomenon and also in developing new technologies employing this extraordinary feature. In this review article, we exclusively discussed the most competent candidate, Ag, among all plasmonic materials. Although Au is an equally accepted metal, we have emphasized the choice of Ag over Au. The outstanding plasmonic response in the desired spectral range and the cost-effectiveness have overpowered the issues of biocompatibility and the stability of Ag nanoparticles. Different facile synthesis methods have also played a vital role in making it more accessible. Here, major techniques for AgNP chemical synthesis and physical deposition have been discussed and reviewed to date. Recent developments in the field of controlled synthesis and the assembly of Ag nanostructures involving nanoskiving and DNA origami are still motivating researchers to unravel the new dimensions of plasmonics and its applications.

The localized plasmonic electric field resonating with external light has enriched spectroscopic techniques, e.g., Raman, fluorescence and UV-vis. Therefore, metal nanoparticles enable surface-enhanced spectroscopy such as SERS, SEF and SEIRS, which are emerging as techniques for the sensing and detection of chemicals and biomolecules. They have been found to have promising applications in numerous fields. In this article, we tried to cover the applications of silver nanostructured substrates in the above techniques with major emphasis on SPR and SERS-based plasmonic applications, e.g., biosensing, environmental monitoring, security and food safety. It is now well understood that the limits and commercial applications of plasmonic sensors can be pushed with an in-depth understanding of the underlying theoretical principles of each of these techniques. Each technique has its own limitations and advantages, which must be carefully taken into account to develop the final prototype for a particular application. The commercialization of SPR-based sensors is promoted due to their high sensitivity and the miniaturized devices facilitated by optical fiber substrates. However, the realization is still limited by their reusability and durability due to the requirement of designing the specific sensing layer. Fluorescence-enhancement mechanism caused by metals provides a boost to the well-established field of fluorescence and FRET-based sensing, improving their sensitivity and providing low-cost, easy-to-handle, readily available commercial systems. However, the low quantum yield of visible fluorophores and tuning the metal NP–fluorophore distance in such systems remain challenges to be addressed in order to carefully avoid fluorescence quenching and to realize enhancement. Similarly, SEIRA sensors have gained advantages from the enhancement of IR signals by metals up to the order of 10^4, but they are still limited by the inherited properties of IR spectra, such as interference from aqueous solutions due to the strong IR signal from water. Among all the surface-enhanced spectroscopies, SERS has proven to be the most reliable and sensitive technique, as it can provide the direct fingerprint of target chemical species, avoiding any interference possibilities. This has enormously guided researchers all over the globe to explore SERS-based sensors for a plethora of applications. As Raman spectrometers are generally large and expensive, this has limited the point-of-care devices based on SERS as compared to the compact instrumentation required for SPR and SEF studies. Thus, now the research is focused on miniaturizing SERS-based systems by developing paper-based substrates, microfluidic platforms and palm-sized spectrometers to commercialize SERS as a promising detection tool.

Thus, plasmonics has emerged as an important fundamental science for resourceful technologies. Apart from sensing, many new applications for metal NPs are rapidly emerging in the fields of plasmonic catalysis, circuitry and quantum computing. The plasmon decay in metals generates highly energetic electrons (hot electrons) and localized heating, which can be utilized in many ways, such as modulating various chemical reactions and the conversion of solar energy to chemical energy. Water splitting, hydrogen, oxygen generation, CO_2 reduction and many more make up the attraction to plasmonics nowadays. The fast-paced growth of plasmonics is evidence of its versatility in next-generation technological solutions for society, and it still inspires researchers worldwide with open questions.

Author Contributions: Writing—original draft preparation, S.K.G. and A.P.; writing—review and editing, B.D.G., A.P. and S.K.G. All authors have read and agreed to the published version of the manuscript.

Funding: This research received no external funding.

Institutional Review Board Statement: Not applicable.

Informed Consent Statement: Not applicable.

Data Availability Statement: Not applicable.

Conflicts of Interest: The authors declare no conflict of interest.

References

1. Brongersma, M.L.; Shalaev, V.M. The case for plasmonics. *Science* **2010**, *328*, 440. [CrossRef] [PubMed]
2. Loiseau, A.; Asila, V.; Boitel-Aullen, G.; Lam, M.; Salmain, M.; Boujday, S. Silver-based plasmonic nanoparticles for and their use in biosensing. *Biosensors* **2019**, *9*, 78. [CrossRef] [PubMed]
3. Rycenga, M.; Cobley, C.M.; Zeng, J.; Li, W.; Moran, C.H.; Zhang, Q.; Qin, D.; Xia, Y. Controlling the synthesis and assembly of silver nanostructures for plasmonic applications. *Chem. Rev.* **2011**, *111*, 3669–3712. [CrossRef] [PubMed]
4. Faraday, M. Experimental relations of gold (and other metals) to light. *Philos. Trans. R. Soc. Lond.* **1857**, *147*, 145.
5. Liang, C.; Lu, Z.A.; Wu, J.; Chen, M.X.; Zhang, Y.; Zhang, B.; Gao, G.L.; Li, S.; Xu, P. Recent Advances in Plasmon-Promoted Organic Transformations Using Silver-Based Catalysts. *ACS Appl. Mater. Interfaces* **2020**, *12*, 54266–54284. [CrossRef]
6. Shen, H.; Bienstman, P.; Maes, B. Plasmonic absorption enhancement in organic solar cells with thin active layers. *J. Appl. Phys.* **2009**, *106*, 073109. [CrossRef]
7. Fang, N.; Lee, H.; Sun, C.; Zhang, X. Sub-diffraction-limited optical imaging with a silver superlens. *Science* **2005**, *308*, 534–537. [CrossRef]
8. Yan, R.; Gargas, D.; Yang, P. Nanowire photonics. *Nat. Photonics* **2009**, *3*, 569–576. [CrossRef]
9. Maier, S.A.; Atwater, H.A. Plasmonics: Localization and guiding of electromagnetic energy in metal/dielectric structures. *J. Appl. Phys.* **2005**, *98*, 10. [CrossRef]
10. Bergman, D.J.; Stockman, M.I. Surface plasmon amplification by stimulated emission of radiation: Quantum generation of coherent surface plasmons in nanosystems. *Phys. Rev. Lett.* **2003**, *90*, 027402. [CrossRef]
11. Tominaga, J.; Mihalcea, C.; Büchel, D.; Fukuda, H.; Nakano, T.; Atoda, N.; Kikukawa, T. Local plasmon photonic transistor. *Appl. Phys. Lett.* **2001**, *78*, 2417–2419. [CrossRef]
12. Stewart, M.E.; Anderton, C.R.; Thompson, L.B.; Maria, J.; Gray, S.K.; Rogers, J.A.; Nuzzo, R.G. Nanostructured plasmonic sensors. *Chem. Rev.* **2008**, *108*, 494–521. [CrossRef]
13. Khurana, K.; Jaggi, N. Localized surface plasmonic properties of Au and Ag nanoparticles for sensors: A review. *Plasmonics* **2021**, *16*, 981–999. [CrossRef]
14. Cobley, C.M.; Skrabalak, S.E.; Campbell, D.J.; Xia, Y. Shape-controlled synthesis of silver nanoparticles for plasmonic and sensing applications. *Plasmonics* **2009**, *4*, 171–179. [CrossRef]
15. Kneipp, J.; Kneipp, H.; Kneipp, K. SERS—A single-molecule and nanoscale tool for bioanalytics. *Chem. Soc. Rev.* **2008**, *37*, 1052–1060. [CrossRef]
16. Li, M.; Cushing, S.K.; Wu, N. Plasmon-enhanced optical sensors: A review. *Analyst* **2015**, *140*, 386–406. [CrossRef]
17. Sebastian, S. SERS microscopy: Nanoparticle probes and biomedical applications. *ChemPhysChem* **2009**, *10*, 1344–1354.
18. Fothergill, S.M.; Joyce, C.; Xie, F. Metal enhanced fluorescence biosensing: From ultra-violet towards second near-infrared window. *Nanoscale* **2018**, *10*, 20914–20929. [CrossRef]
19. Osawa, M. Surface-enhanced infrared absorption. S. Kawata (Ed.): Near-field optics and surface plasmon polaritons. *Top. Appl. Phys.* **2001**, *81*, 163–187.
20. Maier, S.A. *Plasmonics: Fundamentals and Applications*; Springer: New York, NY, USA, 2007; Volume 1.
21. Ru, E.L.; Etchegoin, P. *Principles of Surface Enhanced Raman Spectroscopy*; Elsevier: Oxford, UK, 2009.
22. Wang, H.; Tam, F.; Grady, N.K.; Halas, N.J. Cu nanoshells: effects of interband transitions on the nanoparticle plasmon resonance. *J. Phys. Chem. B* **2005**, *109*, 18218–18222. [CrossRef]
23. Lee, K.-S.; El-Sayed, M.A. Gold and silver nanoparticles in sensing and imaging: sensitivity of plasmon response to size, shape, and metal composition. *J. Phys. Chem. B* **2006**, *110*, 19220–19225. [CrossRef] [PubMed]
24. Murphy, C.J.; Sau, T.K.; Gole, A.M.; Orendorff, C.J.; Gao, J.; Gou, L.; Hunyadi, S.E.; Li, T. Anisotropic metal nanoparticles: synthesis, assembly, and optical applications. *J. Phys. Chem. B* **2005**, *109*, 13857–13870. [CrossRef] [PubMed]
25. Reguera, J.; Langer, J.; Aberasturi, D.J.; Liz-Marzán, L.M. Anisotropic metal nanoparticles for surface enhanced Raman scattering. *Chem. Soc. Rev.* **2017**, *46*, 3866–3885. [CrossRef] [PubMed]
26. Sun, Y.; Xia, Y. Shape-controlled synthesis of gold and silver nanoparticles. *Science* **2002**, *298*, 2176–2179. [CrossRef]
27. Wiley, B.; Sun, Y.; Xia, Y. Synthesis of silver nanostructures with controlled shapes and properties. *Acc. Chem. Res.* **2007**, *40*, 1067–1076. [CrossRef]
28. Zeng, J.; Zheng, Y.; Rycenga, M.; Tao, J.; Li, Z.Y.; Zhang, Q.; Zhu, Y.; Xia, Y. Controlling the shapes of silver nanocrystals with different capping agents. *J. Am. Chem. Soc.* **2010**, *132*, 8552–8553. [CrossRef]
29. Tsuji, M.; Nishizawa, Y.; Matsumoto, K.; Kubokawa, M.; Miyamae, N.; Tsuji, T. Effects of chain length of polyvinylpyrrolidone for the synthesis of silver anostructures by a microwave-polyol. *Mater. Lett.* **2006**, *60*, 834–838. [CrossRef]
30. Sun, Y.; Mayers, B.; Herricks, T.; Xia, Y. Polyol synthesis of uniform silver nanowires: a plausible growth mechanism and the supporting evidence. *Nano Lett.* **2003**, *3*, 955–960. [CrossRef]

31. Skrabalak, S.E.; Wiley, B.J.; Kim, M.; Formo, E.; Xia, Y. On the polyol synthesis of silver nanostructures: Glycolaldehyde as a reducing agent. *Nano Lett.* **2008**, *8*, 2077–2081. [CrossRef]
32. Sun, Y.; Mayers, B.; Xia, Y. Transformation of silver nanospheres into nanobelts and triangular nanoplates through a thermal process. *Nano Lett.* **2003**, *3*, 675–679. [CrossRef]
33. Rocha, T.C.R.; Zanchet, D. Structural defects and their role in the growth of Ag triangular nanoplates. *Phys. Chem. C* **2007**, *111*, 6989–6993. [CrossRef]
34. Kilin, D.S.; Prezhdo, O.V.; Xia, Y. Shape-controlled synthesis of silver nanoparticles: Ab initio study of preferential surface coordination with citric acid. *Chem. Phys. Lett.* **2008**, *458*, 113–116. [CrossRef]
35. Zhang, Q.; Li, N.; Goebl, J.; Lu, Z.; Yin, Y. A systematic study of the synthesis of silver nanoplates: Is citrate a "Magic" reagent? *J. Am. Chem. Soc.* **2011**, *133*, 18931–18939. [CrossRef]
36. Song, H.; Zhang, H.; Sun, Z.; Ren, Z.; Yang, X.; Wang, Q. Triangular silver nanoparticle U-bent fiber sensor based on localized surface plasmon resonance. *AIP Adv.* **2019**, *9*, 085307. [CrossRef]
37. Chen, Z.; Balankura, T.; Fichthorn, K.A.; Rioux, R.M. Revisiting the polyol synthesis of silver nanostructures: Role of chloride in nanocube formation. *ACS Nano* **2019**, *13*, 1849–1860. [CrossRef]
38. Im, S.H.; Lee, Y.T.; Wiley, B.; Xia, Y. Large-scale synthesis of silver nanocubes: The role of HCl in promoting cube perfection and monodispersity. *Angew. Chem. Int. Ed.* **2005**, *44*, 2154–2157. [CrossRef]
39. Wiley, B.; Sun, Y.; Xia, Y. Polyol synthesis of silver nanostructures: control of product morphology with Fe(II) or Fe(III) species. *Langmuir* **2005**, *21*, 8077–8080. [CrossRef]
40. Wu, F.; Wang, W.; Xu, Z.; Li, F. Bromide (Br−) based synthesis of Ag nanocubes with high-yield. *Sci. Rep.* **2015**, *5*, 10772. [CrossRef]
41. Zhu, J.J.; Kan, C.X.; Wan, J.G.; Han, M.; Wang, G.H. High-yield synthesis of uniform Ag nanowires with high aspect ratios by introducing the long-chain PVP in an improved polyol process. *J. Nanomater.* **2012**, *2011*, 982547. [CrossRef]
42. Korte, K.E.; Skrabalak, S.E.; Xia, Y. Rapid synthesis of silver nanowires through a CuCl- or $CuCl_2$-mediated polyol process. *J. Mater. Chem.* **2008**, *18*, 437–441. [CrossRef]
43. Cong, F.Z.; Wei, H.; Tian, X.R.; Xu, H.X. A facile synthesis of branched silver nanowire structures and its applications in surface-enhanced Raman scattering. *Front. Phys.* **2012**, *7*, 521–526. [CrossRef]
44. Lee, P.C.; Meisel, D. Adsorption and surface-enhanced Raman of dyes on silver and gold sols. *J. Phys. Chem.* **1982**, *86*, 3391–3395. [CrossRef]
45. Pillai, Z.S.; Kamat, P.V. What factors control the size and shape of silver nanoparticles in the citrate ion reduction method? *J. Phys. Chem. B* **2004**, *108*, 945–951. [CrossRef]
46. Jin, R.; Cao, Y.W.; Mirkin, C.A.; Kelly, K.L.; Schatz, G.C.; Zheng, J.G. Photoinduced conversion of silver nanospheres to nanoprisms. *Science* **2001**, *294*, 1901–1903. [CrossRef] [PubMed]
47. Wu, X.; Redmond, P.L.; Liu, H.; Chen, Y.; Steigerwald, M.; Brus, L. Photovoltage mechanism for room light conversion of citrate stabilized silver nanocrystal seeds to large nanoprisms. *J. Am. Chem. Soc.* **2008**, *130*, 9500–9506. [CrossRef] [PubMed]
48. Dong, X.; Ji, X.; Wu, H.; Zhao, L.; Li, J.; Yang, W. Shape control of silver nanoparticles by stepwise citrate reduction. *J. Phys. Chem. C* **2009**, *113*, 6573–6576. [CrossRef]
49. Caswell, K.K.; Bender, C.M.; Murphy, C.J. Seedless, surfactantless wet chemical synthesis of silver nanowires. *Nano Lett.* **2003**, *3*, 667–669. [CrossRef]
50. Murph, S.E.H.; Murphy, C.J.; Leach, A.; Gall, K.A. Possible oriented attachment growth mechanism for silver nanowire formation. *Cryst. Growth Des.* **2015**, *15*, 1968–1974. [CrossRef]
51. Koh, A.L.; Bao, K.; Khan, I.; Smith, W.E.; Kothleitner, G.; Nordlander, P.; Maier, S.A.; McComb, D.W. Electron energy-loss spectroscopy (EELS) of surface plasmons in single silver nanoparticles and dimers: Influence of beam damage and mapping of dark modes. *ACS Nano* **2009**, *3*, 3015–3022. [CrossRef]
52. Camden, J.P.; Dieringer, J.A.; Wang, Y.; Masiello, D.J.; Marks, L.D.; Schatz, G.C.; Duyne, R.P.V. Probing the structure of single-molecule surface-enhanced Raman scattering hot spots. *J. Am. Chem. Soc.* **2008**, *130*, 12616–12617. [CrossRef]
53. Mohapatra, B.; Kuriakose, S.; Mohapatra, S. Rapid green synthesis of silver nanoparticles and nanorods using Piper nigrum extract. *J. Alloys Compd.* **2015**, *637*, 119–126. [CrossRef]
54. Al-Dhabi, N.A.; Ghilan, A.K.M.; Arasu, M.V.; Duraipandiyan, V. Green biosynthesis of silver nanoparticles produced from marine Streptomyces sp. Al-Dhabi-89 and their potential applications against wound infection and drug resistant clinical pathogens. *J. Photochem. Photobiol. B* **2018**, *189*, 176–184. [CrossRef]
55. Katta, V.K.M.; Dubey, R.S. Green synthesis of silver nanoparticles using Tagetes erecta plant and investigation of their structural, optical, chemical and morphological properties. *Mater. Today Proc.* **2021**, *45*, 794–798. [CrossRef]
56. Rasheed, T.; Bilal, M.; Iqbal, H.M.N.; Li, C. Green biosynthesis of silver nanoparticles using leaves extract of Artemisia Vulgaris and their potential biomedical applications. *Colloids Surf. B* **2017**, *158*, 408–415. [CrossRef]
57. Santos, E.B.; Madalossi, N.V.; Sigoli, F.A.; Mazali, I.O. Silver nanoparticles: Green synthesis, self-assembled nanostructures and their application as SERS substrates. *New J. Chem.* **2015**, *39*, 2839–2846. [CrossRef]
58. Pietrobon, B.; Kitaev, V. Photochemical synthesis of monodisperse size-controlled silver decahedral nanoparticles and their remarkable optical properties. *Chem. Mater.* **2008**, *20*, 5186–5190. [CrossRef]

59. Zhang, Q.; Li, W.; Moran, C.; Chen, J.; Wen, L.-P.; Xia, Y. Seed-mediated synthesis of Ag nanocubes with controllable edge lengths in the range of 30–200 nm and comparison of their optical properties. *J. Am. Chem. Soc.* **2010**, *132*, 11372–11378. [CrossRef]
60. Seo, D.; Yoo, C.I.; Jung, J.; Song, H. Ag−Au−Ag heterometallic nanorods formed through directed anisotropic growth. *J. Am. Chem. Soc.* **2008**, *130*, 2940–2941. [CrossRef]
61. Shemer, G.; Krichevski, O.; Markovich, G.; Molotsky, T.; Lubitz, I.; Kotlyar, A.B. Chirality of silver nanoparticles synthesized on DNA. *J. Am. Chem. Soc.* **2006**, *128*, 11006–11007. [CrossRef]
62. Maillard, M.; Giorgio, S.; Pileni, M.-P. Tuning the size of silver nanodisks with similar aspect ratios: synthesis and optical properties. *J. Phys. Chem. B* **2003**, *107*, 2466–2470. [CrossRef]
63. Thomas, A.; Goettmann, F.; Antonietti, M. Hard templates for soft materials: Creating nanostructured organic materials. *Chem. Mater.* **2008**, *20*, 738–755. [CrossRef]
64. Yang, R.; Sui, C.; Gong, J.; Qu, L. Silver nanowires prepared by modified AAO template method. *Mater. Lett.* **2007**, *61*, 900–903. [CrossRef]
65. Zheng, X.; Xu, W.; Corredor, C.; Xu, S.; An, J.; Zhao, B.; Lombardi, J.R. Laser-induced growth of monodisperse silver nanoparticles with tunable surface plasmon resonance properties and a wavelength self-limiting effect. *J. Phys. Chem. C* **2007**, *111*, 14962–14967. [CrossRef]
66. Mahmoud, M.A.; El-Sayed, M.A. Substrate effect on the plasmonic sensing ability of hollow nanoparticles of different shapes. *J. Phys. Chem. B* **2013**, *117*, 4468–4477. [CrossRef]
67. Cela, S.A.; Romero, S.C.; Coelho, J.P.; Martínez, A.G. Recent progress on colloidal metal nanoparticles as signal enhancers in nanosensing. *Adv. Colloid Interface Sci.* **2016**, *233*, 255–270. [CrossRef]
68. Chen, Y. Nanofabrication by electron beam lithography and its applications: A review. *Microelectron. Eng.* **2015**, *135*, 57–72. [CrossRef]
69. Kong, D.S.; Varsanik, J.S.; Griffith, S.; Jacobson, J.M. Conductive nanostructure fabrication by focused ion beam direct-writing of silver nanoparticles. *J. Vac. Sci. Technol. B Nanotechnol. Microelectron.* **2004**, *22*, 2987–2991. [CrossRef]
70. Manoccio, M.; Esposito, M.; Passaseo, A.; Cuscunà, M.; Tasco, V. Focused ion beam processing for 3D chiral photonics nanostructures. *Micromachines* **2020**, *12*, 6. [CrossRef] [PubMed]
71. Fischer, U.C.; Zingsheim, H.P. Submicroscopic pattern replication with visible light. *J. Vac. Sci. Technol.* **1981**, *19*, 881. [CrossRef]
72. Deckman, H.W.; Dunsmuir, J.H. Natural lithography. *Appl. Phys. Lett.* **1982**, *41*, 377. [CrossRef]
73. Hulteen, J.C.; Treichel, D.A.; Smith, M.T.; Duval, M.L.; Jensen, T.R.; Van Duyne, R.P. Nanosphere lithography: Size-tunable silver nanoparticle and surface cluster arrays. *J. Phys. Chem. B* **1999**, *103*, 3854–3863. [CrossRef]
74. Bradley, L.; Ye, D.; Luong, H.M.; Zhao, Y. Transition from discrete patches to plasmonic nanohole array by glancing angle deposition on nanosphere monolayers. *Nanotechnology* **2020**, *31*, 205301. [CrossRef]
75. Zhu, C.; Meng, G.; Zheng, P.; Huang, Q.; Li, Z.; Hu, X.; Wang, X.; Huang, Z.; Li, F.; Wu, N. Hierarchically ordered array of silver-nanorod bundles for surface-enhanced Raman scattering detection of phenolic pollutants. *Adv. Mater.* **2016**, *28*, 4871–4876. [CrossRef]
76. Lee, S.H.; Bantz, K.C.; Lindquist, N.C.; Oh, S.H.; Haynes, C.L. Self-assembled plasmonic nanohole arrays. *Langmuir* **2009**, *25*, 13685–13693. [CrossRef]
77. Ingram, W.; Larsen, S.; Carlson, D.; Zhao, Y. Ag–Cu mixed phase plasmonic nanostructures fabricated by shadow nanosphere lithography and glancing angle co-deposition. *Nanotechnology* **2017**, *28*, 015301. [CrossRef]
78. Wang, Y.; Choi, I.; Zhang, K.; Yang, Y.; Ao, S.; Xue, X.; Fu, W.; Zhang, Z.; Zhao, Y. Highly conductive nanograting–nanohole structures with tunable and dual-band spectral transparency. *ACS Appl. Electron. Mater.* **2021**, *3*, 3489–3500. [CrossRef]
79. Ai, B.; Yu, Y.; Möhwald, H.; Wang, L.; Zhang, G. Resonant optical transmission through topologically continuous films. *ACS Nano* **2014**, *8*, 1566–1575. [CrossRef]
80. Ai, B.; Wang, L.; Möhwald, H.; Yu, Y.; Zhang, G. Confined surface plasmon sensors based on strongly coupled disk-in-volcano arrays. *Nanoscale* **2015**, *7*, 2317–2324.
81. Lu, Y.C.; Hsueh, C.H. Fabrication of periodic Ag tetrahedral nanopyramids via H_2O_2-assisted nanosphere lithography for plasmonic applications. *Colloids Surf. A Physicochem. Eng. Asp.* **2021**, *628*, 127278. [CrossRef]
82. Wang, Z.; Ai, B.; Möhwald, H.; Zhang, G. Colloidal lithography meets plasmonic nanochemistry. *Adv. Opt. Mater.* **2018**, *6*, 1800402. [CrossRef]
83. Rino Kaneko, R.; Ichikawa, H.; Hosaka, M.; Sone, Y.; Imura, Y.; Wang, K.H.; Kawai, T. Hole, Convex, and silver nanoparticle patterning on polystyrene nanosheets by colloidal photolithography at air–water interfaces. *Langmuir* **2022**, *38*, 8153–8159. [CrossRef]
84. Qiu, T.; Akinoglu, E.M.; Luo, B.; Konarova, M.; Yun, J.H.; Gentle, I.R.; Wang, L. Nanosphere lithography: A versatile approach to develop transparent conductive films for optoelectronic applications. *Adv. Mater.* **2022**, *34*, 2103842. [CrossRef]
85. Li, X.; Xu, W.; Zhang, J.; Jia, H.; Yang, B.; Zhao, B.; Li, B.; Ozaki, Y. Self-assembled metal colloid films: two approaches for preparing new SERS active substrates. *Langmuir* **2004**, *20*, 1298–1304. [CrossRef]
86. Pathak, A.; Mishra, S.K.; Gupta, B.D. Fiber-optic ammonia sensor using Ag/SnO 2 thin films: Optimization of thickness of SnO2 film using electric field distribution and reaction factor. *Appl. Opt.* **2015**, *54*, 8712–8721. [CrossRef]
87. Hawkeye, M.M.; Brett, M.J. Glancing angle deposition: Fabrication, properties, and applications of micro- and nanostructured thin films. *J. Vac. Sci. Technol. A* **2007**, *25*, 1317–1335. [CrossRef]

88. Taschuk, M.T.; Hawkeye, M.M.; Brett, M.J. Glancing angle deposition. In *Handbook of Deposition Technologies for Films and Coatings*, 3rd ed.; Martin, P.M., Ed.; Elsevier: Oxford, UK, 2010; pp. 621–651.
89. He, Y.; Fu, J.; Zhao, Y. Oblique angle deposition and its applications in plasmonics. *Front. Phys.* **2014**, *9*, 47–59. [CrossRef]
90. Gibbs, J.G.; Mark, A.G.; Lee, T.C.; Eslami, S.; Schamel, D.; Fischer, P. Nanohelices by shadow growth. *Nanoscale* **2014**, *6*, 9457–9466. [CrossRef]
91. Barranco, A.; Borras, A.; Gonzalez-Elipe, A.R.; Palmero, A. Perspectives on oblique angle deposition of thin films: From fundamentals to devices. *Prog. Mat. Sci.* **2016**, *76*, 59–153. [CrossRef]
92. Singh, J.H.; Nair, G.; Ghosh, A.; Ghosh, A. Wafer scale fabrication of porous three-dimensional plasmonic metamaterials for the visible region: Chiral and beyond. *Nanoscale* **2013**, *5*, 7224–7228. [CrossRef]
93. Gahlaut, S.K.; Yadav, K.; Sharan, C.; Singh, J.P. Quick and selective dual mode detection of H_2S gas by mobile app employing silver nanorods array. *Anal. Chem.* **2017**, *89*, 13582–13588. [CrossRef]
94. Gahlaut, S.K.; Kalyani, N.; Sharan, C.; Mishra, P.; Singh, J.P. Smartphone based dual mode in situ detection of viability of bacteria using Ag nanorods array. *Biosens. Bioelectron.* **2019**, *126*, 478–484. [PubMed]
95. Gahlaut, S.K.; Devi, P.; Singh, J.P. Self-sustainable and recyclable Ag nanorods for developing Ag-Ag_2S nano heterostructures using sewage gas: Applications in photocatalytic water purification, hydrogen evolution, SERS and antibacterial activity. *Appl. Surf. Sci.* **2020**, *528*, 147037. [CrossRef]
96. Rajput, A.; Kumar, S.; Singh, J.P. Vertically standing nanoporous Al–Ag zig-zag silver nanorod arrays for highly active SERS substrates. *Analyst* **2017**, *142*, 3959–3966. [CrossRef] [PubMed]
97. Jen, Y.J.; Lin, Y.W.; Huang, Y.J.; Chan, T.L. The effect of glancing angle deposition conditions on the morphology of a silver nanohelix array. *Coatings* **2017**, *7*, 140. [CrossRef]
98. Goel, P.; Singh, K.; Singh, J.P. Polarization dependent diffraction from anisotropic Ag nanorods grown on DVD grating templates by oblique angle deposition. *RSC Adv.* **2014**, *4*, 11130–11135.
99. Wang, X.J.; Haider, A.M.; Abell, J.L.; Zhao, Y.P.; Zhang, Z.M. Anisotropic diffraction from inclined silver nanorod arrays on grating templates. *Nanoscale Microscale Thermophys. Eng.* **2012**, *16*, 18–36. [CrossRef]
100. Zhang, X.; Li, Z.; Ye, S.; Wu, S.; Zhang, J.; Cui, L.; Li, A.; Wang, T.; Li, S.; Yang, B. Elevated Ag nanohole arrays for high performance plasmonic sensors based on extraordinary optical transmission. *J. Mater. Chem.* **2012**, *22*, 8903–8910. [CrossRef]
101. Wang, Y.; Ao, S.; Yang, F.; Zhang, Z.; Zhao, Y.P. Coupling between surface plasmon modes of single-layer complex silver nanohole arrays and enhancing index sensing. *ACS Appl. Nano Mater.* **2022**, *5*, 9761–9770.
102. Giallongo, G.; Pilot, R.; Durante, C.; Rizzi, G.A.; Signorini, R.; Bozio, R.; Gennaro, A.; Granozzi, G. Silver nanoparticle arrays on a DVD-derived template: An easy & cheap SERS substrate. *Plasmonics* **2011**, *6*, 725–733.
103. Avella-Oliver, M.; Puchades, R.; Wachsmann-Hogiu, S.; Maquieira, A. Label-free SERS analysis of proteins and exosomes with large-scale substrates from recordable compact disks. *Sens. Actuat. B Chem.* **2017**, *252*, 657–662. [CrossRef]
104. Karabacak, T.; Singh, J.P.; Zhao, Y.P.; Wang, G.C.; Lu, T.M. Scaling during shadowing growth of isolated nanocolumns. *Phys. Rev. B* **2003**, *68*, 125408. [CrossRef]
105. He, Y.; Larsen, G.K.; Ingram, W.; Zhao, Y. Tunable three-dimensional helically stacked plasmonic layers on nanosphere monolayers. *Nano Lett.* **2014**, *14*, 1976–1981. [CrossRef] [PubMed]
106. Kumar, S.; Gahlaut, S.K.; Singh, J.P. Sculptured thin films: Overcoming the limitations of surface enhanced Raman scattering substrates. *Figshare*, 2022; *preprint*. [CrossRef]
107. Ai, B.; Zhao, Y. Glancing angle deposition meets colloidal lithography: A new evolution in the design of nanostructures. *Nanophotonics* **2019**, *8*, 1–26. [CrossRef]
108. Lee, C.; Lawrie, B.; Pooser, R.; Lee, K.G.; Rockstuhl, C.; Tame, M. Quantum plasmonic sensors. *Chem. Rev.* **2021**, *121*, 4743–4804. [CrossRef]
109. Shalabney, A.; Abdulhalim, I. Sensitivity-enhancement methods for surface plasmon sensors. *Laser Photon. Rev.* **2011**, *5*, 571–606. [CrossRef]
110. Mejía-Salazar, J.R.; Oliveira, O.N., Jr. Plasmonic biosensing focus review. *Chem. Rev.* **2018**, *118*, 10617–10625. [CrossRef]
111. Gupta, B.D.; Kant, R. Recent advances in surface plasmon resonance based fiber optic chemical and biosensors utilizing bulk and nanostructures. *Opt. Laser Technol.* **2018**, *101*, 144–161. [CrossRef]
112. Gupta, B.D.; Srivastava, S.K.; Verma, R. *Fiber Optic Sensors Based on Plasmonics*; World Scientific Publishing Company: Singapore, 2015.
113. Homola, J. Surface plasmon resonance sensors for detection of chemical and biological species. *Chem. Rev.* **2008**, *108*, 462–493. [CrossRef]
114. Masson, J.F. Portable and field-deployed surface plasmon resonance and plasmonic sensors. *Analyst* **2020**, *145*, 3776–3800. [CrossRef]
115. Masson, J.F. Surface plasmon resonance clinical biosensors for medical diagnostics. *ACS Sens.* **2017**, *2*, 16–30. [CrossRef]
116. Gupta, B.D.; Sharma, A.K. Sensitivity evaluation of a multi-layered surface plasmon resonance-based fiber optic sensor: A theoretical study. *Sens. Actuators B Chem.* **2005**, *107*, 40–46. [CrossRef]
117. Gupta, B.D.; Verma, R.K. Surface plasmon resonance-based fiber optic sensors: Principle, probe designs, and some applications. *J. Sens.* **2009**, *16*, 979761. [CrossRef]
118. Pathak, A.; Gupta, B.D. Fiber-optic plasmonic sensor utilizing CTAB-functionalized ZnO nanoparticle-decorated carbon nanotubes on silver films for the detection of catechol in wastewater. *ACS Appl. Nano Mat.* **2020**, *3*, 2582–2593. [CrossRef]

119. Mercado-Zúñiga, C.; Torres-Torres, C.; Trejo-Valdez, M.; Torres-Martı'nez, M.R.; Cervantes-Sodi, F.; Vargas-Garcıa, J.R. Influence of silver decoration on the nonlinear optical absorption exhibited by multiwall carbon nanotubes. *J. Nanoparticle Res.* **2014**, *16*, 2334. [CrossRef]
120. Young, S.J.; Liu, Y.H.; Lin, Z.D.; Ahmed, K.; Shiblee, M.N.I.; Romanuik, S.; Sekhar, P.K.; Thundat, T.; Nagahara, L.; Arya, S.; et al. Multi-walled carbon nanotubes decorated with silver nanoparticles for acetone gas sensing at room temperature. *J. Electrochem. Soc.* **2020**, *167*, 167519. [CrossRef]
121. Gupta, B.D.; Pathak, A.; Semwal, V. Carbon-based nanomaterials for plasmonic sensors: A review. *Sensors* **2019**, *19*, 3536. [CrossRef]
122. Barbillon, G. Latest novelties on plasmonic and non-plasmonic nanomaterials for SERS sensing. *Nanomaterials* **2020**, *10*, 1200. [CrossRef]
123. Martínez-Hernández, M.E.; Sandúa, X.; Rivero, P.J.; Goicoechea, J.; Arregui, F.J. Self-referenced optical fiber sensor based on LSPR generated by gold and silver nanoparticles embedded in layer-by-layer nanostructured coatings. *Chemosensors* **2022**, *10*, 77. [CrossRef]
124. Sharma, S.; Gupta, B.D. Fiber optic surface-plasmon-resonance-based highly sensitive arsenic sensor prepared using α-Fe_2O_3/SnO_2 core-shellnanostructure with optimized probe parameters. *Appl. Opt.* **2018**, *57*, 10466–10473. [CrossRef]
125. Singh, S.; Mishra, S.K.; Gupta, B.D. SPR based fiber optic biosensor for phenolic compounds using immobilization of tyrosinase in polyacrylamide gel. *Sens. Actuators B Chem.* **2013**, *186*, 388–395. [CrossRef]
126. Bhatia, P.; Yadav, P.; Gupta, B.D. Surface plasmon resonance based fiber optic hydrogen peroxide sensor using polymer embedded nanoparticles. *Sens. Actuators B Chem.* **2013**, *182*, 330–335. [CrossRef]
127. Mishra, S.K.; Tripathi, S.N.; Choudhary, V.; Gupta, B.D. Surface plasmon resonance based fiber optic methane gas sensor utilizing graphene carbon nanotubes-poly(methyl methacrylate) hybrid nanocomposite. *Plasmonics* **2015**, *10*, 1147–1157. [CrossRef]
128. Semwal, V.; Gupta, B.D. LSPR- and SPR-based fiber-optic cholesterol sensor using immobilization of cholesterol oxidase over silver nanoparticles coated graphene oxide nanosheets. *IEEE Sens.* **2018**, *18*, 1039–1046. [CrossRef]
129. Gahlaut, S.K.; Pathak, A.; Gupta, B.D.; Singh, J.P. Portable fiber-optic SPR platform for the detection of NS1-antigen for dengue diagnosis. *Biosens. Bioelectron.* **2022**, *196*, 113720. [CrossRef]
130. Yang, Y.; Murray, J.; Haverstick, J.; Tripp, R.A.; Zhao, Y. Silver nanotriangle array based LSPR sensor for rapid coronavirus detection. *Sens. Actuators B Chem.* **2022**, *359*, 131604. [CrossRef]
131. Usha, S.P.; Shrivastav, A.M.; Gupta, B.D. FO-SPR based dextrose sensor using Ag/ZnO nanorods/ GO_x for insulinoma detection. *Biosens. Bioelectron.* **2016**, *85*, 986–995. [CrossRef]
132. Eletxigerra, U.; Martinez-Perdiguero, J.; Barderas, R.; Pingarron, J.M.; Campuzano, S.; Merino, S. Surface plasmon resonance immunosensor for ErbB2 breast cancer biomarker determination in human serum and raw cancer cell lysates. *Anal. Chim. Acta* **2016**, *905*, 156–162. [CrossRef]
133. Baliyan, A.; Usha, S.P.; Gupta, B.D.; Gupta, R.; Sharma, E.K. Localized surface plasmon resonance–based fiber-optic sensor for the detection of triacylglycerides using silver nanoparticles. *J. Biomed. Opt.* **2017**, *22*, 107001. [CrossRef]
134. Raj, D.R.; Sudarsanakumar, C. Surface plasmon resonance based fiber optic sensor for the detection of cysteine using diosmin capped silver nanoparticles. *Sens. Actuators A Phys.* **2017**, *253*, 41–48.
135. Zandieh, M.; Hosseini, S.N.; Vossoughi, M.; Khatami, M.; Abbasian, S.; Moshaii, A. Label-free and simple detection of endotoxins using a sensitive LSPR biosensor based on silver nanocolumns. *Analyt. Biochem.* **2018**, *548*, 96–101. [CrossRef] [PubMed]
136. Raj, D.R.; Prasanth, S.; Vineeshkumar, T.V.; Sudarsanakumar, C. Surface plasmon resonance based fiber optic dopamine sensor using green synthesized silver nanoparticles. *Sens. Actuators B Chem.* **2016**, *224*, 600–606.
137. Shrivastav, A.M.; Usha, S.P.; Gupta, B.D. A localized and propagating SPR, and molecular imprinting based fiber-optic ascorbic acid sensor using an in situ polymerized polyaniline–Ag nanocomposite. *Nanotechnology* **2016**, *27*, 345501. [CrossRef] [PubMed]
138. Verma, R.; Gupta, B.D. A Novel approach for simultaneous sensing of urea and glucose by SPR based optical fiber multianalyte sensor. *Analyst* **2014**, *139*, 1449–1455. [CrossRef]
139. Balbinot, S.; Srivastav, A.M.; Vidic, J.; Abdulhalim, I.; Manzano, M. Plasmonic biosensors for food control. *Trends Food Sci. Technol.* **2021**, *111*, 128–140. [CrossRef]
140. Shrivastav, A.M.; Usha, S.P.; Gupta, B.D. Highly sensitive and selective erythromycin nanosensor employing fiber optic SPR/ERY imprinted nanostructure: Application in milk and honey. *Biosens. Bioelectron.* **2017**, *90*, 516–524. [CrossRef]
141. Shrivastav, A.M.; Gupta, B.D. SPR and molecular imprinting-based fiber-optic melamine sensor with high sensitivity and low limit of detection. *IEEE J. Sel. Top. Quantum Electron.* **2016**, *22*, 172–178. [CrossRef]
142. Gadea, E.A.; Rodríguez-Canto, P.J.; Martínez-Pastor, J.P.; Lopatynskyi, A.; Chegel, V.; Abargues, R. Molecularly imprinted silver nanocomposites for explosive taggant sensing. *ACS Appl. Polym. Mater.* **2021**, *3*, 2960–2970. [CrossRef]
143. Zou, W.; Liu, W.; Luo, L.; Zhang, S.; Lu, R.; Veser, G. Detection of nitro explosives via LSPR sensitive silver clusters embedded in porous silica. *J. Mater. Chem.* **2012**, *22*, 12474–12478. [CrossRef]
144. Fleischmann, M.; Hendra, P.J.; McQuillan, A.J. Raman spectra of pyridine adsorbed at a silver electrode. *Chem. Phys. Lett.* **1974**, *26*, 163–166. [CrossRef]
145. Kneipp, K.; Wang, Y.; Kneipp, H.; Perelman, L.T.; Itzkan, I.; Dasari, R.R.; Feld, M.S. Single molecule detection using surface-enhanced Raman scattering (SERS). *Phys. Rev. Lett.* **1997**, *78*, 1667. [CrossRef]

146. Shanmukh, S.; Jones, L.; Driskell, J.; Zhao, Y.; Dluhy, R.; Tripp, R.A. Rapid and sensitive detection of respiratory virus molecular signatures using a silver nanorod array SERS substrate. *Nano Lett.* **2006**, *6*, 2630–2636. [CrossRef]
147. Zhang, L.; Wang, B.; Zhu, G.; Zhou, X. Synthesis of silver nanowires as a SERS substrate for the detection of pesticide thiram. *Spectrochim. Acta A Mol. Biomol. Spectrosc.* **2014**, *133*, 411–416. [CrossRef]
148. Kumar, S.; Goel, P.; Singh, J.P. Flexible and robust SERS active substrates for conformal rapid detection of pesticide residues from fruits. *Sens. Actuators B Chem.* **2017**, *241*, 577–583. [CrossRef]
149. Li, D.W.; Zhai, W.L.; Li, Y.T.; Long, Y.T. Recent progress in surface enhanced Raman spectroscopy for the detection of environmental pollutants. *Microchim. Acta* **2014**, *181*, 23–43. [CrossRef]
150. Halvorson, R.A.; Vikesland, P.J. Surface-enhanced Raman spectroscopy (SERS) for environmental analyses. *Environ. Sci. Technol.* **2010**, *44*, 7749–7755. [CrossRef]
151. Novotný, F.; Plutnar, J.; Pumera, M. Plasmonic self-propelled nanomotors for explosives detection via solution-based surface enhanced Raman scattering. *Adv. Funct. Mater.* **2019**, *29*, 1903041. [CrossRef]
152. Srivastava, A.K.; Awasthi, P.; Kanojia, S.; Neeraj, N.S.; Sharma, P.; Mukhopadhyay, K.; Prasad, N.E. Synthesis of silver nanostructures and their application in highly sensitive SERS sensors. *Def. Sci. J.* **2018**, *68*, 98–103. [CrossRef]
153. Kumar, G.; Soni, R.K. Trace-level detection of explosive molecules with triangular silver nanoplates-based SERS substrates. *Plasmonics* **2022**, *17*, 559–573. [CrossRef]
154. Adhikari, S.; Ampadu, E.K.; Kim, M.; Noh, D.; Oh, E.; Lee, D. Detection of explosives by SERS platform using metal nanogap substrates. *Sensors* **2021**, *21*, 5567. [CrossRef]
155. Le Ru, E.C.; Blackie, E.; Meyer, M.; Etchegoin, P.G. Surface enhanced Raman scattering enhancement factors: A comprehensive study. *J. Phys. Chem. C* **2007**, *111*, 13794–13803. [CrossRef]
156. Kumar, S.; Tokunaga, K.; Namura, K.; Fukuoka, T.; Suzuki, M. Experimental evidence of a twofold electromagnetic enhancement mechanism of surface-enhanced Raman scattering. *J. Phys. Chem. C* **2020**, *124*, 21215–21222.
157. Zong, C.; Xu, M.; Xu, L.J.; Wei, T.; Ma, X.; Zheng, X.S.; Hu, R.; Ren, B. Surface-enhanced Raman spectroscopy for bioanalysis: Reliability and challenges. *Chem. Rev.* **2018**, *118*, 4946–4980. [CrossRef] [PubMed]
158. Ding, S.Y.; Yi, J.; Li, J.F.; Ren, B.; Wu, D.Y.; Panneerselvam, R.; Tian, Z.Q. Nanostructure-based plasmon-enhanced Raman spectroscopy for surface analysis of materials. *Nat. Rev. Mater.* **2016**, *1*, 16021. [CrossRef]
159. Han, X.X.; Rodriguez, R.S.; Haynes, C.L.; Ozaki, Y.; Zhao, B. Surface-enhanced Raman spectroscopy. *Nat. Rev. Methods Primers* **2021**, *1*, 87. [CrossRef]
160. Driskell, J.D.; Shanmukh, S.; Liu, Y.; Chaney, S.B.; Tang, X.J.; Zhao, Y.P.; Dluhy, R.A. The use of aligned silver nanorod arrays prepared by oblique angle deposition as surface enhanced Raman scattering substrates. *J. Phys. Chem. C* **2008**, *112*, 895–901. [CrossRef]
161. Liu, Y.J.; Chu, H.Y.; Zhao, Y.P. Silver nanorod array substrates fabricated by oblique angle deposition: Morphological, optical, and SERS characterizations. *J. Phys. Chem. C* **2010**, *114*, 8176–8183. [CrossRef]
162. Yadav, S.; Khanam, R.; Singh, J.P. A purview into highly sensitive magnetic SERS detection of hemozoin biomarker for rapid malaria diagnosis. *Sens. Actuators B Chem.* **2022**, *55*, 131303. [CrossRef]
163. Kumar, S.; Lodhi, D.K.; Goel, P.; Mishra, P.; Singh, J.P. A facile method for fabrication of buckled PDMS silver nanorod arrays as active 3D SERS cages for bacterial sensing. *Chem. Commun.* **2015**, *51*, 12411–12414. [CrossRef]
164. Chen, H.Y.; Lin, M.H.; Wang, C.Y.; Chang, Y.M.; Gwo, S. Large-scale hot spot engineering for quantitative SERS at the single-molecule scale. *J. Am. Chem. Soc.* **2015**, *137*, 13698–13705. [CrossRef]
165. Gahlaut, S.K.; Savargaonkar, D.; Sharan, C.; Yadav, S.; Mishra, P.; Singh, J.P. SERS platform for dengue diagnosis from clinical samples employing a hand held Raman spectrometer. *Anal. Chem.* **2020**, *92*, 2527–2534. [CrossRef]
166. Yadav, S.; Senapati, S.; Desai, D.; Gahlaut, S.; Kulkarni, S.; Singh, J.P. Portable and sensitive Ag nanorods based SERS platform for rapid HIV-1 detection and tropism determination. *Colloids Surf. B Biointerfaces* **2021**, *198*, 111477. [CrossRef]
167. Zhang, Z.; Li, D.; Wang, X.; Wang, Y.; Lin, J.; Jiang, S.; Wu, Z.; He, Y.; Gao, X.; Zhu, Z.; et al. Rapid detection of viruses: Based on silver nanoparticles modified with bromine ions and acetonitrile. *Chem. Eng. J.* **2022**, *438*, 135589. [CrossRef]
168. Tang, H.; Zhu, C.; Meng, G.; Wu, N. Review—Surface-enhanced Raman scattering sensors for food safety and environmental monitoring. *J. Electrochem. Soc.* **2018**, *165*, 3098–3118. [CrossRef]
169. Lin, Z.; He, L. Recent advance in SERS techniques for food safety and quality analysis: A brief review. *Curr. Opin. Food Sci.* **2019**, *28*, 82–87. [CrossRef]
170. Langer, J.; Jimenez de Aberasturi, D.; Aizpurua, J.; Alvarez-Puebla, R.A.; Auguié, B.; Baumberg, J.J.; Bazan, G.C.; Bell, S.E.; Boisen, A.; Brolo, A.G.; et al. Present and future of surface-enhanced Raman scattering. *ACS Nano* **2020**, *14*, 28–117. [CrossRef]
171. Sharma, B.; Frontiera, R.R.; Henry, A.I.; Ringe, E.; Van Duyne, R.P. SERS: Materials, applications, and the future. *Mater. Today* **2012**, *15*, 16–25. [CrossRef]
172. Kumar, S.; Kumar, P.; Das, A.; Pathak, C.S. Surface-enhanced Raman scattering: Introduction and applications. In *Recent Advances in Nanophotonics-Fundamentals and Applications*; IntechOpen: London, UK, 2020; pp. 1–24.
173. Pilot, R.; Signorini, R.; Durante, C.; Orian, L.; Bhamidipati, M.; Fabris, L.A. Review on surface-enhanced Raman scattering. *Biosensors* **2019**, *9*, 57. [CrossRef]
174. Purcell, E.M. Spontaneous Emission Probabilities at Radio Frequencies. In *Confined Electrons and Photons*; NATO ASI Series; Burstein, E., Weisbuch, C., Eds.; Springer: Boston, MA, USA, 1995; Volume 340.

175. Tam, F.; Goodrich, G.P.; Johnson, B.R.; Halas, N.J. Plasmonic enhancement of molecular fluorescence. *Nano Lett.* **2007**, *7*, 496–501. [CrossRef]
176. Lakowicz, J.R. Radiative decay engineering 5: Metal-enhanced fluorescence and plasmon emission. *Anal. Biochem.* **2005**, *337*, 171–194. [CrossRef]
177. Drexhage, K.H. Interaction of light with monomolecular dye layers. *Prog. Opt.* **1974**, *12*, 163–232.
178. Lakowicz, J.R.; Ray, K.; Chowdhury, M.; Szmacinski, H.; Fu, Y.; Zhang, J.; Nowaczyk, K. Plasmon-controlled fluorescence: A new paradigm in fluorescence spectroscopy. *Analyst* **2008**, *133*, 1308–1346. [CrossRef] [PubMed]
179. Dong, J.; Zhang, Z.; Zheng, H.; Sun, M. Recent progress on plasmon-enhanced fluorescence. *Nanophotonics* **2015**, *4*, 472–490. [CrossRef]
180. Zhang, Y.; Dragan, A.; Geddes, C.D. Wavelength dependence of metal-enhanced fluorescence. *J. Phys. Chem. C* **2009**, *113*, 12095–12100. [CrossRef]
181. Chen, Y.; Munechika, K.; Ginger, D.S. Dependence of fluorescence intensity on the spectral overlap between fluorophores and plasmon resonant single silver nanoparticles. *Nano Lett.* **2007**, *7*, 690–696. [CrossRef]
182. Bharill, S.; Chen, C.; Stevens, B.; Kaur, J.; Smilansky, Z.; Mandecki, W.; Gryczynski, I.; Gryczynski, Z.; Cooperman, B.S.; Goldman, Y.E. Enhancement of single-molecule fluorescence signals by colloidal silver nanoparticles in studies of protein translation. *ACS Nano* **2011**, *5*, 399–407. [CrossRef]
183. Dong, J.; Qu, S.X.; Zhang, Z.L.; Liu, M.C.; Liu, G.N.; Yan, X.Q.; Zheng, H.R. Surface enhanced fluorescence on three dimensional silver nanostructure substrate. *J. Appl. Phys.* **2012**, *111*, 093101. [CrossRef]
184. Ray, K.; Badugu, R.; Szmacinski, H.; Lakowicz, J.R. Several hundred-fold enhanced fluorescence from single fluorophores assembled on silver nanoparticle–dielectric–metal substrate. *Chem. Commun.* **2015**, *51*, 15023–15026. [CrossRef]
185. Badshah, M.A.; Koh, N.Y.; Zia, A.W.; Abbas, N.; Zahra, Z.; Saleem, M.W. Recent developments in plasmonic nanostructures for metal enhanced fluorescence-based biosensing. *Nanomaterials* **2020**, *10*, 1749. [CrossRef]
186. Jeong, Y.; Kook, Y.M.; Lee, K.; Koh, W.G. Metal enhanced fluorescence (MEF) for biosensors: General approaches and a review of recent developments. *Biosens. Bioelectron.* **2018**, *111*, 102–116. [CrossRef]
187. Lin, G.; Lewandowska, M. Plasmon-enhanced fluorescence provided by silver nanoprisms for sensitive detection of sulfide. *Sens. Actuators B Chem.* **2019**, *292*, 241–246. [CrossRef]
188. Zhang, F.; Ma, N.; Tang, F.; Cui, Q.L.; He, F.; Li, L.D. pH- and glucose-responsive core-shell hybrid nanoparticles with controllable metal-enhanced fluorescence effects. *ACS Appl. Mater. Interfaces* **2012**, *4*, 1747–1751. [CrossRef]
189. Wang, Y.; Li, H.; Xu, D. Aptamers-based sandwich assay for silver-enhanced fluorescence multiplex detection. *Anal. Chim. Acta* **2016**, *905*, 149–155. [CrossRef]
190. Ji, X.; Xiao, C.; Lau, W.F.; Li, J.; Fu, J. Metal enhanced fluorescence improved protein and DNA detection by zigzag Ag nanorod arrays. *Biosens. Bioelectron.* **2016**, *82*, 240–247. [CrossRef]
191. Badshah, M.A.; Ju, J.; Lu, X.; Abbas, N.; Kim, S. Enhancing the sensitivity of DNA microarrays by metal-enhanced fluorescence using vertical nanorod structures. *Sens. Actuat. B Chem.* **2018**, *274*, 451–457. [CrossRef]
192. Pourreza, N.; Ghomi, M. A novel metal enhanced fluorescence bio probe for insulin sensing based on poly vinyl alcohol-borax hydrogel functionalized by Ag dots. *Sens. Actuators B* **2017**, *251*, 609–616. [CrossRef]
193. Cheng, Q.; He, Y.; Ge, Y.; Zhou, J.; Song, G. Ultrasensitive detection of heparin by exploiting the silver nanoparticle-enhanced fluorescence of graphitic carbon nitride (g-C3N4) quantum dots. *Microchim. Acta* **2018**, *185*, 332. [CrossRef]
194. Li, H.; Chen, C.Y.; Wei, X.; Qiang, W.; Li, Z.; Cheng, Q.; Xu, D. Highly sensitive detection of proteins based on metal-enhanced fluorescence with novel silver nanostructures. *Anal. Chem.* **2012**, *84*, 8656–8662. [CrossRef]
195. Xu, J.; Zhang, B.; Jia, L.; Bi, N.; Zhao, T. Metal-enhanced fluorescence detection and degradation of tetracycline by silver nanoparticle-encapsulated halloysite nano-lumen. *J. Hazard. Mater.* **2020**, *386*, 121630. [CrossRef]
196. Hartstein, A.; Kirtley, J.R.; Tsang, J.C. Enhancement of the infrared absorption from molecular monolayers with thin metal overlayers. *Phys. Rev. Lett.* **1980**, *45*, 201. [CrossRef]
197. Aroca, R.F. *Surface-Enhanced Vibrational Spectroscopy*; John Wiley and Sons Ltd.: New York, NY, USA, 2006.
198. Ataka, K.; Heberle, J. Biochemical applications of surface-enhanced infrared absorption spectroscopy. *Anal. Bioanal. Chem.* **2007**, *388*, 47–54. [CrossRef]
199. Huck, C.; Neubrech, F.; Vogt, J.; Toma, A.; Gerbert, D.; Katzmann, J.; Härtling, T.; Pucci, A. Surface-enhanced infrared spectroscopy using nanometer-sized gaps. *ACS Nano* **2014**, *8*, 4908–4914. [CrossRef]
200. Yilmaz, H.; Mohapatra, S.S.; Culha, M. Surface-enhanced infrared absorption spectroscopy for microorganisms discrimination on silver nanoparticle substrates. *Spectrochim. Acta Part A Mol. Biomol. Spectrosc.* **2022**, *268*, 120699. [CrossRef]
201. Pi, M.; Zheng, C.; Ji, J.; Zhao, H.; Peng, Z.; Lang, J.; Liang, L.; Zhang, Y.; Wang, Y.; Tittel, F.K. Surface-enhanced infrared absorption spectroscopic chalcogenide waveguide sensor using a silver island film. *ACS Appl. Mater. Interfaces* **2021**, *13*, 32555–32563. [CrossRef]
202. Kato, Y.; Kikugawa, M.; Sudo, E. Attenuated total reflection surface-enhanced infrared absorption (ATR SEIRA) spectroscopy for the analysis of fatty acids on silver nanoparticles. *Appl. Spectrosc.* **2017**, *71*, 2083–2091. [CrossRef]

203. Zhou, H.; Li, D.; Hui, X.; Mu, X. Infrared metamaterial for surface-enhanced infrared absorption spectroscopy: Pushing the frontier of ultrasensitive on-chip sensing. *Int. J. Optomechat.* **2021**, *15*, 97–119. [CrossRef]
204. Pereira, C.F.; Viegas, I.M.A.; Sobrinha, I.G.S.; Pereira, G.; Pereira, G.A.L.; Krebs, P.; Mizaikoff, B. Surface-enhanced infrared absorption spectroscopy using silver selenide quantum dots. *J. Mater. Chem. C* **2020**, *8*, 10448–10455. [CrossRef]

biosensors

MDPI

Review

Biomedical Applications of an Ultra-Sensitive Surface Plasmon Resonance Biosensor Based on Smart MXene Quantum Dots (SMQDs)

Seyyed Mojtaba Mousavi [1], Seyyed Alireza Hashemi [2], Masoomeh Yari Kalashgrani [3], Vahid Rahmanian [4], Ahmad Gholami [3], Wei-Hung Chiang [1,*] and Chin Wei Lai [5,*]

1 Chemical Engineering Department, National Taiwan University of Science and Technology, Taipei City 106335, Taiwan
2 Nano-Materials and Polymer Nano-Composites Laboratory, School of Engineering, University of British Columbia, Kelowna, BC V1V 1V7, Canada
3 The Center of Biotechnology Research, Shiraz University of Medical Science, Shiraz 71468-64685, Iran
4 The Centre of Molecular and Macromolecular Studies, Polish Academy of Sciences, Sienkiewicza 112, 90-363 Lodz, Poland
5 Nanotechnology & Catalysis Research Centre (NANOCAT), Level 3, Block A, Institute for Advanced Studies (IAS), Universiti Malaya (MU), Kuala Lumpur 50603, Malaysia
* Correspondence: whchiang@mail.ntust.edu.tw (W.-H.C.); cwlai@um.edu.my (C.W.L.)

Abstract: In today's world, the use of biosensors occupies a special place in a variety of fields such as agriculture and industry. New biosensor technologies can identify biological compounds accurately and quickly. One of these technologies is the phenomenon of surface plasmon resonance (SPR) in the development of biosensors based on their optical properties, which allow for very sensitive and specific measurements of biomolecules without time delay. Therefore, various nanomaterials have been introduced for the development of SPR biosensors to achieve a high degree of selectivity and sensitivity. The diagnosis of deadly diseases such as cancer depends on the use of nanotechnology. Smart MXene quantum dots (SMQDs), a new class of nanomaterials that are developing at a rapid pace, are perfect for the development of SPR biosensors due to their many advantageous properties. Moreover, SMQDs are two-dimensional (2D) inorganic segments with a limited number of atomic layers that exhibit excellent properties such as high conductivity, plasmonic, and optical properties. Therefore, SMQDs, with their unique properties, are promising contenders for biomedicine, including cancer diagnosis/treatment, biological sensing/imaging, antigen detection, etc. In this review, SPR biosensors based on SMQDs applied in biomedical applications are discussed. To achieve this goal, an introduction to SPR, SPR biosensors, and SMQDs (including their structure, surface functional groups, synthesis, and properties) is given first; then, the fabrication of hybrid nanoparticles (NPs) based on SMQDs and the biomedical applications of SMQDs are discussed. In the next step, SPR biosensors based on SMQDs and advanced 2D SMQDs-based nanobiosensors as ultrasensitive detection tools are presented. This review proposes the use of SMQDs for the improvement of SPR biosensors with high selectivity and sensitivity for biomedical applications.

Keywords: surface plasmon resonance; biosensor; smart MXene quantum dots; biomedical

Citation: Mousavi, S.M.; Hashemi, S.A.; Kalashgrani, M.Y.; Rahmanian, V.; Gholami, A.; Chiang, W.-H.; Lai, C.W. Biomedical Applications of an Ultra-Sensitive Surface Plasmon Resonance Biosensor Based on Smart MXene Quantum Dots (SMQDs). *Biosensors* **2022**, *12*, 743. https://doi.org/10.3390/bios12090743

Received: 28 July 2022
Accepted: 7 September 2022
Published: 9 September 2022

1. Introduction

The rapid improvement of technology, in line with the growing needs of modern societies in today's world, has paved the way for the development of the factors involved and lifestyles. Sensors are devices that help researchers in various fields such as industry, agriculture, etc. to achieve certain goals by measuring predetermined parameters. Biosensors are a special category of sensors used to study and detect chemical and biological parameters. The diagnosis of diseases, the discovery of new drugs, and the identification of contamination by biological factors such as DNA, proteins, antibodies, enzymes, and viruses are performed by systems called biosensors. Based on their operation, these sensors are classified into mechanical, chemical, electrical, and optical groups. Apart from these, optical biosensors are divided into labeled and label-free sensors [1–3]. Optical biosensors have several advantages, including high sensitivity and insensitivity to electromagnetic interference. Optical biosensors have also covered a wide range of substrates, including SPR, localized surface plasmon resonance (LSPR), interferometers, ring amplifiers, etc. [4]. There are certain types of biosensors that use light sources and light guidance based on different methods to achieve detection and detection objectives. One of the most common types of these sensors are biosensors that use plasmonic sensing in the design of their structure and are referred to as surface plasmon resonance biosensors. SPR biosensors stimulate the phenomenon of the oscillation of electrons in the metal-dielectric junction when their rate of motion matches the rate of motion of the incident light. This category of sensors is of interest to many researchers and scientists in this field due to their small size and optimal sensitivity [5,6]. Therefore, the development of SPR biosensors can be an important area of research to find chemical and biological substances that cause diseases or have negative consequences [7–9]. One of the most important aspects in the development of SPR biosensors may be the accurate detection of 'target molecules' to prevent the occurrence of disease and facilitate early medical therapy. Ultimately, this will accelerate therapeutic efficacy [9–11]. To this end, high selectivity and sensitivity are the most important properties to be considered in the development of SPR biosensors. Numerous nanomaterials, including metal NP and transition metal dichalcogenide (TMD) NP, are being investigated for the development of SPR biosensors [12–15]. Although metal NPs have been more commonly used in the past [16–19], after the discovery of carbon nanomaterials, such as graphene, these nanomaterials showed a more efficient performance than current metal NPs [20–23], and their biocompatibility can render them suitable for monitoring cell-size conditions [24–26]. However, the development of new nanomaterials for SPR biosensors is possible because the demand for nanomaterials with exceptional properties and efficient performance is constantly increasing. 2D nanomaterials, such as smart MXene, are becoming increasingly popular due to their special properties, such as physical, electrical, and chemical properties [27,28]. The term 'smart MXene' has been used by a number of researchers for MXene-based hybrid materials, indicating their unique application-related properties. Such unique compounds make MXene a potential candidate for the fabrication of transparent conductors. MXenes offer transmittance up to 95% in the visible and UV regions with very low sheet resistance (up to 0.01 kO per square). Due to their excellent mechanical properties and tunable optical properties, MXenes can be used as transparent conductive electrodes for touchscreen applications, various sensors, light-emitting diodes, and flexible displays [29–31]. Among the nanomaterials for SPR biosensors, SMQDs attract much attention due to their great potential and exclusive properties in developing SPR biosensors [32–34]. SMQDs are 2D inorganic compounds composed of transition metal carbides and possess a significant atomic layer thickness. Their exceptional properties include high conductivity and plasmonic and optical qualities [35–38]. SMQDs can be used in biomedical applications due to their biocompatibility. [39–42]. This new nanomaterial is now the best option for the development of SPR biosensors in biomedical applications, based on the present research to improve SPR biosensors. The main objective of this review is an ultra-sensitive plasmon resonance nano biosensor on a surface based on SMQDs for biomedical applications. This article is divided into three topics: an introduction to SPR and SPR biosensors; explanation and characteristics of SMQDs; and SPR biosensors based on SMQDs. Accordingly, this review clearly presents the characteristics of SMQDs for the development of SPR biosensors and their biomedical applications. In summary, the team believes that this article can highlight current research directions as well as ways to utilize SMQDs for the efficient improvement of SPR biosensors with a high selectivity and sensitivity in biomedical applications.

2. SPR

SPR refers to the collective oscillations of electrons on the surface of metal nanostructures that occur in response to an external stimulus such as light or a charge. When the particle size reaches the order of nanometers, the electron can spontaneously accelerate on the surface of the particle and absorb electromagnetic waves of a certain wavelength; a schematic of this phenomenon is shown in Figure 1a [43–45]. The solutions of nanoparticles of rare metals such as silver and gold (which have a high conductivity) often show a strong absorption band in the visible spectrum. When the solution of these nanoparticles with the same size range is exposed to electromagnetic radiation, part of the radiation is scattered, and part of it is accelerated by the free electrons of the nanoparticles; therefore, in this phenomenon, certain frequencies are absorbed, resulting in an enhancement of the electron resonance and appearing as a strong peak in the visible region. The shape and frequency of the resonance spectrum depend on the size, shape, and distance between the nanoparticles and their dielectric properties but, most importantly, on the dielectric properties of the environment in which the nanoparticles are located [46–48]. Figure 1b shows the steps of a typical SPR method. Each sensor-level measurement begins with the selection of a suitable buffer solution, which is the most basic task before starting the association process. At this stage, the sensor surface contains active ligands ready to accept the target analytes. By injecting the solution containing the analytes, the association cycle begins. If the correct ligands are not selected, special bonds can form between the ligands and the analytes after the solution has passed, leading to the instability and detachment of the ligands from the surface. In this step, the kinetic energy resulting from the interaction of analytes and ligands is measured in real time. In the next step, a solution is brought into contact with the sensor surface to regenerate the initial state. As shown in Figure 1b, this step destroys the non-specific binding elements so that the mass accumulated on the surface can be recovered from the sensor reaction. At this stage, analyte dissociation also begins, and the kinetic energy of the dissociation process can be studied. Finally, the regeneration solution is injected, breaking the link between ligands and analytes. If the ligands are properly placed on the surface, they will remain on the surface after passing the regeneration solution as the analytes are gradually removed [49,50].

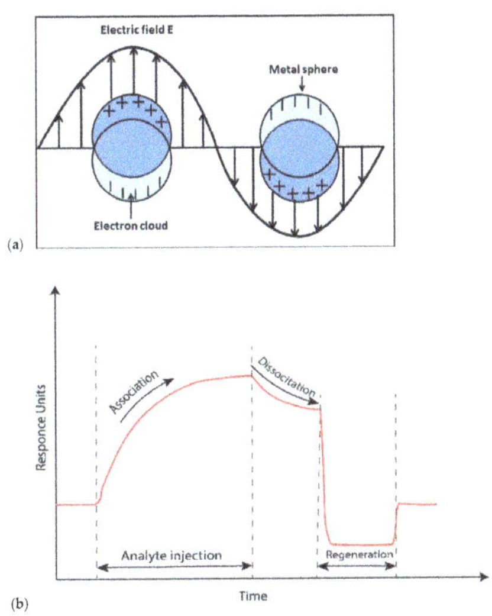

Figure 1. (**a**) Schematic of the surface plasmon (electronic cloud) resonated due to the electric field. (Reprinted with permission [51]. Copyright © 2016, The Author(s). Licensee: IntechOpen.) (**b**) Steps in a typical SPR method. (Reprinted with permission [49] Copyright © 2020 by the authors. Licensee MDPI, Basel, Switzerland).

Biosensor Using SPR

The light beam propagates in a medium with a larger refractive index $n1$ and reaches the common section of a material that has a lower index with a refractive form than the first medium, i.e., n_2 ($n_1 > n_2$). At an incidence angle greater than the limit angle (θ), the light is completely reflected and returns to the environment with a higher refractive index [52–55]. Additionally, no energy is lost during the reflection of the beam, and the light beam causes the penetration of an electric field intensity into the material with a low refractive index, which is introduced as an evanescent wave. The 'P-polarized' component of the evanescent subject can penetrate the metal layer and excite electromagnetic surface plasmon waves that are propagated inside the conductive surface associated with a material with a low refractive index if the total internal reflection interface is covered with a layer of appropriate conductive materials, such as metal with an acceptable thickness. This 'surface plasmon wave' has P polarization for a non-magnetic metal, namely, 'gold', and because of its electromagnetic properties and diffusion surface, it generates an amplified evanescent wave in comparison to incident electromagnetic waves. If the size and directions of the 'photon wave' vector kx and the plasmon wave vector ksp are equal for waves of the same frequency, an amplified evanescent field is produced. When this condition occurs at the landing angle θ, the 'intensity of reflection' at the angle θ can be zero due to the conversion of the energy into a 'surface' electric field. With increasing penetration into the thinner material n2, the loss in this evanescent field wave's amplitude is approximately half the wavelength of its resonance away from the surface. To put it another way, the field loss for visible light is of the order of several hundred nanometers. As a result, just the quenching zone is used to investigate analyte molecules. The SPR biosensor is a group of optical biosensors that have advantages such as real-time detection, a short response time, the simultaneous detection of several types of analytes, and non-labeled sensors [56–59]. Exciting surface plasmon waves and their characteristics depend on the electromagnetic properties of the dielectric metal interface. Resonance coupling causes a valley in the reflection spectrum at the SPR resonance angle. SPR biosensing can be obtained by the absorption of 'target analytes' on the metal surface and dependent changes in the wavelength and intensity in reflected light. These optical changes can rely on alterations in the refractive index due to the phenomenon of surface absorption. Figure 2 shows the basis of SPR biosensors. The high sensitivity to alterations in the features of the dielectric is caused by the transfer of incident light energy to the 'surface plasmon wave' and the resulting high density of the electromagnetic field in the dielectric near the metal layer. The gold metal layer's penetration depth of 200–300 nanometers offers the chance to detect minute variations in the thickness or 'refractive index' of layers on the surface of the metal [60–62]. The resolution limit of SPR biosensors provides the possibility of detecting surfaces with an approximate coverage of 1 picogram/mm [63,64]. Currently, SPR-based biosensors are the most commercialized type of optical biosensors; they usually have large dimensions and high prices and are suitable for laboratory use [65–67]. The technology of surface plasmon resonance sensing as a detector or diagnostic has developed rapidly and has now become an effective tool for direct monitoring and, especially, the analysis of biomolecular interactions. It is also widely used for interactions of biological molecules such as protein–protein supplements, drugs–protein, nucleic acid–protein, nuclear acceptor–DNA, and DNA–DNA. Its fields of application include immunodiagnosis, signal transduction, drug screening, antibody conjugation, and protein conformational changes (Table 1) [56,68].

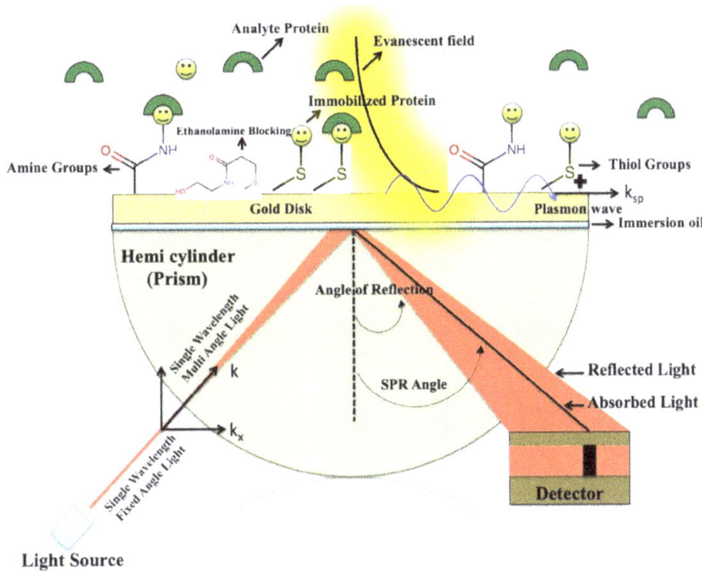

Figure 2. The functionality of SPR biosensors. When the analytes interact with the ligands fixed on the 'surface', the dielectric index on the gold 'surface' changes, and the reflection maximum is seen at a different angle. This angular shift is the result of the interactions between the analyte and the ligand. (Reprinted with permission [69]. Copyright © 2020, Authors. Exclusive licensee: Bio-protocol LLC.)

Table 1. Applying surface plasmon resonance biosensors in medical diagnosis.

Field	Detection	Species	Ref.
Medical diagnostics	Virus marker	Ebola, Hepatitis B virus	[70,71]
	Cardiac marker	Myoglobin	[72]
	Drug	Warfarin, Morphine	[73,74]
	Cancer marker	Interleukin 8, Prostate-specific antigen	[75,76]

3. SMQDs Structure

The aging of the 'MAX -phase' 'A' layers leads to a layered structure of SMQDs, a new class of 2D materials. These MAX phases consist of a large family of nitrides as well as carbonitrides with the chemical formula nAX M_{n+1}, where 'M' stands for the primary transition metal (such as Sc, Zr, HF, V, Nb, Mo, Ta, Cr) in layer n + 1, 'A' stands for an element from the periodic table (usually group 13 or even 14), and 'X' stands for carbon as well as 'nitrogen' in layers X. [77–80]. In Figure 3, all constituent elements of the phase MAX are marked with different colors.

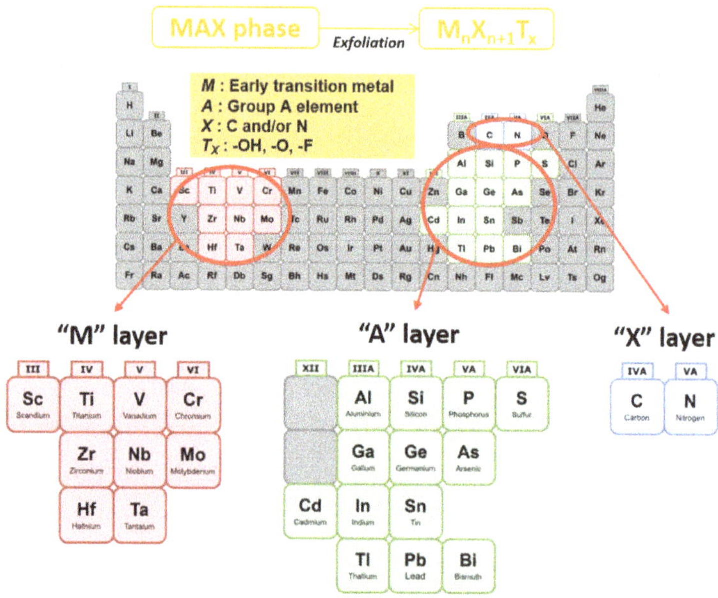

Figure 3. Periodic table of the elements that make up the 'MAX' phases as well as SMQDs: M: metals with early transition, I: 'Group A' element, X: C as well as Tx, and N: 'surface function group'. (Reprinted with permission [81]. Copyright © 2021, The Author(s).)

Here, 'n' can take the numbers 1, 2, and 3. By changing n from '1' to '3', SMQDs contain layers between three and seven layers of atoms for 'M2×', 'M3 × 2', and 'M4 × 3', respectively [82,83]. As shown in Figure 4, during the etching process, the group A element from the MAX phase is replaced by surface groups such as oxygen (−O), hydroxyl (−OH), and fluorine without destroying the MX layers by suitable chemicals [84,85]. In several studies, $Ti_3C_2T_x$ with a surface termination group of -Cl has also been observed, and the general formula is $M_{n+1}X_nT_x$, where T is the symbol for the surface groups [86–88]. MAX has 'layered' structures in which the bonds between the layers are weaker than the bonds in the layer [89,90]. In other words, the bonds between M and X are a mixture of ionic and covalent bonds, which can be much stronger than the bonds between M and A [91,92]. As a result, the bond between M and A is decomposed at high temperatures, and the 2D structure $M_{n+1}X_n$ is formed. Upon deformation, they become laminated and exhibit a combination of unusual and sometimes unique properties that are intermediate between those of ceramics and metals. For example, like metals, they are capable of conducting electricity and heat, and they can be hard, brittle, and heat-resistant [93,94]. In addition, they are resistant to chemical agents and thermal shocks. However, these ceramics are fabricated as 3D materials, and one of the first experiences with their 2D fabrication is due to '2D Ti_3C_2' nanoplatelets. Researchers attempted to remove aluminum from titanium aluminum carbide (Ti_3AlC_2) powder by placing it in hydrofluoric acid. Through a chemical process called exfoliation, 2D Ti3C2 nanoplatelets were thus obtained [95,96]. The interesting thing about SMQDs is the naming of this substance. This material is produced from a bulk crystal called Max with the suffix 'ene' added to the end, similar to graphene [97,98].

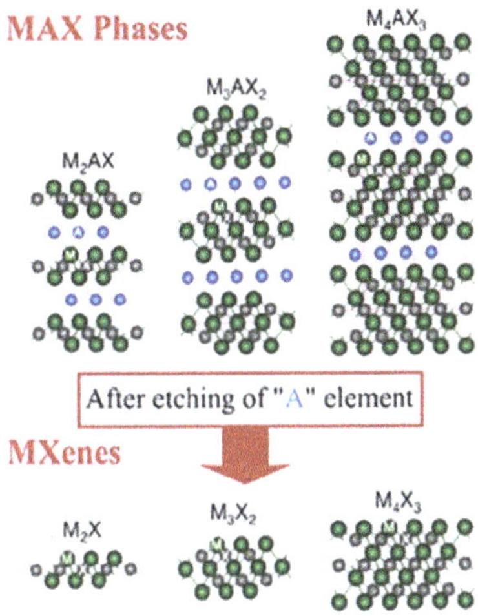

Figure 4. Etching of the MAX phase and creation of SMQDs with surface groups. (Reprinted with permission [99] Copyright © 2021 by the authors. Licensee MDPI, Basel, Switzerland.)

3.1. Functional Group on the Surface of SMQDs

Surface end groups that are −OH or −O replace the A layers by the chemical etching of the MAX phase to produce SMQDs. In these materials, two to four M layers are interspersed with layers of C (carbon) or N (nitrogen) in the clever MXene QD structure. Unlike graphene, the surfaces of these materials have functional groups, −O or −OH, that make them hydrophilic. These surface groups are strongly dependent on the etching technique [100,101]. For example, $Ti_3C_2T_x$ etched with HF has four times more 'F-functional groups' than the material etched with a LiF mixture [102,103]. These functional groups also have a great impact on the detection of the electronic functions of SMQDs. For example, it has been shown that both −F and −OH functional groups on the surface of Ti_3C_2 lead to a semiconductor behavior with a band gap of 0.05 to 0.1 ev, while Ti_3C_2 without surface termination exhibits a metallic behavior [104,105]. Moreover, surface functional groups can influence the energy storage application of SMQDs. For example, density functional theory studies confirm that Ti_3C_2 without functional groups stores more lithium ions than its counterpart with a fluorine functional group (Ti_3C_2F) because the surface functional groups block lithium adsorption [106,107].

3.2. SMQDs Synthesis

The synthesis of SMQDs by precursors is called the top-down method. Depending on the type of precursor, i.e., either MAX or not MAX, this method is divided into two subgroups [100,102]. The most common SMQD precursor is a part of 3D layered carbides as well as nitrides, called the SMQD phase [108,109]. In the precursor materials of the MAX phase, such as $M_{n+1}AlX_n$ or $M_{n+1}SiX_n$, various etching methods are used to break the bonds within the layers and replace the individual elements Al and Si with surface groups [110,111]. Layered materials, where the 'layer-to-layer' bonding is not significantly stronger than the bonding between layers, are divided into one or more atomic layers to produce 2D materials. Ghidiu and co-workers [107] argued in 2014 that MAX can be etched with a solution of lithium fluoride and hydrochloric acid or with various amounts of hydrofluoric acid. Table 2 lists various approaches for the synthesis of SMQDs. In general, experimental factors such as the etching time, the particle size of the MAX phase, and the acid concentration used affect the better performance in the preparation of higher-quality SMQDs [33]. Recently, some non-Mex phase precursors have been used to fabricate MXs. For example, $Zr_3Al_3C_5$ has been used as a precursor, although its constituents are similar to those of Al-bonded MAX phase precursors; however, in this precursor, an 'Al-C' layer is etched instead of a pure 'Al' layer to produce MXene Zr_3C_2 [112,113]. Considering that both the constituents of the composition and the surface end groups can be changed, the properties and characteristics of SMQDs can also be easily modified [114,115].

Table 2. A summary of the different SMQDs synthesis methods.

SMQDs	Functionalization(s)	Synthesis Method of SMQDs	Ref.
Ta_4C_3	Manganese oxide (MnOx), 'soy bean' phospholipid (SP)	HF etching	[116]
Ti_3C_2	Poly 'lactic-co- glycolic acid' (PLGA), SP, IONPs	HF etching, TPAOH intercalation	[117]
TiO_2–Ti_3C_2	Hemoglobin (Hb), Nafion	Hydrothermal synthesis	[118,119]
Ti_3C_2	Cobalt nanowires (CoNWs), Dox	LiF + HCl etching	[120]
Ti_3C_2 QDs	–	Hydrothermal synthesis	[121]
Ti_2N QDs	SP	KF + HCl etching, sonication in NMP	[122]
Nb_2C QDs	–	HF etching, TPAOH sonication (ultrasoundassisted)	[123]

3.3. *Characteristics and Features of SMQDs*

The SMQD material has very interesting properties; for example, although it falls into the category of ceramics, unlike many others, it has good electrical conductivity, which makes it suitable for biomedical applications. The electronic properties of SMQDs are of particular importance because they can be tailored by changing the 'elemental' composition of the SMQDs or the surface functional groups. Other factors such as the band gap can also affect the electronic properties of SMQDs. Unlike graphene, SMQD is hydrophilic, which can be very advantageous in many applications. It is also flexible, pliable, and soft. Because of these properties, it can be formed into complex shapes (its use in the form of a tube or a sheet for materials with a conductivity as high as that of metals is very undesirable) [124,125].

3.4. *Preparing Hybrid NPs Using SMQDs*

A SMQDs composite with tiny magnetic Fe_3O_4 NPs with a size of about ~4.9 nm ($Ti_3C_2T_x/Fe_3O_4/TiO_2$) was prepared in an 'aqueous' solution of vitamin C and Fe^{3+} salt for 5 h at 150 °C in a stainless steel autoclave with Teflon coating by the hydrothermal method. It is also possible to selectively enrich different biomolecules/antigens based on affinity interactions through these hybrid magnetic NPs. Another interesting alternative for nanocomposites is the combination of SMQDs sheets and metal NPs, which are modified by crosslinkers to detect target molecules due to their strong affinity for SMQDs or other biomolecules [126–128]. SMQDs/metal nanoparticle-based nanocomposites can be prepared using an external reducing agent such as $NaBH_4$ or the reduction of noble metal salts. To form particles showing surface-enhanced Raman spectroscopy (SERS), the spontaneous reduction of metal salts such as silver, gold, and palladium is applied to $Ti_3C_2T_x$ SMQDs sheets [129] to form NPs. In addition, it is possible to increase the detection sensitivity of oncomarkers such as microRNA using an AuNP/SMQDs composite [130]. The composite has also been used to detect important small bioactive compounds [131] and electrochemical catalysis [132]. The formation of a composite with SMQDs is also possible using graphite oxide as another 2D material, and such a composite for sensor-based applications leads to the maintenance of the biological activity of hemoglobin even after inkjet printing, as well as the stable and efficient electrochemical detection of H_2O_2 [133].

4. SPR Biosensors Based on SMQDs

Along with other 2D materials, SMQDs are a potential biosensor application material. In biosensing, the unique benefits of SMQDs include their biocompatibility and minimal cytotoxicity. In addition, MXenes provide a wide adsorption range for optical detection and enhanced DNA interaction [134,135]. MXenes are also related with metallic conductivity, intrinsic surface functionalization, and hydrophilic characteristics, all of which may increase the efficacy of SPR biosensors based on MXenes. The features of MXene that make it necessary for biosensors are summarized in Figure 5a. Ti_3C_2, among other compounds, has been extensively documented. Few studies have been conducted on additional MXenes and their composites with metallic nanoparticles, particularly in immunosensing. Ti_3C_2, a member of the MXenes and titanium families, is used in a variety of applications, including SPR biosensors. Numerous publications on the diverse uses of Ti_3C_2 MXenes in electrochemical and optical smart biosensors have been published [136–140]. Although MXenes have been widely investigated, the biosensor applications of Ti_2C-MXenes, particularly their composites with nanoparticles, have received less attention (e.g., Au, Ag, etc.). Wang et al. described the production and optical characteristics of Ti_2C@Au core-shell nanosheets for photonic applications [141]. Zhu et al. built a bifunctional smart nanosensor platform based on Au-Ag nanoshuttles (NSs), utilizing Ti_2C for the electrochemical and SERS measurement of ultratrace carbendazim (CBZ) residues in tea and rice for environmental monitoring [142]. The mechanisms behind SPR biosensors based on SMQDs usually utilize the exclusive 'electrocatalytic' properties of the MXene sheet with respect to the relationship of the 'target signal' (Figure 5b) [143]. The electronic properties and current signal change when biological targets are attached to SMQDs films. The 2D layered nanostructure provides a large surface area to accommodate biological materials. The electrocatalytic properties change and lead to a linear response when biological components can be immobilized by functional groups on SMQDs nanocomposites. SPR biosensors built on smart MXene QDs have impeccable repeatability, stability, and reproducibility. The use of functional groups enriched on the surface of SMQDs material could be a potential solution, since non-covalent interactions and physical adsorption are not sustainable for some biomedical applications. This would allow for surface bonding in new and controllable ways to alter surface properties [144–148].

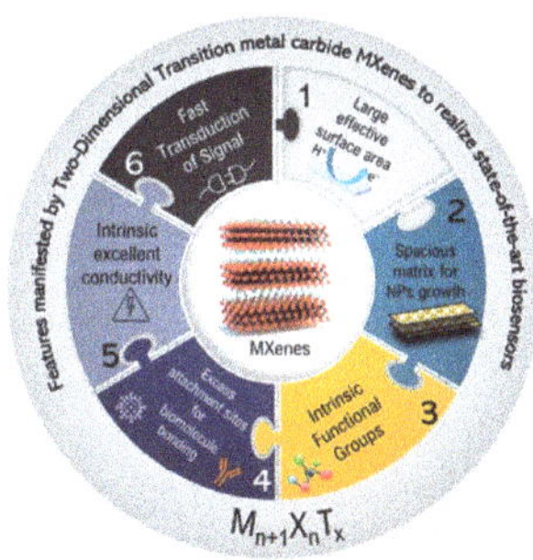

(a)

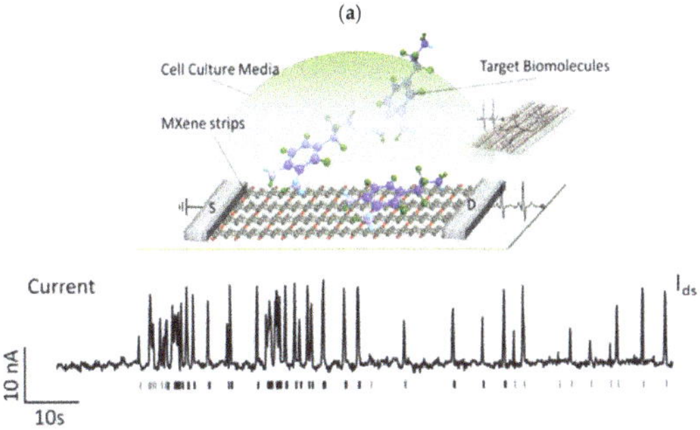

(b)

Figure 5. (**a**) Schematic representation of the main features of MXenes with regard to their application in biosensors. (Reprinted with permission [149]. This journal is © The Royal Society of Chemistry 2022.) (**b**) Mechanisms of the SPR biosensors based on SMQDs groups. (Reprinted with permission [150]. Copyright © 2020, Author(s). Published by IOP Publishing Ltd.)

5. Advanced 2D SMQDs-Based SPR Nanobiosensors as Ultra-Sensitive Detection Gadgets

The use of biosensing platforms that use nanomaterials or nanostructures with exceptional optical, magnetic, electrical, mechanical, and electrocatalytic capabilities promotes the link between advancing detection and routine testing. Incorporating new multifunctional nanoscale structures, morphologies, and controlled structures and a large surface-to-volume ratio enables immobilization in bioreceptors while maintaining biostability, biocompatibility, and biodistribution [151]. Therefore, the SPR sensing strategy using nanomaterials can not only be used as an effective tool for the detection of difficult-to-detect molecules in the concentration range between pmol and amol, but it also facilitates the improvement of sensing properties [152]. It is expected that the design of SPR biosensors is promising for the ultrasensitive and selective detection of cancer. 2D layered materials such as SMQDs have anisotropic electron transport behavior and a large surface area, which makes them potential transducer materials for biosensing applications [153–155]. The results of Wu et al. show an increase in the sensitivity of an SPR biosensor by about 25% with ten graphene layers [156]. Gupta et al. also investigated an SPR biosensor with graphene and silicon to increase the sensitivity [157], and their results showed a maximum sensitivity of ~134.6°/RIU. Ouyang et al. investigated an SPR biosensor with MoS_2 and silicon to increase the sensitivity [158], and the highest sensitivity was ~125.44°/RIU. Wu et al. investigated a novel SPR biosensor with Ti_3C_2Tx-MXene multilayers to increase the sensitivity. According to their results, the sensitivity can reach 224.5°/RIU [159]. SMQDs nanomaterials exhibit a unique combination of excellent mechanical properties, an ease of functionalization, an excellent electrical conductivity, an extremely thin 2D sheet-like morphology, etc. compared to other 2D materials such as graphite carbon nitride, MoS_2, and graphene [160,161]. Among the properties that significantly affect the strength, sensitivity, and selectivity of a biosensor are the inherent properties of the bioreceptor, including its tendency to be structurally stable during the operation of the biosensor, the analyte, and the method used to stabilize the bioreceptor on the surface of the transducer. The bioreceptor component is often attached to a surface, placing it in close proximity to the transducer. Additional requirements that must be met for improved biosensor performance include the interfacial density of the bioreceptor and the distance between the bioreceptor and the transducer (surface). Aptamers, antibodies, enzymes, and protein molecules can be used to influence the design of biosensors based on 2D SMQDs nanomaterials to improve biocompatibility and increase the transporter surface area of the biosensor in conjunction with the increased activity of the catalyst [162–164]. In addition, the implementation of SMQDs as next-generation diagnostic devices requires a significant improvement in the stability of SMQDs against oxidation. Biosensors are small, portable analytical instruments that convert a biochemical process into a quantitative, analytical signal. Because of their high 'specificity', small size, and ease of use, biosensors are the preferred instruments for biological components and chemical detection. Biosensors consist of two parts: a bio-detection component that uses a biological element (enzymes, antibodies, nucleic acids, etc.) that interacts with an analyte in a specific biochemical manner, and transducers in which the interaction is converted into quantifiable signals. The integration of the bio-receptor into a suitable matrix for the interaction between analytes and such receptors are the two main obstacles to the improvement of biosensors [165]. Chen et al. designed a new SPR biosensor using thiol-functionalized niobium carbide MXene QDs (referred to as Nb_2C- SH QDs) as a bio-platform for the N58 aptamer targeting the N gene. As shown in Figure 6, this biosensor was investigated for the sensitive detection of the N gene in various complex environments (e.g., human serum). By the solvothermal method, Nb_2C-QDs were obtained from Nb_2C-MXene nanosheets and then modified with thiol groups (Figure 6a). The Nb_2C- SH QDs were homogeneously distributed on the surface of the chip due to the self-assembly effect between Nb2C- SH QD and the SPR gold chip, and the N58 aptamer was stabilized by hydrogen bonding, π-π* stacking, and electrostatic adsorption. In the presence of SARS-CoV-2, it is also possible to form a G-quadruplex between the N58 aptamer and the N gene of SARS-CoV-2. Thus, upon binding to the N gene, the structure of the aptamer strands is altered, resulting in an increase in the contact area or the distance between the probe molecule and the chip. These changes were then translated into changes in the SPR signal for the detection of the SARS-CoV-2 N gene (Figure 6b) [166].

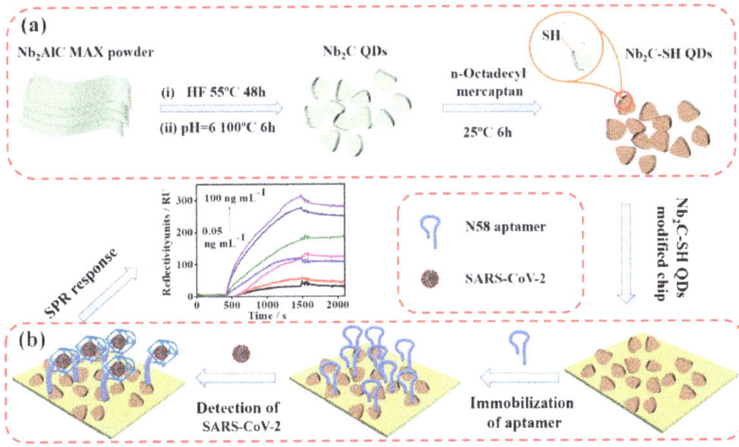

Figure 6. (**a**) Synthesis of Nb_2C-SH QDs. (**b**) Fabrication of an Nb_2C-SH QD-based SPR aptasensor for SARS-CoV-2 N-gene detection. (Reprinted by permission [166]. Copyright © 2022, Springer Nature Switzerland AG. Part of Springer Nature.)

5.1. MXene-Based Electro-Chemical SPR Nanobiosensors

Electro-chemical biosensors can be promising selective tools for detecting cancer diseases in the early stages [167]. SWV (square wave voltammetry), CV (cyclic voltammetry), DPV (differential-pulse-voltammetry), and EIS (electro-chemical impedance spectroscopy) are among the electro-chemical methods [168–170]. 'Lab-on-chip' biosensors have been miniaturized instruments used in the biomarker research of tumors, leading to potential clinical properties. The small volume of analytes, the direct miniaturization, and the optically absorbing and fluorescent compounds are among the attractive features of biosensors that use surface nano-architectures with this type of detection. Kumar et al. investigated the covalent binding of bioreceptors to f-Ti_3C_2 SMQDs for the electro-chemical detection of carcinoembryonic antigen (CEA) as a cancer detector. Single-layer SMQDs (Ti_3C_2) nano-sheets were used with 3-aminopropyl tri-ethoxy-silane. The enhancement of antibodies anchoring and faster access to analytes are possible by ultra-thin 2D nano-sheets of single/multilayer Ti_3C_2 SMQDs. According to the findings, the synthesized biofunctional Ti_3C_2 SMQDs have a linear detection range of 0.0001–2000 n.gm L^{-1}, with a sensitivity to approximately 37.9 Ang^{-1} mL cm^{-2} per 10 years [160]. A conductive support for the immobilization of aptamer probes is also employed in 2D SMQDs because of their outstanding electrical conductivity and sizable particular surface areas by a variety of possible binding sites. Lorenkova et al. investigated the electrochemical performance of Ti3C2Tx-MXenes as sensors [162]. The results obtained showed that the detection limit of 0.7 nm is comparable to the best result obtained so far, which is 0.3 nm [171]. However, there are few reports on SPR sensors integrated with MXene. A recent theoretical study of an MXene-based SPR sensor showed that the coating layers on the gold film can increase the sensitivity of the gold-based SPR sensor. An RI sensitivity of 160 was achieved with four layers of coated gold film at an excitation wavelength of 633 nm, while it was 137 for the devoid setup [159,172].

5.2. SMQDs-Based Optical SPR Nanobiosensors

An important technique for the in situ detection of the affinity of various biomolecules that do not require enzymatic labeling is SPR. SPR optical sensing technology is also useful for biomolecule detection. To make the SPR optical biosensor specific for the analytes of interest, they need to be functionalized by bio-recognition molecules (such as proteins, RNA, DNA, cells, etc.). The adhesion of biomolecules to the optical surface is generally achieved by chemical bonds such as (3-aminopropyl) triethox-ysilane and N-succinimidyl-4-maleimidobutyrate [173–175]. In recent years, 2D transition metal dichalcogenides (TMDs), especially MoS_2, have attracted the attention of researchers in various scientific fields due to their high optical absorption efficiency, high electron conductivity, and tunable band gap [176,177]. The distinctive features of MoS_2 that make it a potential material for the development of biosensor interfaces include the presence of free sulfur atoms, its hydrophobic nature, and its large surface area [178,179]. In addition, MoS_2 layers are also used to inhibit the oxidation of metal layers such as aluminum in SPR biosensors [180]. Additionally, improved operating parameters using nanomaterials have the potential to develop SPR biosensors [181]. The SPR detection platform offers useful advantages such as the ease of miniaturization, 'label-free' and 'real-time' detections, and rapid detection for bioassays. $Ti_3C_2T_x$ SMQDs multilayers improve the applicability of SPR biosensors due to their absorption [159]. The 'gold layer' SMQDs/WS_2 'phosphorus'-based platform, using a monolayer of each nanomaterial, was shown to be a new SPR 'sensing material' with an increased sensitivity of 15.6% compared to bare 'metal films' [182]. SMQDs-based composites such as g-C_3N_4/SMQDs AgNPs containing g-C_3N_4 as a photocatalyst, SMQDs, and AgNPs as electron mediators enhance the photocatalytic activity. In the interface modified with the nanocomposite, the decrease in the band gap energy and the increase in the optical absorption can be observed thanks to the SPR effect of the 'deposited silver NP' [183]. The label-free detection of the bovine serum albumin (BSA) protein using an alternative method of fiber optic SPR probe activation with antibodies was evaluated by Kaushik et al. In this new method, gold-coated fibers were first modified with molybdenum disulfide (MoS_2) nanosheets. The developed technique enables the direct and chemical-free binding of representative antibodies through hydrophobic interactions and also allows for the amplification of SPR signals by the synergistic effects of MoS_2 and the gold metal thin film. The results showed that the sensitivity of the modified MoS_2 sensing probe was improved with a detection limit of 0.29 μg/mL compared to the optical fiber SPR biosensor without MoS_2 coating [184].

According to Wu et al., employing composites constructed of SMQDs, such as g-C3N4/SMQDs AgNPs, which include g-C3N4 as a photocatalyst, SMQDs, and AgNPs as an electron mediator, increases the photocatalytic activity. The band gap energy is decreased and the optical absorbance is raised at the nanocomposite modified interface as a result of the deposited silver NPs SPR influence. As a signal amplifier, amino-functionalized N-Ti_3C_2-MXene-hollow gold NPs (HGNPs)—staphylococcal protein A—were employed for the detection of CEA with a L.O.D of 0.15 fM (linear range of 0.001 to1000 p.M) in SPR (Figure 7) [185].

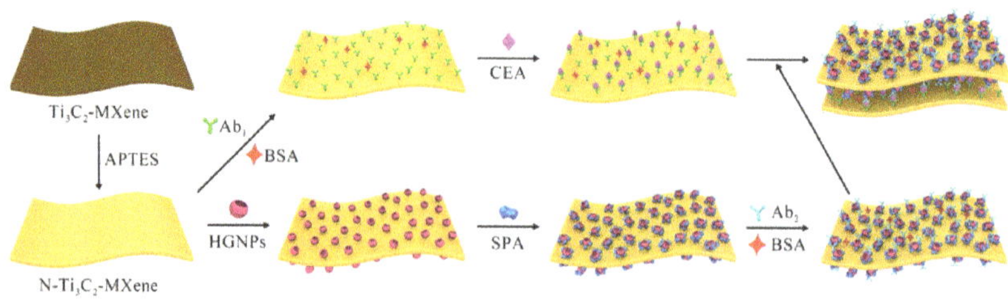

Figure 7. Plan of the prepared 'SPR biosensor' detection method. (Reprinted with permission [185]. Copyright © 2020, The American Chemical Society.)

6. Biomedical Applications of SPR Biosensors Based on SMQDs

SPR biosensors provide a label-free, sensitive, specific, and rapid detection method that is preferred for chemical analysis and medical diagnostics [186]. Over the last three decades, since their beginnings in 1982 as gas sensors [187], SPR biosensors based on 2D nanomaterials such as SMQDs have emerged as suitable sensing platforms for a wide range of applications, e.g., in medicine. Various SPR-based configurations have also been investigated for medical and environmental applications, including SPR biosensors based on SMQDs and fiber-optic SPR sensors [188–190]. Thus, VDW (Weak van der Waals) forces combined with strong 'hydrogen bonding interactions' between 'surface functional groups' cause SMQDs to assemble into stacked 2D layers [191]. Chemical reactivity and functionalization ability are among the properties of surface functional groups. In biomedical studies, the level of SMQDs is adapted to various materials suitable for cancer treatment and diagnosis, biosensing, antigen detection, drug delivery, and antimicrobial activity (Figure 8) [81]. The medical applications of SMQDs are shown in Table 3.

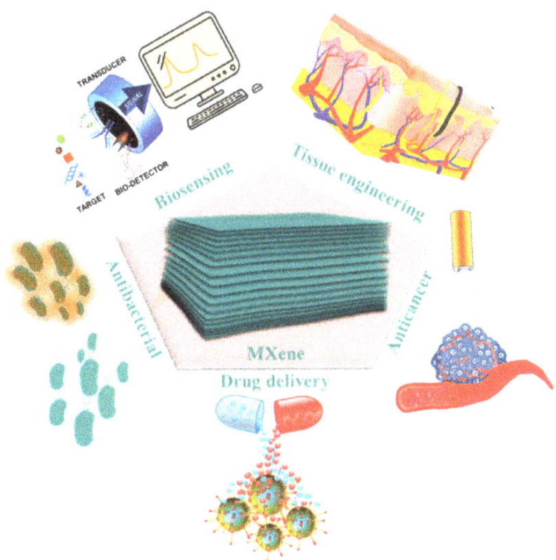

Figure 8. Biomedical applications of SMQDs. (Reprinted with permission [192] Copyright © 2022 by the authors. Licensee MDPI, Basel, Switzerland.)

Table 3. The medical applications of SMQDs.

SMQDs	Applications	Ref.
Ti_3C_2	Detection of curcumin and hypochlorite (ClO^-)	[193]
Ti_3C_2	Glutathione detection and photoelectrochemical biosensing	[194]
V_2C Quantum dots	(Bio)imaging, photothermal therapy, and tumor detection	[195]
Ti_3C_2	Bioimaging, macrophage labeling, and Cu^{2+} detection	[196]
2D Nb_2C-MXenes	Photothermal therapy	[197]
$Ti_3C_2T_x$-SP	Drug delivery	[198]

6.1. Detection of Cancer Biomarkers

SMQDs as new 2D nano-materials have the potential to affect aspects of biosensing such as SPR biosensors in medical applications. Therefore, to detect cancer biomarkers in 'blood', SPR biosensors based on SMQDs offer sufficient sensitivity up to ng·m^{-1} or better. In order to simultaneously immobilize biomolecules while resisting non-specific protein binding, much effort should be devoted to finding suitable decoration strategies for SMQDs. These criteria state that, due to their distinct physical and chemical features, SMQDs-based SPR biosensors may be employed to assess complicated substances such as plasma or blood serum (Table 4) [142]. Additionally, the Ti_3C_2 MXene-based SPR biosensor in human serum samples exhibits an ultrasensitive cancer biomarker response with a high recovery, good reproducibility, and good selectivity [161]. Additionally, SMQDs with a high density of functional groups have an ultrathin 2D nano-sheet morphology that can optimize biomolecule loading and speed up access to the analyte. In addition to enabling a larger density of bound biomarkers, which improves biosensor performance, the covalent immobilization of bioreceptors including enzymes, DNA, and proteins can also enhance homogeneity and dispersion [199]. Sundaram et al. studied the engineering of MXenes nitrides and 2D transition metal carbides for the therapy and diagnostics of cancer. The findings show that electro-chemical devices based on MXene have the ability to detect cancer biomarkers and have an extraordinarily high sensitivity in identifying the target analyte [200]. To determine the osteosarcoma cancer biomarker by a microgap dielectrode sensor, the MXene surface on multiple connection triangles was investigated. The detection limit and sensitivity were found to be one fM by having good regression co-efficient values ($y = 1.0036\dot{x} + 0.525$; $R^2 = 0.978$), and a current increase was found when raising the target DNA concentration. Based on the results of detecting the levels of operating system complications and the quantification of the survivin gene at a lower level, it can be said that the microgap device with the dielectric surface of multiple connection triangles modified with MXene is useful [201].

Table 4. SPR nanobiosensor-based SMQDs to detect cancer biomarkers.

MXene-Based Biosensors	Target Biomarker	LOD	Diagnosis Method	Ref.
ssRNA, MoS_2, AuNPs, Ti_3C_2, GCE, and BSA	miRNA-182	0.43 fM	Electrochemical/DPV	[202]
PMo12/PPy@$Ti_3C_2T_x$/Apt/AE	OPN	0.98 fg m·L^{-1}	Electrochemical/EIS	[203]
CD63 aptamer that has been tagged with Cy3 and Ti_3C_2 MXenes	Exosomes	1.4×10^3 particles m·L^{-1}	Ratiometric fluorescence resonance	[204]
M B, DNA, H T, HP 1, AuNPs, Ti_3C_2, $BiVO_4$, and GCE	$VEGF_{165}$	3.3 f.M	Photoelectro-chemical	[205]
MXene/IDE HRP-Au-Ab2-PSA-Ab1	PSA	0.031 ng m·L^{-1}	Electrochemical/EIS,CV	[35]
N-$Ti_3C_2T_x$-MXene	CEA	1.7 pg m·L^{-1}	SPR	[206]

6.2. Detecting an Exosome as a Supply of Biomarkers of Cancer by Applying 2D SMQDs

Exosome signals transmit in intercellular communications. Additionally, exosomes have the ability to deliver cargo that affects nearby cells and can form pre-metastatic cavities. Exosomes are responsible for the initiation, development, and progression of local malignancies, as well as the formation of metastatic lesions. Exosomes themselves are a popular choice for cancer diagnosis since tumor cells produce more exosomes than normal/healthy cells due to their significantly increased cellular activity [168]. Due to its quick response time, minimal background signal, and high sensitivity, electrochemiluminescence (ECL) has been extensively employed for biomarker research [169]. Because 2D Ti_3C_2 MXenes nanosheets have a great conductivity, a large surface area, and catalytic characteristics, Zhang et al. studied the possibility of using them as ECL nanoprobes to create a sensitive ECL biosensor to detect exosomes. The results showed that the limit is about 124 μL^{-1} particles, which is more than 100-fold lower than that of the current enzyme-linked immunosorbent assay (ELISA) method (Figure 9) [207].

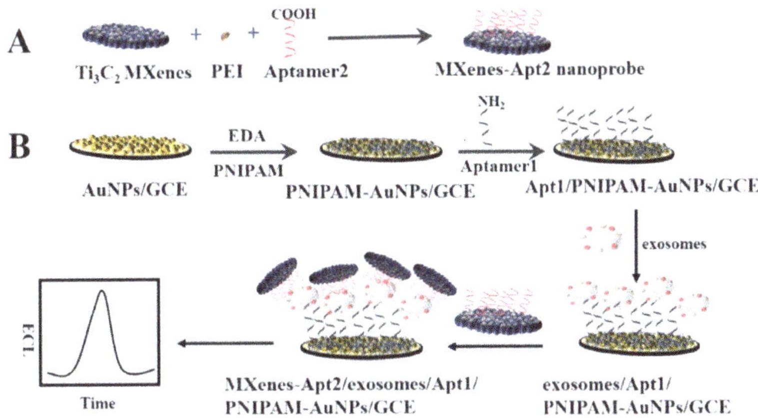

Figure 9. ECL biosensor principle for the signal amplification strategy of exosome activity detection. (Reprinted with permission [207]. © 2018 Elsevier B.V. All rights are reserved.)

6.3. Detection of Carcinoembryonic Antigen

Carcino-embryonic antigen (CEA) can be one of the cancer markers considered for cancer diagnosis [161]. Ti_3C_2 SMQDs that are monolayer- or multilayer-coated with an acceptor amino group for covalently immobilizing the carcinoembryonic monoclonal antibody for cancer biomarker detection are the first SMQDs-based CEA detectors. SPR technology, which enables chemical molecules as well as refractive index measurements, has been introduced as a technology to increase the sensitivity of the CEA biosensor [185]. However, due to a lack of quick and accurate diagnostic techniques, it is difficult to identify CEA-related tumors at an early stage, which is essential for effective treatment. SPR biosensor technologies can thus be crucial in reaching this objective [171,172]. While ELISA is traditionally used in scientific settings, SPR biosensors based on SMQDs provide a label-free and real-time detection approach. [208]. SPR biosensors according to SMQDs using genetics [209,210] present in tissue have been used for other ailments that take place at high incidence levels [211]. Liu et al. evaluated the detecting growth differentiation factor-11(GDF11) anti-body using an SPR fiber biosensor based on Ti_3C_2 MXenes. They found that the detection of GDF11 after activation with the GDF11 antibody is performed by the fiber SPR sensor, and the sensitivity of the fiber SPR sensor increases to 4804.64 nm/RIU. Likewise, the limit of detection in comparison with the single-molecule ELISA procedure could reach 0.577 pg/L, which is 100-fold lower in comparison with that of the single-molecule ELISA procedure [212]. Altintas et al. studied carcinoembryonic antigen cancer biomarker detection. The results showed that a detection limit of 3 ng/mL CEA was achieved with sustainable detections with a correlation co-efficiency of 1 as well as 0.99 for rabbit anti-mouse (RAM) recording assays [213]. Wu et al. used an SPR biosensor based on 2D transition metal carbide MXene for ultrasensitive CEA detection. They also found that Ti_3C_2 MXene, as a novel class of 2D transition metal carbides, provides a large compatible hydrophilic surface that is ideal for SPR biosensing. Based on the results, the dynamic range and detection limit for determining CEA is from 2×10^{-16} to 2×10^{-8} M and 0.07 fM, respectively. Additionally, the results showed that this biosensor approach shows good reproducibility and high specificity for CEA in real serum samples, which provides a promising procedure for evaluating CEA in human serum for the early detection and monitoring of cancer [161].

7. Conclusions and Futures Outlooks

In the development of SPR biosensors to achieve high sensitivity and selectivity, numerous nanomaterials have been synthesized and used due to their inherent properties such as extreme conductivity and plasmonic nature. SMQDs attract a lot of attention in developing SPR biosensors due to their exceptional properties. Accordingly, their potential for biosensor development has been widely investigated since the first reports on SMQDs. Moreover, recent research on the improvement of SPR biosensors based on SMQDs has confirmed that, among various nanomaterials, SMQDs may be the best candidates for the development of various types of biosensors, including fluorescent, optical, and electrochemical biosensors. Moreover, there is still much room for progress in the development of SPR biosensor systems and other next-generation biosensors. These views are supported by recent research on the properties of SMQDs and SPR biosensors based on them. SMQDs improve the performance of SPR biosensors and help in the development of SPR biosensors, as explained in this article. The practical application of these SPR biosensors based on SMQDs faces several challenges, such as the reproducibility of these SPR biosensors and their potential for mass production. However, the commercialization of various SMQDs and the development of SPR biosensors based on SMQDs will depend on ongoing research to develop new synthesis techniques or new SMQD architectures. In addition, it is expected that this ongoing research will lead to a more efficient method of combining SMQDs with other nanomaterials to improve the intrinsic properties of new SMQDs that will be developed in the near future.

Author Contributions: S.M.M. and A.G. developed the idea and structure of the review article. V.R., S.A.H. and M.Y.K. wrote the manuscript and collected the materials from databases. C.W.L. and A.G. revised and improved the manuscript. A.G. and W.-H.C. supervised the manuscript. All authors have read and agreed to the published version of the manuscript.

Funding: This work is sponsored by the Ministry of Science and Technology, Taiwan (grant number: MOST 110-2628-E-011-003, MOST 109-2923-E-011-003-MY, MOST 111-NU-E-011-001-NU).

Institutional Review Board Statement: Not applicable.

Informed Consent Statement: Not applicable.

Data Availability Statement: All data generated or analyzed during this study are included in this published article.

Acknowledgments: This research work was financially supported by the Fundamental Research Grant 529 Scheme FRGS/1/2020/TK0/UM/02/8 (No. FP023-2020) and the Global Collaborative Program—SATU 530 Joint Research Scheme (No. ST004-2021).

Conflicts of Interest: The authors declare no conflict of interest.

References

1. Mehrotra, P. Biosensors and their applications—A review. *J. Oral Biol. Craniofacial Res.* **2016**, *6*, 153–159. [CrossRef] [PubMed]
2. Koyun, A.; Ahlatcolu, E.; Koca, Y.; Kara, S. Biosensors and their principles. *A Roadmap Biomed. Eng. Milest.* **2012**, 117–142. [CrossRef]
3. Pandey, S. *Advance Nanomaterials for Biosensors*; MDPI: Basel, Switzerland, 2022; Volume 12, p. 219.
4. Fan, X.; White, I.M.; Shopova, S.I.; Zhu, H.; Suter, J.D.; Sun, Y. Sensitive optical biosensors for unlabeled targets: A review. *Anal. Chim. Acta* **2008**, *620*, 8–26. [CrossRef] [PubMed]
5. Chen, Y.; Ming, H. Review of surface plasmon resonance and localized surface plasmon resonance sensor. *Photonic Sens.* **2012**, *2*, 37–49. [CrossRef]
6. Cao, J.; Sun, T.; Grattan, K.T. Gold nanorod-based localized surface plasmon resonance biosensors: A review. *Sens. Actuators B Chem.* **2014**, *195*, 332–351. [CrossRef]
7. Kirsch, J.; Siltanen, C.; Zhou, Q.; Revzin, A.; Simonian, A. Biosensor technology: Recent advances in threat agent detection and medicine. *Chem. Soc. Rev.* **2013**, *42*, 8733–8768. [CrossRef]
8. Park, Y.M.; Ahn, J.; Choi, Y.S.; Jeong, J.-M.; Lee, S.J.; Lee, J.J.; Choi, B.G.; Lee, K.G. Flexible nanopillar-based immunoelectrochemical biosensor for noninvasive detection of Amyloid beta. *Nano Converg.* **2020**, *7*, 29. [CrossRef]
9. Alipour, A.; Kalashgarani, M.Y. Nano Protein and Peptides for Drug Delivery and Anticancer Agents. *Adv. Appl. NanoBio-Technol.* **2022**, *3*, 60–64.
10. Roointan, A.; Mir, T.A.; Wani, S.I.; Hussain, K.K.; Ahmed, B.; Abrahim, S.; Savardashtaki, A.; Gandomani, G.; Gandomani, M.; Chinnappan, R. Early detection of lung cancer biomarkers through biosensor technology: A review. *J. Pharm. Biomed. Anal.* **2019**, *164*, 93–103. [CrossRef]

11. Mousavi, S.M.; Hashemi, S.A.; Kalashgrani, M.Y.; Omidifar, N.; Bahrani, S.; Vijayakameswara Rao, N.; Babapoor, A.; Gholami, A.; Chiang, W.-H. Bioactive Graphene Quantum Dots Based Polymer Composite for Biomedical Applications. *Polymers* **2022**, *14*, 617. [CrossRef]
12. Nunna, B.B.; Mandal, D.; Lee, J.U.; Singh, H.; Zhuang, S.; Misra, D.; Bhuyian, M.N.U.; Lee, E.S. Detection of cancer antigens (CA-125) using gold nano particles on interdigitated electrode-based microfluidic biosensor. *Nano Converg.* **2019**, *6*, 3. [CrossRef] [PubMed]
13. Liu, Y.; Yu, D.; Zeng, C.; Miao, Z.; Dai, L. Biocompatible graphene oxide-based glucose biosensors. *Langmuir* **2010**, *26*, 6158–6160. [CrossRef] [PubMed]
14. Kazemi, K.; Ghahramani, Y.; Kalashgrani, M.Y. Nano biofilms: An emerging biotechnology applications. *Adv. Appl. NanoBio-Technol.* **2022**, *3*, 8–15.
15. Mousavi, S.M.; Hashemi, S.A.; Gholami, A.; Kalashgrani, M.Y.; Vijayakameswara Rao, N.; Omidifar, N.; Hsiao, W.W.-W.; Lai, C.W.; Chiang, W.-H. Plasma-Enabled Smart Nanoexosome Platform as Emerging Immunopathogenesis for Clinical Viral Infection. *Pharmaceutics* **2022**, *14*, 1054. [CrossRef]
16. Mohammadniaei, M.; Yoon, J.; Lee, T.; Choi, J.-W. Spectroelectrochemical detection of microRNA-155 based on functional RNA immobilization onto ITO/GNP nanopattern. *J. Biotechnol.* **2018**, *274*, 40–46. [CrossRef]
17. Elahi, N.; Kamali, M.; Baghersad, M.H.; Amini, B. A fluorescence Nano-biosensors immobilization on Iron (MNPs) and gold (AuNPs) nanoparticles for detection of *Shigella* spp. *Mater. Sci. Eng. C* **2019**, *105*, 110113. [CrossRef]
18. Mousavi, S.M.; Hashemi, S.A.; Kalashgrani, M.Y.; Gholami, A.; Omidifar, N.; Babapoor, A.; Vijayakameswara Rao, N.; Chiang, W.-H. Recent Advances in Plasma-Engineered Polymers for Biomarker-Based Viral Detection and Highly Multiplexed Analysis. *Biosensors* **2022**, *12*, 286. [CrossRef]
19. Kalashgrani, M.Y.; Javanmardi, N. Multifunctional Gold nanoparticle: As novel agents for cancer treatment. *Adv. Appl. NanoBio-Technol.* **2022**, *3*, 43–48.
20. Hussein, M.A.; El-Said, W.A.; Abu-Zied, B.M.; Choi, J.-W. Nanosheet composed of gold nanoparticle/graphene/epoxy resin based on ultrasonic fabrication for flexible dopamine biosensor using surface-enhanced Raman spectroscopy. *Nano Converg.* **2020**, *7*, 15. [CrossRef]
21. Wee, Y.; Park, S.; Kwon, Y.H.; Ju, Y.; Yeon, K.-M.; Kim, J. Tyrosinase-immobilized CNT based biosensor for highly-sensitive detection of phenolic compounds. *Biosens. Bioelectron.* **2019**, *132*, 279–285. [CrossRef]
22. Mousavi, S.M.; Hashemi, S.A.; Rahmanian, V.; Kalashgrani, M.Y.; Gholami, A.; Omidifar, N.; Chiang, W.-H. Highly Sensitive Flexible SERS-Based Sensing Platform for Detection of COVID-19. *Biosensors* **2022**, *12*, 466. [CrossRef] [PubMed]
23. Kalashgrani, M.Y.; Nejad, F.F.; Rahmanian, V. Carbon Quantum Dots Platforms: As nano therapeutic for Biomedical Applications. *Adv. Appl. NanoBio-Technol.* **2022**, *3*, 38–42.
24. Wang, X.; Liu, A.; Xing, Y.; Duan, H.; Xu, W.; Zhou, Q.; Wu, H.; Chen, C.; Chen, B. Three-dimensional graphene biointerface with extremely high sensitivity to single cancer cell monitoring. *Biosens. Bioelectron.* **2018**, *105*, 22–28. [CrossRef] [PubMed]
25. Mousavi, S.M.; Hashemi, S.A.; Yari Kalashgrani, M.; Kurniawan, D.; Gholami, A.; Rahmanian, V.; Omidifar, N.; Chiang, W.-H. Recent Advances in Inflammatory Diagnosis with Graphene Quantum Dots Enhanced SERS Detection. *Biosensors* **2022**, *12*, 461. [CrossRef] [PubMed]
26. Kalashgrani, M.Y.; Harzand, F.V.; Javanmardi, N.; Nejad, F.F.; Rahmanian, V. Recent Advances in Multifunctional magnetic nano platform for Biomedical Applications: A mini review. *Adv. Appl. NanoBio-Technol.* **2022**, *3*, 31–37.
27. Zhang, H.; Cheng, H.-M.; Ye, P. 2D nanomaterials: Beyond graphene and transition metal dichalcogenides. *Chem. Soc. Rev.* **2018**, *47*, 6009–6012. [CrossRef]
28. Abootalebi, S.N.; Mousavi, S.M.; Hashemi, S.A.; Shorafa, E.; Omidifar, N.; Gholami, A. Antibacterial effects of green-synthesized silver nanoparticles using Ferula asafoetida against Acinetobacter baumannii isolated from the hospital environment and assessment of their cytotoxicity on the human cell lines. *J. Nanomater.* **2021**, *2021*, 6676555. [CrossRef]
29. Sreenilayam, S.P.; Ahad, I.U.; Nicolosi, V.; Brabazon, D. Mxene materials based printed flexible devices for healthcare, biomedical and energy storage applications. *Mater. Today* **2021**, *43*, 99–131. [CrossRef]
30. Hantanasirisakul, K.; Zhao, M.Q.; Urbankowski, P.; Halim, J.; Anasori, B.; Kota, S.; Ren, C.E.; Barsoum, M.W.; Gogotsi, Y. Fabrication of Ti3C2Tx MXene transparent thin films with tunable optoelectronic properties. *Adv. Electron. Mater.* **2016**, *2*, 1600050. [CrossRef]
31. Jiang, X.; Kuklin, A.V.; Baev, A.; Ge, Y.; Ågren, H.; Zhang, H.; Prasad, P.N. Two-dimensional MXenes: From morphological to optical, electric, and magnetic properties and applications. *Phys. Rep.* **2020**, *848*, 1–58. [CrossRef]
32. Kalambate, P.K.; Gadhari, N.S.; Li, X.; Rao, Z.; Navale, S.T.; Shen, Y.; Patil, V.R.; Huang, Y. Recent advances in MXene–based electrochemical sensors and biosensors. *TrAC Trends Anal. Chem.* **2019**, *120*, 115643. [CrossRef]
33. Sinha, A.; Zhao, H.; Huang, Y.; Lu, X.; Chen, J.; Jain, R. MXene: An emerging material for sensing and biosensing. *TrAC Trends Anal. Chem.* **2018**, *105*, 424–435. [CrossRef]
34. Tech, J.E.T. Investigating the activity of antioxidants activities content in Apiaceae and to study antimicrobial and insecticidal activity of antioxidant by using SPME Fiber assembly carboxen/polydimethylsiloxane (CAR/PDMS). *J. Environ. Treat. Tech.* **2020**, *8*, 214–224.
35. Chen, J.; Tong, P.; Huang, L.; Yu, Z.; Tang, D. Ti3C2 MXene nanosheet-based capacitance immunoassay with tyramine-enzyme repeats to detect prostate-specific antigen on interdigitated micro-comb electrode. *Electrochim. Acta* **2019**, *319*, 375–381. [CrossRef]

36. Chen, X.; Sun, X.; Xu, W.; Pan, G.; Zhou, D.; Zhu, J.; Wang, H.; Bai, X.; Dong, B.; Song, H. Ratiometric photoluminescence sensing based on Ti 3 C 2 MXene quantum dots as an intracellular pH sensor. *Nanoscale* **2018**, *10*, 1111–1118. [CrossRef]
37. Sarycheva, A.; Makaryan, T.; Maleski, K.; Satheeshkumar, E.; Melikyan, A.; Minassian, H.; Yoshimura, M.; Gogotsi, Y. Two-dimensional titanium carbide (MXene) as surface-enhanced Raman scattering substrate. *J. Phys. Chem. C* **2017**, *121*, 19983–19988. [CrossRef]
38. Mousavi, S.M.; Hashemi, S.A.; Parvin, N.; Gholami, A.; Ramakrishna, S.; Omidifar, N.; Moghadami, M.; Chiang, W.-H.; Mazraedoost, S. Recent biotechnological approaches for treatment of novel COVID-19: From bench to clinical trial. *Drug Metab. Rev.* **2021**, *53*, 141–170. [CrossRef]
39. Lin, H.; Chen, Y.; Shi, J. Insights into 2D MXenes for versatile biomedical applications: Current advances and challenges ahead. *Adv. Sci.* **2018**, *5*, 1800518. [CrossRef]
40. Huang, K.; Li, Z.; Lin, J.; Han, G.; Huang, P. Two-dimensional transition metal carbides and nitrides (MXenes) for biomedical applications. *Chem. Soc. Rev.* **2018**, *47*, 5109–5124. [CrossRef]
41. Dai, C.; Lin, H.; Xu, G.; Liu, Z.; Wu, R.; Chen, Y. Biocompatible 2D titanium carbide (MXenes) composite nanosheets for pH-responsive MRI-guided tumor hyperthermia. *Chem. Mater.* **2017**, *29*, 8637–8652. [CrossRef]
42. Ahmadi, S.; Fazilati, M.; Nazem, H.; Mousavi, S.M. Green synthesis of magnetic nanoparticles using Satureja hortensis essential oil toward superior antibacterial/fungal and anticancer performance. *BioMed Res. Int.* **2021**, *2021*, 8822645. [CrossRef] [PubMed]
43. Geldhauser, T.; Ikegaya, S.; Kolloch, A.; Murazawa, N.; Ueno, K.; Boneberg, J.; Leiderer, P.; Scheer, E.; Misawa, H. Visualization of near-field enhancements of gold triangles by nonlinear photopolymerization. *Plasmonics* **2011**, *6*, 207–212. [CrossRef]
44. Galarreta, B.C.; Harté, E.; Marquestaut, N.; Norton, P.R.; Lagugné-Labarthet, F. Plasmonic properties of Fischer's patterns: Polarization effects. *Phys. Chem. Chem. Phys.* **2010**, *12*, 6810–6816. [CrossRef] [PubMed]
45. Hashemi, S.A.; Mousavi, S.M.; Naderi, H.R.; Bahrani, S.; Arjmand, M.; Hagfeldt, A.; Chiang, W.-H.; Ramakrishna, S. Reinforced polypyrrole with 2D graphene flakes decorated with interconnected nickel-tungsten metal oxide complex toward superiorly stable supercapacitor. *Chem. Eng. J.* **2021**, *418*, 129396. [CrossRef]
46. Krutyakov, Y.A.; Kudrinsky, A.; Olenin, A.Y.; Lisichkin, G. Synthesis of highly stable silver colloids stabilized with water soluble sulfonated polyaniline. *Appl. Surf. Sci.* **2010**, *256*, 7037–7042. [CrossRef]
47. Hashemi, S.A.; Mousavi, S.M.; Faghihi, R.; Arjmand, M.; Rahsepar, M.; Bahrani, S.; Ramakrishna, S.; Lai, C.W. Superior X-ray radiation shielding effectiveness of biocompatible polyaniline reinforced with hybrid graphene oxide-iron tungsten nitride flakes. *Polymers* **2020**, *12*, 1407. [CrossRef]
48. Kelly, K.L.; Coronado, E.; Zhao, L.L.; Schatz, G.C. The Optical Properties of Metal Nanoparticles: The Influence of Size, Shape, and Dielectric Environment. *J. Phys. Chem. B* **2003**, *107*, 668–677. [CrossRef]
49. Kamal Eddin, F.B.; Fen, Y.W. The principle of nanomaterials based surface plasmon resonance biosensors and its potential for dopamine detection. *Molecules* **2020**, *25*, 2769. [CrossRef]
50. Schasfoort, R.B. *Handbook of Surface Plasmon Resonance*; Royal Society of Chemistry: London, UK, 2017.
51. Sun, L.; Chen, P.; Lin, L. Enhanced Molecular Spectroscopy via Localized Surface Plasmon Resonance. In *Applications of Molecular Spectroscopy to Current Research in the Chemical and Biological Sciences*; IntechOpen: London, UK, 2016.
52. Homola, J.; Yee, S.S.; Gauglitz, G. Surface plasmon resonance sensors. *Sens. Actuators B Chem.* **1999**, *54*, 3–15. [CrossRef]
53. Daghestani, H.N.; Day, B.W. Theory and applications of surface plasmon resonance, resonant mirror, resonant waveguide grating, and dual polarization interferometry biosensors. *Sensors* **2010**, *10*, 9630–9646. [CrossRef]
54. Homola, J. Electromagnetic theory of surface plasmons. In *Surface Plasmon Resonance Based Sensors*; Springer: Berlin/Heidelberg, Germany, 2006; pp. 3–44.
55. Mousavi, S.M.; Low, F.W.; Hashemi, S.A.; Lai, C.W.; Ghasemi, Y.; Soroshnia, S.; Savardashtaki, A.; Babapoor, A.; Pynadathu Rumjit, N.; Goh, S.M. Development of graphene based nanocomposites towards medical and biological applications. *Artif. Cells Nanomed. Biotechnol.* **2020**, *48*, 1189–1205. [CrossRef] [PubMed]
56. Yao, Y.; Yi, B.; Xiao, J.; Li, Z. Surface plasmon resonance biosensors and its application. In Proceedings of the 2007 1st International Conference on Bioinformatics and Biomedical Engineering, Wuhan, China, 6–8 July 2007; pp. 1043–1046.
57. Zeng, Y.; Hu, R.; Wang, L.; Gu, D.; He, J.; Wu, S.-Y.; Ho, H.-P.; Li, X.; Qu, J.; Gao, B.Z. Recent advances in surface plasmon resonance imaging: Detection speed, sensitivity, and portability. *Nanophotonics* **2017**, *6*, 1017–1030. [CrossRef]
58. Martín-Becerra, D.; Armelles, G.; González, M.; García-Martín, A. Plasmonic and magnetoplasmonic interferometry for sensing. *New J. Phys.* **2013**, *15*, 085021. [CrossRef]
59. Ahmadi, S.; Fazilati, M.; Mousavi, S.M.; Nazem, H. Anti-bacterial/fungal and anti-cancer performance of green synthesized Ag nanoparticles using summer savory extract. *J. Exp. Nanosci.* **2020**, *15*, 363–380. [CrossRef]
60. Skorobogatiy, M.; Kabashin, A.V. Photon crystal waveguide-based surface plasmon resonance biosensor. *Appl. Phys. Lett.* **2006**, *89*, 143518. [CrossRef]
61. Arasu, P.; Al-Qazwini, Y.; Onn, B.I.; Noor, A. Fiber Bragg grating based surface plasmon resonance sensor utilizing FDTD for alcohol detection applications. In Proceedings of the 2012 IEEE 3rd International Conference on Photonics, Pulau Pinang, Malaysia, 1–3 October 2012; pp. 93–97.
62. Mousavi, S.; Esmaeili, H.; Arjmand, O.; Karimi, S.; Hashemi, S. Biodegradation study of nanocomposites of phenol novolac epoxy/unsaturated polyester resin/egg shell nanoparticles using natural polymers. *J. Mater.* **2015**, *2015*, 131957. [CrossRef]

63. Hoa, X.D.; Kirk, A.; Tabrizian, M. Towards integrated and sensitive surface plasmon resonance biosensors: A review of recent progress. *Biosens. Bioelectron.* **2007**, *23*, 151–160. [CrossRef]
64. Mousavi, S.; Arjmand, O.; Hashemi, S.; Banaei, N. Modification of the epoxy resin mechanical and thermal properties with silicon acrylate and montmorillonite nanoparticles. *Polym. Renew. Resour.* **2016**, *7*, 101–113. [CrossRef]
65. Samudrala, P.K. Alumina Waveguide Characterization and SPARROW Biosensor Modeling. Graduate Theses, Dissertations, and Problem Reports, 1786. Master's Thesis, West Virginia University, Morgantown, WV, USA, 2006.
66. Nguyen, H.H.; Park, J.; Kang, S.; Kim, M. Surface plasmon resonance: A versatile technique for biosensor applications. *Sensors* **2015**, *15*, 10481–10510. [CrossRef]
67. Mousavi, S.; Aghili, A.; Hashemi, S.; Goudarzian, N.; Bakhoda, Z.; Baseri, S. Improved morphology and properties of nanocomposites, linear low density polyethylene, ethylene-co-vinyl acetate and nano clay particles by electron beam. *Polym. Renew. Resour.* **2016**, *7*, 135–153. [CrossRef]
68. Mousavi, S.M.; Hashemi, S.A.; Zarei, M.; Gholami, A.; Lai, C.W.; Chiang, W.H.; Omidifar, N.; Bahrani, S.; Mazraedoost, S. Recent progress in chemical composition, production, and pharmaceutical effects of kombucha beverage: A complementary and alternative medicine. *Evid. -Based Complementary Altern. Med.* **2020**, *2020*, 4397543. [CrossRef] [PubMed]
69. Rath, P.P.; Anand, G.; Agarwal, S. Surface Plasmon Resonance Analysis of the Protein-protein Binding Specificity Using Autolab ESPIRIT. *Bio-protocol* **2020**, *10*, e3519. [CrossRef] [PubMed]
70. Wang, X.; Li, Y.; Wang, H.; Fu, Q.; Peng, J.; Wang, Y.; Du, J.; Zhou, Y.; Zhan, L. Gold nanorod-based localized surface plasmon resonance biosensor for sensitive detection of hepatitis B virus in buffer, blood serum and plasma. *Biosens. Bioelectron.* **2010**, *26*, 404–410. [CrossRef] [PubMed]
71. Sharma, P.K.; Kumar, J.S.; Singh, V.V.; Biswas, U.; Sarkar, S.S.; Alam, S.I.; Dash, P.K.; Boopathi, M.; Ganesan, K.; Jain, R. Surface plasmon resonance sensing of Ebola virus: A biological threat. *Anal. Bioanal. Chem.* **2020**, *412*, 4101–4112. [CrossRef]
72. Gnedenko, O.V.; Mezentsev, Y.V.; Molnar, A.A.; Lisitsa, A.V.; Ivanov, A.S.; Archakov, A.I. Highly sensitive detection of human cardiac myoglobin using a reverse sandwich immunoassay with a gold nanoparticle-enhanced surface plasmon resonance biosensor. *Anal. Chim. Acta* **2013**, *759*, 105–109. [CrossRef] [PubMed]
73. Ke, H.; Du, X.; Wang, L.; Wang, X.; Zhu, J.; Gao, Y.; Peng, B.; Hao, H.; Cai, N. Detection of morphine in urine based on a surface plasmon resonance imaging immunoassay. *Anal. Methods* **2020**, *12*, 3038–3044. [CrossRef]
74. Fitzpatrick, B.; O'Kennedy, R. The development and application of a surface plasmon resonance-based inhibition immunoassay for the determination of warfarin in plasma ultrafiltrate. *J. Immunol. Methods* **2004**, *291*, 11–25. [CrossRef]
75. Yang, C.-Y.; Brooks, E.; Li, Y.; Denny, P.; Ho, C.-M.; Qi, F.; Shi, W.; Wolinsky, L.; Wu, B.; Wong, D.T. Detection of picomolar levels of interleukin-8 in human saliva by SPR. *Lab A Chip* **2005**, *5*, 1017–1023. [CrossRef]
76. Ertürk, G.; Özen, H.; Tümer, M.A.; Mattiasson, B.; Denizli, A. Microcontact imprinting based surface plasmon resonance (SPR) biosensor for real-time and ultrasensitive detection of prostate specific antigen (PSA) from clinical samples. *Sens. Actuators B Chem.* **2016**, *224*, 823–832. [CrossRef]
77. Naguib, M.; Kurtoglu, M.; Presser, V.; Lu, J.; Niu, J.; Heon, M.; Hultman, L.; Gogotsi, Y.; Barsoum, M.W. Two-dimensional nanocrystals produced by exfoliation of Ti_3AlC_2. *Adv. Mater.* **2011**, *23*, 4248–4253. [CrossRef]
78. Naguib, M.; Gogotsi, Y. Synthesis of two-dimensional materials by selective extraction. *Acc. Chem. Res.* **2015**, *48*, 128–135. [CrossRef] [PubMed]
79. Halim, J. *An X-ray Photoelectron Spectroscopy Study of Multilayered Transition Metal Carbides (MXenes)*; Drexel University: Philadelphia, PA, USA, 2016.
80. Amani, A.M.; Hashemi, S.A.; Mousavi, S.M.; Abrishamifar, S.M.; Vojood, A. Electric field induced alignment of carbon nanotubes: Methodology and outcomes. In *Carbon Nanotubes-Recent Progress*; IntechOpen: London, UK, 2017.
81. Zamhuri, A.; Lim, G.P.; Ma, N.L.; Tee, K.S.; Soon, C.F. MXene in the lens of biomedical engineering: Synthesis, applications and future outlook. *Biomed. Eng. Online* **2021**, *20*, 33. [CrossRef] [PubMed]
82. Anasori, B.; Lukatskaya, M.R.; Gogotsi, Y. 2D metal carbides and nitrides (MXenes) for energy storage. *Nat. Rev. Mater.* **2017**, *2*, 16098. [CrossRef]
83. Hashemi, S.A.; Mousavi, S.M.; Arjmand, M.; Yan, N.; Sundararaj, U. Electrified single-walled carbon nanotube/epoxy nanocomposite via vacuum shock technique: Effect of alignment on electrical conductivity and electromagnetic interference shielding. *Polym. Compos.* **2018**, *39*, E1139–E1148. [CrossRef]
84. Xu, S.; Liu, W.; Liu, X.; Kuang, X.; Wang, X. A MXene based all-solid-state microsupercapacitor with 3D interdigital electrode. In Proceedings of the 2017 19th International Conference on Solid-State Sensors, Actuators and Microsystems (TRANSDUCERS), Kaohsiung, Taiwan, 18–22 June 2017; pp. 706–709.
85. Mousavi, S.M.; Hashemi, S.A.; Arjmand, M.; Amani, A.M.; Sharif, F.; Jahandideh, S. Octadecyl amine functionalized Graphene oxide towards hydrophobic chemical resistant epoxy Nanocomposites. *ChemistrySelect* **2018**, *3*, 7200–7207. [CrossRef]
86. Yang, S.; Zhang, P.; Wang, F.; Ricciardulli, A.G.; Lohe, M.R.; Blom, P.W.; Feng, X. Fluoride-free synthesis of two-dimensional titanium carbide (MXene) using a binary aqueous system. *Angew. Chem.* **2018**, *130*, 15717–15721. [CrossRef]
87. Sun, W.; Shah, S.; Chen, Y.; Tan, Z.; Gao, H.; Habib, T.; Radovic, M.; Green, M. Electrochemical etching of Ti_2AlC to Ti_2CT_x (MXene) in low-concentration hydrochloric acid solution. *J. Mater. Chem. A* **2017**, *5*, 21663–21668. [CrossRef]

88. Mousavi, S.M.; Hashemi, S.A.; Ramakrishna, S.; Esmaeili, H.; Bahrani, S.; Koosha, M.; Babapoor, A. Green synthesis of supermagnetic Fe_3O_4–MgO nanoparticles via Nutmeg essential oil toward superior anti-bacterial and anti-fungal performance. *J. Drug Deliv. Sci. Technol.* **2019**, *54*, 101352. [CrossRef]
89. Nicolosi, V.; Chhowalla, M.; Kanatzidis, M.G.; Strano, M.S.; Coleman, J.N. Liquid exfoliation of layered materials. *Science* **2013**, *340*, 1226419. [CrossRef]
90. Mousavi, S.M.; Zarei, M.; Hashemi, S.A.; Ramakrishna, S.; Chiang, W.-H.; Lai, C.W.; Gholami, A.; Omidifar, N.; Shokripour, M. Asymmetric membranes: A potential scaffold for wound healing applications. *Symmetry* **2020**, *12*, 1100. [CrossRef]
91. Abdelmalak, M.N. *MXenes: A New Family of Two-Dimensional Materials and Its Application as Electrodes for Li-ion Batteries*; Drexel University: Philadelphia, PA, USA, 2014.
92. Hashemi, S.A.; Mousavi, S.M. Effect of bubble based degradation on the physical properties of Single Wall Carbon Nanotube/Epoxy Resin composite and new approach in bubbles reduction. *Compos. Part A Appl. Sci. Manuf.* **2016**, *90*, 457–469. [CrossRef]
93. Naguib, M.; Mashtalir, O.; Carle, J.; Presser, V.; Lu, J.; Hultman, L.; Gogotsi, Y.; Barsoum, M.W. Two-dimensional transition metal carbides. *ACS Nano* **2012**, *6*, 1322–1331. [CrossRef] [PubMed]
94. Mousavi, S.M.; Hashemi, S.A.; Jahandideh, S.; Baseri, S.; Zarei, M.; Azadi, S. Modification of phenol novolac epoxy resin and unsaturated polyester using sasobit and silica nanoparticles. *Polym. Renew. Resour.* **2017**, *8*, 117–132. [CrossRef]
95. Verger, L.; Xu, C.; Natu, V.; Cheng, H.-M.; Ren, W.; Barsoum, M.W. Overview of the synthesis of MXenes and other ultrathin 2D transition metal carbides and nitrides. *Curr. Opin. Solid State Mater. Sci.* **2019**, *23*, 149–163. [CrossRef]
96. Mousavi, S.M.; Hashemi, S.A.; Amani, A.M.; Saed, H.; Jahandideh, S.; Mojoudi, F. Polyethylene terephthalate/acryl butadiene styrene copolymer incorporated with oak shell, potassium sorbate and egg shell nanoparticles for food packaging applications: Control of bacteria growth, physical and mechanical properties. *Polym. Renew. Resour.* **2017**, *8*, 177–196. [CrossRef]
97. Barsoum, M.W. *MAX Phases: Properties of Machinable Ternary Carbides and Nitrides*; John Wiley & Sons: Hoboken, NJ, USA, 2013.
98. Mousavi, S.M.; Hashemi, S.A.; Salahi, S.; Hosseini, M.; Amani, A.M.; Babapoor, A. *Development of Clay Nanoparticles toward Bio and Medical Applications*; IntechOpen: London, UK, 2018.
99. Shi, Z.; Khaledialidusti, R.; Malaki, M.; Zhang, H. MXene-based materials for solar cell applications. *Nanomaterials* **2021**, *11*, 3170. [CrossRef]
100. Verger, L.; Natu, V.; Carey, M.; Barsoum, M.W. MXenes: An introduction of their synthesis, select properties, and applications. *Trends Chem.* **2019**, *1*, 656–669. [CrossRef]
101. Mousavi, S.M.; Hashemi, S.A.; Esmaeili, H.; Amani, A.M.; Mojoudi, F. Synthesis of Fe_3O_4 nanoparticles modified by oak shell for treatment of wastewater containing Ni (II). *Acta Chim. Slov.* **2018**, *65*, 750–756. [CrossRef]
102. Hope, M.A.; Forse, A.C.; Griffith, K.J.; Lukatskaya, M.R.; Ghidiu, M.; Gogotsi, Y.; Grey, C.P. NMR reveals the surface functionalisation of Ti_3C_2 MXene. *Phys. Chem. Chem. Phys.* **2016**, *18*, 5099–5102. [CrossRef]
103. Mousavi, M.; Hashemi, A.; Arjmand, O.; Amani, A.M.; Babapoor, A.; Fateh, M.A.; Fateh, H.; Mojoudi, F.; Esmaeili, H.; Jahandideh, S. Erythrosine adsorption from aqueous solution via decorated graphene oxide with magnetic iron oxide nano particles: Kinetic and equilibrium studies. *Acta Chim. Slov.* **2018**, *65*, 882–894. [CrossRef]
104. Ahmed, B.; Anjum, D.H.; Gogotsi, Y.; Alshareef, H.N. Atomic layer deposition of SnO_2 on MXene for Li-ion battery anodes. *Nano Energy* **2017**, *34*, 249–256. [CrossRef]
105. Mousavi, S.; Zarei, M.; Hashemi, S. Polydopamine for biomedical application and drug delivery system. *Med. Chem.* **2018**, *8*, 218–229. [CrossRef]
106. Berdiyorov, G. Effect of surface functionalization on the electronic transport properties of Ti_3C_2 MXene. *EPL (Europhys. Lett.)* **2015**, *111*, 67002. [CrossRef]
107. Mousavi, S.M.; Hashemi, S.A.; Zarei, M.; Bahrani, S.; Savardashtaki, A.; Esmaeili, H.; Lai, C.W.; Mazraedoost, S.; Abassi, M.; Ramavandi, B. Data on cytotoxic and antibacterial activity of synthesized Fe_3O_4 nanoparticles using Malva sylvestris. *Data Brief* **2020**, *28*, 104929. [CrossRef] [PubMed]
108. Sokol, M.; Natu, V.; Kota, S.; Barsoum, M.W. On the chemical diversity of the MAX phases. *Trends Chem.* **2019**, *1*, 210–223. [CrossRef]
109. Azhdari, R.; Mousavi, S.M.; Hashemi, S.A.; Bahrani, S.; Ramakrishna, S. Decorated graphene with aluminum fumarate metal organic framework as a superior non-toxic agent for efficient removal of Congo Red dye from wastewater. *J. Environ. Chem. Eng.* **2019**, *7*, 103437. [CrossRef]
110. Barsoum, M.W.; El-Raghy, T.; Farber, L.; Amer, M.; Christini, R.; Adams, A. The topotactic transformation of Ti_3SiC_2 into a partially ordered cubic Ti ($C_{0.67}Si_{0.06}$) phase by the diffusion of si into molten cryolite. *J. Electrochem. Soc.* **1999**, *146*, 3919. [CrossRef]
111. Hashemi, S.A.; Mousavi, S.M.; Ramakrishna, S. Effective removal of mercury, arsenic and lead from aqueous media using Polyaniline-Fe3O4-silver diethyldithiocarbamate nanostructures. *J. Clean. Prod.* **2019**, *239*, 118023. [CrossRef]
112. Zhou, J.; Zha, X.; Chen, F.Y.; Ye, Q.; Eklund, P.; Du, S.; Huang, Q. A two-dimensional zirconium carbide by selective etching of Al3C3 from nanolaminated $Zr_3Al_3C_5$. *Angew. Chem. Int. Ed.* **2016**, *55*, 5008–5013. [CrossRef]
113. Ravanshad, R.; Karimi Zadeh, A.; Amani, A.M.; Mousavi, S.M.; Hashemi, S.A.; Savar Dashtaki, A.; Mirzaei, E.; Zare, B. Application of nanoparticles in cancer detection by Raman scattering based techniques. *Nano Rev. Exp.* **2018**, *9*, 1373551. [CrossRef]

114. Halim, J.; Cook, K.M.; Naguib, M.; Eklund, P.; Gogotsi, Y.; Rosen, J.; Barsoum, M.W. X-ray photoelectron spectroscopy of select multi-layered transition metal carbides (MXenes). *Appl. Surf. Sci.* **2016**, *362*, 406–417. [CrossRef]
115. Mousavi, S.M.; Low, F.W.; Hashemi, S.A.; Samsudin, N.A.; Shakeri, M.; Yusoff, Y.; Rahsepar, M.; Lai, C.W.; Babapoor, A.; Soroshnia, S. Development of hydrophobic reduced graphene oxide as a new efficient approach for photochemotherapy. *RSC Adv.* **2020**, *10*, 12851–12863. [CrossRef]
116. Dai, C.; Chen, Y.; Jing, X.; Xiang, L.; Yang, D.; Lin, H.; Liu, Z.; Han, X.; Wu, R. Two-dimensional tantalum carbide (MXenes) composite nanosheets for multiple imaging-guided photothermal tumor ablation. *ACS Nano* **2017**, *11*, 12696–12712. [CrossRef] [PubMed]
117. Lin, H.; Wang, X.; Yu, L.; Chen, Y.; Shi, J. Two-dimensional ultrathin MXene ceramic nanosheets for photothermal conversion. *Nano Lett.* **2017**, *17*, 384–391. [CrossRef] [PubMed]
118. Wang, F.; Yang, C.; Duan, M.; Tang, Y.; Zhu, J. TiO_2 nanoparticle modified organ-like Ti_3C_2 MXene nanocomposite encapsulating hemoglobin for a mediator-free biosensor with excellent performances. *Biosens. Bioelectron.* **2015**, *74*, 1022–1028. [CrossRef] [PubMed]
119. Liu, H.; Duan, C.; Yang, C.; Shen, W.; Wang, F.; Zhu, Z. A novel nitrite biosensor based on the direct electrochemistry of hemoglobin immobilized on MXene-Ti_3C_2. *Sens. Actuators B Chem.* **2015**, *218*, 60–66. [CrossRef]
120. Liu, Y.; Han, Q.; Yang, W.; Gan, X.; Yang, Y.; Xie, K.; Xie, L.; Deng, Y. Two-dimensional MXene/cobalt nanowire heterojunction for controlled drug delivery and chemo-photothermal therapy. *Mater. Sci. Eng. C* **2020**, *116*, 111212. [CrossRef]
121. Xue, Q.; Zhang, H.; Zhu, M.; Pei, Z.; Li, H.; Wang, Z.; Huang, Y.; Huang, Y.; Deng, Q.; Zhou, J. Photoluminescent Ti_3C_2 MXene quantum dots for multicolor cellular imaging. *Adv. Mater.* **2017**, *29*, 1604847. [CrossRef]
122. Shao, J.; Zhang, J.; Jiang, C.; Lin, J.; Huang, P. Biodegradable titanium nitride MXene quantum dots for cancer phototheranostics in NIR-I/II biowindows. *Chem. Eng. J.* **2020**, *400*, 126009. [CrossRef]
123. Yang, G.; Zhao, J.; Yi, S.; Wan, X.; Tang, J. Biodegradable and photostable Nb_2C MXene quantum dots as promising nanofluorophores for metal ions sensing and fluorescence imaging. *Sens. Actuators B Chem.* **2020**, *309*, 127735. [CrossRef]
124. Novoselov, K.S.; Colombo, L.; Gellert, P.R.; Schwab, M.G.; Kim, K. A roadmap for graphene. *Nature* **2012**, *490*, 192–200. [CrossRef]
125. Hashemi, S.A.; Mousavi, S.M.; Bahrani, S.; Ramakrishna, S. Integrated polyaniline with graphene oxide-iron tungsten nitride nanoflakes as ultrasensitive electrochemical sensor for precise detection of 4-nitrophenol within aquatic media. *J. Electroanal. Chem.* **2020**, *873*, 114406. [CrossRef]
126. Rakhi, R.; Nayak, P.; Xia, C.; Alshareef, H.N. Novel amperometric glucose biosensor based on MXene nanocomposite. *Sci. Rep.* **2016**, *6*, 36422. [CrossRef] [PubMed]
127. Jiang, Y.; Zhang, X.; Pei, L.; Yue, S.; Ma, L.; Zhou, L.; Huang, Z.; He, Y.; Gao, J. Silver nanoparticles modified two-dimensional transition metal carbides as nanocarriers to fabricate acetycholinesterase-based electrochemical biosensor. *Chem. Eng. J.* **2018**, *339*, 547–556. [CrossRef]
128. Song, D.; Jiang, X.; Li, Y.; Lu, X.; Luan, S.; Wang, Y.; Li, Y.; Gao, F. Metal-organic frameworks-derived MnO_2/Mn_3O_4 microcuboids with hierarchically ordered nanosheets and Ti_3C_2 MXene/Au NPs composites for electrochemical pesticide detection. *J. Hazard. Mater.* **2019**, *373*, 367–376. [CrossRef] [PubMed]
129. Satheeshkumar, E.; Makaryan, T.; Melikyan, A.; Minassian, H.; Gogotsi, Y.; Yoshimura, M. One-step solution processing of Ag, Au and Pd@ MXene hybrids for SERS. *Sci. Rep.* **2016**, *6*, 32049. [CrossRef] [PubMed]
130. Mohammadniaei, M.; Koyappayil, A.; Sun, Y.; Min, J.; Lee, M.-H. Gold nanoparticle/MXene for multiple and sensitive detection of oncomiRs based on synergetic signal amplification. *Biosens. Bioelectron.* **2020**, *159*, 112208. [CrossRef] [PubMed]
131. Lorencova, L.; Bertok, T.; Filip, J.; Jerigova, M.; Velic, D.; Kasak, P.; Mahmoud, K.A.; Tkac, J. Highly stable $Ti_3C_2T_x$ (MXene)/Pt nanoparticles-modified glassy carbon electrode for H_2O_2 and small molecules sensing applications. *Sens. Actuators B Chem.* **2018**, *263*, 360–368. [CrossRef]
132. Filip, J.; Zavahir, S.; Lorencova, L.; Bertok, T.; Yousaf, A.B.; Mahmoud, K.A.; Tkac, J.; Kasak, P. Tailoring electrocatalytic properties of Pt nanoparticles grown on $Ti_3C_2T_x$ MXene surface. *J. Electrochem. Soc.* **2019**, *166*, H54. [CrossRef]
133. Zheng, J.; Diao, J.; Jin, Y.; Ding, A.; Wang, B.; Wu, L.; Weng, B.; Chen, J. An inkjet printed Ti3C2-GO electrode for the electrochemical sensing of hydrogen peroxide. *J. Electrochem. Soc.* **2018**, *165*, B227. [CrossRef]
134. Li, R.; Zhang, L.; Shi, L.; Wang, P. MXene Ti_3C_2: An effective 2D light-to-heat conversion material. *ACS Nano* **2017**, *11*, 3752–3759. [CrossRef] [PubMed]
135. Manzanares-Palenzuela, C.L.; Pourrahimi, A.M.; Gonzalez-Julian, J.; Sofer, Z.; Pykal, M.; Otyepka, M.; Pumera, M. Interaction of single-and double-stranded DNA with multilayer MXene by fluorescence spectroscopy and molecular dynamics simulations. *Chem. Sci.* **2019**, *10*, 10010–10017. [CrossRef] [PubMed]
136. Ramanavicius, S.; Ramanavicius, A. Progress and insights in the application of MXenes as new 2D nano-materials suitable for biosensors and biofuel cell design. *Int. J. Mol. Sci.* **2020**, *21*, 9224. [CrossRef]
137. Wu, X.; Ma, P.; Sun, Y.; Du, F.; Song, D.; Xu, G. Application of MXene in electrochemical sensors: A review. *Electroanalysis* **2021**, *33*, 1827–1851. [CrossRef]
138. Li, X.; Lu, Y.; Liu, Q. Electrochemical and optical biosensors based on multifunctional MXene nanoplatforms: Progress and prospects. *Talanta* **2021**, *235*, 122726. [CrossRef]
139. Thenmozhi, R.; Maruthasalamoorthy, S.; Nirmala, R.; Navamathavan, R. MXene Based Transducer for Biosensor Applications. *J. Electrochem. Soc.* **2021**, *168*, 117507. [CrossRef]

140. Shahzad, F.; Zaidi, S.A.; Naqvi, R.A. 2D transition metal carbides (MXene) for electrochemical sensing: A review. *Crit. Rev. Anal. Chem.* **2022**, *52*, 848–864. [CrossRef]
141. Wang, Y.; Liu, S.; Zhu, F.; Gan, Y.; Wen, Q. MXene Core-Shell Nanosheets: Facile Synthesis, Optical Properties, and Versatile Photonics Applications. *Nanomaterials* **2021**, *11*, 1995. [CrossRef] [PubMed]
142. Zhu, X.; Liu, P.; Xue, T.; Ge, Y.; Ai, S.; Sheng, Y.; Wu, R.; Xu, L.; Tang, K.; Wen, Y. A novel graphene-like titanium carbide MXene/Au–Ag nanoshuttles bifunctional nanosensor for electrochemical and SERS intelligent analysis of ultra-trace carbendazim coupled with machine learning. *Ceram. Int.* **2021**, *47*, 173–184. [CrossRef]
143. Xu, B.; Zhu, M.; Zhang, W.; Zhen, X.; Pei, Z.; Xue, Q.; Zhi, C.; Shi, P. Ultrathin MXene-micropattern-based field-effect transistor for probing neural activity. *Adv. Mater.* **2016**, *28*, 3333–3339. [CrossRef]
144. Amstad, E.; Textor, M.; Reimhult, E. Stabilization and functionalization of iron oxide nanoparticles for biomedical applications. *Nanoscale* **2011**, *3*, 2819–2843. [CrossRef] [PubMed]
145. Chimene, D.; Alge, D.L.; Gaharwar, A.K. Two-dimensional nanomaterials for biomedical applications: Emerging trends and future prospects. *Adv. Mater.* **2015**, *27*, 7261–7284. [CrossRef]
146. Li, X.; Shan, J.; Zhang, W.; Su, S.; Yuwen, L.; Wang, L. Recent advances in synthesis and biomedical applications of two-dimensional transition metal dichalcogenide nanosheets. *Small* **2017**, *13*, 1602660. [CrossRef] [PubMed]
147. Svenson, S.; Tomalia, D.A. Dendrimers in biomedical applications—Reflections on the field. *Adv. Drug Deliv. Rev.* **2012**, *64*, 102–115. [CrossRef]
148. Thanh, N.T.; Green, L.A. Functionalisation of nanoparticles for biomedical applications. *Nano Today* **2010**, *5*, 213–230. [CrossRef]
149. Babar, Z.U.D.; Della Ventura, B.; Velotta, R.; Iannotti, V. Advances and emerging challenges in MXenes and their nanocomposites for biosensing applications. *RSC Adv.* **2022**, *12*, 19590–19610. [CrossRef]
150. Xu, B.; Zhi, C.; Shi, P. Latest advances in MXene biosensors. *J. Phys. Mater.* **2020**, *3*, 031001. [CrossRef]
151. Zamora-Galvez, A.; Morales-Narváez, E.; Mayorga-Martinez, C.C.; Merkoçi, A. Nanomaterials connected to antibodies and molecularly imprinted polymers as bio/receptors for bio/sensor applications. *Appl. Mater. Today* **2017**, *9*, 387–401. [CrossRef]
152. Zeng, S.; Baillargeat, D.; Ho, H.-P.; Yong, K.-T. Nanomaterials enhanced surface plasmon resonance for biological and chemical sensing applications. *Chem. Soc. Rev.* **2014**, *43*, 3426–3452. [CrossRef]
153. Taniselass, S.; Arshad, M.M.; Gopinath, S.C. Graphene-based electrochemical biosensors for monitoring noncommunicable disease biomarkers. *Biosens. Bioelectron.* **2019**, *130*, 276–292. [CrossRef]
154. Majd, S.M.; Salimi, A.; Ghasemi, F. An ultrasensitive detection of miRNA-155 in breast cancer via direct hybridization assay using two-dimensional molybdenum disulfide field-effect transistor biosensor. *Biosens. Bioelectron.* **2018**, *105*, 6–13. [CrossRef] [PubMed]
155. Cai, S.; Xiao, W.; Duan, H.; Liang, X.; Wang, C.; Yang, R.; Li, Y. Single-layer Rh nanosheets with ultrahigh peroxidase-like activity for colorimetric biosensing. *Nano Res.* **2018**, *11*, 6304–6315. [CrossRef]
156. Wu, L.; Chu, H.-S.; Koh, W.S.; Li, E.-P. Highly sensitive graphene biosensors based on surface plasmon resonance. *Opt. Express* **2010**, *18*, 14395–14400. [CrossRef] [PubMed]
157. Verma, R.; Gupta, B.D.; Jha, R. Sensitivity enhancement of a surface plasmon resonance based biomolecules sensor using graphene and silicon layers. *Sens. Actuators B Chem.* **2011**, *160*, 623–631. [CrossRef]
158. Ouyang, Q.; Zeng, S.; Dinh, X.-Q.; Coquet, P.; Yong, K.-T. Sensitivity enhancement of MoS_2 nanosheet based surface plasmon resonance biosensor. *Procedia Eng.* **2016**, *140*, 134–139. [CrossRef]
159. Wu, L.; You, Q.; Shan, Y.; Gan, S.; Zhao, Y.; Dai, X.; Xiang, Y. Few-layer $Ti_3C_2T_x$ MXene: A promising surface plasmon resonance biosensing material to enhance the sensitivity. *Sens. Actuators B Chem.* **2018**, *277*, 210–215. [CrossRef]
160. Kumar, S.; Lei, Y.; Alshareef, N.H.; Quevedo-Lopez, M.; Salama, K.N. Biofunctionalized two-dimensional Ti_3C_2 MXenes for ultrasensitive detection of cancer biomarker. *Biosens. Bioelectron.* **2018**, *121*, 243–249. [CrossRef]
161. Wu, Q.; Li, N.; Wang, Y.; Xu, Y.; Wei, S.; Wu, J.; Jia, G.; Fang, X.; Chen, F.; Cui, X. A 2D transition metal carbide MXene-based SPR biosensor for ultrasensitive carcinoembryonic antigen detection. *Biosens. Bioelectron.* **2019**, *144*, 111697. [CrossRef]
162. Lorencova, L.; Bertok, T.; Dosekova, E.; Holazova, A.; Paprckova, D.; Vikartovska, A.; Sasinkova, V.; Filip, J.; Kasak, P.; Jerigova, M. Electrochemical performance of $Ti_3C_2T_x$ MXene in aqueous media: Towards ultrasensitive H_2O_2 sensing. *Electrochim. Acta* **2017**, *235*, 471–479. [CrossRef]
163. Gajdosova, V.; Lorencova, L.; Prochazka, M.; Omastova, M.; Micusik, M.; Prochazkova, S.; Kveton, F.; Jerigova, M.; Velic, D.; Kasak, P. Remarkable differences in the voltammetric response towards hydrogen peroxide, oxygen and $Ru(NH_3)^{63+}$ of electrode interfaces modified with HF or LiF-HCl etched $Ti_3C_2T_x$ MXene. *Microchim. Acta* **2020**, *187*, 52. [CrossRef]
164. Lorencova, L.; Gajdosova, V.; Hroncekova, S.; Bertok, T.; Blahutova, J.; Vikartovska, A.; Parrakova, L.; Gemeiner, P.; Kasak, P.; Tkac, J. 2D MXenes as perspective immobilization platforms for design of electrochemical nanobiosensors. *Electroanalysis* **2019**, *31*, 1833–1844. [CrossRef]
165. Grieshaber, D.; MacKenzie, R.; Vörös, J.; Reimhult, E. Electrochemical biosensors-sensor principles and architectures. *Sensors* **2008**, *8*, 1400–1458. [CrossRef]
166. Chen, R.; Kan, L.; Duan, F.; He, L.; Wang, M.; Cui, J.; Zhang, Z.; Zhang, Z. Surface plasmon resonance aptasensor based on niobium carbide MXene quantum dots for nucleocapsid of SARS-CoV-2 detection. *Microchim. Acta* **2021**, *188*, 316. [CrossRef] [PubMed]

167. Bertok, T.; Lorencova, L.; Chocholova, E.; Jane, E.; Vikartovska, A.; Kasak, P.; Tkac, J. Electrochemical impedance spectroscopy based biosensors: Mechanistic principles, analytical examples and challenges towards commercialization for assays of protein cancer biomarkers. *ChemElectroChem* **2019**, *6*, 989–1003. [CrossRef]
168. Lorencova, L.; Bertok, T.; Bertokova, A.; Gajdosova, V.; Hroncekova, S.; Vikartovska, A.; Kasak, P.; Tkac, J. Exosomes as a source of cancer biomarkers: Advances in electrochemical biosensing of exosomes. *ChemElectroChem* **2020**, *7*, 1956–1973. [CrossRef]
169. Reddy, K.K.; Bandal, H.; Satyanarayana, M.; Goud, K.Y.; Gobi, K.V.; Jayaramudu, T.; Amalraj, J.; Kim, H. Recent trends in electrochemical sensors for vital biomedical markers using hybrid nanostructured materials. *Adv. Sci.* **2020**, *7*, 1902980. [CrossRef] [PubMed]
170. Gajdosova, V.; Lorencova, L.; Kasak, P.; Tkac, J. Electrochemical nanobiosensors for detection of breast cancer biomarkers. *Sensors* **2020**, *20*, 4022. [CrossRef]
171. Xiao, F.; Zhao, F.; Zhang, Y.; Guo, G.; Zeng, B. Ultrasonic electrodeposition of gold-platinum alloy nanoparticles on ionic liquid− chitosan composite film and their application in fabricating nonenzyme hydrogen peroxide sensors. *J. Phys. Chem. C* **2009**, *113*, 849–855. [CrossRef]
172. Xu, Y.; Ang, Y.S.; Wu, L.; Ang, L.K. High sensitivity surface plasmon resonance sensor based on two-dimensional MXene and transition metal dichalcogenide: A theoretical study. *Nanomaterials* **2019**, *9*, 165. [CrossRef]
173. Liu, C.; Cai, Q.; Xu, B.; Zhu, W.; Zhang, L.; Zhao, J.; Chen, X. Graphene oxide functionalized long period grating for ultrasensitive label-free immunosensing. *Biosens. Bioelectron.* **2017**, *94*, 200–206. [CrossRef]
174. Kaushik, S.; Tiwari, U.; Saini, T.; Pandey, A.; Paul, A.; Sinha, R. Long Period Grating Modified with Fe-Metal Organic Frameworks for Detection of Isopropanol. In Proceedings of the International Conference on Fibre Optics and Photonics, Kanpur, India, 4–8 December 2016.
175. Kaushik, S.; Tiwari, U.; Nilima; Prashar, S.; Das, B.; Sinha, R.K. Label-free detection of E scherichia coli bacteria by cascaded chirped long period gratings immunosensor. *Rev. Sci. Instrum.* **2019**, *90*, 025003. [CrossRef] [PubMed]
176. Singh, E.; Kim, K.S.; Yeom, G.Y.; Nalwa, H.S. Two-dimensional transition metal dichalcogenide-based counter electrodes for dye-sensitized solar cells. *RSC Adv.* **2017**, *7*, 28234–28290. [CrossRef]
177. Mishra, A.K.; Mishra, S.K.; Verma, R.K. Graphene and beyond graphene MoS_2: A new window in surface-plasmon-resonance-based fiber optic sensing. *J. Phys. Chem. C* **2016**, *120*, 2893–2900. [CrossRef]
178. Tuteja, S.K.; Duffield, T.; Neethirajan, S. Liquid exfoliation of 2D MoS_2 nanosheets and their utilization as a label-free electrochemical immunoassay for subclinical ketosis. *Nanoscale* **2017**, *9*, 10886–10896. [CrossRef]
179. Lee, J.; Dak, P.; Lee, Y.; Park, H.; Choi, W.; Alam, M.A.; Kim, S. Two-dimensional layered MoS_2 biosensors enable highly sensitive detection of biomolecules. *Sci. Rep.* **2014**, *4*, 7352. [CrossRef]
180. Xu, Y.; Wu, L.; Ang, L.K. MoS_2-Based Highly Sensitive Near-Infrared Surface Plasmon Resonance Refractive Index Sensor. *IEEE J. Sel. Top. Quantum Electron.* **2018**, *25*, 2868795.
181. Kumar, R.; Pal, S.; Verma, A.; Prajapati, Y.; Saini, J. Effect of silicon on sensitivity of SPR biosensor using hybrid nanostructure of black phosphorus and MXene. *Superlattices Microstruct.* **2020**, *145*, 106591. [CrossRef]
182. Srivastava, A.; Verma, A.; Das, R.; Prajapati, Y. A theoretical approach to improve the performance of SPR biosensor using MXene and black phosphorus. *Optik* **2020**, *203*, 163430. [CrossRef]
183. Ding, X.; Li, C.; Wang, L.; Feng, L.; Han, D.; Wang, W. Fabrication of hierarchical g-C_3N_4/MXene-AgNPs nanocomposites with enhanced photocatalytic performances. *Mater. Lett.* **2019**, *247*, 174–177. [CrossRef]
184. Kaushik, S.; Tiwari, U.K.; Deep, A.; Sinha, R.K. Two-dimensional transition metal dichalcogenides assisted biofunctionalized optical fiber SPR biosensor for efficient and rapid detection of bovine serum albumin. *Sci. Rep.* **2019**, *9*, 6987. [CrossRef]
185. Wu, Q.; Li, N.; Wang, Y.; Xu, Y.; Wu, J.; Jia, G.; Ji, F.; Fang, X.; Chen, F.; Cui, X. Ultrasensitive and selective determination of carcinoembryonic antigen using multifunctional ultrathin amino-functionalized Ti_3C_2-MXene nanosheets. *Anal. Chem.* **2020**, *92*, 3354–3360. [CrossRef]
186. Shrivastav, A.M.; Mishra, S.K.; Gupta, B.D. Fiber optic SPR sensor for the detection of melamine using molecular imprinting. *Sens. Actuators B Chem.* **2015**, *212*, 404–410. [CrossRef]
187. Liedberg, B.; Nylander, C.; Lunström, I. Surface plasmon resonance for gas detection and biosensing. *Sens. Actuators* **1983**, *4*, 299–304. [CrossRef]
188. Abdulhalim, I.; Zourob, M.; Lakhtakia, A. Surface plasmon resonance for biosensing: A mini-review. *Electromagnetics* **2008**, *28*, 214–242. [CrossRef]
189. Singh, P. SPR biosensors: Historical perspectives and current challenges. *Sens. Actuators B Chem.* **2016**, *229*, 110–130. [CrossRef]
190. Gupta, B.D.; Verma, R.; Srivastava, S.K. *Fiber Optic Sensors Based on Plasmonics*; World Scientific: Singapore, 2015.
191. Zhang, C.J.; Pinilla, S.; McEvoy, N.; Cullen, C.P.; Anasori, B.; Long, E.; Park, S.-H.; Seral-Ascaso, A.; Shmeliov, A.; Krishnan, D. Oxidation stability of colloidal two-dimensional titanium carbides (MXenes). *Chem. Mater.* **2017**, *29*, 4848–4856. [CrossRef]
192. Koyappayil, A.; Chavan, S.G.; Roh, Y.-G.; Lee, M.-H. Advances of MXenes; Perspectives on Biomedical Research. *Biosensors* **2022**, *12*, 454. [CrossRef]
193. Liu, M.; Bai, Y.; He, Y.; Zhou, J.; Ge, Y.; Zhou, J.; Song, G. Facile microwave-assisted synthesis of Ti_3C_2 MXene quantum dots for ratiometric fluorescence detection of hypochlorite. *Microchim. Acta* **2021**, *188*, 15. [CrossRef]

194. Chen, X.; Li, J.; Pan, G.; Xu, W.; Zhu, J.; Zhou, D.; Li, D.; Chen, C.; Lu, G.; Song, H. Ti_3C_2 MXene quantum dots/TiO_2 inverse opal heterojunction electrode platform for superior photoelectrochemical biosensing. *Sens. Actuators B Chem.* **2019**, *289*, 131–137. [CrossRef]
195. Cao, Y.; Wu, T.; Zhang, K.; Meng, X.; Dai, W.; Wang, D.; Dong, H.; Zhang, X. Engineered exosome-mediated near-infrared-II region V_2C quantum dot delivery for nucleus-target low-temperature photothermal therapy. *Acs Nano* **2019**, *13*, 1499–1510. [CrossRef]
196. Guan, Q.; Ma, J.; Yang, W.; Zhang, R.; Zhang, X.; Dong, X.; Fan, Y.; Cai, L.; Cao, Y.; Zhang, Y. Highly fluorescent Ti_3C_2 MXene quantum dots for macrophage labeling and Cu^{2+} ion sensing. *Nanoscale* **2019**, *11*, 14123–14133. [CrossRef]
197. Han, X.; Jing, X.; Yang, D.; Lin, H.; Wang, Z.; Ran, H.; Li, P.; Chen, Y. Therapeutic mesopore construction on 2D Nb_2C MXenes for targeted and enhanced chemo-photothermal cancer therapy in NIR-II biowindow. *Theranostics* **2018**, *8*, 4491. [CrossRef]
198. Han, X.; Huang, J.; Lin, H.; Wang, Z.; Li, P.; Chen, Y. 2D ultrathin MXene-based drug-delivery nanoplatform for synergistic photothermal ablation and chemotherapy of cancer. *Adv. Healthc. Mater.* **2018**, *7*, 1701394. [CrossRef]
199. Wang, X.D.; Rabe, K.S.; Ahmed, I.; Niemeyer, C.M. Multifunctional silica nanoparticles for covalent immobilization of highly sensitive proteins. *Adv. Mater.* **2015**, *27*, 7945–7950. [CrossRef]
200. Sundaram, A.; Ponraj, J.S.; Wang, C.; Peng, W.K.; Manavalan, R.K.; Dhanabalan, S.C.; Zhang, H.; Gaspar, J. Engineering of 2D transition metal carbides and nitrides MXenes for cancer therapeutics and diagnostics. *J. Mater. Chem. B* **2020**, *8*, 4990–5013. [CrossRef]
201. Zhou, D.; Gopinath, S.C.; Saheed, M.S.M.; Sangu, S.S.; Lakshmipriya, T. MXene surface on multiple junction triangles for determining osteosarcoma cancer biomarker by dielectrode microgap sensor. *Int. J. Nanomed.* **2020**, *15*, 10171. [CrossRef]
202. Liu, L.; Wei, Y.; Jiao, S.; Zhu, S.; Liu, X. A novel label-free strategy for the ultrasensitive miRNA-182 detection based on MoS_2/Ti_3C_2 nanohybrids. *Biosens. Bioelectron.* **2019**, *137*, 45–51. [CrossRef]
203. Zhou, S.; Gu, C.; Li, Z.; Yang, L.; He, L.; Wang, M.; Huang, X.; Zhou, N.; Zhang, Z. $Ti_3C_2T_x$ MXene and polyoxometalate nanohybrid embedded with polypyrrole: Ultra-sensitive platform for the detection of osteopontin. *Appl. Surf. Sci.* **2019**, *498*, 143889. [CrossRef]
204. Zhang, Q.; Wang, F.; Zhang, H.; Zhang, Y.; Liu, M.; Liu, Y. Universal Ti_3C_2 MXenes based self-standard ratiometric fluorescence resonance energy transfer platform for highly sensitive detection of exosomes. *Anal. Chem.* **2018**, *90*, 12737–12744. [CrossRef]
205. Liu, Y.; Zeng, H.; Chai, Y.; Yuan, R.; Liu, H. $Ti_3C_2/BiVO_4$ Schottky junction as a signal indicator for ultrasensitive photoelectrochemical detection of VEGF 165. *Chem. Commun.* **2019**, *55*, 13729–13732. [CrossRef]
206. Wen, W.; Qiong, W.; Sheng-Nan, R.; Zhuo, L.; Fang-Fang, C. Ti_3C_2-MXene-assisted signal amplification for sensitive and selective surface plasmon resonance biosensing of biomarker. *Chin. J. Anal. Chem.* **2022**, *50*, 13–18.
207. Zhang, H.; Wang, Z.; Zhang, Q.; Wang, F.; Liu, Y. Ti_3C_2 MXenes nanosheets catalyzed highly efficient electrogenerated chemiluminescence biosensor for the detection of exosomes. *Biosens. Bioelectron.* **2019**, *124*, 184–190. [CrossRef] [PubMed]
208. Pichersky, E.; Dudareva, N.; Lee, G.Y.; Lim, C.T.; Hatti-Kaul, R.; Törnvall, U.; Gustafsson, L.; Börjesson, P.; Healy, D.A.; Hayes, C.J. 93 Multifunctional cargo systems for biotechnology. *Biotechnology* **2007**, *25*, 89–138.
209. Tombelli, S.; Minunni, M.; Luzi, E.; Mascini, M. Aptamer-based biosensors for the detection of HIV-1 Tat protein. *Bioelectrochemistry* **2005**, *67*, 135–141. [CrossRef]
210. Gao, Z.; Zhang, J.; Ting, B.P. A doubly amplified electrochemical immunoassay for carcinoembryonic antigen. *Biosens. Bioelectron.* **2009**, *24*, 1825–1830. [CrossRef]
211. Quershi, A.; Gurbuz, Y.; Kang, W.P.; Davidson, J.L. A novel interdigitated capacitor based biosensor for detection of cardiovascular risk marker. *Biosens. Bioelectron.* **2009**, *25*, 877–882. [CrossRef]
212. Liu, C.; Wang, R.; Shao, Y.; Chen, C.; Wu, P.; Wei, Y.; Gao, Y. Detection of GDF11 by using a Ti_3C_2-MXene-based fiber SPR biosensor. *Opt. Express* **2021**, *29*, 36598–36607. [CrossRef]
213. Altintas, Z.; Uludag, Y.; Gurbuz, Y.; Tothill, I.E. Surface plasmon resonance based immunosensor for the detection of the cancer biomarker carcinoembryonic antigen. *Talanta* **2011**, *86*, 377–383. [CrossRef]

biosensors

MDPI

Review

Recent Advances in DNA Nanotechnology for Plasmonic Biosensor Construction

Jeong Ah Park [1,†], Chaima Amri [2,†], Yein Kwon [1], Jin-Ho Lee [2,3,*] and Taek Lee [1,*]

1 Department of Chemical Engineering, Kwangwoon University, Seoul 01897, Korea; m3m33@kw.ac.kr (J.A.P.); ijmr9126@kw.ac.kr (Y.K.)
2 Department of Convergence Medical Sciences, School of Medicine, Pusan National University, Yangsan 50612, Korea; a.chaima@pusan.ac.kr
3 School of Biomedical Convergence Engineering, Pusan National University, Yangsan 50612, Korea
* Correspondence: leejh@pusan.ac.kr (J.-H.L.); tlee@kw.ac.kr (T.L.); Tel.: +82-51-510-8547 (J.-H.L.); +82-2-940-5771 (T.L.)
† These authors contributed equally to this work.

Abstract: Since 2010, DNA nanotechnology has advanced rapidly, helping overcome limitations in the use of DNA solely as genetic material. DNA nanotechnology has thus helped develop a new method for the construction of biosensors. Among bioprobe materials for biosensors, nucleic acids have shown several advantages. First, it has a complementary sequence for hybridizing the target gene. Second, DNA has various functionalities, such as DNAzymes, DNA junctions or aptamers, because of its unique folded structures with specific sequences. Third, functional groups, such as thiols, amines, or other fluorophores, can easily be introduced into DNA at the 5′ or 3′ end. Finally, DNA can easily be tailored by making junctions or origami structures; these unique structures extend the DNA arm and create a multi-functional bioprobe. Meanwhile, nanomaterials have also been used to advance plasmonic biosensor technologies. Nanomaterials provide various biosensing platforms with high sensitivity and selectivity. Several plasmonic biosensor types have been fabricated, such as surface plasmons, and Raman-based or metal-enhanced biosensors. Introducing DNA nanotechnology to plasmonic biosensors has brought in sight new horizons in the fields of biosensors and nanobiotechnology. This review discusses the recent progress of DNA nanotechnology-based plasmonic biosensors.

Keywords: biosensors; DNA nanotechnology; DNAzyme; plasmonic biosensor; aptamer; DNA origami

Citation: Park, J.A.; Amri, C.; Kwon, Y.; Lee, J.-H.; Lee, T. Recent Advances in DNA Nanotechnology for Plasmonic Biosensor Construction. *Biosensors* **2022**, *12*, 418. https://doi.org/10.3390/bios12060418

Received: 23 May 2022
Accepted: 13 June 2022
Published: 15 June 2022

1. Introduction

Nucleic acids are regarded as the one of the most vital molecules in living organisms; they have been credited with the origin of life on Earth [1,2]. Our understanding of nucleic acids has led to immense advances in medicine, pharmacology, biology, and biotechnology [3–5]. In particular, the convergence of biotechnology and nanotechnology has resulted in the development of innovative technologies for biosensors, drug delivery, and bioimaging [6–8]. Among these, DNA nanotechnology has received the most attention from researchers interested in the material rather than genetic characteristics of DNA [9,10]. DNA nanotechnology has brought forth a new paradigm for DNA research, one that employs the genetic functionality of DNA for engineering applications in biosensors [11,12], nanoarchitecture [13,14], drug delivery [15,16], and computations [17,18]. This integrated approach has resulted in the development of a tremendous line of products, which would have been impossible with conventional nucleic acid research or nanobiotechnology alone.

Prior to the rise of DNA nanotechnology, DNA was regarded as a molecule studied exclusively in the fields of life sciences and medicine. DNA nanotechnology has shifted its focus to the valuable material characteristics of DNA, which provides unusual stability, flexibility, complementary binding, and tailored functionality [19–21]. Several researchers have reported on DNA-based computations [22], DNA origami structures [23], DNA-based biosensors [24], DNA-based vaccines [25], and drug development using DNA [26]. Of note are the DNA aptamers [27–29] and DNAzymes (deoxy → ribozymes) [30–32] that mimic the characteristics of natural antibodies and enzymes, respectively. A DNA aptamer can bind a specific target with high binding affinity. Usually, the DNA aptamer and DNAzyme can be produced by the systematic evolution of ligands by exponential enrichment (SELEX) that can bind specifically designed targets [33]. An aptamer is nucleic acid or peptide sequence that can mimic the functions of an antibody. The terms aptamer and SELEX were reported by different research groups [34,35].

Several groups reported on the application of aptamers in medicines [28], pharmacology [36], and molecular biology research [27]. Several types of aptamers are reported to have biosensor and diagnostic applications [37–41]. The DNA or RNA aptamer is a great bioreceptor that can alter an antibody with the manufacturing cost. Aptamers can also be produced via chemical synthesis without requiring experiments on animals for antibody production. Moreover, compared to antibodies, chemical production can easily be scaled up in aptamers because of the short production time. DNAzymes (Deoxyribozyme), also known as catalytic DNA, act as enzymes that can catalyze specific reactions [41,42]. These DNAzymes can be produced by the SELEX method [28]. Unlike natural enzyme, DNAzymes were reported to have several functionalities such as DNA self-modification [43], RNA cleavage [44] and H_2O_2 reduction [45]. These unique characteristics make DNAzymes a good material for biosensor construction as the bioprobe [43–46].

In particular, the study of DNA nanotechnology has led to innovative advances in biosensors. Here too, the structural flexibility and malleability of the structure of DNA allows the nucleic acid to be employed as the bioprobe [23,47,48]. For example, the DNA 3-way junction (3WJ) structure has added one more sequence to the typical double-stranded DNA that can add extra functional groups to the structure [49,50]. Likewise, Y-shaped DNA gives a similar effect [51,52]. Based on this concept, DNA 4-way junctions (4WJ) provide additional arms to improve the functionality of biosensor fabrication [38,53]. G-quadruplex DNA-Based Biosensors are a good example of the structural application of DNA nanotechnology [54,55]. Conventional single-stranded DNA can bind complementary sequences; this principle is used as the basis of a PCR-based in vitro diagnostic (IVD) apparatuses or other genetic biosensor constructions. DNA origami, in contrast, provides unprecedented DNA structures that expand the various structural bioprobes. Thus, DNA nanotechnology offers new methods for biosensor development.

Incorporating DNA nanotechnology with the plasmonic biosensors provided several merits compared to conventional protein-based biosensors. (1) The multi-functional bioprobe can minimize the bioprobe construction. For example, the DNA can be easily manufactured by chemical synthesis that means we can easily introduce the chemical linker or fluorescence dye at the end of each terminal. (2) The production cost of aptamer is cheap compared to a protein or antibody. (3) DNA hybridization with nanomaterial provides the new sensing platform for plasmonic biosensor. Based on those advantages of DNA nanotechnology, the focus of this review will be on the recent progress of nucleic acid engineering for plasmonic biosensor construction. The applications of both simple aptameric structures and origami-shaped structures in SPR, SERS, and MEF will be discussed.

Meanwhile, researchers in the field of biosensing have shifted their attention to plasmonic biosensors. This propriety can be harnessed through multiple applications in the field of biosensing [56]. For instance, to generate surface plasmon resonance (SPR), an incident light produces energetic electrons that propagate through the surface [57]. In contrast, in local surface Plasmon resonance (LSPR), coherent oscillations are confined to a metallic nanostructure and cannot propagate thus generating an electromagnetic field (EM) [58]. These properties can be harnessed for multiple applications in the field of biosensing [56]. For instance, when a molecule, such as a protein or an aptamer, is adsorbed on to the plasmonic surface, the refractive index changes and such values can be measured. SPR/LSPR is a highly sensitive, simple, low cost, and label-free detection method [59,60]. Raman scattering is the inelastic scattering of photons caused by a high-intensity incident light. This effect generates a spectrum fingerprint proper to the molecule of interest [61]. However, such measurements are not precise at low concentrations or for complex structures such as proteins [62]. To overcome this issue, surface-enhanced Raman scattering (SERS) combines a major LSPR EM-induced enhancement and a lesser chemical enhancement induced by Raman reporters. Thus, leading to a stronger peak detection signal in SERS that can provide information on the chemical species and structure of a wide range of molecules [63]. Metal-enhanced fluorescence (MEF), also referred to as radiative decay engineering, is another application of LSPR EM that has been extensively studied. It is known that radiating surface plasmons can affect the spectral properties of a fluorophore [64], i.e., the combination of a metal and a fluorescent molecule at optimal distance has been shown to increase fluorescence emission, prevent photobleaching, and increase the sensitivity of detection, among other things [65].

As the field of nucleic acid-based Plasmon biosensors is continuously evolving, a regular reviewing of the latest breakthroughs and the most trending technologies is necessary. To highlight the recent progress of nucleic acid engineering for Plasmonic biosensor construction, this review will be chiefly focused on research outcomes published within the last five years. The applications of both simple aptameric structures and origami-shaped structures in SPR/LSPR, SERS, and MEF will be discussed.

2. Aptameric Structure-Based Plasmon Biosensor

Despite the promising results shown by antibody-bound plasmon-generating structures in sensing a multitude of molecules from proteins to whole cells, the biosensor trend is progressively shifting toward aptameric detection probes for numerous reasons. Not only do aptasensors have a low cost of production and are easy to use, but they also have repeatedly demonstrated a high sensitivity to an extensive range of targets.

2.1. SPR/LSPR-Based Sensing

Using broad-spectrum antibiotics can significantly affect both the environment and the living organisms involuntarily exposed to them by generating drug resistant bacteria. Oxytetracycline (OTC), a semi-synthetic antibiotic, is commonly used in animal feeds and is often released into the environment. In Huang et al.'s experiment, a liquid crystal catalytic amplification-nanogold SPR aptamer absorption assay was developed for the detection of trace OTC [66]. The team developed a novel SPR spectrophotometric analysis method by combining 4-heptylbenzoic acid (HPB) and $HAuCl_4$. The reduction of $HAuCl_4$ to gold NPs was hindered by the adsorbed OTC-binding aptamers. When in the presence of the target molecule, the aptamer would be released and the NPs would catalyze the reaction. This sensing mechanism achieved a detection limit of 0.50 ng/mL for OTC [66]. In their research, Écija-Arenas et al., developed an aptamer-based SPR biosensor for the determination of kanamycin, an aminoglycoside bacteriocidal antibiotic, residue in food [67]. The team tested the combination of a gold and a graphene film through two methods: self-assembly of reduced-graphene (rGO) and chemical vapor deposition (CVD) of graphene (GO) (Figure 1a). The former step consisted of a 1-pyrenebutyric acid (PBA) modification via π-stacking before binding the kanamycin specific aptamer to the graphene surface. As a proof of concept, the sensor was tested on kanamycin spiked commercial cow milk samples. Overall, the GO-based sensor had a detection limit of 285 nmol L^{-1}, seven-fold lower than the rGO-based platform. This was most likely due to the homogeneity of the GO monolayer when directly deposited through CVD [67]. As the sensitivity of SPR sensors can be affected by the homogeneity of the probe surface, producing a homogeneous deposition is one of the challenges to overcome. In their study, Chen et al., constructed an SPR aptasensor for the detection of SARS-CoV-2 based on a gold chip coated with thiol-modified niobium carbide MXene quantum dots (Nb2C-SH QDs) [68]. This was followed by the immobilization of a detection aptamer (N58) through π-π stacking, electrostatic adsorption and hydrogen bond on to the chip. The multifunctional proprieties of the chosen nanoparticles allowed for a homogeneous deposition leading to a detection limit of 4.9 pg/mL for the N-gene of SARS-CoV-2. Furthermore, the sensor displayed good selectivity in the presence of other viruses and a great stability when tested in other environments such as human serum [68].

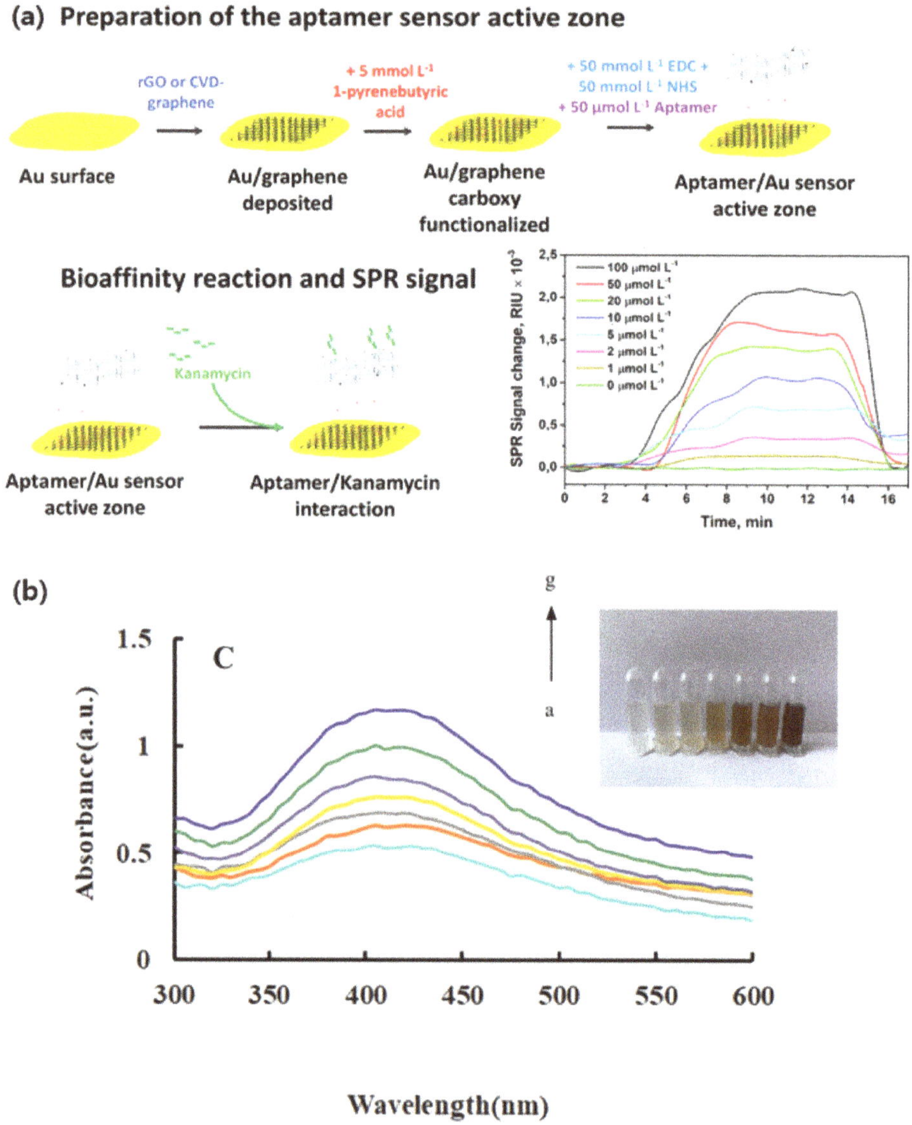

Figure 1. Aptamer-based SPR detection mechanism: (**a**) Kanamycin sensing platform. (**b**) Surface plasmon resonance absorption spectra of the nanosilver sol systems with a–g as As^{3+} concentrations of 0; 0.0945; 0.189; 0.2835; 0.378; 0.4725 and 0.567 µg/L. (Reproduced with permission from [67], published by Elsevier 2021; reproduced with permission from [69], published by Elsevier 2020).

Arsenic is a naturally occurring poisonous non-metallic element. The standard method, silver diethyldithiocarbamate spectrophotometry (SDDC), for determining the arsenic concentration in water is simple but has a detection limit of 0.5 mg/L. Considering that ingesting 1 mg of arsenic is highly toxic for the human body [69,70], the development of a more sensitive detection method is necessary. Through a catalytic amplification method based on colored nanosilver surface plasmon resonance (SPR), Zhang et al., achieved a detection limit of 0.01 μg/L [69]. The sensing mechanism was based on the catalytic effect of gold-doped carbon dots (CDAu) in the reductive reaction of $AgNO_3$ with glucose. In the initial stage, aptamers were adsorbed on to the CDAu surface, stopping the catalysis. Only in the presence of arsenic (As^{3+}), which forms a conjugate with the aptamer, would the carbon dots be released. The ensuing reaction would lead to the synthesis of yellow spherical silver nanosols exhibiting an intense SPR absorption peak (Figure 1b) [69].

Both SPR and LSPR aptameric sensors offer great advantages when compared to other sensing platforms. However, due to the nature of the interaction between the aptamer, the target, and the LSPR/SPR generating structure, certain factors must be taken into account. Some detection probes are solely based on the adsorption of a target detection aptamer onto a metallic nanostructure which in the presence of the target would detach and subsequently modulate the aggregation of the nanoparticles. Yet, in certain cases, the target itself has been shown to adsorb onto the nanostructures. For instance, Ochratoxin A (OTA), one of the most common food contaminants, an extremely toxic molecule for living organisms, can also adsorb onto AuNPs, thus making accurate detection difficult [71,72]. This phenomenon was first described by Liu et al., during their attempt to widen the detection range of OTA by doubling the calibration curve of their gold NP aptamer-based localized surface plasmon resonance (LSPR) sensor [71]. The team established that at low concentration, OTA would bind to its aptamer inducing the aggregation of the AuNPs but as the concentration increases the free OTA would directly bind to the AuNPs and protect them from aggregating. By developing a double calibration curve the sensing mechanism achieved a detection limit of $10^{-10.5}$ g/mL. More importantly, the same effect was observed for different analytes such as adenosine triphosphate (ATP) and 17β-estradiol (EST) [71].

2.2. *Surface-Enhanced Raman Scattering-Based Sensing*

Herbicides have detrimental side effects on human health and their long term exposure can lead to chronic disease and childbirth defects [73]. To achieve the detection of glyphosate (GLY) at very low concentrations, Liu et al., constructed a simple and sensitive quantitative analysis based on gold nanoplasmon resonance Rayleigh scattering (RRS) and SERS [74]. In this study, gold nanoparticles (AuNPs) were generated through the catalytic effect of a covalent organic framework (COF) on a glycol and chloroauric acid reaction. The COF was prepared by the polycondensation of melamine and p-benzaldehyde, and could interact with a GLY-binding aptamer through hydrogen bonds (HB), hydrophobic interactions (HI), and intermolecular forces (IF). However, in the presence of GLY, the aptamers would detach from the CPF and allow for the catalysis of AuNP synthesis in a proportional manner (Figure 2a,b). The detection of GLY was possible through RRS, but as the blank was high, SERS analysis showed a better signal with a peak at 1617 cm^{-1} using Victoria blue B molecular probes. The team was able to achieve a detection limit of 0.002 nmol/L for GLY [74].

Urea, a protein decomposition product, is another harmful molecule that needs to be monitored [75]. Yao et al., used a similar RRS/SERS dual-mode method to detect urea concentrations as low as 0.06 nM and 0.03 nM, respectively, in milk [76]. The slow reaction of Citrate-Ag (I) can be catalyzed using a polystyrene nanosphere (PN) to produce yellow silver NPs (AgNPs). Similar to the previous GLY detection mechanism, PN adsorbed urea-binding aptamers would weaken the RRS and SERS signal (Figure 2c,d). When in the presence of their target, the aptamers would release the catalytic NPs in a linear manner, thus generating a stronger signal [76].

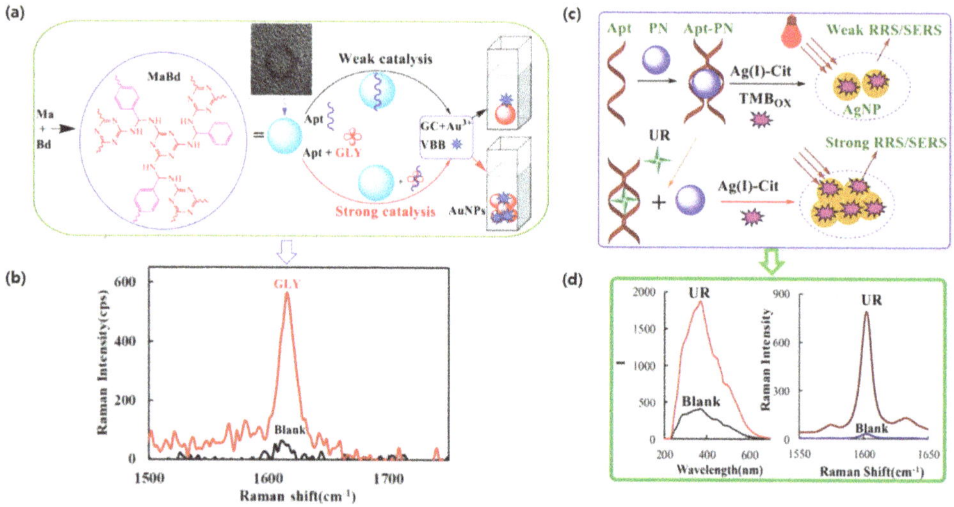

Figure 2. SERS aptameric catalysis mechanism: (**a**) Aptamer-mediated MaBd catalysis of GC-HAuCl4 nanogold reaction for the detection of glyphosate. (**b**) Raman spectra of glyphosate detection peak. (**c**) Aptamer-mediated PN-$AgNO_3$-Cit catalytic amplification reaction for the detection of urea. (**d**) Raman spectra of RRS and SERS dual-mode urea detection. (Reproduced with permission from [74], published by Elsevier 2021; reproduced with permission from [76], published by Elsevier 2022).

Cancer detection is also a great application of SERS-based plasmonic sensors due to the low concentration of most biomarkers. In a study by Bhamidipati et al., the combination of gold nanostarts and truncated aptamers allowed for the quantification of epithelial cell adhesion molecules (EpCAM), a common cancer biomarker, at the single cell level with concentrations in the subnanomolar range [77]. As the expression of cancer markers can vary at the cellular level, the quantification of EpCAM density in individual cells can help clinicians monitor the phenotypic evolution of cancer cells. Herein, the biomarker expression of both MCF-7 and PC-3 cancer cell lines, with high and low EpCAM expression, respectively, was successfully measured with aptamer-functionalized gold nanostarts (Figure 3a,b). Interestingly, when compared to a longer, 48-base-pair oligonucleotide, a truncated EpCAM aptamer seemed to have better sensitivity, most likely due to a reduction in the number of conformations possible at room temperature [77]. Ning et al., exploited an aptamer-bound gold–silver bimetallic nanotrepang structure in their SERS analysis for the simultaneous detection of multiple cancer-related exosomes [78]. By alternating between a gold nanorod (GNR) core, Raman reports and silver layers, external interference was prevented, and the SERS signal was 151.7-fold stronger when compared to GNRs only. Furthermore, this structure provides an outer shell for functional modifications without the traditional competitive binding of Raman reports. Although the sensor required streptavidin-modified magnetic beads as capture substrates, which may be viewed as an extra step compared to previous sensors, it was successfully tested on real blood samples of cancer patients [78].

Multiplexed detection of biomarkers can also be applied in the diagnosis of neurodegenerative diseases, such as Alzheimer's disease (AD). Zhang et al., engineered a SERS detection platform that detected both Aβ (1–42) oligomers and Tau protein simultaneously [79]. To do so, the team combined different Raman dye-coded polyA aptamers with AuNPs. The simplicity of the system was based on two types of Raman dye-coded polyA aptamer-AuNPs. In the presence of the protein of interest, the polyA aptamer would detach from the AuNPs which would induce the aggregation of the NPs. This would activate a strong SERS signal and a colorimetric change in the solution (Figure 3c,d) [79].

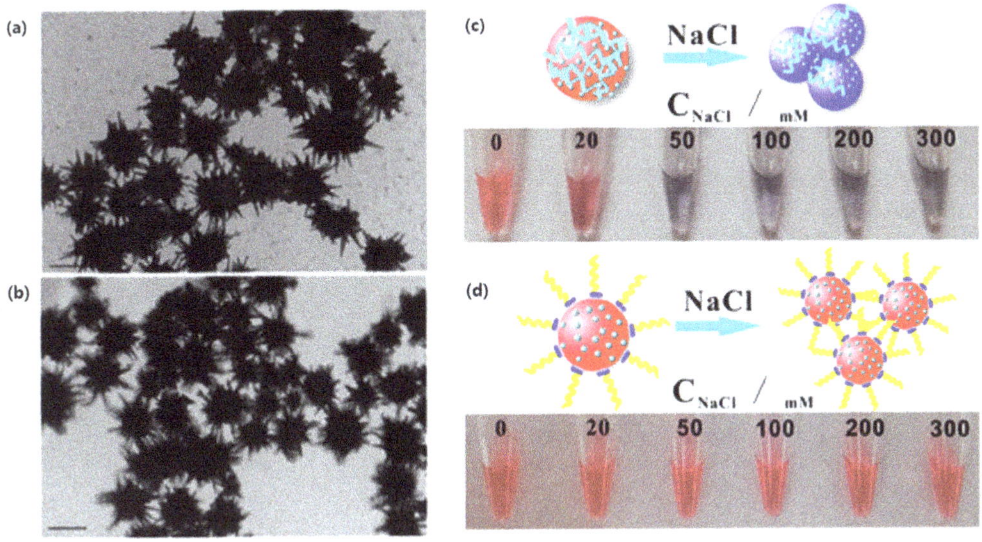

Figure 3. Gold nanoparticles aptamer binding: (**a**) TEM of synthesized gold nanostars and (**b**) 4-aminothiophenol and EpCAM aptamer-functionalized gold nanostars. (Scale bars are 100 nm) (**c**) Photo of the ionic strength-dependent color change of the non-polyA-Tau apt/DTNB/AuNPs. (**d**) Photo of the ionic strength-dependent color change of the polyA-Tau apt/DTNB/AuNPs. (Reproduced with permission from [77], published by American Chemical Society 2018; reproduced with permission from [79], published by American Chemical Society 2019).

2.3. MEF-Based Sensing

Metal-enhanced fluorescence (MEF) is a promising application of aptameric plasmon sensors. As the intensity of MEF can be affected significantly by the distance between the fluorophore and the localized surface plasmon resonance-generating structure, aptamers present a convenient method to optimize such factors [80]. Hence, Choi et al., took advantage of this aspect to design a rapid, simple, and one-step technique for the real-time monitoring of intracellular proteolytic enzymes, such as Caspase-3 (Figure 4a) [81]. The nanobiosensor was based on an AuNP with a double connection to a fluorophore molecule, fluorescein isothiocyanate (FITC). The first link was made of peptides and was meant to keep the FITC molecule close to the AuNP and quenched in the absence of caspase-3. The second link, single-stranded oligonucleotides, was designed to allow an optimal MEF of FITC, in the presence of a proteolytic enzyme that would degrade the peptide connection. The short reaction time of 1 h and the low detection limit of 10 pg/mL allowed for the successful detection of preapoptotic cells [81].

Jiang et al., developed a highly sensitive fluorescence-enhanced aptasensor based on a polyA_n-aptamer nanostructure for the detection of adenosine triphosphate (ATP) [82]. Their approach involved optimizing the distance between fluorophores, in this case fluorescein (FAM), and the AuNP to enhance fluorescence by adjusting the number of adenosine bases at the 5′ end of a polyA_n-aptamer. A FAM-bound aptamer would hybridize with the AuNP-bound polyA_n-aptamer and induce stronger fluorescence. However, in the presence of targeted ATP, the FAM-aptamer would be released, leading to weak fluorescence (Figure 4b). Through this mechanism, a detection limit of 200 pM was successfully achieved when the polyA_n-aptamer reached a length of 9 nm, a seven-fold sensitivity improvement compared to more common detection methods [82]. Since ATP is a very important molecule in living organisms, it is not surprising to find many studies aiming to improve its detection. Zheng et al., designed a sensing mechanism based on structure-switching aptamer-triggering metal-enhanced fluorescence for Cy7 [83]. Gold nanobipyramids were synthesized by a seed-mediated growth method, then separated and functionalized with an aptamer probe that can bind ATP (Figure 4c,d). The team reported the first hybridization chain reaction-induced MEF, but despite the novelty of their method, the detection limit achieved was not as low as other biosensors with a marked 35 nm [83].

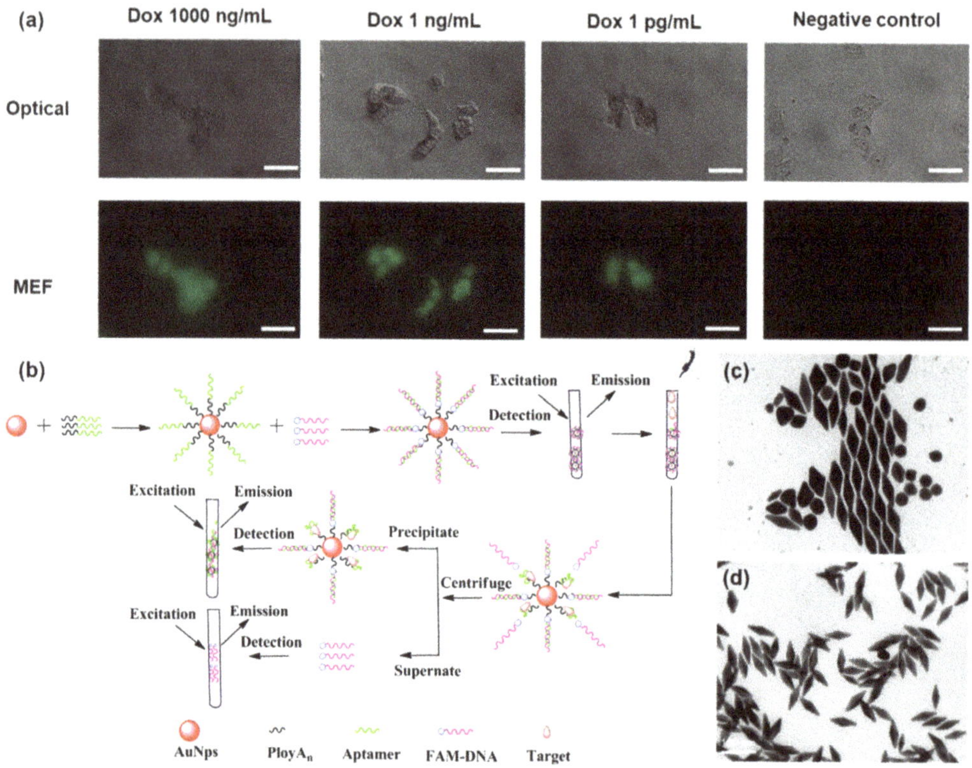

Figure 4. Metal-enhanced fluorescence biosensors: (**a**) Fluorescence imaging of Dox-treated MCF-7 cells. (Scale bars are 100 μm) (**b**) AuNPs@polyAn-aptamer@FAM-DNA nanostructure for ATP detection. (**c**) TEM image of synthetized AuNBPs with LPRW at 751 nm before separation. (**d**) TEM image of AuNBPs with LPRW at 751 nm after separation. (Reproduced with permission from [81], published by American Chemical Society 2020; reproduced with permission from [81], published by Elsevier 2019; reproduced with permission from [83], published by Elsevier 2020).

MEF aptasensors were also proven to be efficient in the detection of viral DNA. Furui et al., constructed another MEF microarray for the detection of hepatitis B with a LOD of 50 fM [84]. The enzyme free detection mechanism was based on the sandwich binding of Tag-A and Tag-B, both aptamer-functionalized silver nanoparticles (AgNPs), through a capture probe immobilized on to the chip with the viral target serving as a bridge. The detection assay demonstrated up to 120-fold enhancement factor due to the increasing number of fluorophores and the MEF caused by the AgNPs [84].

3. DNA Origami-Based Plasmonic Biosensor

Due to its versatile structure modifications, DNA origami can easily be applied in the development of biosensors [85,86]. Biosensors designed with synthetic DNA origami are rolled to two specific goals. First, the specific structure of DNA origami was applied to a biosensor platform. For example, the DNA tetrahedron or DNA branch structure can be introduced to bioprobe [85,87]. Second, each arm is designed to have unique functionalities including target recognition, signal generation or immobilization. Conventional single-stranded or double-stranded DNA only has two ends. However, DNA origami, such as Y shape or 3WJ DNA, has an additional arm to serve as a multi-functional bioprobe [53,88]. Furthermore, the X shape or 4WJ can improve functionality significantly. Assembled, multi-functional DNA origami can reduce the fabrication and labeling process for biosensors.

Lee et al., developed the LSPR biosensor which contained a multi-functional DNA 3WJ on the hollow Au spike-like NP-modified substrate for detecting avian influenza [89]. In this report, DNA 3WJ was designed to provide several functionalities, such as target recognition, immobilization, and signal enhancement, simultaneously. Each end of the DNA arm has a functional moiety. The DNA 3WJ-a arm was tagged to the hemagglutinin (HA)-binding aptamer as the bioprobe, the DNA 3WJ-b arm was connected to the FAM dye for signal enhancement function, and the DNA 3WJ-c arm was modified to the thiol group for anchoring the surface. Hollow Au spike-like NPs provided the plasmonic platform for target detection (Figure 5a,b). In this paper, Hollow Au spike-like NPs provided more surface roughness compared to normal Au nanoparticle that can provide more bioprobe immobilization number. Additionally, Hollow Au spike structure enhanced the LSPR effect compared to normal Au NP. Based on this DNA 3WJ and NP combination, the HA protein can be detected in PBS buffer and diluted chicken serum. The detection range of HA protein is 1 pM to 100 nM with high selectivity (Figure 5c,d). Thus, DNA 3WJ is a great tool for LSPR biosensor fabrication.

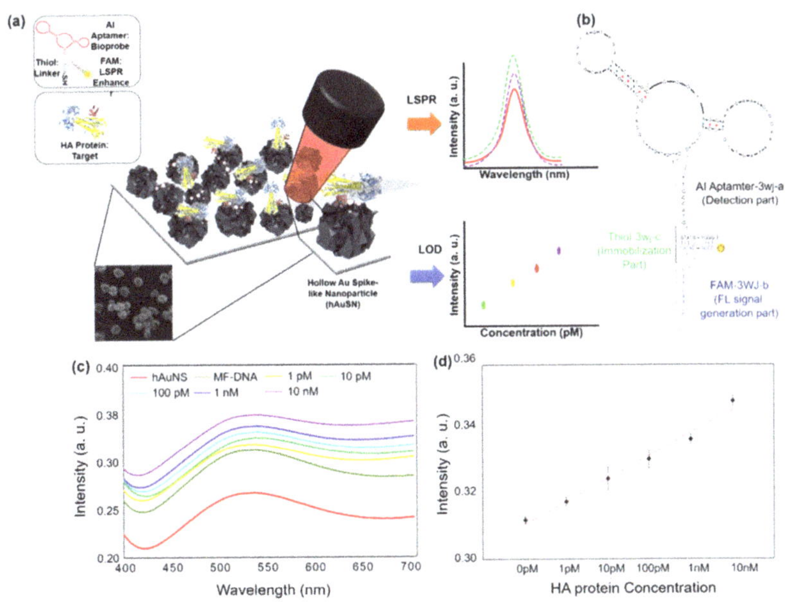

Figure 5. (**a**) Schematic diagram of multi-functional DNA 3WJ for AIV H5N1 detection through LSPR, (**b**) Expected 2D structure of multi-functional DNA 3WJ, HA protein, spike protein. (**c**) Detection of HA protein in 10% chicken serum on the DNA 3WJ-based localized surface plasmon resonance (LSPR) biosensor. Absorbance increases from different HA concentrations in 10% chicken serum (1 pM to 10 nM) of (**d**) Calibration characteristics of the different concentration of HA protein range from 1 pM to 10 nm with correlation coefficient (R2) of 0.9796. (Reproduced with permission from [89], published by Elsevier 2019).

Puchkova et al., demonstrated fluorescence enhancement using a DNA origami nanoantenna technology and proposed an MEF sensor [90]. The researchers attached 100 nm colloidal gold NPs to columnar DNA origami and a fluorophore was incorporated into the inter-particle spacing. The gold NPs were functionalized with thiol groups and bound to DNA origami to induce a particle distance between 12 and 17 nm. Furthermore, the authors quenched the fluorophore with $NiCl_2$, employing the phenomenon that effective quantum yield improves if the intrinsic quantum yield of the fluorophore is lower (>10). As a result, a fluorescence enhancement index of 306 was obtained, which is the highest value reported to date (Figure 6a,b). In this contribution, the researchers showed substantial improvements in the plasmonic structure by improving DNA origami, reducing the interparticle distance, and introducing a quencher to reduce the fluorophore quantum yield. Since nanoantennas are shown to be capable of detecting individual molecules at concentrations as low as 25 uM, DNA nanotechnology creates the possibility of placing molecules in self-assembled plasmonic nanoantennas.

Thacker et al., showed that DNA origami and two gold NPs are powerful tools for generating SERS active NPs [91]. Strong plasmonic binding was induced by holding two gold NPs (40 nm in diameter) at a distance of 3.3 ± 1 nm, which is one of the shortest intervals that can be made by DNA origami assembly. Moreover, the attachment of individual ssDNA-coated gold NPs to DNA origami sheets allowed for the characterization of the red shift and the subsequent extraction of the effective refractive index of DNA origami. Consequently, they showed the efficiency of dimer structures for SERS measurements using enhancement factors of up to 7000 times. It showed local field enhancement through the detection of short single-stranded DNA oligonucleotides as well as a small number of dye molecules. This demonstrated the effectiveness of the combination of DNA origami and SERS, suggesting that it has great potential in various biosensing fields.

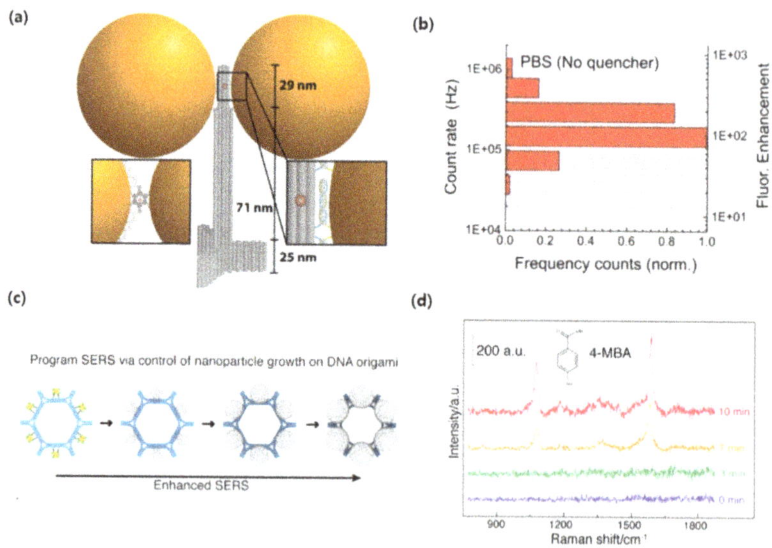

Figure 6. (**a**) Sketch of the DNA origami pillar (gray) employed to build the optical nanoantenna with two 100 nm Au nanoparticles together with a top-view (lower-left inset), (**b**) Photon count rate histogram for the dimer nanoantennas with no quencher, (**c**) Design of DNA origami-templated metamolecules with programmable surface-enhanced Raman scattering (**d**) Single-particle Raman spectra of Raman tag (4-mercaptobenzoic acid, 4-MBA) after adsorbed onto Ag@Au hexagon monomer metamolecules of varied silver growth time. (Reproduced with permission from [90], published by American Chemical Society 2015; reproduced with permission from [92], published by Elsevier 2020).

There is also a case where DNA origami became a template and SERS was programmed. Zhou et al., improved the complexity of SERS programming using DNA origami to increase the number of NPs in SERS metamolecules and arrange them into elaborate configurations [92]. The template was designed based on DNA origami hexagon tiles (DHT), and six gold NPs (10 nm in diameter) were fixed on the outer surface of the hexagon, and Ag–Au core−shell NPs were formed through subsequent Ag growth. The longer the silver growth time, the longer the color wavelength, and the metamolecules containing the largest NPs had the strongest interparticle electromagnetic field and Raman enhancement effect (Figure 6c,d). In general, SERS metamolecules are assembled into clusters of a small number (2–4) of nanoparticles and have limited programmability, but in this study, they expanded the structural complexity by increasing the number of nanoparticles. In other words, new plasmon research was made possible using DNA origami templates to precisely control the structural composition of metal NP clusters. This technology, it was shown, has several applications in fields such as photonics and sensing.

Jung et al., confirmed that DNA nanotechnology capable of calculating molecular information, called toehold-mediated DNA strand displacement (TMSD), could be linked to in vitro transcription and act as an information-processing unit for cell-free biosensors [93]. They developed design rules for interfacing small molecule sensing platforms with toehold-mediated strand displacements to construct hybrid RNA-DNA circuits that allow fine-tuning of reaction kinetics. Additionally, a cell-free biosensor platform, called an RNA output sensor, could be activated by ligand induction and combined with the computational power of TMSD to construct 12 different circuits that realize seven logical functions. Thus, a circuit capable of estimating the concentration range of a target compound in the sample was designed and verified. As a result, they extended their functions by interfacing with toehold-mediated strand displacement circuits through programmable interactions between nucleic acid strands. This platform presented the potential for different types of molecular computations in cell-free systems and of TMSD circuits that improve cell-free biosensing technology.

Dass et al., presented recent studies of plasmon sensing using DNA-based nanostructures. Using DNA origami for the coordinated arrangement of plasmonic nanoparticles holds great promise in the field of biosensing. Plasmonic particles can also be detected, ranging from fluorescence enhancement to enabling visualization of nanoscale dynamic behavior. Among prevalent technologies, DNA-based nanotechnology is the most successful strategy, and plasmonics coupled with DNA origami provide a novel and more effective approach [94]. This capability provides a suitable platform for the fabrication of plasmonic sensors for single-molecule measurements. From this point of view, the authors discussed recent developments and future research directions for plasmonic sensing with DNA origami, such as fluorescence-based plasmon sensing, SERS sensing, and chiral sensing. It was concluded that the fabrication of DNA origami-based plasmonic sensors is a new technology platform for engineering molecular scale sensors and has reached a level that can be applied in the field.

Recently, the DNAzyme is also used as the bioprobe for constructing plasmon biosensor. Conventionally, in the area of biosensors, the DNAzyme has been used to construct electrochemical biosensor beause of thier catalytic effect including H_2O_2 reduction or DNA cleavage [95–97].

Other than aptamer or origami-based plasmonic biosensors, few researchers have worked on developing plasmonic biosensors. Only recently have scientists introduced this unique material for plasmon biosensor fabrication. The group developed a DNAzyme integrated with magnetic beads (MBs) to create an *Escherichia coli* specific plasmonic biosensor [98]. They used DNAzyme because its RNA-cleaving characteristics would offer accurate target recognition. The MBs provide the plasmonic effect for the biosensor. When the DNAzyme-MB reacted with *E. coli* lysate, DNA cleavage was activated. Based on this, signal amplification was processed by a hybridization chain reaction with silver etching via an enzyme-triggered reaction. This detection method can selectively detect *E. coli* compared to other samples including *M. subtilis, M. peli* etc. *E. coli* was even detected in apple juice or skim milk using the DNAzyme-MB system. The detection range of *E. coli* was reportedly 50 to 5 × 106 cfu/mL.

Lee et al., also reported the use of DNAzyme conjugated with gold NPs for detecting *Salmonella choleraesuis* [99]. This study used multi-component DNAzymes for performing various functions in plasmonic biosensors. Multi-component DNAzyme can cleave the target, make the color change, and can be attached with gold NPs for plasmonic detection. The proposed biosensor can detect the 16 rRNA real target sample in the range of 50 nM to 500 nM. Moreover, Wu et al., fabricated a lead (II) ion detecting SPR biosensor composed of DNAzyme-gold NP hybrids [100]. The substrate cleavage characteristic with Pb^{+2} ions using the DNAzyme can be combined with the gold NP for SPR biosensor. LOD was determined to 80 p.m. in drinking water. Thus, DNAzyme is also a versatile bioprobe for plasmonic biosensor development.

4. Conclusions

This review covered the most recent progress in nucleic acid-based plasmonic biosensor fabrication (Table 1). From cancer marker sensing to toxin detection, the combination of plasmon resonance-generating structures with oligonucleotides, such as aptamers, or more complex constructions, such as origami folded aptamers and DNAzymes, has certainly proven to be efficient. Not only do nucleic acid-mediated sensors have better affinity and stability, they are also easier to produce. Furthermore, surface plasmon resonance-based sensing presents numerous advantages, such as lower detection limits due to their increased sensitivity and a shorter sensing time. Yet, certain aspects, such as the nonspecific adsorption of unwanted molecules and the production of homogeneous plasmon surfaces should be considered for further improvements. Only then can the potential of plasmonic biosensors for application in real sample analysis be achieved. Nonetheless, based on recent studies, DNA nanotechnology for plasmonic biosensors offers a more sensitive and less costly alternative to other traditional sensing methods.

Table 1. DNA nanotechnology for plasmonic biosensor construction.

Detection Method	Oligonucleotide Component	Plasmonic Component	Target	LOD	Reference
SPR [1]	Aptamer	AuNPs	Oxytetracycline	0.50 ng/mL	[66]
		Graphene coated Au chip	Kanamycin	285 nmol/L	[67]
		niobium carbide MXene quantum dots coated Au chip	N-gene of SARS-CoV-2	4.9 pg/mL	[68]
		Ag nanosols	Arsenic (As^{3+})	0.01 μg/L	[69]
LSPR [2]	Aptamer	AuNPs	Ochratoxin A, triphosphate, 17β-estradiol and oxytetracycline hydrochloride	$10^{-10.5}$ g/mL	[71]
	DNA 3 Way Junction	Hollow Au nanospike	Avian influenza	1 pM	[89]

Table 1. *Cont.*

Detection Method	Oligonucleotide Component	Plasmonic Component	Target	LOD	Reference
SERS [3]	Aptamer	Au nanosols	Glyphosate	0.002 nmol/L	[74]
		Ag nanosols	Urea	0.03 nM	[76]
		Au nanostars	Epithelial cell adhesion molecule (EpCAM)	100 nM to 500 nM.	[77]
		Au–Ag bimetallic nanotrepangs	PSMA, Her2 and AFP proteins expressing exosomes derived from LNCaP, SKBR3 and HepG2 cell lines	6 particles/μL, 72 particles/μL and 35 particles/μL, respectively	[78]
		AuNPs	Tau protein and Aβ(1–42)	3.7×10^{-2} nM	[79]
	DNA origami	AuNPs	N/A N/A	N/A N/A	[90] [91]
		AuNPs and AgNPs	N/A	N/A	[92]
MEF [4]	Aptamer	AuNPs	Caspase-3	10 pg/mL	[81]
			ATP	200 pM	[82]
		Au nanobipyramids	ATP	35 nM	[83]
		AgNPs aggregates	Hepatitis B virus DNA	50 fM	[84]

[1] Surface Plasmon Resonance. [2] Localized Surface Plasmon Resonance. [3] Surface-Enhanced Raman Scattering. [4] Metal-Enhanced Fluorecence.

Author Contributions: T.L. and J.-H.L. conceived and designed the manuscript; J.A.P. and Y.K. wrote the origami-based biosensor section; C.A. wrote the aptamer-based biosensor section. All authors have read and agreed to the published version of the manuscript.

Funding: This work was supported by the National Research Foundation of Korea (NRF) grant funded by the Korea government (MSIT) (No. 2021R1C1C1005583) and by Korea Environment Industry & Technology Institute (KEITI) through the program for the management of aquatic ecosystem health, funded by Korea Ministry of Environment (MOE). (2020003030001) and by the National Research Foundation of Korea (NRF) grant by the Korea government (MSIT) (2020R1C1C1011876) and by the Research Grant of Kwangwoon University in 2022.

Institutional Review Board Statement: Not applicable.

Informed Consent Statement: Not applicable.

Data Availability Statement: Data sharing not applicable to this article.

Conflicts of Interest: The authors declare no conflict of interest.

References

1. Albert, B.; Johnson, A.; Lewis, J.; Reff, M.; Roberts, K.; Walter, P. DNA, Chromosomes and Genomes. In *Molecular Biology of the Cell*, 6th ed.; Garland Pub: Spokane, WA, USA, 2014; Chapter 4; ISBN 970-0-8153-4432-2.
2. Birney, E.; Stamatoyannopoulos, J.A.; Dutta, A.; Guigó, R.; Gingeras, T.R.; Margulies, E.H.; Weng, Z.; Snyder, M.; Dermitzakis, E.T.; Thurman, R.E. Identification and Analysis of Functional Elements in 1% of the Human Genome by the ENCODE Pilot Project. *Nature* **2007**, *447*, 799–816. [PubMed]
3. Roberts, T.C.; Langer, R.; Wood, M.J.A. Advances in Oligonucleotide Drug Delivery. *Nat. Rev. Drug Discov.* **2020**, *19*, 673–694. [CrossRef] [PubMed]
4. Kulkarni, J.A.; Witzigmann, D.; Thomson, S.B.; Chen, S.; Leavitt, B.R.; Cullis, P.R.; van der Meel, R. The Current Landscape of Nucleic Acid Therapeutics. *Nat. Nanotechnol.* **2021**, *16*, 630–643. [CrossRef] [PubMed]
5. Li, Y.; Zhao, D. *Basics of Molecular Biology*; Springer: Berlin/Heidelberg, Germany, 2013; pp. 541–601.
6. Fakruddin, M.; Hossain, Z.; Afroz, H. Prospects and Applications of Nanobiotechnology: A Medical Perspective. *J. Nanobiotechnol.* **2012**, *10*, 31. [CrossRef] [PubMed]

7. Naresh, V.; Lee, N. A Review on Biosensors and Recent Development of Nanostructured Materials-Enabled Biosensors. *Sensors* **2021**, *21*, 1109. [CrossRef]
8. Nagamune, T. Biomolecular Engineering for Nanobio/Bionanotechnology. *Nano Converg.* **2017**, *4*, 1–56. [CrossRef]
9. Chen, Y.J.; Groves, B.; Muscat, R.A.; Seelig, G. DNA Nanotechnology from the Test Tube to the Cell. *Nat. Publ. Group* **2015**, *10*, 748–760. [CrossRef]
10. Keller, A.; Linko, V. Challenges and Perspectives of DNA Nanostructures in Biomedicine. *Angew. Chem. Int. Ed.* **2020**, *59*, 15818–15833. [CrossRef]
11. Chao, J.; Zhu, D.; Zhang, Y.; Wang, L.; Fan, C. DNA Nanotechnology-Enabled Biosensors. *Biosens. Bioelectron.* **2016**, *76*, 15818–15833. [CrossRef]
12. Shen, L.; Wang, P.; Ke, Y. DNA Nanotechnology-Based Biosensors and Therapeutics. *Adv. Healthc. Mater.* **2021**, *10*, 2002205. [CrossRef]
13. Ke, Y.; Castro, C.; Choi, J.H. Structural DNA Nanotechnology: Artificial Nanostructures for Biomedical Research. *Annu. Rev. Biomed. Eng.* **2018**, *20*, 375–401. [CrossRef]
14. Pinheiro, A.V.; Han, D.; Shih, W.M.; Yan, H. Challenges and Opportunities for Structural DNA Nanotechnology. *Nat. Publ. Group* **2011**, *6*, 763–772. [CrossRef]
15. La, T.H.; Nguyen, T.T.T.; Pham, V.P.; Nguyen, T.M.H.; Le, Q.H. Using DNA Nanotechnology to Produce a Drug Delivery System. *Adv. Nat. Sci. Nanosci. Nanotechnol.* **2013**, *4*, 015002. [CrossRef]
16. Hu, Q.; Li, H.; Wang, L.; Gu, H.; Fan, C. DNA Nanotechnology-Enabled Drug Delivery Systems. *Chem. Rev.* **2019**, *119*, 6459–6506. [CrossRef]
17. Song, T.; Eshra, A.; Shah, S.; Bui, H.; Fu, D.; Yang, M.; Mokhtar, R.; Reif, J. Fast and Compact DNA Logic Circuits Based on Single-Stranded Gates Using Strand-Displacing Polymerase. *Nat. Nanotechnol.* **2019**, *14*, 1075–1081. [CrossRef]
18. Zhang, C.; Zhao, Y.; Xu, X.; Xu, R.; Li, H.; Teng, X.; Du, Y.; Miao, Y.; Lin, H.C.; Han, D. Cancer Diagnosis with DNA Molecular Computation. *Nat. Nanotechnol.* **2020**, *15*, 709–715. [CrossRef]
19. Dey, S.; Fan, C.; Gothelf, K.V.; Li, J.; Lin, C.; Liu, L.; Liu, N.; Nijenhuis, M.A.D.; Saccà, B.; Simmel, F.C.; et al. DNA Origami. *Nat. Rev. Methods Primers* **2021**, *1*, 13. [CrossRef]
20. Hong, F.; Zhang, F.; Liu, Y.; Yan, H. DNA Origami: Scaffolds for Creating Higher Order Structures. *Chem. Rev.* **2017**, *117*, 12584–12640. [CrossRef]
21. Liu, W.; Duan, H.; Zhang, D.; Zhang, X.; Luo, Q.; Xie, T.; Yan, H.; Peng, L.; Hu, Y.; Liang, L.; et al. Concepts and Application of DNA Origami and DNA Self-Assembly: A Systematic Review. *Appl. Bionics Biomech.* **2021**, *2021*, 9112407. [CrossRef]
22. Amir, Y.; Ben-Ishay, E.; Levner, D.; Ittah, S.; Abu-Horowitz, A.; Bachelet, I. Universal Computing by DNA Origami Robots in a Living Animal. *Nat. Publ. Group* **2014**, *9*, 353–357. [CrossRef]
23. Saran, R.; Wang, Y.; Li, I.T.S. Mechanical Flexibility of Dna: A Quintessential Tool for Dna Nanotechnology. *Sensors* **2020**, *20*, 7019. [CrossRef] [PubMed]
24. Raveendran, M.; Lee, A.J.; Sharma, R.; Wälti, C.; Actis, P. Rational Design of DNA Nanostructures for Single Molecule Biosensing. *Nat. Commun.* **2020**, *11*, 4384. [CrossRef] [PubMed]
25. Liu, S.; Jiang, Q.; Zhao, X.; Zhao, R.; Wang, Y.; Wang, Y.; Liu, J.; Shang, Y.; Zhao, S.; Wu, T.; et al. A DNA Nanodevice-Based Vaccine for Cancer Immunotherapy. *Nat. Mater.* **2021**, *20*, 421–430. [CrossRef] [PubMed]
26. Weiden, J.; Bastings, M.M.C. DNA Origami Nanostructures for Controlled Therapeutic Drug Delivery. *Curr. Opin. Colloid Interface Sci.* **2021**, *52*, 101411. [CrossRef]
27. Adachi, T.; Nakamura, Y. Aptamers: A Review of Their Chemical Properties. *Molecules* **2019**, *24*, 4229. [CrossRef]
28. Keefe, A.D.; Pai, S.; Ellington, A. Aptamers as Therapeutics. *Nat. Rev. Drug Discov.* **2010**, *9*, 537–550. [CrossRef]
29. Afrasiabi, S.; Pourhajibagher, M.; Raoofian, R.; Tabarzad, M.; Bahador, A. Therapeutic Applications of Nucleic Acid Aptamers in Microbial Infections. *J. Biomed. Sci.* **2020**, *27*, 6. [CrossRef]
30. Achenbach, J.; Chiuman, W.; Cruz, R.; Li, Y. DNAzymes: From Creation In Vitro to Application In Vivo. *Curr. Pharm. Biotechnol.* **2005**, *5*, 321–336. [CrossRef]
31. Ma, L.; Liu, J. Catalytic Nucleic Acids: Biochemistry, Chemical Biology, Biosensors, and Nanotechnology. *Iscience* **2020**, *23*, 100815. [CrossRef]
32. Morrison, D.; Rothenbroker, M.; Li, Y. DNAzymes: Selected for Applications. *Small Methods* **2018**, *2*, 1700319. [CrossRef]
33. Sun, H.; Zhu, X.; Lu, P.Y.; Rosato, R.R.; Tan, W.; Zu, Y. Oligonucleotide Aptamers: New Tools for Targeted Cancer Therapy. *Am. Soc. Gene Cell Ther.* **2014**, *3*, e182. [CrossRef]
34. Ellington, A.D.; Szostak, J.W. In vitro selection of RNA molecules that bind specific ligands. *Nature* **1990**, *34*, 818–822. [CrossRef]
35. Tuerk, C.; Gold, L. Systematic Evolution of Ligands by Exponential Enrichment: RNA Ligands to Bacteriophage T4 DNA Polymerase. *Science* **1990**, *249*, 505–510. [CrossRef]
36. Ng, E.W.M.; Shima, D.T.; Calias, P.; Cunningham, E.T.; Guyer, D.R.; Adamis, A.P. Pegaptanib, a Targeted Anti-VEGF Aptamer for Ocular Vascular Disease. *Nat. Rev. Drug Discov.* **2006**, *5*, 123–132. [CrossRef]
37. Lee, M.; Park, S.J.; Kim, G.; Park, C.; Lee, M.H.; Ahn, J.H.; Lee, T. A Pretreatment-Free Electrical Capacitance Biosensor for Exosome Detection in Undiluted Serum. *Biosens. Bioelectron.* **2022**, *199*, 113872. [CrossRef]
38. Kim, J.; Noh, S.W.; Park, C.; Lee, J.H.; Cho, H.Y.; Min, J.; Lee, T. Fabrication of Electrochemical Biosensor Composed of Multi-Functional DNA 4 Way Junction for TNF-α Detection in Human Serum. *Bioelectrochemistry* **2021**, *142*, 107939. [CrossRef]

39. Pultar, J.; Sauer, U.; Domnanich, P.; Preininger, C. Aptamer-Antibody on-Chip Sandwich Immunoassay for Detection of CRP in Spiked Serum. *Biosens. Bioelectron.* **2009**, *24*, 1456–1461. [CrossRef]
40. Abbaspour, A.; Norouz-Sarvestani, F.; Noori, A.; Soltani, N. Aptamer-Conjugated Silver Nanoparticles for Electrochemical Dual-Aptamer-Based Sandwich Detection of Staphylococcus Aureus. *Biosens. Bioelectron.* **2015**, *68*, 149–155. [CrossRef]
41. Kaur, H.; Bruno, J.G.; Kumar, A.; Sharma, T.K. Aptamers in the Therapeutics and Diagnostics Pipelines. *Theranostics* **2018**, *8*, 4016. [CrossRef]
42. Huo, W.; Li, X.; Wang, B.; Zhang, H.; Zhang, J.; Yang, X.; Jin, Y. Recent Advances of DNAzyme-Based Nanotherapeutic Platform in Cancer Gene Therapy. *Biophys. Rep.* **2020**, *6*, 256–265. [CrossRef]
43. Wang, Y.; Nguyen, K.; Spitale, R.C.; Chaput, J.C. A Biologically Stable DNAzyme That Efficiently Silences Gene Expression in Cells. *Nat. Chem.* **2021**, *13*, 319–326. [CrossRef]
44. McConnell, E.M.; Cozma, I.; Mou, Q.; Brennan, J.D.; Lu, Y.; Li, Y. Biosensing with DNAzymes. *Chem. Soc. Rev.* **2021**, *50*, 8954–8994. [CrossRef]
45. Khan, S.; Burciu, B.; Filipe, C.D.M.; Li, Y.; Dellinger, K.; Didar, T.F. DNAzyme-Based Biosensors: Immobilization Strategies, Applications, and Future Prospective. *ACS Nano* **2021**, *15*, 13943–13969. [CrossRef]
46. Lee, T.; Park, S.Y.; Jang, H.; Kim, G.H.; Lee, Y.; Park, C.; Mohammadniaei, M.; Lee, M.H.; Min, J. Fabrication of Electrochemical Biosensor Consisted of Multi-Functional DNA Structure/Porous Au Nanoparticle for Avian Influenza Virus (H5N1) in Chicken Serum. *Mater. Sci. Eng. C* **2019**, *99*, 511–519. [CrossRef]
47. Ramakrishnan, S.; Krainer, G.; Grundmeier, G.; Schlierf, M.; Keller, A. Structural Stability of DNA Origami Nanostructures in the Presence of Chaotropic Agents. *Nanoscale* **2016**, *8*, 10398–10405. [CrossRef]
48. Rutten, I.; Daems, D.; Lammertyn, J. Boosting Biomolecular Interactions through DNA Origami Nano-Tailored Biosensing Interfaces. *R. Soc. Chem.* **2020**, *8*, 3606–3615. [CrossRef]
49. Assenberg, R.; Weston, A.; Cardy, D.L.N.; Fox, K.R. Sequence-Dependent Folding of DNA Three-Way Junctions. *Nucleic Acids Res.* **2002**, *30*, 5142–5150. [CrossRef]
50. Lee, T.; Lee, Y.; Park, S.Y.; Hong, K.; Kim, Y.; Park, C.; Chung, Y.H.; Lee, M.H.; Min, J. Fabrication of Electrochemical Biosensor Composed of Multi-Functional DNA Structure/Au Nanospike on Micro-Gap/PCB System for Detecting Troponin I in Human Serum. *Colloids Surf. B Biointerfaces* **2019**, *175*, 343–350. [CrossRef]
51. Li, J.; Liu, Q.; Xi, H.; Wei, X.; Chen, Z. Y-Shaped DNA Duplex Structure-Triggered Gold Nanoparticle Dimers for Ultrasensitive Colorimetric Detection of Nucleic Acid with the Dark-Field Microscope. *Anal. Chem.* **2017**, *89*, 12850–12856. [CrossRef]
52. Zhou, L.; Wang, Y.; Yang, C.; Xu, H.; Luo, J.; Zhang, W.; Tang, X.; Yang, S.; Fu, W.; Chang, K.; et al. A Label-Free Electrochemical Biosensor for MicroRNAs Detection Based on DNA Nanomaterial by Coupling with Y-Shaped DNA Structure and Non-Linear Hybridization Chain Reaction. *Biosens. Bioelectron.* **2019**, *126*, 657–663. [CrossRef]
53. Kim, J.; Park, J.A.; Yim, G.; Jang, H.; Kim, T.H.; Sohn, H.; Lee, T. Fabrication of an Electrochemical Biosensor Composed of Multi-Functional Ag Ion Intercalated DNA Four-Way Junctions/Rhodium Nanoplate Heterolayer on a Micro-Gap for C-Reactive Protein Detection in Human Serum. *R. Soc. Chem.* **2021**, *146*, 2085–2402. [CrossRef] [PubMed]
54. Xi, H.; Juhas, M.; Zhang, Y. G-Quadruplex Based Biosensor: A Potential Tool for SARS-CoV-2 Detection. *Biosens. Bioelectron.* **2020**, *167*, 112494. [CrossRef] [PubMed]
55. Peng, Y.; Wang, X.; Xiao, Y.; Feng, L.; Zhao, C.; Ren, J.; Qu, X. I-Motif Quadruplex DNA-Based Biosensor for Distinguishing Single- and Multiwalled Carbon Nanotubes. *J. Am. Chem. Soc* **2009**, *131*, 13813–13818. [CrossRef] [PubMed]
56. Mochán, W.L. Plasmons. *Ref. Modul. Mater. Sci. Mater. Eng.* **2016**, 1–13.
57. Li, H.; Zhang, L. Photocatalytic Performance of Different Exposed Crystal Facets of BiOCl. *Curr. Opin. Green Sustain. Chem.* **2017**, *6*, 48–56. [CrossRef]
58. Petryayeva, E.; Krull, U.J. Localized Surface Plasmon Resonance: Nanostructures, Bioassays and Biosensing—A Review. *Anal. Chim. Acta* **2011**, *706*, 8–24. [CrossRef]
59. Szunerits, S.; Boukherroub, R. Sensing Using Localised Surface Plasmon Resonance Sensors. *Chem. Commun.* **2012**, *48*, 8999–9010. [CrossRef]
60. Duan, Q.; Liu, Y.; Chang, S.; Chen, H.; Chen, J.H. Surface Plasmonic Sensors: Sensing Mechanism and Recent Applications. *Sensors* **2021**, *21*, 5262. [CrossRef]
61. Chatterjee, B.; Kalyani, N.; Das, S.; Anand, A.; Sharma, T.K. Nano-Realm for Point-of-Care (POC) Bacterial Diagnostics. *Methods Microbiol.* **2019**, *46*, 19–42.
62. Langer, J.; de Aberasturi, D.J.; Aizpurua, J.; Alvarez-Puebla, R.A.; Auguié, B.; Baumberg, J.J.; Bazan, G.C.; Bell, S.E.J.; Boisen, A.; Brolo, A.G.; et al. Present and Future of Surface-Enhanced Raman Scattering. *ACS Nano* **2020**, *14*, 28–117. [CrossRef]
63. Bodelón, G.; Montes-García, V.; Pérez-Juste, J.; Pastoriza-Santos, I. Surface-Enhanced Raman Scattering Spectroscopy for Label-Free Analysis of *P. Aeruginosa* Quorum Sensing. *Front. Cell. Infect. Microbiol.* **2018**, *8*, 143. [CrossRef]
64. Jeong, Y.; Kook, Y.M.; Lee, K.; Koh, W.G. Metal Enhanced Fluorescence (MEF) for Biosensors: General Approaches and a Review of Recent Developments. *Biosens. Bioelectron.* **2018**, *111*, 102–116. [CrossRef]
65. Fothergill, S.M.; Joyce, C.; Xie, F. Metal-Enhanced Fluorescence Biosensing: From Ultra-Violet towards Second near-Infrared Window. *Nanoscale* **2018**, *10*, 20914–20929. [CrossRef]
66. Huang, H.; Li, J.; Pan, S.; Wang, H.; Liang, A.; Jiang, Z. A Novel Small Molecular Liquid Crystal Catalytic Amplification-Nanogold SPR Aptamer Absorption Assay for Trace Oxytetracycline. *Talanta* **2021**, *233*, 122528. [CrossRef]

67. Écija-Arenas, Á.; Kirchner, E.M.; Hirsch, T.; Fernández-Romero, J.M. Development of an Aptamer-Based SPR-Biosensor for the Determination of Kanamycin Residues in Foods. *Anal. Chim. Acta* **2021**, *1169*, 338631. [CrossRef]
68. Chen, R.; Kan, L.; Duan, F.; He, L.; Wang, M.; Cui, J.; Zhang, Z.; Zhang, Z. Surface Plasmon Resonance Aptasensor Based on Niobium Carbide MXene Quantum Dots for Nucleocapsid of SARS-CoV-2 Detection. *Microchim. Acta* **2021**, *188*, 316. [CrossRef]
69. Zhang, Z.H.; Lei, K.N.; Li, C.N.; Luo, Y.H.; Jiang, Z.L. A New and Facile Nanosilver SPR Colored Method for Ultratrace Arsenic Based on Aptamer Regulation of Au-Doped Carbon Dot Catalytic Amplification. *Spectrochim. Acta-Part A Mol. Biomol. Spectrosc.* **2020**, *232*, 118174. [CrossRef]
70. Sodhi, K.K.; Kumar, M.; Agrawal, P.K.; Singh, D.K. Perspectives on Arsenic Toxicity, Carcinogenicity and Its Systemic Remediation Strategies. *Environ. Technol. Innov.* **2019**, *16*, 100462. [CrossRef]
71. Liu, B.; Huang, R.; Yu, Y.; Su, R.; Qi, W.; He, Z. Gold Nanoparticle-Aptamer-Based LSPR Sensing of Ochratoxin A at a Widened Detection Range by Double Calibration Curve Method. *Front. Chem.* **2018**, *6*, 94. [CrossRef]
72. Sun, D.; Wu, Y.; Chang, S.J.; Chen, C.J.; Liu, J.T. Investigation of the Recognition Interaction between Glycated Hemoglobin and Its Aptamer by Using Surface Plasmon Resonance. *Talanta* **2021**, *222*, 121466. [CrossRef]
73. Hu, R.; Huang, X.; Huang, J.; Li, Y.; Zhang, C.; Yin, Y.; Chen, Z.; Jin, Y.; Cai, J.; Cui, F. Long- and Short-Term Health Effects of Pesticide Exposure: A Cohort Study from China. *PLoS ONE* **2015**, *10*, e0128766. [CrossRef] [PubMed]
74. Liu, Q.; Zhang, R.; Yu, B.; Liang, A.; Jiang, Z. A Highly Sensitive Gold Nanosol SERS Aptamer Assay for Glyphosate with a New COF Nanocatalytic Reaction of Glycol-Au(III). *Sens. Actuators B Chem.* **2021**, *344*, 130288. [CrossRef]
75. Austin, J. Urea Toxicity and Its Prevention. In *Urea as a Protein Supplement*; Briggs, M.H., Ed.; Elsevier: Amsterdam, The Netherlands, 1967; Chapter 9; pp. 173–184.
76. Yao, D.; Wang, H.; Lu, S.; Li, C.; Liang, A.; Wen, G.; Jiang, Z. On-Signal Amplification of Silver Nanosol RRS/SERS Aptamer Detection of Ultratrace Urea by Polystyrene Nanosphere Catalyst. Spectrochim. *Acta-Part A Mol. Biomol. Spectrosc.* **2022**, *265*, 120353. [CrossRef] [PubMed]
77. Bhamidipati, M.; Cho, H.Y.; Lee, K.B.; Fabris, L. SERS-Based Quantification of Biomarker Expression at the Single Cell Level Enabled by Gold Nanostars and Truncated Aptamers. *Bioconjug. Chem.* **2018**, *29*, 2970–2981. [CrossRef] [PubMed]
78. Ning, C.F.; Wang, L.; Tian, Y.F.; Yin, B.C.; Ye, B.C. Multiple and Sensitive SERS Detection of Cancer-Related Exosomes Based on Gold-Silver Bimetallic Nanotrepangs. *Analyst* **2020**, *145*, 2795–2804. [CrossRef]
79. Zhang, X.; Liu, S.; Song, X.; Wang, H.; Wang, J.; Wang, Y.; Huang, J.; Yu, J. Robust and Universal SERS Sensing Platform for Multiplexed Detection of Alzheimer's Disease Core Biomarkers Using PAapt-AuNPs Conjugates. *ACS Sens.* **2019**, *4*, 2140–2149. [CrossRef]
80. Zhang, Y.; Yang, C.; Zhang, G.; Peng, Z.; Yao, L.; Wang, Q.; Cao, Z.; Mu, Q.; Xuan, L. Distance-Dependent Metal Enhanced Fluorescence by Flowerlike Silver Nanostructures Fabricated in Liquid Crystalline Phase. *Opt. Mater.* **2017**, *72*, 289–294. [CrossRef]
81. Choi, J.H.; Choi, J.W. Metal-Enhanced Fluorescence by Bifunctional Au Nanoparticles for Highly Sensitive and Simple Detection of Proteolytic Enzyme. *Nano Lett.* **2020**, *20*, 7100–7107. [CrossRef]
82. Jiang, L.; Hang, X.; Zhang, P.; Zhang, J.; Wang, Y.; Wang, W.; Ren, L. A Highly Sensitive Fluorescence-Enhanced Aptasensor Based on PolyA_n-Aptamer Nanostructure. *Microchem. J.* **2019**, *148*, 285–290. [CrossRef]
83. Zheng, M.; Kang, Y.; Liu, D.; Li, C.; Zheng, B.; Tang, H. Detection of ATP from "Fluorescence" to "Enhanced Fluorescence" Based on Metal-Enhanced Fluorescence Triggered by Aptamer Nanoswitch. *Sens. Actuators B Chem.* **2020**, *319*, 128263. [CrossRef]
84. Jin, F.; Li, H.; Xu, D. Enzyme-Free Fluorescence Microarray for Determination of Hepatitis B Virus DNA Based on Silver Nanoparticle Aggregates-Assisted Signal Amplification. *Anal. Chim. Acta* **2019**, *1077*, 297–304. [CrossRef]
85. Wang, S.; Zhou, Z.; Yang, S.; Li, K.; Teng, C.; Ke, Y.; Tian, Y. DNA Origami-Enabled Biosensors. *Sensors* **2020**, *20*, 6899. [CrossRef]
86. Bellassai, N.; D'Agata, R.; Spoto, G. Novel Nucleic Acid Origami Structures and Conventional Molecular Beacon–Based Platforms: A Comparison in Biosensing Applications. *Anal. Bioanal. Chem.* **2021**, *413*, 6063–6077. [CrossRef]
87. Lu, J.; Wang, J.; Hu, X.; Gyimah, E.; Yakubu, S.; Wang, K.; Wu, X.; Zhang, Z. Electrochemical Biosensor Based on Tetrahedral DNA Nanostructures and G-Quadruplex-Hemin Conformation for the Ultrasensitive Detection of MicroRNA-21 in Serum. *Anal. Chem.* **2019**, *91*, 7353–7359. [CrossRef]
88. Lee, T.; Mohammadniaei, M.; Zhang, H.; Yoon, J.; Choi, H.K.; Guo, S.; Guo, P.; Choi, J.W. Single Functionalized PRNA/Gold Nanoparticle for Ultrasensitive MicroRNA Detection Using Electrochemical Surface-Enhanced Raman Spectroscopy. *Adv. Sci.* **2020**, *7*, 1902477. [CrossRef]
89. Lee, T.; Kim, G.H.; Kim, S.M.; Hong, K.; Kim, Y.; Park, C.; Sohn, H.; Min, J. Label-Free Localized Surface Plasmon Resonance Biosensor Composed of Multi-Functional DNA 3 Way Junction on Hollow Au Spike-like Nanoparticles (HAuSN) for Avian Influenza Virus Detection. *Colloids Surf. B Biointerfaces* **2019**, *182*, 110341. [CrossRef]
90. Puchkova, A.; Vietz, C.; Pibiri, E.; Wünsch, B.; Sanz Paz, M.; Acuna, G.P.; Tinnefeld, P. DNA Origami Nanoantennas with over 5000-Fold Fluorescence Enhancement and Single-Molecule Detection at 25 μM. *Nano Lett.* **2015**, *15*, 8354–8359. [CrossRef]
91. Thacker, V.V.; Herrmann, L.O.; Sigle, D.O.; Zhang, T.; Liedl, T.; Baumberg, J.J.; Keyser, U.F. DNA Origami Based Assembly of Gold Nanoparticle Dimers for Surface-Enhanced Raman Scattering. *Nat. Commun.* **2014**, *5*, 3448. [CrossRef]
92. Zhou, C.; Yang, Y.; Li, H.; Gao, F.; Song, C.; Yang, D.; Xu, F.; Liu, N.; Ke, Y.; Su, S.; et al. Programming Surface-Enhanced Raman Scattering of DNA Origami-Templated Metamolecules. *Nano Lett.* **2020**, *20*, 3155–3159. [CrossRef]
93. Jung, J.K.; Archuleta, C.M.; Alam, K.K.; Lucks, J.B. Programming Cell-Free Biosensors with DNA Strand Displacement Circuits. *Nat. Chem. Biol.* **2022**, *18*, 385–393. [CrossRef]

94. Dass, M.; Gür, F.N.; Kołątaj, K.; Urban, M.J.; Liedl, T. DNA Origami-Enabled Plasmonic Sensing. *J. Phys. Chem.* **2021**, *125*, 5969–5981. [CrossRef] [PubMed]
95. Zhao, Y.; Chu, X.; Yang, B. Electrochemical Behavior of Hemin Binding with Human Centrin 3. *Bioelectrochemistry* **2017**, *117*, 15–22. [CrossRef] [PubMed]
96. Peng, H.; Newbigging, A.M.; Wang, Z.; Tao, J.; Deng, W.; Le, X.C.; Zhang, H. DNAzyme-Mediated Assays for Amplified Detection of Nucleic Acids and Proteins. *Anal. Chem.* **2018**, *90*, 190–207. [CrossRef] [PubMed]
97. Hasick, N.J.; Ramadas, R.; Todd, A.V. Subzymes: Regulating DNAzymes for Point of Care Nucleic Acid Sensing. *Sens. Actuators B Chem.* **2019**, *297*, 126704. [CrossRef]
98. Yu, F.; Li, Y.; Li, M.; Tang, L.; He, J.J. DNAzyme-Integrated Plasmonic Nanosensor for Bacterial Sample-to-Answer Detection. *Biosens. Bioelectron.* **2017**, *89*, 880–885. [CrossRef]
99. Lee, W.; Hwang, B.H. Plasmonic Biosensor Controlled by DNAzyme for On-Site Genetic Detection of Pathogens. *Biotechnol. J.* **2020**, *15*, 1900329. [CrossRef]
100. Wu, H.; Wang, S.; Li, S.F.Y.; Bao, Q.; Xu, Q. A label-free lead(II) ion sensor based on surface plasmon resonance and DNAzyme-gold nanoparticle conjugates. *Anal. Bioanal. Chem.* **2020**, *412*, 7525–7533. [CrossRef]

biosensors

Review

Recent Trends in SERS-Based Plasmonic Sensors for Disease Diagnostics, Biomolecules Detection, and Machine Learning Techniques

Reshma Beeram, Kameswara Rao Vepa and Venugopal Rao Soma *

Advanced Centre of Research in High Energy Materials (ACRHEM), DRDO Industry Academia—Centre of Excellence (DIA-COE), University of Hyderabad, Hyderabad 500046, Telangana, India
* Correspondence: soma_venu@uohyd.ac.in or soma_venu@yahoo.ac.in

Abstract: Surface-enhanced Raman spectroscopy/scattering (SERS) has evolved into a popular tool for applications in biology and medicine owing to its ease-of-use, non-destructive, and label-free approach. Advances in plasmonics and instrumentation have enabled the realization of SERS's full potential for the trace detection of biomolecules, disease diagnostics, and monitoring. We provide a brief review on the recent developments in the SERS technique for biosensing applications, with a particular focus on machine learning techniques used for the same. Initially, the article discusses the need for plasmonic sensors in biology and the advantage of SERS over existing techniques. In the later sections, the applications are organized as SERS-based biosensing for disease diagnosis focusing on cancer identification and respiratory diseases, including the recent SARS-CoV-2 detection. We then discuss progress in sensing microorganisms, such as bacteria, with a particular focus on plasmonic sensors for detecting biohazardous materials in view of homeland security. At the end of the article, we focus on machine learning techniques for the (a) identification, (b) classification, and (c) quantification in SERS for biology applications. The review covers the work from 2010 onwards, and the language is simplified to suit the needs of the interdisciplinary audience.

Keywords: biosensing; SERS; plasmonics; disease diagnosis; biomolecules; microorganisms; COVID-19; biohazardous molecules; cancer

Citation: Beeram, R.; Vepa, K.R.; Soma, V.R. Recent Trends in SERS-Based Plasmonic Sensors for Disease Diagnostics, Biomolecules Detection, and Machine Learning Techniques. *Biosensors* **2023**, *13*, 328. https://doi.org/10.3390/bios13030328

Received: 31 January 2023
Revised: 20 February 2023
Accepted: 24 February 2023
Published: 27 February 2023

1. Introduction

Plasmonics is the study of electron oscillations in metal nanostructures and their interaction with electromagnetic radiation. Since its conception in the 1950s, researchers have been interested in studying the fundamentals of the effects of shape, surrounding medium, material, and their interaction with light of different wavelengths [1]. With this well-established knowledge, plasmonics is witnessing an enormous potential for applications in different fields, including forensics [2]; environmental safety [3]; biosensing [4–11], e.g., SARS-CoV-2 detection [12]; and homeland security [13]. The applications of plasmonics majorly rely on surface plasmon resonance (SPR) or localized surface plasmon resonance (LSPR) effects [14]. Some of the significant techniques that were developed using these include higher-order harmonic generation, microscopy, drug delivery, photovoltaics, surface-enhanced Raman spectroscopy (SERS) and fluorescence, and surface-enhanced infrared absorption spectroscopy (SEIAS) and waveguides. The use of plasmonics in these techniques has significantly improved their efficiency over existing conventional techniques, offering flexibility, signal enhancement, and ease of use [15]. Advents in plasmonics have led to the emergence of SERS with impressive signal enhancements over traditional Raman spectroscopy [16]. SERS-based sensing is being widely used for the trace detection of different molecules, such as explosives [17], pesticides [18,19], food adulterants [20,21], drugs [22], biomolecules [23–27], medicine [28–30], and microorganisms [31].

SERS typically utilizes localized surface plasmon resonances in metal nanostructures to enhance the weak Raman signal significantly. The phenomenon was first observed by Fleischmann in 1974 while studying pyridine adsorbed on a roughened silver electrode [32]. However, the enhancement was attributed to increased surface area for adsorption. It took further experiments in 1977 by two independent groups, Jeanmaire and van Duyne [33] and Albrecht and Creighton [34], to understand the origin of the enhancement. Now it is established that the enhancement predominantly comes from two mechanisms: electromagnetic enhancement (EE) and chemical enhancement (CE) [35]. The electromagnetic enhancement in SERS is a two-step process, and the total enhancement is multiplicative. When a molecule of interest is in the vicinity of a plasmonic nanostructure, it experiences an enhanced field called local field enhancement (LFE). The molecule then radiates with increased efficiency, referred to as radiation enhancement [36,37]. In addition, there is chemical enhancement which occurs because of charge-transfer mechanisms between nanoparticles and the analyte. Figure 1 summarizes the two enhancement mechanisms in SERS. The type of the plasmonic material, choice of wavelength, surface coverage of the molecules, and concentration of the analyte are the factors that influence SERS's efficiency [38]. This technique is label-free, rapid, non-destructive, and water compatible and offers the fingerprint of the molecule, making it suitable for numerous applications. Nobel metals such as Au, Ag, and Cu and their alloys are the widely used materials for SERS for their tunability in the visible and IR region, inertness, sensitivity, and compatibility [39,40]. Despite the superior performance of Ag owing to its high-quality resonance in the visible region, Au is the preferred material, as it is known to be biocompatible and non-reactive in an oxygen atmosphere. The near-field enhancement in SERS is dependent on the shape and size of the nanostructures, in addition to the distance between the nanoparticles and distribution of probe molecules around the nanoparticles [41]. The different morphologies of nanoparticles, such as core–shell, rods, spherical, triangular, stars, and nanopyramids, are synthesized by widely reported chemical routes in bottom-up or top-down approaches [42]. Anisotropic nanostructures such as dendrites, rods, stars, and triangle are considered highly desirable for SERS since they enable lower detection limits owing to the lightening-rod effect [43,44]. The performance of SERS is also dependent on the choice of wavelength, and most biological tissues are transparent in the IR region, making it a preferred choice [45]. Recently, there is also growing interest in the UV and deep UV SERS for applications concerning biomolecules such as amino acids and DNA bases because they have electronic transitions in the UV region [39].

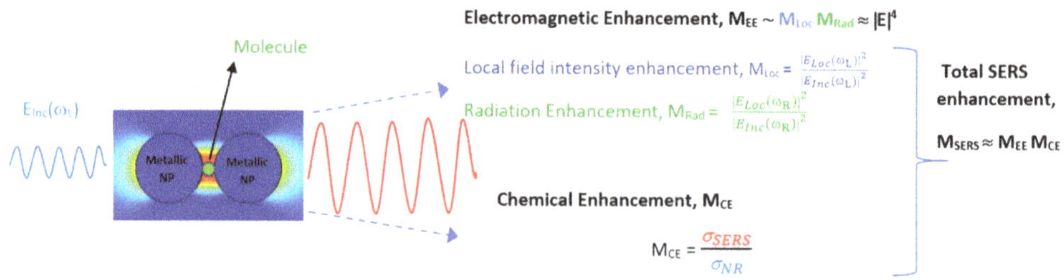

Figure 1. Schematic of total enhancement in SERS via electromagnetic and chemical enhancement mechanisms.

With a growing population and, consequently, the diseases worldwide, there is a need to develop point-of-care (POC) devices that are easy to use, reliable, rapid, and low cost. Over the years, SERS has been proven to possess all of these advantages, including trace detection with sub-picomolar sensitivity. Particularly, there are many reasons for the surge of using SERS for biosensing. Firstly, given the low scattering cross-section of water, SERS is extremely compatible with liquid samples, paving the way for use in biology applications, including liquid biopsy [46,47]. SERS has been widely used for disease diagnosis using urine, blood, serum, plasma, saliva, breath, and tear samples, establishing its compatibility. Measurements in SERS can be performed using liquids, gases, solids, and powders, unlike traditional tests. Secondly, SERS gives specific molecular information, which is often a vibrational fingerprint of the molecule or cell under study. Biomarkers that are Raman active are extensively used for the identification of different diseases, using SERS [48]. Frequently, when the variations are unrecognizable to the human eye, machine learning techniques are used to extract the patterns and discriminate the samples [49]. This was successfully used to classify normal and cancer cells [50], identify microorganism species [51], and monitor disease progression [52]. Thirdly, SERS is a rapid technique, can accomplish trace detection, and has a test time of three to five minutes [53]. Combined with recent developments in flexible SERS sensors, it also offers easy sample-collection methods, such as swabbing from an uneven surface [54]. Lastly, advances in portable instrumentation and low-cost lasers leveraged the usage of SERS for real-world applications [55]. The easy availability of IR lasers that have a low damage threshold with biology samples, as well as quench fluorescence, has favored the development of SERS for biosensing. All of these advantages have made SERS a popular choice for biosensing recently.

There have been many review articles concerning the applications of plasmonics for biosensing and biosensors over the years. Salazar et al. and Han et al. reviewed different techniques, including LSPR, Chiral Plasmonic Biosensors, Magnetoplasmonic Biosensors, and Quantum Plasmonics Biosensors [56,57]. Anand et al. published a comprehensive review on plasmonic biosensors for the detection of viruses, with a special focus on COVID-19. They have focused on LSPR, SPR, SERS, SEF, and SEIAS techniques [58]. There are reviews and book chapters elaborating specifically on various SPR [59–61] and LSPR [62] techniques that are currently being used for biosensing. Similarly, Sarah et al. focused exclusively on LSPR techniques and associated challenges in the detection [63]. Alexandre reviewed the future of plasmonic biosensing with a goal of single-molecule and single-particle sensing [64]. Juanjuan et al. discussed on the challenges and future of using plasmonic materials for point-of-need applications [65]. However, although significant work has been performed using SERS for biology applications, no reviews of the literature can be found in this area. Here, we present a review of the work conducted in SERS for biosensing and the recent developments, with a special focus on machine learning techniques that are being used for the same. The article covers work from 2010 onwards and is organized into different sections, as shown in the index. Figure 2a,b illustrate the statistics of publications in different areas discussed in this review article. The data presented in the figure indicate that there is a growing interest in the usage of SERS for cancer-related applications and the usage of machine learning techniques for biosensing using SERS. There is also relative growth in using SERS for respiratory disease diagnosis recently owing to the COVID-19 situation.

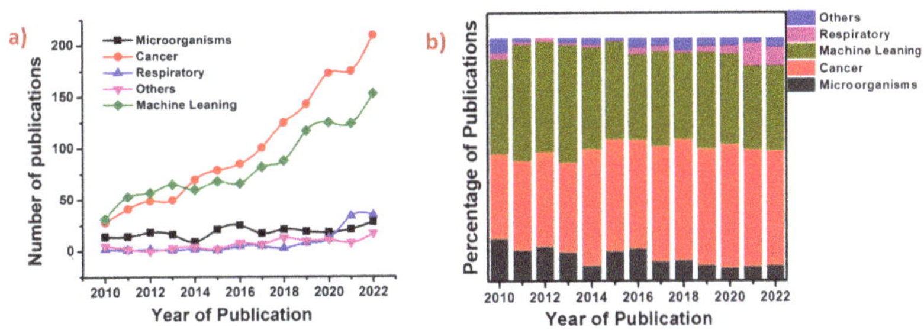

Figure 2. (**a**) Trends in research on SERS based plasmonic applications in the detection of microorganisms, cancers, respiratory diseases, other diseases such as heart ailments and diabetes and the use of different machine learning techniques for SERS based biosensing. (**b**) Bar chart with percentage contribution from each area shown on the label for the past 12 years. Source: Scopus search with the keywords mentioned in both the panels as on 5 January 2023.

2. SERS for Disease Diagnosis

With growing zoonotic diseases, cancers, diabetes, and other ailments, there is a pressing need to develop low-cost and POC identification techniques. Early and rapid diagnosis is the key to saving a life and prevent the rapid transmission of diseases. A trace detection technique such as SERS will aid in tracking the minute changes in cells or biomarkers, thus enabling early diagnosis. SERS is being extensively used for the same in both labeled and label-free approaches, often targeting specific biomarkers of the disease expression [30]. In the label-free approach, the sample is directly studied in contact with the plasmonic material, whereas in the labeled approach, a Raman reporter, such as fluorophores, antibodies, or ligands, is attached to the sample for detection and imaging [66,67]. Different biomarkers, such as proteins, antibodies, miRNAs, exosomes, and DNA, are used as indicators for the presence of the disease. In our observation, where full cells, tissues, or body fluids are studied, a machine learning algorithm is used hand in hand for accurate identification. SERS has been used for the detection of conditions such as Alzheimer's [68–71], PCOS [72], diabetes [73,74], inflammation [74], Crohn's disease [75], and single Hb molecule [76], to name a few. Here we review the progress on SERS for the diagnosis of (a) cancer, paying special attention to lung and breast cancer, as they are the leading causes of deaths due to cancer; and (b) respiratory viruses, including COVID-19.

2.1. *Cancer Diagnosis and Theranostics*

Cancer is the new pandemic and a leading cause of deaths in the modern world [77]. There is an increase in the incidence of various types of cancers, including mouth, gastric, lungs, ovaries, skin, and blood cancer. Numerous factors, such as environment, diet, lifestyle, and smoking, can trigger cancer. The early diagnosis of cancer is extremely important, as it is lifesaving with existing treatment protocols. Conventional cancer diagnosis is often performed using imaging techniques such as X-ray, computerized tomography scan (CT), positron emission tomography (PET), ultrasound, and magnetic resonance imaging (MRI). These techniques are often destructive, posing the risk of radiation ionization, and are often not compatible with patients with pre-existing conditions and medical devices such as pacemakers [78]. These are also expensive, involve sophisticated instruments, are time-consuming, and are often performed with multiple tests to avoid ambiguity [79]. Recently, there has been an increase in using plasmonic biosensing for cancer diagnosis and therapy, with review articles summarizing the progress in the same [80–86]. They are established to be minimally invasive, rapid, low cost, and offer point-of-care testing [87,88]. Of all plasmonic-based detection techniques, SERS is being extensively used for cancer identification, monitoring, and other theranostics, including imaging and chemo/photothermal therapy [89–96]. Figure 2 also indicates the growing interest in the last decade for the use of SERS-based plasmonic techniques for cancer diagnosis. SERS facilitates liquid biopsy [96] by using urine, saliva, and serum, thus making it low cost and enabling easier frequent sampling compared to the existing tissue-biopsy techniques, which are often destructive [97]. Different cancer biomarkers, such as miRNA [98,99], proteins, exosomes [100,101], circulating tumor DNA (ctDNA), genes [102], peptides [103], and blood plasma [104], are studied using SERS for disease identification. SERS tags that specifically bind to the targets under study are widely used for analyzing cancer samples [105–109]. Machine learning algorithms are used to analyze complex patterns and recognize buried signals overcoming noise from undesirable constituents of cells and other bio-fluids. Here, we focus only on SERS-based plasmonic biosensing for cancer-related applications in recent times, focusing on lung and breast cancers.

2.1.1. Lung Cancer

Lung cancer is known as the most fatal and frequently diagnosed cancer of all [110]. The cited report projected 2.89 million cases of lung cancer by 2030. Smoking, the presence of carcinogenic substances in the environment, and lifestyle are considered to be the main causes of lung cancer [88]. There are two kinds of lung cancers, non-small cell (NSCLC) and small cell lung cancer (SCLC). NSCLC is the most common kind of lung cancer, accounting for 80% of the cases. SCLC is the most fatal and fast-spreading cancer, and it is often diagnosed only at the later stages. SERS has been successfully used for the diagnosis of both kinds of lung cancers, with a prospect of developing point-of-care and rapid testing.

Using DNA-based complexes as SERS tags and miRNA as a biomarker, Mao et al. have developed a lateral-flow-assay-based SERS substrate for the rapid detection and quantification of lung cancer biomarkers in less than 30 min [111]. Two types of biomarkers, miR-21 and miR-196a-5p, were detected simultaneously and with comparable accuracy with the existing qRT-PCR techniques. Similar studies were carried out with a flexible filter paper substrate [112] and with different biomarkers [113,114]. They extended their studies with ctDNA as the biomarker, thus implying the versatility of SERS [115]. Similarly, miRNA has been used for the detection of lung cancer by using circular exponential amplification reaction (EXPAR)-based SERS [116]. With a combination of asymmetric PCR and SERS, in regard to mutation genes in ctDNA, Guo et al. achieved a highly specific (100%) and sensitive (75%) lung-cancer-detection method in blood samples. Asymmetric PCA was performed to obtain single-stranded DNA, followed by the SERS-based detection using specifically labeled Au substrates [117]. With exosomes as biomarkers, lung cancers at different stages were identified accurately (~90%) by using SERS and a deep learning algorithm to classify healthy and malignant samples, exhibiting potential for early diagnosis, as shown in Figure 3 [118]. Serum samples of normal and lung-cancer patients were analyzed using PCA and PLS analysis to discriminate and identify the cancer samples with SERS spectra and achieved an accuracy of 92% [119]. Similarly, using a core-satellite type of plasmonic materials and SERS, serum samples of healthy, benign, and malignant cases of lung cancers were classified with a combination of principal component analysis (PCA) and support vector machines (SVMs) [120]. CtDNA-based identification of lung cancer using a DNA-rN1-DNA-mediated SERS frequency shift method was developed to achieve sub-femtomolar sensitivity [121]. Similarly, exosomes derived from bronchoalveolar fluid were used for the detection [122]. Whole-exosome SERS spectra have been analyzed using PCA to classify lung cancer and normal samples with 95.3% sensitivity and 97.3% specificity [123]. Lung-cancer biomarkers (aldehydes) and cells were identified rapidly using renewable porous $CuFeSe_2$/Au nanostructures achieving an LOD of 1 ppb [124]. Challenging gaseous biomarkers called volatile organic compounds, which serve as indicators for lung cancer, were detected using ZIF-8-coated gold superparticles for the sensitive identification of lung cancer [125]. Pleural effusions of lung cancer and normal samples were studied using SERS and machine learning techniques to achieve a classification accuracy of 85% [126]. A combination of PCA and LDA has been used to classify lung cancer and normal samples with SERS analysis of serum samples, achieving sensitivity and specificity of 100% and 90%, respectively [127]. The same technique was used for SERS-based classification of lung-cancer-tissue slices [128]. Using common protein carcinoembryonic antigen (CEA) and a-fetoprotein (AFP) as biomarkers, SERS-based detection was performed for the diagnosis of lung cancer [129]. A non-destructive photothermal therapy targeting lung cancer cells (A549 cells) was developed using NIR radiation and Ag-Au shell–core structures were used for the SERS-based detection of the A549 cells. These nanostructures are highly specific and have different affinities for cancerous and non-cancerous cells, thus helping in tagging the cells. Based on the SERS activity of R6G molecules, the detection and phototherapy can be monitored [130]. Similar studies were carried out using reduced graphene oxide plasmonic substrates [131]. A multivariate analysis (SVM and PCA) of SERS data was used to identify and classify different types of lung cancers with an accuracy of 95% [132]. Choosing aldehydes in exhaled breath as biomarkers, highly sensitive, portable detection was performed, achieving LOD of 1.35 nM [133]. Chemometric techniques coupled with slippery liquid-infused porous surface-enhanced Raman spectroscopy were used for concentrating blood samples in a small area and thus enhancing the SERS signal for trace detection [134]. A gap mode plasmonic SERS substrate with a combination of Ag nanocubes and Au nanorods was used for the identification of lung-cancer-related exosomes [135]. A SERS analysis of saliva samples was performed to classify healthy and cancerous samples, using SVM and random forest, with a sensitivity of 95% and 97%, respectively [136]. With adenosine as a biomarker, urine samples were analyzed using SERS with Fe_3O_4/Au/Ag-based substrates, achieving good reproducibility, stability, and sensitivity of 10^{-10} M [137].

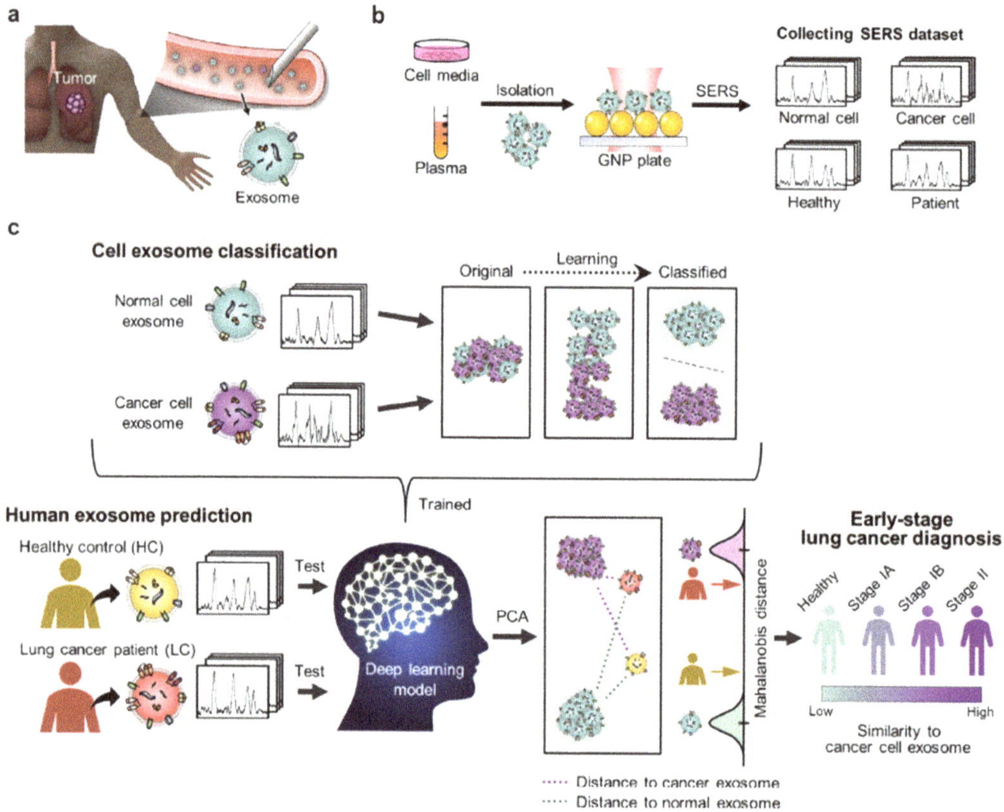

Figure 3. Schematic of work created by Hyunku et al. in lung cancer identification via a combination of SERS and deep learning. (**a**) Exosomes that were used as biomarkers for sensing. (**b**) Sample preparation and data collection. (**c**) Deep learning model used for the classification of normal and cancer cells exosomes using SERS spectra. Reproduced with permission from [118]. Copyright (2020), American Chemical Society.

2.1.2. Breast Cancer

Breast cancer is considered to be the second leading cause of death among women after lung cancer [138]. Breast cancer is often diagnosed by a mammogram, ultrasound, MRI, or biopsy. Furthermore, it is often concluded by a histopathological test, which is unfortunately time-consuming and is highly prone to human interpretation error. In addition to identification using urine, serum [139], and tear samples [140], SERS has also been used to understand the drug carrier mechanism [141] and classification of different stages of breast cancer [142], as shown in Figure 4.

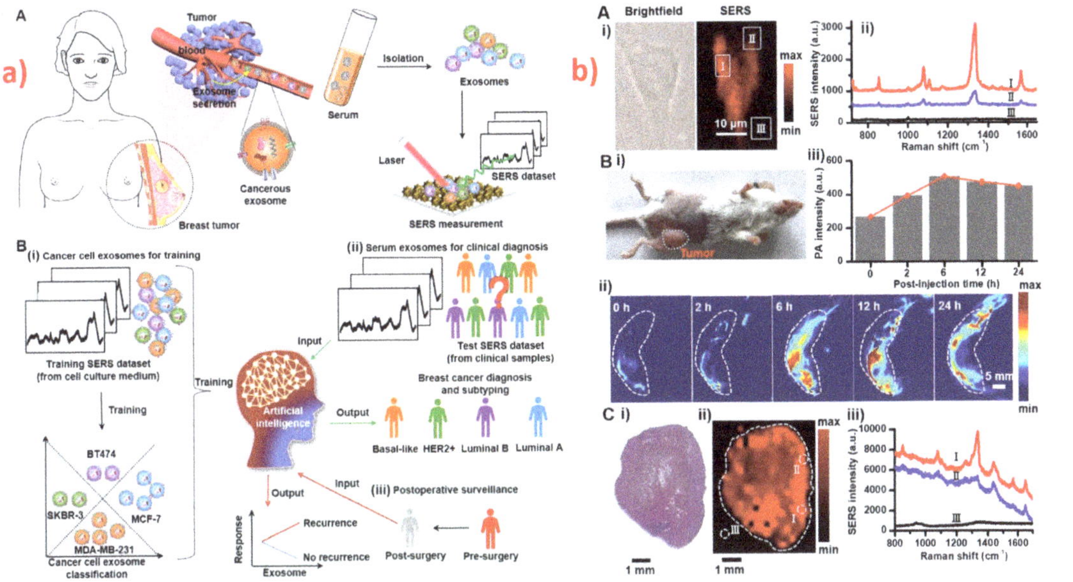

Figure 4. Breast cancer detection using SERS. (**a**) Schematic of workflow for (**A**) sample collection and (**B**) deep learning model-based breast cancer detection with exosomes-based SERS sensor. Reproduced with permission from [143]. Copyright (2022), American Chemical Society. (**b**) SERS and Photoacoustic (PA) imaging of breast cancer cells (**A**): (**i**) brightfield microscopic images and SERS mapping area, and (**ii**) corresponding SERS spectra in different regions labeled in (**i**). (**B**) (**i**) Photo of mouse with tumor, (**ii**) corresponding representative PA images for different post injection times, and (**iii**) PA intensity at 750 nm. (**C**) (**i**) Optical image of tumor, (**ii**) SERS image of tumor, and (**iii**) corresponding spectra for different regions in the image. Reproduced with permission from [144]. Copyright (2021), American Chemical Society.

Using the epidermal growth factor receptor as a biomarker, a gold-nanorods-based SERS tool that can identify and image the spatial and temporal distribution of breast cancer cells was developed by Xiao et al. [145]. Sialic acid, with its specificity towards a phenyboronic-acid-based nanoprobe, was used as biomarker for identification and imaging of breast cancer in human cells and saliva [146,147]. The miRNA of breast cancer was detected with a high sensitivity of 10^{-10} M, using a hybrid SERS substrate of GaN nanostructures with Au/Ag [148]. Functionalized SERS substrates with specific tags were used for the simultaneous isolation and detection of breast cancer cell lines [149]. Zheng et al. developed a SERS-based microfluidic channel for detection and quantification of prominent breast cancer biomarkers in real samples [150]. A combination of SERS and electrochemical biosensor was developed to monitor the drug response of DNA associated with breast cancer cells [52]. Hameed et al. have worked on fabricating anisotropic gold nano-stars that showed specific affinity to breast cancer cells compared to normal cells, aiding in the detection of the same [151]. Multiple SERS tags were used for understanding the drug-carrier mechanisms in breast cancer cells for the antineoplastic drug tamoxifen [141]. Similar studies were carried out with estrogen receptor alpha (ER-α) as the biomarker [152]. ER-α-based SERS has also been used for understanding cellular uptake mechanisms in breast cancer [153]. Labeled hollow silica-encapsulated gold nano-spheres were used for identifying and quantifying breast cancer biomarkers [154]. Choi et al. developed SERS nanotags by using Ag-Au hollow nanospheres that are durable, reproducible, and sensitive for the detection of various biomarkers for SERS [155]. A SERS-based 3D holograph was developed to detect and quantify nine miRNAs corresponding to breast cancer. Hairpin-like

DNA was used as SERS tags along with Raman reporters for each miRNA that are spatially separated on the SERS substrate [156]. Similar studies were carried out by Weng et al. [157], Li et al. [158], and Lee et al. [159]. A ratio-type method was developed for the discrimination of breast cancer and non-cancer cells, using the SERS technique. A plasmonic material with Rh6G as a tag for breast cancer biomarker (MMP-2), along with a standard (2-NT), was used for analyzing the live cells based on the ratio of SERS signals in standard and R6G [160]. Li et al. also performed similar studies [161] and was also used for quantitative molecular phenotyping in a different study [162]. Recently, ratiometric SERS has been used for the identification of breast cancer using Au@Ag and GO nanostructures [163]. Cell suspensions of normal and breast cancer cells were analyzed using SERS coupled with Random Forest classifier to understand the differences. It was found that breast cancer cells have high cholesterol, lipids, proteins, and nucleic acids relative to the normal cells, and the classification accuracy was nearly 78% [164]. A comparison was made between SERS and Raman performance for the classification of different stages of breast cancers, using PCA and PLS-LDA, and found that SERS leads to better accuracy (94%) relative to the Raman (83%) method [142]. PLS-LDA was also used by Zheng et al. for the identification of breast-cancer biomarkers, using HAp [165]. PLS-SVM, PLS-LDA, and PCA-LDA were used for the classification of breast cancer and the normal group [166–176]. An exosome-based CNN model was developed for the classification of breast cancer and normal samples, with an accuracy of 95% [177]. A systematic analysis of SERS spectra obtained from urine and serum samples was performed, and it was found that the urine samples demonstrated better accuracy in the classification [170]. Biomarkers tracking the epithelial–mesenchymal transition in the plasma samples of breast cancer cells were identified using SERS immunoassay [178]. SERS-based cancer cell imaging was performed using gold nanoparticles based on the specific affinity of phenylalanine [179]. Different methods for the preparation of plasmonic Ag nanoparticles and their effects in SERS signal were discussed by Beata et al. [180]. Photothermal therapy and SERS-based identification of breast cancer were performed using gold nanorods [181] and gold nanobipyramids [182]. NIR and SERS-based phototherapy and detection were also performed [183]. A three-in-one tool consisting of photoacoustic imaging, thermosurgery, and SERS was developed to address the concern of residual microtumors in breast cancer [144]. By combining artificial intelligence and SERS, researchers developed a label-free detection method of breast cancer exosomes with 100% accuracy. This was also used to assess the outcomes of the surgeries [143]. Au/HCP-PS nanospheres were used for the SERS-based detection of breast cancer, using tears from asymptomatic patients, along with chemometric analysis [140]. A Pt-based SERS template was developed using cost-effective methods for the detection of breast cancer exosomes that achieved a sensitivity of 83.3% and a specificity of 95.8% [184]. A combination of 2D graphene and plasmonic gold nanostars was used for trace identification of exosomes [185]. There are many reports of researchers using exosomes as biomarkers for the identification of cancer, using SERS [186]. A highly sensitive (EF ~10^5) and reproducible (2.7%) method was developed using Au@Ag nanospheres for the detection of breast-cancer-based extracellular metabolites [187]. Systematic experiments were performed to understand the effects of laser power and acquisition time on the reproducibility in immune-SERS microscopy and found that a longer acquisition time and higher laser power lead to poor reproducibility [188].

2.1.3. Miscellaneous

With the mechanism for detection being the same, SERS has been extensively used for the detection of several other cancers, including gastric [189], oral, liver, ovarian [190,191], and prostate cancers [192–197]. Gastric cancer diagnoses have been performed using different plasmonic materials by analyzing SERS spectra of serum samples [198–200], blood plasma [201], exosomes [202], extracellular vesicles [203], telomerase [204], saliva [205], and ctDNA [206]. In a breakthrough study, a breath analysis based on SERS was performed to identify different stages of gastric cancers by analyzing the Raman bands [207,208]. Different chemometric techniques such as PCA [209,210], PCA-LDA [201,211–213], SVM [214],

ANN [215], and PCA-QDA [216] were also used for classification and identification of gastric cancers. Li et al. used a combination of classification algorithms, such as PCA-LDA, PCA-SVM, and PCA-CART, for identifying gastric diseases in serum samples [217]. Blood samples from healthy and normal patients were analyzed for different cancers, such as liver cancer, colonic cancer, esophageal cancer, nasopharyngeal cancer, gastric cancer along with PCA-SVM and achieved an accuracy of 96% [218]. Based on the SERS profiling of urine samples, bladder cancer was studied using machine learning algorithms with miRNA as a biomarker [219]. PCA, random forest, KNN, and naive Bayes algorithms were used for the identification of renal cancer, with the SERS profiling of serum samples achieving accuracy greater than 75% [139]. Taking advantage of the coffee-ring effect, the serum samples of lung and prostate cancer patients were identified with 100% accuracy, using PLS-SVM algorithms on SERS data [220]. Gaussian-based CNNs were used for the same application elsewhere [221]. Recently, there was a review article specifically focusing on SERS-based biosensing for liver cancer detection applications [222]. Zhang et al. elaborated on the existing literature for oral cancer diagnosis and therapy with gold nanoparticles, highlighting the current progress and challenges [223]. A similar review article was also published for the case of ovarian cancer [224]. Oral cancer was studied using saliva samples and the miRNA of normal and cancer patients with the SERS technique [225,226]. SVM in combination with SERS has been used for the early detection of oral cancer among patients using serum and saliva samples and achieved an accuracy of 80% [227]. Prostate cancer has been extensively studied and successfully identified using different techniques, such as serum analysis combined with PCA-SVM [228]; detection of prostate specific antigens [229–232]; EVs combined with CNN [230]; miRNAs [233]; different multivariate techniques, e.g., PCA-LDA and PCA-SVM [234]; and urine profiling [235].

2.2. SARS-CoV-2 and Other Respiratory Diseases

With the onset of the pandemic and the fast-spreading variants, there was a need to rapidly identify, detect, and quarantine the infected population. Surveying the presence of antibodies in large populations, often called a serological survey, was important to access the percentage of population infected and to monitor community transmission [236]. The dominant existing technique for the identification of SARS-CoV was PCR, which relies on analyzing the genetic material of the virus [237]. However, the test is expensive, thus preventing wide usage and also is time consuming. The Raman spectrum of a whole organism, including viruses, is contributed to by the proteins, carbohydrates, and nucleic acids that make up the organism [238]. The expression of these building blocks is controlled by the genetic material of the organism, hence helping in the unique identification [239]. SERS has enabled trace, point-of-care (POC), sample-collection-friendly, rapid, flexible, and cost-effective covid detection alternatives with the use of diverse nanomaterials [10,240–242]. In addition, both portable and handheld systems have indeed enabled point-of-care testing based on Raman spectroscopy [56,243]. SERS has also been widely used for the detection of other respiratory zoonotic diseases, such as H1N1, H7N9, H3N2, and H5N1; and other coronaviruses, such as MERS-CoV [244,245]. Often, machine learning algorithms are used in combination to enable the identification of patterns that are not apparent to the human eye [246,247]. The availability of large data and the ease of collection have accelerated the potential of machine learning algorithms in identifying viruses and their variants with reliable accuracies for POC devices [248,249]. In addition to trace identification, SERS has also enabled quantification of viral load to access the severity of the infection [250,251].

SERS in combination with LDA has been used for the rapid (2 min) identification of respiratory viruses, including SARS-CoV-2, human adenovirus type 7, and H1N1, using label-free silver nanoparticles [252]. Fe_3O_4@Ag nanoparticles tagged with specific antibodies were used for the detection of adenovirus and influenza virus [253]. Eleven different respiratory pathogens were identified using SERS, with nanoparticles tagged with nucleic acids achieving remarkable LODs in the sub-picomolar range [254]. Gold nanoparticles functionalized with a specific enzyme were used for the detection of S protein expressed by the COVID-19 viruses with SERS-based sensing in water [255]. Trace S protein detection has also been performed with SERS substrates enabling both chemical and electromagnetic enhancement [256] and using DNA-aptamer-based substrates, achieving a 0.7 fg mL^{-1} LOD [257]. Influenza-infected cells were identified based on proteins, using SERS and PCA [258]. Influenza and covid viruses were detected in human nasal fluid and saliva, using SERS [259], and also in untreated saliva [260]. A portable breath analyzer for covid detection based on the presence of organic volatile compounds was developed, achieving a sensitivity greater than 95% with less than 5 min of detection time [53]. A lateral-flow-immunoassay-based SERS was proposed for the quantitative detection of SARS-CoV-2 [261]. Similar work was performed for the trace detection of SARS-CoV-2 antibodies and spike proteins [262–264]. Li et al. optimized the silver nanostructures to increase the LOD for SARS-CoV-2 detection [265]. In a unique study, Kim et al. studied the efficacy of the Oxford–AstraZeneca vaccine by using SERS studies on tear samples and achieved excellent reproducibility and LOD in the femtomolar regime [266]. Machine learning algorithms such as PCA and SVM were used for the classification of normal and SARS-CoV-2 saliva samples with SERS data, with an accuracy of 95% [267]. Different respiratory viruses and their variants were identified using a silver-nanorods-based SERS sensor [268]. Different respiratory syncytial viruses have been identified and classified using SERS and classification algorithms such as PCA and HCA [269]. A deep-learning-based on-site SERS detection was developed to detect the SARS-CoV-2 virus based on the spike protein with 87% accuracy. This work also studied Raman modes of the spike protein theoretically and established a database [270]. Different variants of the SARS-CoV-2 virus, including wild-type, Alpha, Delta, and Omicron, were successfully identified using specific antibody-tagged 3D porous Ag-based SERS substrates [271]. SERS has also shown the potential of simultaneous detection of influenza virus (H1N1), SARS-CoV-2, and respiratory syncytial virus by using magnetic-tags-based SERS substrates with extended studies in throat swabs [272]. Label-free SERS was performed on serum samples of patients after 4 to 16 days of testing positive for COVID-19, and chemometric techniques were used to find significant difference in the SERS spectral features [273]. Figure 5 summarizes different techniques that are used for the SERS-based detection of SARS-CoV-2.

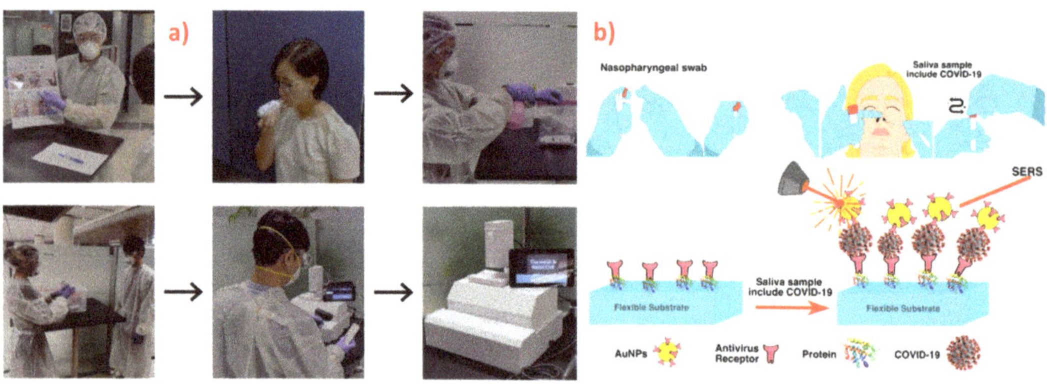

Figure 5. (**a**) Optical images of the workflow used for SERS-based detection of COVID-19 disease using breath analysis with a detection time of 5 min, achieving sensitivity >95% in nearly 500 participants, establishing the rapidness and specificity of SERS. Reproduced with permission from [53]. Copyright (2022), American Chemical Society. (**b**) Schematic of nasopharyngeal-swab-based Covid detection using SERS with flexible substrates enabling sensitive detection. Reproduced with permission from [238]. Copyright (2022), MDPI.

3. SERS-Based Detection of Microorganisms

3.1. Bacteria Sensing

A bacterium is a living cell and falls under the class of prokaryotic microorganisms. Bacteria come in different shapes, including spheres, rods, spiral, and comma, and have a typical size of few micrometers [274]. Bacterial cells are omnipresent, as they are found in water, food, soil, air, and the human body, and, interestingly, the human body contains 10 times more bacterial cells than human cells. However, only 3% of the bacteria are pathogenic, while the other 97% are essential for the survival of different life forms on the earth [275]. The identification of bacteria is important to assess the quality and contamination of food, soil, and water as a measure of public health. In some cases, the presence of bacteria is also desirable to ensure the decomposition of undesirable contaminants through a process called bioremediation [276–278]. Conventionally, PCR, plate culture, and flow cytometry are used for the detection of bacteria. However, all of them are time-consuming and need 2 to 3 days to arrive at conclusions [279]. SERS-based sensing for bacteria is extensively used for its proven advantages of being specific, sensitive [280–282], rapid [283], and water compatible to perform in situ measurements [284], as well as having the ability to quantify [285–287] and potential for trace detection [288–293]. Point-of-care devices for detection of bacteria can also be realized through SERS [294,295]. The sensitivity of SERS even enabled the detection of the single bacterium [295]. It is even possible to distinguish between live and dead bacteria cells by using SERS [296]. With the use of appropriate machine learning techniques, researchers achieved strain-level distinction using SERS spectra [297].

A SERS biosensor using aptamer (aptamer–Fe3O4@Au) and antibiotic (Vancomycin–Au@MBA) molecules has been used for the detection and quantification of pathogenic bacteria achieving a LOD of 3 cells/mL [298]. Vancomycin tagged NPs were also used in fabricating a sandwich such as SERS substrate for identification and photothermal elimination of bacteria in blood samples [299]. Different bacteria species such as *S. typhi*, *E. coli*, and *L. mono* were identified using SERS with Fe_3O_4@Au magnetic nanoparticles and demonstrated good accuracy in real world samples such as beef, saliva, and urine [300]. Wang et al. have also used magnetic nanoparticles for the detection of *S. aureus* [301,302]. Inspired by polyphenolic chemistry, SERS substrates with metal phenolic networks were designed for the detection of *E. coli* and *S. aureus* [303]. In addition to *E. coli* detection, antibiotic susceptibility was studied using core–shell Au@Ag nanorods. This study was also extended to mice blood, implying practical usage [304]. Bacteria present in serum and human blood sampl was identified using SERS based sensing [305,306]. Polymer mats prepared by force spinning were used for the detection of *S. aureus*, *P. aeruginosa*, and *S. Typhimurium* in blood plasma [307]. Using external magnetic field and plasmonic magnetic nanoparticles, the sensitive detection of Gram-negative bacteria was performed by concentrating the sample to a small area [308]. Similar work was accomplished using a microfluidic device to analyze drinking water for bacterial contamination [309]. The quantification of Salmonella typhimurium was performed using 3D DNA-based SERS substrates [310]. SERS-based immunoassay was used for the ultrasensitive and quantitative detection of different bacteria species simultaneously [311]. Multiplexing was also demonstrated by Hayleigh et al. [312] and Gracie et al., who then went on to conduct quantification in multiplexing [313]. A ceramic-filter-based SERS substrate, along with metal nanoparticles, was used for the detection of *E. coli* and Shewanella putrefaciens [314]. Nine different species of *E. coli* were studied using a SERS microfluidic device and discriminated with 92% accuracy, using support vector machine analysis [315]. The label-free and portable detection of various foodborne bacteria was studied using SERS and different chemometric techniques, e.g., PCA and PLS-DA [316]. Silver nanoparticles synthesized using leaf extract were used for the detection of two bacteria species [317]. SERS, in combination with deep learning techniques, was used for the accurate identification of Staphylococcus aureus to achieve an accuracy of ~98% [318].

3.2. Sensing of Biohazardous Molecules for Homeland Security

Bioterrorism is the new threat facing the world and is equally potent to cause large-scale destruction of civil, animal, and plant life. Often, biological agents are easy to prepare and scale up; can be contaminated in food, water, and soil; and are easy to carry, making them the future weapons. Many countries keep them in their military stockpiles despite the regulations [319]. According to the Centre for Disease Control and Prevention (CDC), a biohazardous material is defined as any infectious agent or biological material that poses a threat to human health, the environment, and animals. A review by Lister et al. summarized different biological agents that concern homeland security [320]. Different pathogens and biological agents, such as toxins, venom, and allergens, are some examples of biohazardous materials. Nerve agents are a big concern owing to their high solubility, high toxicity, and durability, with the Tokyo event in 1995 being an example [321,322]. Nerve agents can be classified into G-series, representing agents developed by Germans; V-series for venomous agents; GV series for the combination of G- and V-series; and Novichock series [323]. It is imperative to have a detection system that is sensitive, rapid, portable, and functional for different background media, such as liquids and gases, for the detection of these nerve agents. Plasmonic sensors are widely used for the detection of chemical and biological war threats [324,325]. Of all, SERS has its own advantages for the reasons discussed in the Introduction section and hence is widely used for the detection of biological threats, with a potential for field applications using portable devices [326] and chemometrics [327]. Here we focus on nerve agents, risk-grade-two and -three bacteria species, or their biomarkers' sensing, using SERS, with an interest in homeland security.

A sensitive and selective identification of the nerve agents Tabun, Cyclosarin, and VX was performed using gold- and silver-coated Si nanostructures both without [328] and with a tag (antidote) [329] in two different studies. VX and its hydrolysis products were studied elsewhere, too [330,331]. Sarin, an organophosphorus nerve agent, was detected using plasmonic Si nanocone structures [332]. Three nerve agents, i.e., isopropyl methylphosphonofluoridate (GB), pinacolyl methylphosphonofluoridate (GD), and cyclohexyl methylphosphonofluoridate (GF), were identified, and their hydrolysis degradation was distinguished using SERS [322]. A mustard simulant, pathogenic bacteria, and cyanide were detected using SERS [333]. A reproducible (7%), rapid (30 s), and sensitive (1 ppb) was used for the detection of a nerve simulant, pinacolyl methyl phosphonic acid (PMPA) [334]. Gaseous warfare agents such as dimethyl methylphosphonate were identified using SERS on LiCl microlenses [335]. Various G-series and VX nerve agents were identified using novel pinhole shell-isolated Au nanoparticles substrates achieving sensitivity of 10 ng/L and 20 ng/L, respectively [336]. Using plasmonic 3D fractal structures, a G-series nerve agent called dimethyl methylphosphonate (DMMP) was detected in the gaseous state, with a sensitivity of 12 ppmV [337]. Bacillus anthracis is a highly infectious bacteria that causes the fatal disease anthrax in humans. It is a cause for concern because of its recent usage as a biowarfare agent by many countries [338]. Farrell et al. summarized different anthrax biomarkers and existing detection techniques [339]. Plasmonic metal decorated anisotropic Ni nanostructures were used for detection of dipicolinic acid (DPA), a biomarker for anthrax [340]. Specifically, tagged SERS substrates were used for the detection of anthrax protective antigens, achieving a remarkable LOD of 1 pg/mL [341]. A magnetic microfluidic SERS sensor using specifically tagged Au nanoparticles was used for the detection of the anthrax biomarker poly-γ-D-glutamic acid, with an LOD of 100 pg/mL [342]. Reusable and sensitive laser-ablated Au nanostructures were used for the detection of dipicolinic acid (DPA) with a LOD of 0.83 pg/L and signal enhancement of ~10^{12} [343]. A selective SERS substrate that can discriminate between different strains of bacteria by specifically binding to Bacillus anthracis was designed with DPA as a biomarker [344]. Gold nanorods were also employed for the sensitive detection of DPA and anthrax-protective antigen [345,346]. The trace detection of DPA, equivalent to nearly 18 spores, was achieved using super-hydrophobic SERS sensors [347]. The effects of aggregation of NPs and pH on the SERS performance for the detection of components of cell wall and endospores of

Bacillus thuringiensis were studied extensively [348]. Different chemical and biological warfare agents were classified using techniques such as PCA, PLS-DA, as well as hierarchical classification techniques based on the SERS spectra [328,349].

4. Machine Learning in SERS-Based Biosensing

4.1. Introduction to Machine Learning

In recent times, machine learning is widely being used for many applications including spectroscopy for both data pre- and postprocessing. Machine learning (ML), as the name suggests, is a technique in which the algorithm learns patterns from the existing data and will attempt to make accurate predictions on the unknown based on the trained data. The potential for its ability to find complex patterns from big data sets has given an opportunity to extract and model data purposefully. There are different existing algorithms, both supervised and unsupervised, depending on the problem at hand. Deep learning is a subdomain of machine learning inspired by the human brain that uses multilayered neural networks for modeling data. Throughout this article, machine learning also implies deep learning techniques. Advances in computation facilities and with increasing availability and complexity of big data, deep learning, which is a kind of machine learning, has found its place. Some popular and relevant examples of ML being classification of emails as span and not span, identifying cancer in early stage using medical images, face recognition and weather prediction. ML algorithms can be broadly classified into three types, namely supervised for labeled observations, unsupervised for unlabeled observations, and reinforcement learning for models that learn from the errors to improve accuracy [350], as summarized in the Figure 6 below.

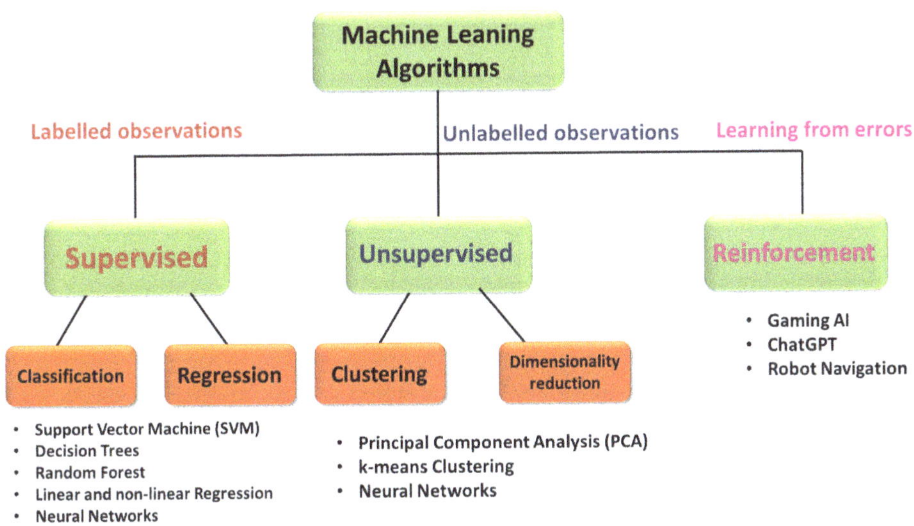

Figure 6. Flow chart illustrating the classification of different machine learning algorithms as supervised, unsupervised, and reinforcement models.

With the ease of data collection and availability of open source Raman spectroscopy data, SERS has also seen a surge in machine learning models [49,351,352]. The trend is welcoming and desirable as the nature of existing challenges in SERS involving trace detection, signal fluctuations, quantification and identification are complex with many variables calling for an analytical tool that has the ability to capture the patterns devoid of experts [353]. Trace detection implies identifying signal from a noisy background where ML could be aided. SERS is also known to have inherent signal fluctuations owing to localization of hotspots. Especially in the case of bio samples, they have background contribution from different undesirable components thus interfering with the signal and need ML algorithms to extract the useful information [2,354–357]. The process of data collection, identification of chemical composition and quantification is non-linear and is highly dependent on human intelligence making it a barrier to carry the benefits of SERS to onsite [358]. Some of the widely used techniques include Principal Component Analysis (PCA), Support Vector Machine (SVM), Partial Least Squares (PLS), Decision Trees (DTs) and Convolutional Neural Networks (CNNs). PCA is a dimensionality reduction technique where components representative of the data with large variance are preserved. This is extensively used a preprocessing step in order to reduce complexity of the models or also as a classification technique [359–361]. SVM is a nonlinear ML technique that can be used for both regression and classification [360]. It works by finding a hyperplane that distinguishes two or more classes using a kernel function [362]. If the data set is small and the number of variables is large, PLS is useful for its ability to still extract useful information and is often used for quantitative studies [363,364]. DTs are widely used for classification of the data using a method bootstrapping [365]. CNNs are a kind of neural networks which employ filters and pooled layers in the architecture and often used if the size of the data set is large enough and if images are involved in the modeling [366]. Specifically, in the field of biophotonics, machine learning models using SERS can be efficiently classified into three domains: identification, classification, and quantification, with interests such as disease and molecular diagnosis [367,368]; microorganism classification, identification, etc. [369–372]; and cancer diagnosis [373], as shown in Figure 7. In addition, machine learning was also used to improve data collection to overcome signal fluctuations and enhance the usability on site [374], to estimate the effect of scattering [375] and for the SERS signal enhancement itself [376]. In further sections, we discuss different ML techniques that were used in SERS for biology applications.

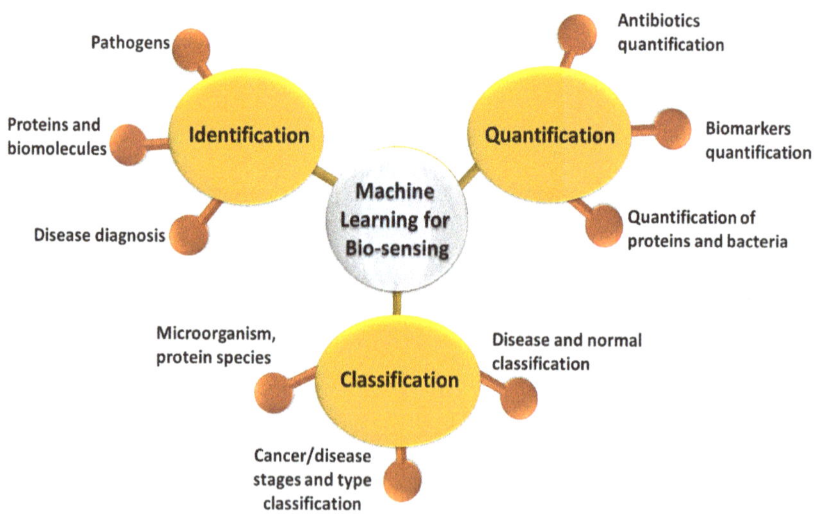

Figure 7. Schematic of applications of machine learning for biosensing using SERS based plasmonic sensors.

4.2. Identification

SERS provides the vibrational fingerprint of many biomolecules, including amino acids, peptides, carbohydrates, pathogens, and nuclei acids [377]. It is also label free and non-destructive, making it desirable for in situ and rapid identification. Often in real-world situations of biology sample analysis, there are undesirable effects from background cell signals or with the similarity of spectra from two subspecies. Machine learning models can be successfully trained to capture these complex differences and distinguish two similar spectra devoid of the background helping in identification of the sample. Figure 8 summarizes the work so far in using ML for identification applications in biosensing with SERS. CNNs were used for identification of cancer using SERS with gold multi-branched nanoparticles (AuMs), functionalized with different chemical groups, and achieved 100% accuracy in identifying the structural changes [378]. Drug-sensitive and drug-resistant bacterial strains were identified using SERS with a combination of CNNs and achieved 100% accuracy [379]. Different classification algorithms such as LDA, SVM, and KNN were used for the classification of bacterial extracellular vesicles for *E. coli* by strain and culture time using label-free approach of SERS [380]. SVM was successfully used for the identification of different drugs in human urine at trace levels with an accuracy greater than 92% [381]. A SERS chip was designed to identify a cancer marker, TIMP-1, and combined it with ML to identify lung and colon cancer in patients [382]. A label-free SERS, in combination with different machine learning algorithms, such as random forest, PCA-LDA, and decision trees, was used for the identification of colon cancer using serum samples. It was found that the random forest model outperformed the other two models in terms of accuracy and specificity [383]. SERS combined with ANN was used for the identification of different pollen samples despite many spectral contributions using Au NPs [384]. A microfluidic-chip-based SERS substrate with Au nanoparticles was used for the identification of l Jurkat, THP-1, and MONO-MAC-6 leukemia cell lysates, using SVM, and achieved 99% accuracy [385]. A lab-on-chip SERS device was fabricated and used for the successful identification of different species of mycobacteria [386]. The machine learning models PLS-DA and CNN were used to identify different stages of kidney malfunction in dialysis patients by using serum analysis by SERS. The CNN model achieved an accuracy of 96%, which is better than that of PLS-DA, with 84% [387]. The SVM outperformed other techniques in the identification of cyanobacteria, using SERS spectra of mutant and wild-type strains [388]. Using a dimensionality reduction technique, followed by a probabilistic ML model, SARS-CoV-2 identification was performed with an accuracy of ~85% [389]. SERS coupled with SVM was also used for the identification of lung cancers [96].

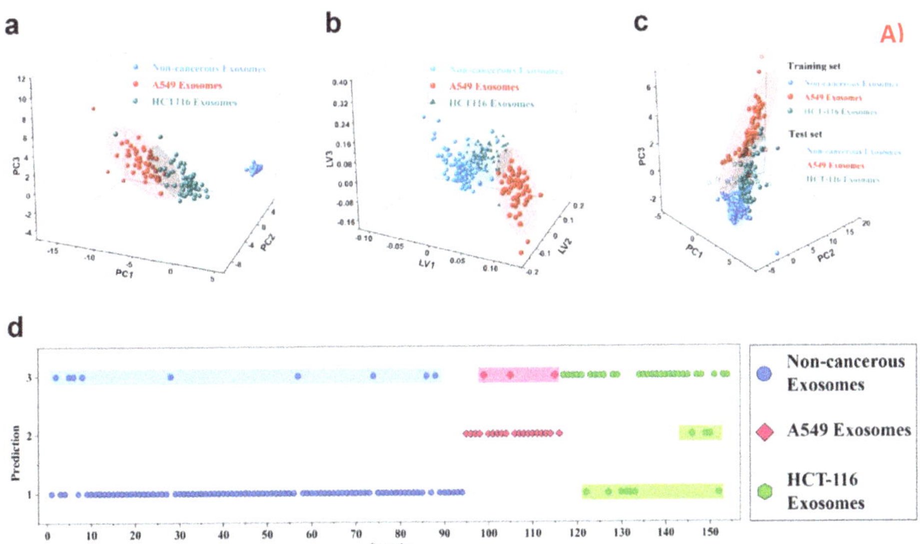

Figure 8. *Cont.*

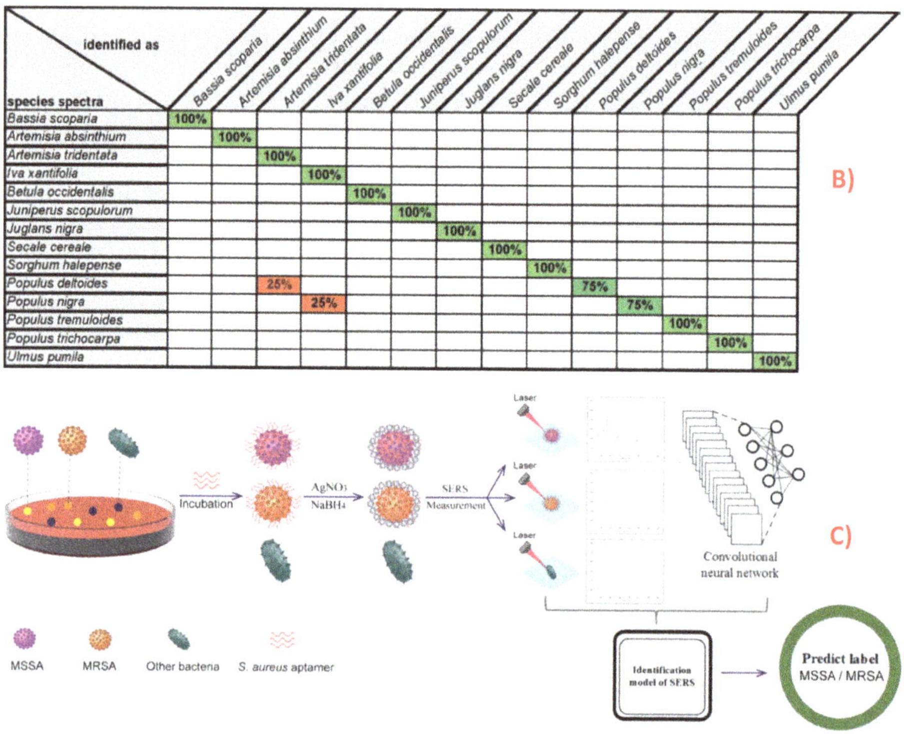

Figure 8. Applications of machine learning techniques used in the identification of biological samples, using the SERS technique for the (**A**) identification of lung and colon cancers from exosomes HCT-116 (colon cancer biomarker), A549 (lung cancer biomarker), and non-cancerous samples performed with (**a**) PCA, (**b**) PLSDA, and (**c**) SVM, with 60% of training set and 40% of test set. (**d**) Predicted labels for the test set using the SVM model. Value 1 is a prediction for normal plasma, Value 2 is a prediction for A549 (lung cancer) exosomes, and Value 3 is a prediction for HCT-116 (colon cancer) exosomes. The highlighted portion shows labels that are wrongly identified. Reproduced with permission from [382]. Copyright (2022), Elsevier. (**B**) Identification of 14 commercially available pollen species using SERS spectra combined with an artificial neural network, using a winner-takes-all (WTA) method. Reproduced with permission from [384]. Copyright (2015), Wiley. (**C**) Schematic for SERS-based ML model used in the identification of methicillin-susceptible Staphylococcus aureus (MSSA) and methicillin-resistant Staphylococcus aureus (MRSA) bacteria, using a CNN model. Reproduced with permission from [379]. Copyright (2021), RSC.

4.3. *Quantification*

One of the interests of using SERS for sensing also lies in its ability to detect trace and ultra-trace molecules. The intensity and concentration relation for a peak of choice in the SERS spectrum is often non-linear due to many factors, such as the inhomogeneous distribution of hotspots, non-uniform adsorption of molecules, and localization of the hotspots [361,390]. This calls for machine learning models that have the ability to capture non-linear patterns of intensity and concentration relation and further predict the unknown concentration. As the problem demands, regression ML models such as PCR, PLSR, SVR, and XGBR are used for the quantification of trace biomolecules.

A quantitative analysis of antibiotics and a mixture of antibiotics was performed using PLSR with an accuracy of 96% [391]. An SERS-based lateral flow assay was used for the quantification of *E. coli* in milk and beef, using the Bayesian ridge regression (BRR), support vector regression (SVR), elastic net regression (ENR), and extreme gradient boosting regression (XGBR) algorithm, as shown in Figure 9 [392]. A SERS substrate with plasmonic nanogaps was fabricated and used for the trace sensing of pyocyanin, a secondary metabolite of Pseudomonas aeruginosa, from a complex background. Furthermore, using machine learning algorithms, the quantification of pyocyanin was performed with an accuracy until five significant digits, using PLS [393]. The quantification of very low concentrations of fumonicins in maize was performed using different chemometric techniques such as PCR and PLSR and achieved an accuracy above 90% [394]. Thiols found in the whole blood of umbilical cords were quantified using a PLSR model on SERS spectra collected using silver nanoparticles as plasmonic substrates [395]. PCA, followed by SVR, was used for the quantification of histamine, an allergen, in seafood, using spectral data from a combination of TLS and SERS [396].

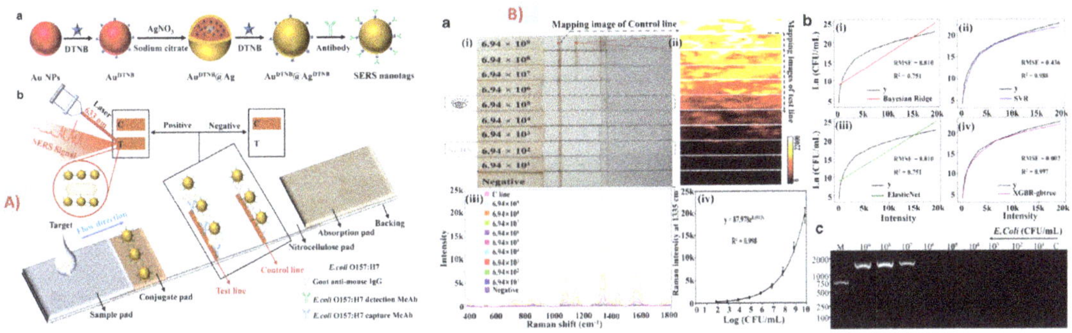

Figure 9. (**A**) Schematic of ML-based quantification of *E. coli* O157:H7, using (**a**) SERS nanotags and (**b**) lateral flow assay along with different regression models, including SVR, BNR, and XGBR. Reproduced with permission from [392]. Copyright (2020), Springer. (**B**) (**a**) (**i**) Optical images of the lateral flow strips and (**ii**) SERS mapping region of the prominent peak in the SERS intensity profile. (**iii**) Corresponding SERS spectra of the test lines. (**iv**) Intensity and concentration fit. (**b**) Machine-learning-based regression fits for (**i**) Bayesian ridge regression (BRR), (**ii**) support vector regression (SVR), and the (**iii**) elastic net regression (ENR) and (**iv**) extreme gradient boosting regression (XGBR). (**c**) PCR image for the *E. coli* detection. Reproduced with permission from [392]. Copyright (2020), Springer.

4.4. Classification

The goal of the classification algorithms employed for data analysis in SERS for biosensing is often differentiating different classes, species, and spectra corresponding to different stages of the disease or different diseases themselves. So far, classification algorithms such as SVM, KNN, and PCA; and different neural networks, such as CNN, were used for the problems stated.

Different bacteria species were classified and identified using SVM, with an accuracy of 87% by using SERS with bacterial cellulose nanocrystals (BCNCs) decorated with Au nanoparticles [397]. K-nearest neighbor and decision trees were used for the classification of SERS-based liquid biopsy assay to identify five protein biomarkers (CA19-9, HE4, MUC4, MMP7, and mesothelin) in pancreatic cancer patients, ovarian cancer patients, pancreatitis patients, and healthy individuals [398]. The direct serum analysis of liver cancer samples is performed using Au-Ag nano complex-decorated ZnO nanopillars on paper for the classification of different stages of cancer using CNNs. This method achieved an accuracy of 97.78% [399]. SERS combined with machine learning was also used for the screening of PCOS, using classification algorithms on SERS data. Samples of follicular fluids and plasma from healthy and PCOS patients were successfully classified, with an accuracy of 89%, using stacked models for both [400]. Protein species with similar spectral profiles were classified using principal component analysis (PCA) applied to SERS spectra [401]. CNNs without any preprocessing steps were used for the classification of different grades of bladder cancer tissue, using Raman spectra, and different species of *E. coli*, using SERS spectra. Different classification algorithms, such as KNN, PCA, SVM, and ANN, were used, but CNN was found to outperform the others in terms of accuracy [402]. Using Non-Structural Protein 1 (NS1) as a biomarker for dengue, extreme learning machine and PCA models were used for the classification of dengue patients with 100% accuracy towards a goal of early diagnosis [403]. Bacterial endotoxins of twelve different species were identified and classified using SERS spectra and machine learning algorithms such as KNN, RF, SVM, and RamanNet. While the other algorithms achieved accuracy greater than 90%, RamanNet outperformed them, with 100% accuracy [404]. With a goal to identify cancer at an early stage, a point-of-care diagnosis system using a novel hydrophobic SERS substrate combined with machine learning techniques was used, as shown in Figure 10 [50]. The SERS spectra of serum samples collected from nearly 690 patients, including normal and different cancers (breast cancer, leukemia, and hepatitis B virus), were collected and analyzed using deep learning techniques to achieve 100% accuracy in successfully classifying the data. They performed external testing with an accuracy of 98%, indicating potential usage in the real world.

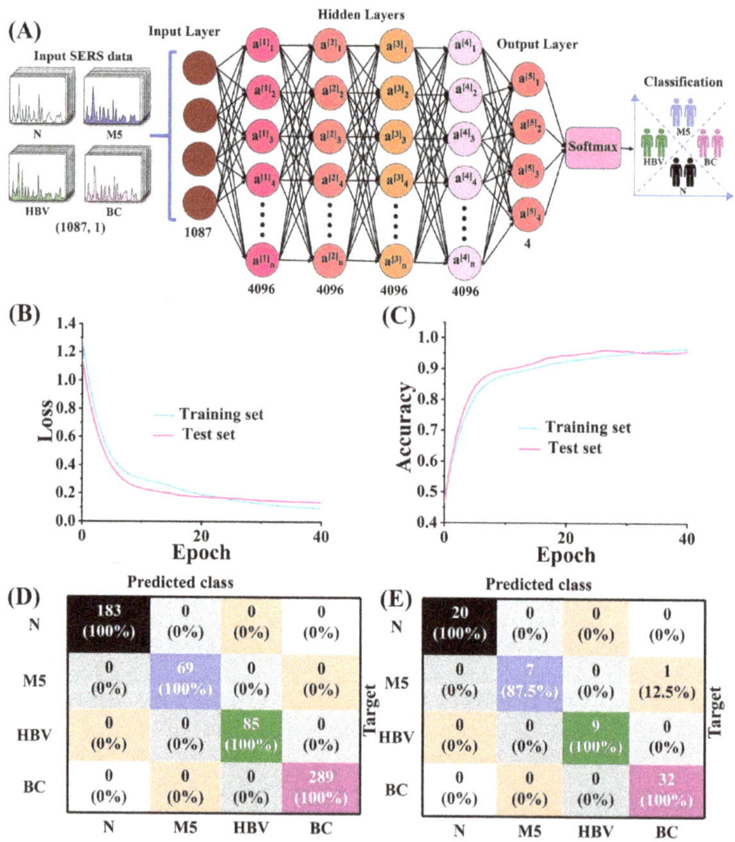

Figure 10. (**A**) Schematic of the architecture used for classification of different cancers and normal serum samples using SERS spectra collected from nearly 695 patients. (**B**) Learning curves for the model implemented with loss and (**C**) accuracy as metrics. (**D**) Confusion matrix for the training (**E**) test data sets communicating good accuracy. Reproduced with permission from [50]. Copyright (2021), Wiley.

5. Conclusions and Scope

The vast existing literature and continued interest in the SERS technique for biosensing is a promising sign to realize point-of-care devices based on SERS. This would revolutionize disease diagnosis due to its ability to identify traces, enabling early detection, cost effectiveness, and rapid diagnosis. Under optimized conditions, a single bacterial cell was also detected using SERS, thus demonstrating its sensitivity [295]. Using SERS, it is possible to identify disease biomarkers in a variety of bio-fluids, such as urine, saliva, plasma, and blood, as well as in volatile compounds and gases. Unlike many commercial techniques, SERS is reagent free and does not need sequential procedures for the identification of disease biomarkers. Machine learning techniques are extensively being used in SERS for their ability to recognize complex and intricate patterns devoid of background noise. Different models, such as PCA, SVM, ANN, CNN, KNN, and PLS, were used for identification, quantification, and classification of microorganisms and different diseases, including cancers. In regard to cancer diagnosis using SERS, the distinguishment between normal and cancerous samples, including cells and liquids, and the discrimination of different stages of cancer have also been performed. These methods are cost efficient, rapid, and sensitive, as opposed to the existing cancer-screening techniques. In response to the pandemic situation, SERS has been extensively used for the detection of novel COVID-19 virus and also for tracking the efficiency of the vaccines [266]. Furthermore, SERS has been widely used for the detection of various nerve agents and other bio-warfare agents, thus expanding its application in homeland security. Commercialization of SERS is already underway, with many lateral flow and point-of-care devices that have been developed in response to the pandemic [261] and diagnosis of other diseases with equal and par performance as existing commercial techniques [150]. For example, it is established that SERS performs better than the commercial enzyme-linked immunosorbent assay (ELISA) test kits in cancer detection, allowing multiplexing with very less sample volume [405]. It was also shown to achieve a lower LOD than radioimmunoassay (RIA) and ELISA in a different study [406]. In a recent study, SERS was compared with a clinically available method for quantification of glucose in blood sugar and shown to perform equally good [407]. It was found that SERS is 16-to-32-times more sensitive than the commercial lateral flow assay and >400-times more sensitive than the ELISA with the same reagents for the detection of covid [264]. The major components of a Raman system consist of a laser source, a probe for excitation and signal collection, and a detection system with a spectrometer [408]. Recent advances on all of these fronts for miniaturizing and reducing the cost are enabling the widespread usage of SERS-based detection with portable systems.

Despite its merits, there are few challenges that stand in the way of scaling up SERS for biosensing in the real world. Firstly, there are reasons inherent to the SERS enhancement mechanisms that turn out to be undesirable, often causing signal fluctuations and poor reproducibility. Due to the localization of dense field enhancement areas ("hotspots") and metal-sample adsorption artefacts, SERS signals are known to fluctuate. Upon laser illumination, these hotspots are also known to diffuse or transform, thus adding further to the poor reproducibility. A substrate with homogeneous field enhancement promises good reproducibility but comes at the cost of enhancement and eventually limiting trace detection [374]. Secondly, SERS substrates majorly comprise Au or Ag nanoparticles/nanostructures. These nanostructures are generally not stable for long durations, with a risk of rapid oxidation upon exposure to ambient atmosphere. Owing to their large surface charge, they also tend to aggregate to form clusters. Often, aggregation and oxidation are prevented by the addition of capping agents or ligands which could affect the SERS signal and compatibility with the bio-samples. In view of commercialization, there is also a question of the reusability of the SERS substrates. Thirdly, SERS substrates that are used in the lab are optimized under specific instrument conditions, such as laser wavelength, acquisition time, power, and focusing conditions. In regard to point-of-care applications, it is a challenge to have the same experimental conditions, thus limiting the substrate efficiency. Field applications also call for cost-effective and miniature devices that are easy to operate by a non-expert [39]. Specifically in regard to biosensing, SERS-based detection in the field is a challenge because of the lack of disease specificity. In most of the scenarios, the biomarkers for disease detection, such as proteins and antigens, are not disease specific and need further evaluation in order for researchers to arrive at conclusions [409]. In the case of whole-organism or cell/tissue studies, it is difficult to ascertain the peaks because of contribution from various components, such as proteins, DNA bases, lipids, and other cell components. Figure 11 summarizes some best practices for quantification and qualitative analysis in SERS at various stages of experiments. The current research in SERS for biosensing is moving in the right direction to overcome these challenges with exploration in the direction of instrumentation, standoff detection, and the usage of ML techniques to improve data collection and identification without expertise. We believe that, in the coming years, these challenges will be successfully met and SERS will realize its full potential in real-world low-cost biosensing.

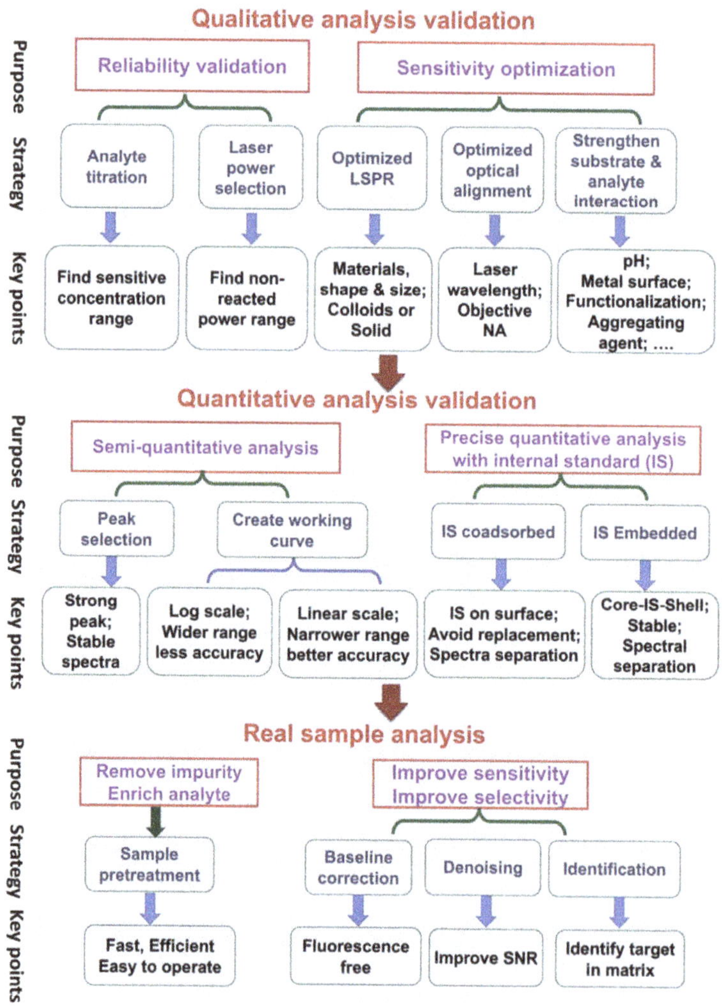

Figure 11. Steps for best practices for quantitative and qualitative detection using SERS to overcome the challenges at each stage with a goal of real-world applications. Reproduced with permission from [390]. Copyright (2020), RSC.

Author Contributions: Conceptualization, R.B. and V.R.S.; methodology, R.B. and V.R.S.; resources, V.R.S. and K.R.V.; writing—original draft preparation, R.B.; writing—review and editing, V.R.S. and K.R.V.; visualization, R.B. and V.R.S.; supervision, V.R.S. and K.R.V.; project administration, V.R.S.; funding acquisition, V.R.S. All authors have read and agreed to the published version of the manuscript.

Funding: The authors thank DRDO for the funding [Project #ERIP/ER/1501138/M/01/319/D(R&D)]. V.R. Soma also thanks the University of Hyderabad (UoH) for financial support through the Institute of Eminence (IoE) project [No. UOH/IOE/RC1/RC1-20-016]. The IoE scheme was granted to the UoH by the Ministry of Education, Government of India, vide MHRD notification F11/9/2019-U3(A).

Conflicts of Interest: The authors declare no conflict of interest.

References

1. Stewart, M.E.; Anderton, C.R.; Thompson, L.B.; Maria, J.; Gray, S.K.; Rogers, J.A.; Nuzzo, R.G. Nanostructured Plasmonic Sensors. *Chem. Rev.* **2008**, *108*, 494–521. [CrossRef]
2. Costanzo, H.; Gooch, J.; Frascione, N. Nanomaterials for Optical Biosensors in Forensic Analysis. *Talanta* **2023**, *253*, 123945. [CrossRef]
3. Chen, G.; Chen, Y.; Huang, W.; Shi, Y. Plasmonic Nanobiosensors for Detection of Different Targets. In Proceedings of the Second International Conference on Medical Imaging and Additive Manufacturing (ICMIAM 2022), Xiamen, China, 25–27 February 2022. [CrossRef]
4. Sadani, K.; Nag, P.; Thian, X.Y.; Mukherji, S. Enzymatic Optical Biosensors for Healthcare Applications. *Biosens. Bioelectron. X* **2022**, *12*, 100278. [CrossRef]
5. Erkmen, C.; Selcuk, O.; Unal, D.N.; Kurbanoglu, S.; Uslu, B. Layer-by-Layer Modification Strategies for Electrochemical Detection of Biomarkers. *Biosens. Bioelectron. X* **2022**, *12*, 100270. [CrossRef]
6. Spillman, W.B. *Fiber Optic Biosensors*; Elsevier: Amsterdam, The Netherlands, 2011; Volume 3, ISBN 9780470126844.
7. Kazanskiy, N.L.; Khonina, S.N.; Butt, M.A.; Kaźmierczak, A.; Piramidowicz, R. State-of-the-Art Optical Devices for Biomedical Sensing Applications—A Review. *Electronics* **2021**, *10*, 973. [CrossRef]
8. Ramirez, J.C.; Grajales García, D.; Maldonado, J.; Fernández-Gavela, A. Current Trends in Photonic Biosensors: Advances towards Multiplexed Integration. *Chemosensors* **2022**, *10*, 398. [CrossRef]
9. Chadha, U.; Bhardwaj, P.; Agarwal, R.; Rawat, P.; Agarwal, R.; Gupta, I.; Panjwani, M.; Singh, S.; Ahuja, C.; Selvaraj, S.K.; et al. Recent Progress and Growth in Biosensors Technology: A Critical Review. *J. Ind. Eng. Chem.* **2022**, *109*, 21–51. [CrossRef]
10. Dutta, G. *Next-Generation Nanobiosensor Devices for Point-of-Care Diagnostics*; Springer: Singapore, 2023; ISBN 9789811971303. [CrossRef]
11. Ahangari, A.; Mahmoodi, P.; Mohammadzadeh, A. Advanced Nano Biosensors for Rapid Detection of Zoonotic Bacteria. *Biotechnol. Bioeng.* **2022**, *120*, 41–56. [CrossRef]
12. Taha, B.A.; Al Mashhadany, Y.; Bachok, N.N.; Ashrif, A.; Bakar, A.; Hafiz Mokhtar, M.H.; Dzulkefly Bin Zan, M.S.; Arsad, N. Detection of COVID-19 Virus on Surfaces Using Photonics: Challenges and Perspectives. *Diagnostics* **2021**, *11*, 1119. [CrossRef]
13. Soma, V.R.; Podagatlapalli, G.K.; Hamad, S.; Mechanics, F. Ultrafast Laser Ablation in Liquids for Nanomaterials and Applications. *J. Nanosci. Nanotechnol.* **2014**, *14*, 1364–1388. [CrossRef]
14. Soler, M.; Lechuga, L.M. Principles, Technologies, and Applications of Plasmonic Biosensors. *J. Appl. Phys.* **2021**, *129*, 111102. [CrossRef]
15. Barbillon, G. Plasmonics and Its Applications. *Materials* **2019**, *12*, 1502. [CrossRef] [PubMed]
16. Michaels, A.M.; Jiang, J.; Brus, L. Ag Nanocrystal Junctions as the Site for Surface-Enhanced Raman Scattering of Single Rhodamine 6G Molecules. *J. Phys. Chem. B* **2000**, *104*, 11965–11971. [CrossRef]
17. Golightly, R.S.; Doering, W.E.; Natan, M.J. Surface-Enhanced Raman Spectroscopy and Homeland Security: A Perfect Match? *ACS Nano* **2009**, *3*, 2859–2869. [CrossRef] [PubMed]
18. Vendamani, V.S.; Beeram, R.; Nageswara Rao, S.V.S.; Pathak, A.P.; Soma, V.R. Trace Level Detection of Explosives and Pesticides Using Robust, Low-Cost, Free-Standing Silver Nanoparticles Decorated Porous Silicon. *Opt. Express* **2021**, *29*, 30045. [CrossRef]
19. Liu, C.; Xu, D.; Dong, X.; Huang, Q. A Review: Research Progress of SERS-Based Sensors for Agricultural Applications. *Trends Food Sci. Technol.* **2022**, *128*, 90–101. [CrossRef]
20. Zhang, D.; Pu, H.; Huang, L.; Sun, D.W. Advances in Flexible Surface-Enhanced Raman Scattering (SERS) Substrates for Nondestructive Food Detection: Fundamentals and Recent Applications. *Trends Food Sci. Technol.* **2021**, *109*, 690–701. [CrossRef]
21. Li, Y.; Liu, X.; Lin, Z. Recent Developments and Applications of Surface Plasmon Resonance Biosensors for the Detection of Mycotoxins in Foodstuffs. *Food Chem.* **2012**, *132*, 1549–1554. [CrossRef]
22. Dies, H.; Raveendran, J.; Escobedo, C.; Docoslis, A. Rapid Identification and Quantification of Illicit Drugs on Nanodendritic Surface-Enhanced Raman Scattering Substrates. *Sens. Actuators B Chem.* **2018**, *257*, 382–388. [CrossRef]
23. Vendamani, V.S.; Beeram, R.; Rao, S.V.S.N.; Rao, S.V. Protocol for Designing AuNP-Capped Ag Dendrites as Surface-Enhanced Raman Scattering Sensors for Trace Molecular Detection Protocol for Designing AuNP-Capped Ag Dendrites as Surface-Enhanced Raman Scattering Sensors for Trace Molecular Detection. *STAR Protoc.* **2023**, *4*, 102068. [CrossRef]

24. He, L.; Rodda, T.; Haynes, C.L.; Deschaines, T.; Strother, T.; Diez-Gonzalez, F.; Labuza, T.P. Detection of a Foreign Protein in Milk Using Surface-Enhanced Raman Spectroscopy Coupled with Antibody-Modified Silver Dendrites. *Anal. Chem.* **2011**, *83*, 1510–1513. [CrossRef]
25. Jebakumari, K.A.E.; Murugasenapathi, N.K. Engineered Two-Dimensional Nanostructures as SERS Substrates for Biomolecule Sensing: A Review. *Biosensors* **2023**, *13*, 102. [CrossRef]
26. Bantz, K.C.; Meyer, A.F.; Wittenberg, N.J.; Im, H.; Kurtuluş, Ö.; Lee, S.H.; Lindquist, N.C.; Oh, S.H.; Haynes, C.L. Recent Progress in SERS Biosensing. *Phys. Chem. Chem. Phys.* **2011**, *13*, 11551–11567. [CrossRef] [PubMed]
27. Vendamani, V.S.; Nageswara Rao, S.V.S.; Venugopal Rao, S.; Kanjilal, D.; Pathak, A.P. Three-Dimensional Hybrid Silicon Nanostructures for Surface Enhanced Raman Spectroscopy Based Molecular Detection. *J. Appl. Phys.* **2018**, *123*, 014301. [CrossRef]
28. Huang, Z.; Zhang, A.; Zhang, Q.; Cui, D. Nanomaterial-Based SERS Sensing Technology for Biomedical Application. *J. Mater. Chem. B* **2019**, *7*, 3755–3774. [CrossRef]
29. Szaniawska, A.; Kudelski, A. Applications of Surface-Enhanced Raman Scattering in Biochemical and Medical Analysis. *Front. Chem.* **2021**, *9*, 664134. [CrossRef]
30. Chen, Y.; An, Q.; Teng, K.; Liu, C.; Sun, F.; Li, G. Application of SERS in In-Vitro Biomedical Detection. *Chem. Asian J.* **2022**, *18*, e202201194. [CrossRef]
31. Hegde, M.; Pai, P.; Gangadhar Shetty, M.; Sundara Babitha, K. Gold Nanoparticle Based Biosensors for Rapid Pathogen Detection: A Review. *Environ. Nanotechnol. Monit. Manag.* **2022**, *18*, 100756. [CrossRef]
32. Fleischmann, M.; Hendra, P.J.; McQuillan, A.J. Raman Spectra of Pyridine Adsorbed at a Silver Electrode. *Chem. Phys. Lett.* **1974**, *26*, 163–166. [CrossRef]
33. Jeanmaire, D.L.; Duyne, R.P.V.A.N. Surface Raman spectroelectrochemistry: Part I. Heterocyclic, aromatic, and aliphatic amines adsorbed on the anodized silver electrode. *J. Electroanal. Chem. Interfacial Electrochem.* **1977**, *84*, 1–20. [CrossRef]
34. Albrecht, M.G.; Creighton, J.A. Anomalously Intense Raman Spectra of Pyridine at a Silver Electrode. *J. Am. Chem. Soc.* **1977**, *99*, 5215–5217. [CrossRef]
35. Pilot, R.; Signorini, R.; Durante, C.; Orian, L.; Bhamidipati, M.; Fabris, L. A Review on Surface-Enhanced Raman Scattering. *Biosensors* **2019**, *9*, 57. [CrossRef] [PubMed]
36. Le Ru, E.; Etchegoin, P. *Principles of Surface Enhanced Raman Spectroscopy and Related Plasmonic Effects*; Elseveir: Amsterdam, The Netherlands, 2008.
37. Etchegoin, P.G.; Ru, E.C.L. Basic Electromagnetic Theory of SERS. In *Surface Enhanced Raman Spectroscopy: Analytical, Biophysical and Life Science Applications*; John Wiley & Sons: Hoboken, NJ, USA, 2011; pp. 1–37. ISBN 9783527325672.
38. Le Ru, E.C.; Blackie, E.; Meyer, M.; Etchegoint, P.G. Surface Enhanced Raman Scattering Enhancement Factors: A Comprehensive Study. *J. Phys. Chem. C* **2007**, *111*, 13794–13803. [CrossRef]
39. Sharma, B.; Frontiera, R.R.; Henry, A.I.; Ringe, E.; Van Duyne, R.P. SERS: Materials, Applications, and the Future. *Mater. Today* **2012**, *15*, 16–25. [CrossRef]
40. Li, Q.; Huo, H.; Wu, Y.; Chen, L.; Su, L.; Zhang, X.; Song, J.; Yang, H. Design and Synthesis of SERS Materials for In Vivo Molecular Imaging and Biosensing. *Adv. Sci.* **2023**, *2023*, 2202051. [CrossRef]
41. Israelsen, N.D.; Hanson, C.; Vargis, E. Nanoparticle Properties and Synthesis Effects on Surface-Enhanced Raman Scattering Enhancement Factor: An Introduction. *Sci. World J.* **2015**, *2015*, 124582. [CrossRef]
42. Wang, A.X.; Kong, X. Review of Recent Progress of Plasmonic Materials and Nano-Structures for Surface-Enhanced Raman Scattering. *Materials* **2015**, *8*, 3024–3052. [CrossRef]
43. Moram, S.S.B.; Byram, C.; Soma, V.R. Gold-Nanoparticle- and Nanostar-Loaded Paper-Based SERS Substrates for Sensing Nanogram-Level Picric Acid with a Portable Raman Spectrometer. *Bull. Mater. Sci.* **2020**, *43*, 53. [CrossRef]
44. Zhang, Z.; Guan, R.; Li, J. Engineering Rational SERS Nanotags for Parallel Detection of Multiple Cancer Circulating Biomarkers. *Chemosensors.* **2023**, *11*, 110. [CrossRef]
45. Pilot, R.; Massari, M. Silver Nanoparticle Aggregates: Wavelength Dependence of Their SERS Properties in the First Transparency Window of Biological Tissues. *Chem. Phys. Impact* **2021**, *2*, 100014. [CrossRef]
46. Zhang, Y.; Mi, X.; Tan, X.; Xiang, R. Recent Progress on Liquid Biopsy Analysis Using Surface-Enhanced Raman Spectroscopy. *Theranostics* **2019**, *9*, 491–525. [CrossRef] [PubMed]
47. Aitekenov, S.; Sultangaziyev, A.; Ilyas, A.; Dyussupova, A.; Boranova, A.; Gaipov, A.; Bukasov, R. Surface-Enhanced Raman Spectroscopy (SERS) for Protein Determination in Human Urine. *Sens. Bio-Sens. Res.* **2022**, *38*, 100535. [CrossRef]
48. Akgönüllü, S.; Denizli, A. Recent Advances in Optical Biosensing Approaches for Biomarkers Detection. *Biosens. Bioelectron. X* **2022**, *12*, 100269. [CrossRef]
49. Lussier, F.; Thibault, V.; Charron, B.; Wallace, G.Q.; Masson, J.F. Deep Learning and Artificial Intelligence Methods for Raman and Surface-Enhanced Raman Scattering. *TrAC Trends Anal. Chem.* **2020**, *124*, 115796. [CrossRef]
50. Lin, X.; Lin, D.; Chen, Y.; Lin, J.; Weng, S.; Song, J.; Feng, S. High Throughput Blood Analysis Based on Deep Learning Algorithm and Self-Positioning Super-Hydrophobic SERS Platform for Non-Invasive Multi-Disease Screening. *Adv. Funct. Mater.* **2021**, *31*, 2103382. [CrossRef]
51. Breuch, R.; Klein, D.; Siefke, E.; Hebel, M.; Herbert, U.; Wickleder, C.; Kaul, P. Differentiation of Meat-Related Microorganisms Using Paper-Based Surface-Enhanced Raman Spectroscopy Combined with Multivariate Statistical Analysis. *Talanta* **2020**, *219*, 121315. [CrossRef]

52. Ilkhani, H.; Hughes, T.; Li, J.; Zhong, C.J.; Hepel, M. Nanostructured SERS-Electrochemical Biosensors for Testing of Anticancer Drug Interactions with DNA. *Biosens. Bioelectron.* **2016**, *80*, 257–264. [CrossRef] [PubMed]
53. Leong, S.X.; Leong, Y.X.; Tan, E.X.; Sim, H.Y.F.; Koh, C.S.L.; Lee, Y.H.; Chong, C.; Ng, L.S.; Chen, J.R.T.; Pang, D.W.C.; et al. Noninvasive and Point-of-Care Surface-Enhanced Raman Scattering (SERS)-Based Breathalyzer for Mass Screening of Coronavirus Disease 2019 (COVID-19) under 5 Min. *ACS Nano* **2022**, *16*, 2629–2639. [CrossRef]
54. Bharati, M.S.S.; Soma, V.R. Flexible SERS Substrates for Hazardous Materials Detection: Recent Advances. *Opto-Electron. Adv.* **2021**, *4*, 210048. [CrossRef]
55. Ali, A.; Nettey-Oppong, E.E.; Effah, E.; Yu, C.Y.; Muhammad, R.; Soomro, T.A.; Byun, K.M.; Choi, S.H. Miniaturized Raman Instruments for SERS-Based Point-of-Care Testing on Respiratory Viruses. *Biosensors* **2022**, *12*, 590. [CrossRef]
56. Mejía-Salazar, J.R.; Oliveira, O.N. Plasmonic Biosensing. *Chem. Rev.* **2018**, *118*, 10617–10625. [CrossRef]
57. Han, X.; Liu, K.; Sun, C. Plasmonics for Biosensing. *Materials* **2019**, *12*, 1411. [CrossRef] [PubMed]
58. Shrivastav, A.M.; Cvelbar, U.; Abdulhalim, I. A Comprehensive Review on Plasmonic-Based Biosensors Used in Viral Diagnostics. *Commun. Biol.* **2021**, *4*, 70. [CrossRef] [PubMed]
59. Wang, Q.; Ren, Z.H.; Zhao, W.M.; Wang, L.; Yan, X.; Zhu, A.S.; Qiu, F.M.; Zhang, K.K. Research Advances on Surface Plasmon Resonance Biosensors. *Nanoscale* **2022**, *14*, 564–591. [CrossRef] [PubMed]
60. Homola, J. Present and Future of Surface Plasmon Resonance Biosensors. *Anal. Bioanal. Chem.* **2003**, *377*, 528–539. [CrossRef]
61. Piliarik, M.; Vaisocherová, H.; Homola, J. Surface Plasmon Resonance Biosensing. In *Biosensors and Biodetection*; Humana Press: Totowa, NJ, USA, 2009; pp. 65–88. [CrossRef]
62. Hong, Y.; Huh, Y.M.; Yoon, D.S.; Yang, J. Nanobiosensors Based on Localized Surface Plasmon Resonance for Biomarker Detection. *J. Nanomater.* **2012**, *2012*, 759830. [CrossRef]
63. Unser, S.; Bruzas, I.; He, J.; Sagle, L. Localized Surface Plasmon Resonance Biosensing: Current Challenges and Approaches. *Sensors* **2015**, *15*, 15684–15716. [CrossRef]
64. Brolo, A.G. Plasmonics for Future Biosensors. *Nat. Photonics* **2012**, *6*, 709–713. [CrossRef]
65. Liu, J.; Jalali, M.; Mahshid, S.; Wachsmann-Hogiu, S. Are Plasmonic Optical Biosensors Ready for Use in Point-of-Need Applications? *Analyst* **2020**, *145*, 364–384. [CrossRef]
66. Moore, T.J.; Moody, A.S.; Payne, T.D.; Sarabia, G.M.; Daniel, A.R.; Sharma, B. In Vitro and in Vivo Sers Biosensing for Disease Diagnosis. *Biosensors* **2018**, *8*, 46. [CrossRef]
67. Alvarez-Puebla, R.A.; Liz-Marzán, L.M. SERS-Based Diagnosis and Biodetection. *Small* **2010**, *6*, 604–610. [CrossRef] [PubMed]
68. Zhang, X.; Liu, S.; Song, X.; Wang, H.; Wang, J.; Wang, Y.; Huang, J.; Yu, J. Robust and Universal SERS Sensing Platform for Multiplexed Detection of Alzheimer's Disease Core Biomarkers Using PAapt-AuNPs Conjugates. *ACS Sens.* **2019**, *4*, 2140–2149. [CrossRef] [PubMed]
69. Park, H.J.; Cho, S.; Kim, M.; Jung, Y.S. Carboxylic Acid-Functionalized, Graphitic Layer-Coated Three-Dimensional SERS Substrate for Label-Free Analysis of Alzheimer's Disease Biomarkers. *Nano Lett.* **2020**, *20*, 2576–2584. [CrossRef]
70. Dang, H.; Joung, Y.; Jeong, C.; Jeon, C.S.; Pyun, S.H.; Park, S.-G.; Choo, J. Nanoplasmonic Assay Platforms for Reproducible SERS Detection of Alzheimer's Disease Biomarker. *Bull. Korean Chem. Soc.* **2023**, *2023*, 1. [CrossRef]
71. Momenpour, A.; Lima, P.D.A.; Chen, Y.-A.; Tzeng, C.-R.; Tsang, B.K.; Anis, H. Surface-Enhanced Raman Scattering for the Detection of Polycystic Ovary Syndrome. *Biomed. Opt. Express* **2018**, *9*, 801. [CrossRef] [PubMed]
72. Lyandres, O.; Shah, N.C.; Yonzon, C.R.; Walsh, J.T.; Glucksberg, M.R.; Van Duyne, R.P. Real-Time Glucose Sensing by Surface-Enhanced Raman Spectroscopy in Bovine Plasma Facilitated by a Mixed Decanethiol/Mercaptohexanol Partition Layer. *Anal. Chem.* **2005**, *77*, 6134–6139. [CrossRef]
73. Qi, G.; Jia, K.; Fu, C.; Xu, S.; Xu, W. A Highly Sensitive SERS Sensor for Quantitative Analysis of Glucose Based on the Chemical Etching of Silver Nanoparticles. *J. Opt.* **2015**, *17*, 114020. [CrossRef]
74. Rong, Z.; Xiao, R.; Xing, S.; Xiong, G.; Yu, Z.; Wang, L.; Jia, X.; Wang, K.; Cong, Y.; Wang, S. SERS-Based Lateral Flow Assay for Quantitative Detection of C-Reactive Protein as an Early Bio-Indicator of a Radiation-Induced Inflammatory Response in Nonhuman Primates. *Analyst* **2018**, *143*, 2115–2121. [CrossRef]
75. Li, B.; Wu, Y.; Wang, Z.; Xing, M.; Xu, W.; Zhu, Y.; Du, P.; Wang, X.; Yang, H. Non-Invasive Diagnosis of Crohn's Disease Based on SERS Combined with PCA-SVM. *Anal. Methods* **2021**, *13*, 5264–5273. [CrossRef]
76. Xu, H.; Bjerneld, E.J.; Käll, M.; Börjesson, L. Spectroscopy of Single Hemoglobin Molecules by Surface Enhanced Raman Scattering. *Phys. Rev. Lett.* **1999**, *83*, 4357–4360. [CrossRef]
77. World Health Organization Report. Available online: https://www.who.int/news-room/fact-sheets/detail/cancer (accessed on 30 January 2023).
78. Blanco-Formoso, M.; Alvarez-Puebla, R.A. Cancer Diagnosis through Sers and Other Related Techniques. *Int. J. Mol. Sci.* **2020**, *21*, 2253. [CrossRef]
79. Guerrini, L.; Pazos-Perez, N.; Garcia-Rico, E.; Alvarez-Puebla, R. Cancer Characterization and Diagnosis with SERS-Encoded Particles. *Cancer Nanotechnol.* **2017**, *8*, 5. [CrossRef]
80. Kaur, B.; Kumar, S.; Kaushik, B.K. Recent Advancements in Optical Biosensors for Cancer Detection. *Biosens. Bioelectron.* **2022**, *197*, 113805. [CrossRef]
81. Thenrajan, T.; Wilson, J. Biosensors for Cancer Theranostics. *Biosens. Bioelectron. X* **2022**, *12*, 100232. [CrossRef]

82. Falkowski, P.; Lukaszewski, Z.; Gorodkiewicz, E. Potential of Surface Plasmon Resonance Biosensors in Cancer Detection. *J. Pharm. Biomed. Anal.* **2021**, *194*, 113802. [CrossRef]
83. Fu, Q.; Zhang, X.; Song, J.; Yang, H. Plasmonic Gold Nanoagents for Cancer Imaging and Therapy. *View* **2021**, *2*, 20200149. [CrossRef]
84. Azzouz, A.; Hejji, L.; Kim, K.H.; Kukkar, D.; Souhail, B.; Bhardwaj, N.; Brown, R.J.C.; Zhang, W. Advances in Surface Plasmon Resonance–Based Biosensor Technologies for Cancer Biomarker Detection. *Biosens. Bioelectron.* **2022**, *197*, 113767. [CrossRef] [PubMed]
85. Sugumaran, S.; Jamlos, M.F.; Ahmad, M.N.; Bellan, C.S.; Schreurs, D. Nanostructured Materials with Plasmonic Nanobiosensors for Early Cancer Detection: A Past and Future Prospect. *Biosens. Bioelectron.* **2018**, *100*, 361–373. [CrossRef] [PubMed]
86. Fattahi, Z.; Khosroushahi, A.Y.; Hasanzadeh, M. Recent Progress on Developing of Plasmon Biosensing of Tumor Biomarkers: Efficient Method towards Early Stage Recognition of Cancer. *Biomed. Pharmacother.* **2020**, *132*, 110850. [CrossRef] [PubMed]
87. Bellassai, N.; D'Agata, R.; Jungbluth, V.; Spoto, G. Surface Plasmon Resonance for Biomarker Detection: Advances in Non-Invasive Cancer Diagnosis. *Front. Chem.* **2019**, *7*, 570. [CrossRef] [PubMed]
88. Usman, F.; Dennis, J.O.; Aljameel, A.I.; Ali, M.K.M.; Aldaghri, O.; Ibnaouf, K.H.; Zango, Z.U.; Beygisangchin, M.; Alsadig, A.; Meriaudeau, F. Plasmonic Biosensors for the Detection of Lung Cancer Biomarkers: A Review. *Chemosensors* **2021**, *9*, 326. [CrossRef]
89. Yin, B.; Ho, W.K.H.; Xia, X.; Chan, C.K.W.; Zhang, Q.; Ng, Y.M.; Lam, C.Y.K.; Cheung, J.C.W.; Wang, J.; Yang, M.; et al. A Multilayered Mesoporous Gold Nanoarchitecture for Ultraeffective Near-Infrared Light-Controlled Chemo/Photothermal Therapy for Cancer Guided by SERS Imaging. *Small* **2023**, *2023*, 2206762. [CrossRef] [PubMed]
90. Constantinou, M.; Hadjigeorgiou, K.; Abalde-cela, S.; Andreou, C. Label-Free Sensing with Metal Nanostructure-Based Surface-Enhanced Raman Spectroscopy for Cancer Diagnosis. *ACS Appl. Nano Mater.* **2022**, *5*, 12276–12299. [CrossRef] [PubMed]
91. Rajput, S.; Pink, D.; Findlay, S.; Woolner, E.; Lewis, J.D.; McDermott, M.T. Application of Surface-Enhanced Raman Spectroscopy to Guide Therapy for Advanced Prostate Cancer Patients. *ACS Sens.* **2022**, *7*, 827–838. [CrossRef]
92. Avci, E.; Yilmaz, H.; Sahiner, N.; Tuna, B.G.; Cicekdal, M.B.; Eser, M.; Basak, K.; Altıntoprak, F.; Zengin, I.; Dogan, S.; et al. Label-Free Surface Enhanced Raman Spectroscopy for Cancer Detection. *Cancers* **2022**, *14*, 5021. [CrossRef]
93. Guerrini, L.; Alvarez-puebla, R.A. Surface-Enhanced Raman Spectroscopy in Cancer Diagnosis, Prognosis and Monitoring. *Cancers* **2019**, *11*, 748. [CrossRef]
94. Pollap, A.; Paweł, S. Recent Advances in Sandwich SERS Immunosensors for Cancer Detection. *Int. J. Mol. Sci.* **2022**, *23*, 4740. [CrossRef]
95. Vendrell, M.; Maiti, K.K.; Dhaliwal, K.; Chang, Y.T. Surface-Enhanced Raman Scattering in Cancer Detection and Imaging. *Trends Biotechnol.* **2013**, *31*, 249–257. [CrossRef]
96. Moisoiu, V.; Iancu, S.D.; Stefancu, A.; Moisoiu, T.; Pardini, B.; Dragomir, M.P.; Crisan, N.; Avram, L.; Crisan, D.; Andras, I.; et al. SERS Liquid Biopsy: An Emerging Tool for Medical Diagnosis. *Colloids Surfaces B Biointerfaces* **2021**, *208*, 112064. [CrossRef]
97. Shanmugasundaram, K.B.; Li, J.; Sina, A.I.; Wuethrich, A.; Trau, M. Toward Precision Oncology: SERS Microfluidic Systems for Multiplex Biomarker Analysis in Liquid Biopsy. *Mater. Adv.* **2022**, *3*, 1459–1471. [CrossRef]
98. Song, C.Y.; Yang, Y.J.; Yang, B.Y.; Sun, Y.Z.; Zhao, Y.P.; Wang, L.H. An Ultrasensitive SERS Sensor for Simultaneous Detection of Multiple Cancer-Related MiRNAs. *Nanoscale* **2016**, *8*, 17365–17373. [CrossRef] [PubMed]
99. Wang, H.N.; Crawford, B.M.; Norton, S.J.; Vo-Dinh, T. Direct and Label-Free Detection of MicroRNA Cancer Biomarkers Using SERS-Based Plasmonic Coupling Interference (PCI) Nanoprobes. *J. Phys. Chem. B* **2019**, *123*, 10245–10251. [CrossRef]
100. Guerrini, L.; Garcia-Rico, E.; O'loghlen, A.; Giannini, V.; Alvarez-Puebla, R.A. Surface-Enhanced Raman Scattering (Sers) Spectroscopy for Sensing and Characterization of Exosomes in Cancer Diagnosis. *Cancers.* **2021**, *13*, 2179. [CrossRef] [PubMed]
101. Lee, J.U.; Kim, S.; Sim, S.J. SERS-Based Nanoplasmonic Exosome Analysis: Enabling Liquid Biopsy for Cancer Diagnosis and Monitoring Progression. *BioChip J.* **2020**, *14*, 231–241. [CrossRef]
102. Vo-Dinh, T.; Allain, L.R.; Stokes, D.L. Cancer Gene Detection Using Surface-Enhanced Raman Scattering (SERS). *J. Raman Spectrosc.* **2002**, *33*, 511–516. [CrossRef]
103. Zhu, D.; Li, A.; Di, Y.; Wang, Z.; Shi, J.; Ni, X.; Wang, Y. Interference-Free SERS Nanoprobes for Labeling and Imaging of MT1-MMP in Breast Cancer Cells. *Nanotechnology* **2022**, *33*, 115702. [CrossRef]
104. Lin, J.; Chen, R.; Feng, S.; Pan, J.; Li, Y.; Chen, G.; Cheng, M.; Huang, Z.; Yu, Y.; Zeng, H. A Novel Blood Plasma Analysis Technique Combining Membrane Electrophoresis with Silver Nanoparticle-Based SERS Spectroscopy for Potential Applications in Noninvasive Cancer Detection. *Nanomed. Nanotechnol. Biol. Med.* **2011**, *7*, 655–663. [CrossRef]
105. Fabris, L. SERS Tags: The Next Promising Tool for Personalized Cancer Detection? *ChemNanoMat* **2016**, *2*, 249–258. [CrossRef]
106. Davis, R.M.; Campbell, J.L.; Burkitt, S.; Qiu, Z.; Kang, S.; Mehraein, M.; Miyasato, D.; Salinas, H.; Liu, J.T.C.; Zavaleta, C. A Raman Imaging Approach Using CD47 Antibody-Labeled SERS Nanoparticles for Identifying Breast Cancer and Its Potential to Guide Surgical Resection. *Nanomaterials* **2018**, *8*, 953. [CrossRef]
107. Yang, J.; Wang, Z.; Zong, S.; Song, C.; Zhang, R.; Cui, Y. Distinguishing Breast Cancer Cells Using Surface-Enhanced Raman Scattering. *Anal. Bioanal. Chem.* **2012**, *402*, 1093–1100. [CrossRef]
108. Dinish, U.S.; Balasundaram, G.; Chang, Y.T.; Olivo, M. Actively Targeted in Vivo Multiplex Detection of Intrinsic Cancer Biomarkers Using Biocompatible SERS Nanotags. *Sci. Rep.* **2014**, *4*, 4075. [CrossRef] [PubMed]

109. Liu, H.; Gao, X.; Xu, C.; Liu, D. SERS Tags for Biomedical Detection and Bioimaging. *Theranostics* **2022**, *12*, 1870–1903. [CrossRef] [PubMed]
110. Report on Lung Cancer. Available online: https://www.chestnet.org/newsroom/chest-news/2020/07/world-lung-cancer-day-2020-fact-sheet (accessed on 30 January 2023).
111. Mao, Y.; Sun, Y.; Xue, J.; Lu, W.; Cao, X. Ultra-Sensitive and High Efficiency Detection of Multiple Non-Small Cell Lung Cancer-Related MiRNAs on a Single Test Line in Catalytic Hairpin Assembly-Based SERS-LFA Strip. *Anal. Chim. Acta* **2021**, *1178*, 338800. [CrossRef] [PubMed]
112. Xia, J.; Liu, Y.; Ran, M.; Lu, D.; Cao, X.; Wang, Y. SERS Platform Based on Bimetallic Au-Ag Nanowires-Decorated Filter Paper for Rapid Detection of MiR-196ain Lung Cancer Patients Serum. *J. Chem.* **2020**, *2020*, 5073451. [CrossRef]
113. Cao, X.; Sun, Y.; Mao, Y.; Ran, M.; Liu, Y.; Lu, D.; Bi, C. Rapid and Sensitive Detection of Dual Lung Cancer-Associated MiRNA Biomarkers by a Novel SERS-LFA Strip Coupling with Catalytic Hairpin Assembly Signal Amplification. *J. Mater. Chem. C* **2021**, *9*, 3661–3671. [CrossRef]
114. Cao, X.; Mao, Y.; Gu, Y.; Ge, S.; Lu, W.; Gu, Y.; Li, Z. Highly Sensitive and Simultaneous Detection of CtDNAs Related to Non-Small Cell Lung Cancer in Serum Using a Catalytic Hairpin Assembly Strategy in a SERS Microfluidic Chip. *J. Mater. Chem. B* **2022**, *10*, 6194–6206. [CrossRef]
115. Cao, X.; Ge, S.; Zhou, X.; Mao, Y.; Sun, Y.; Lu, W.; Ran, M. A Dual-Signal Amplification Strategy Based on Pump-Free SERS Microfluidic Chip for Rapid and Ultrasensitive Detection of Non-Small Cell Lung Cancer-Related Circulating Tumour DNA in Mice Serum. *Biosens. Bioelectron.* **2022**, *205*, 114110. [CrossRef]
116. Ye, L.P.; Hu, J.; Liang, L.; Zhang, C.Y. Surface-Enhanced Raman Spectroscopy for Simultaneous Sensitive Detection of Multiple MicroRNAs in Lung Cancer Cells. *Chem. Commun.* **2014**, *50*, 11883–11886. [CrossRef]
117. Guo, T.; Li, W.; Qian, L.; Yan, X.; Cui, D.; Zhao, J.; Ni, H.; Zhao, X.; Zhang, Z.; Li, X.; et al. Highly-Selective Detection of EGFR Mutation Gene in Lung Cancer Based on Surface Enhanced Raman Spectroscopy and Asymmetric PCR. *J. Pharm. Biomed. Anal.* **2020**, *190*, 113522. [CrossRef]
118. Shin, H.; Oh, S.; Hong, S.; Kang, M.; Kang, D.; Ji, Y.G.; Choi, B.H.; Kang, K.W.; Jeong, H.; Park, Y.; et al. Early-Stage Lung Cancer Diagnosis by Deep Learning-Based Spectroscopic Analysis of Circulating Exosomes. *ACS Nano* **2020**, *14*, 5435–5444. [CrossRef]
119. Lei, J.; Yang, D.; Li, R.; Dai, Z.X.; Zhang, C.; Yu, Z.; Wu, S.; Pang, L.; Liang, S.; Zhang, Y. Label-Free Surface-Enhanced Raman Spectroscopy for Diagnosis and Analysis of Serum Samples with Different Types Lung Cancer. *Spectrochim. Acta Part A Mol. Biomol. Spectrosc.* **2021**, *261*, 120021. [CrossRef]
120. Wang, Z.; Hong, Y.; Yan, H.; Luo, H.; Zhang, Y.; Li, L.; Lu, S.; Chen, Y.; Wang, D.; Su, Y.; et al. Fabrication of Optoplasmonic Particles through Electroless Deposition and the Application in SERS-Based Screening of Nodule-Involved Lung Cancer. *Spectrochim. Acta Part A Mol. Biomol. Spectrosc.* **2022**, *279*, 121483. [CrossRef] [PubMed]
121. Zhang, J.; Dong, Y.; Zhu, W.; Xie, D.; Zhao, Y.; Yang, D.; Li, M. Ultrasensitive Detection of Circulating Tumor DNA of Lung Cancer via an Enzymatically Amplified SERS-Based Frequency Shift Assay. *ACS Appl. Mater. Interfaces* **2019**, *11*, 18145–18152. [CrossRef] [PubMed]
122. Jonak, S.T.; Liu, Z.; Liu, J.; Li, T.; D'Souza, B.V.; Schiaffino, A.; Oh, S.; Xie, Y.-H. Analyzing Bronchoalveolar Fluid Derived Small Extracellular Vesicles Using Single-Vesicle SERS for Non-Small Cell Lung Cancer Detection. *Sens. Diagn.* **2023**, *2*, 90–99. [CrossRef]
123. Park, J.; Hwang, M.; Choi, B.; Jeong, H.; Jung, J.H.; Kim, H.K.; Hong, S.; Park, J.H.; Choi, Y. Exosome Classification by Pattern Analysis of Surface-Enhanced Raman Spectroscopy Data for Lung Cancer Diagnosis. *Anal. Chem.* **2017**, *89*, 6695–6701. [CrossRef]
124. Wen, H.; Wang, H.; Hai, J.; He, S.; Chen, F.; Wang, B. Photochemical Synthesis of Porous CuFeSe 2 /Au Heterostructured Nanospheres as SERS Sensor for Ultrasensitive Detection of Lung Cancer Cells and Their Biomarkers. *ACS Sustain. Chem. Eng.* **2019**, *7*, 5200–5208. [CrossRef]
125. Qiao, X.; Su, B.; Liu, C.; Song, Q.; Luo, D.; Mo, G.; Wang, T. Selective Surface Enhanced Raman Scattering for Quantitative Detection of Lung Cancer Biomarkers in Superparticle@MOF Structure. *Adv. Mater.* **2018**, *30*, 1702275. [CrossRef]
126. Perumal, J.; Lee, P.; Dev, K.; Lim, H.Q.; Dinish, U.S.; Olivo, M. Machine Learning Assisted Real-Time Label-Free SERS Diagnoses of Malignant Pleural Effusion Due to Lung Cancer. *Biosensors* **2022**, *12*, 940. [CrossRef] [PubMed]
127. Zhang, K.; Liu, X.; Man, B.; Yang, C.; Zhang, C.; Liu, M.; Zhang, Y.; Liu, L.; Chen, C. Label-Free and Stable Serum Analysis Based on Ag-NPs/PSi Surface-Enhanced Raman Scattering for Noninvasive Lung Cancer Detection. *Biomed. Opt. Express* **2018**, *9*, 4345. [CrossRef]
128. Zhang, K.; Hao, C.; Huo, Y.; Man, B.; Zhang, C.; Yang, C.; Liu, M.; Chen, C. Label-Free Diagnosis of Lung Cancer with Tissue-Slice Surface-Enhanced Raman Spectroscopy and Statistical Analysis. *Lasers Med. Sci.* **2019**, *34*, 1849–1855. [CrossRef]
129. Chon, H.; Lee, S.; Yoon, S.Y.; Chang, S.I.; Lim, D.W.; Choo, J. Simultaneous Immunoassay for the Detection of Two Lung Cancer Markers Using Functionalized SERS Nanoprobes. *Chem. Commun.* **2011**, *47*, 12515–12517. [CrossRef] [PubMed]
130. Wu, P.; Gao, Y.; Lu, Y.; Zhang, H.; Cai, C. High Specific Detection and Near-Infrared Photothermal Therapy of Lung Cancer Cells with High SERS Active Aptamer-Silver-Gold Shell-Core Nanostructures. *Analyst* **2013**, *138*, 6501–6510. [CrossRef]
131. Chen, Y.W.; Liu, T.Y.; Chen, P.J.; Chang, P.H.; Chen, S.Y. A High-Sensitivity and Low-Power Theranostic Nanosystem for Cell SERS Imaging and Selectively Photothermal Therapy Using Anti-EGFR-Conjugated Reduced Graphene Oxide/Mesoporous Silica/AuNPs Nanosheets. *Small* **2016**, *12*, 1458–1468. [CrossRef] [PubMed]

132. Zhang, Y.; Ye, X.; Xu, G.; Jin, X.; Luan, M.; Lou, J.; Wang, L.; Huang, C.; Ye, J. Identification and Distinction of Non-Small-Cell Lung Cancer Cells by Intracellular SERS Nanoprobes. *RSC Adv.* **2016**, *6*, 5401–5407. [CrossRef]
133. Huang, Y.; Xie, T.; Zou, K.; Gu, Y.; Yang, G.; Zhang, F.; Qu, L.L.; Yang, S. Ultrasensitive SERS Detection of Exhaled Biomarkers of Lung Cancer Using a Multifunctional Solid Phase Extraction Membrane. *Nanoscale* **2021**, *13*, 13344–13352. [CrossRef]
134. Cai, C.; Liu, Y.; Li, J.; Wang, L.; Zhang, K. Serum Fingerprinting by Slippery Liquid-Infused Porous SERS for Non-Invasive Lung Cancer Detection. *Analyst* **2022**, *147*, 4426–4432. [CrossRef]
135. Sivashanmugan, K.; Huang, W.L.; Lin, C.H.; Liao, J.D.; Lin, C.C.; Su, W.C.; Wen, T.C. Bimetallic Nanoplasmonic Gap-Mode SERS Substrate for Lung Normal and Cancer-Derived Exosomes Detection. *J. Taiwan Inst. Chem. Eng.* **2017**, *80*, 149–155. [CrossRef]
136. Qian, K.; Wang, Y.; Hua, L.; Chen, A.; Zhang, Y. New Method of Lung Cancer Detection by Saliva Test Using Surface-Enhanced Raman Spectroscopy. *Thorac. Cancer* **2018**, *9*, 1556–1561. [CrossRef]
137. Yang, T.; Guo, X.; Wu, Y.; Wang, H.; Fu, S.; Wen, Y.; Yang, H. Facile and Label-Free Detection of Lung Cancer Biomarker in Urine by Magnetically Assisted Surface-Enhanced Raman Scattering. *ACS Appl. Mater. Interfaces* **2014**, *6*, 20985–20993. [CrossRef]
138. Breast Cancer Report. Available online: https://www.cancer.org/cancer/breast-cancer/about/how-common-is-breast-cancer.html (accessed on 30 January 2023).
139. Moisoiu, T.; Iancu, S.D.; Burghelea, D.; Dragomir, M.P.; Iacob, G.; Stefancu, A.; Cozan, R.G.; Antal, O.; Bálint, Z.; Muntean, V.; et al. SERS Liquid Biopsy Profiling of Serum for the Diagnosis of Kidney Cancer. *Biomedicines* **2022**, *10*, 233. [CrossRef]
140. Kim, S.; Kim, T.G.; Lee, S.H.; Kim, W.; Bang, A.; Moon, S.W.; Song, J.; Shin, J.H.; Yu, J.S.; Choi, S. Label-Free Surface-Enhanced Raman Spectroscopy Biosensor for On-Site Breast Cancer Detection Using Human Tears. *ACS Appl. Mater. Interfaces* **2020**, *12*, 7897–7904. [CrossRef] [PubMed]
141. Teixeira, R.A.R.; Lataliza, A.A.B.; Raposo, N.R.B.; Costa, L.A.S.; Sant'Ana, A.C. Insights on the Transport of Tamoxifen by Gold Nanoparticles for MCF-7 Breast Cancer Cells Based on SERS Spectroscopy. *Colloids Surf. B Biointerfaces* **2018**, *170*, 712–717. [CrossRef] [PubMed]
142. Nargis, H.F.; Nawaz, H.; Bhatti, H.N.; Jilani, K.; Saleem, M. Comparison of Surface Enhanced Raman Spectroscopy and Raman Spectroscopy for the Detection of Breast Cancer Based on Serum Samples. *Spectrochim. Acta Part A Mol. Biomol. Spectrosc.* **2021**, *246*, 119034. [CrossRef] [PubMed]
143. Xie, Y.; Su, X.; Wen, Y.; Zheng, C.; Li, M. Artificial Intelligent Label-Free SERS Profiling of Serum Exosomes for Breast Cancer Diagnosis and Postoperative Assessment. *Nano Lett.* **2022**, *22*, 7910–7918. [CrossRef] [PubMed]
144. Wen, Y.; Truong, V.X.; Li, M. Real-Time Intraoperative Surface-Enhanced Raman Spectroscopy-Guided Thermosurgical Eradication of Residual Microtumors in Orthotopic Breast Cancer. *Nano Lett.* **2021**, *21*, 3066–3074. [CrossRef]
145. Xiao, L.; Harihar, S.; Welch, D.R.; Zhou, A. Imaging of Epidermal Growth Factor Receptor on Single Breast Cancer Cells Using Surface-Enhanced Raman Spectroscopy. *Anal. Chim. Acta* **2014**, *843*, 73–82. [CrossRef]
146. Liang, L.; Shen, Y.; Zhang, J.; Xu, S.; Xu, W.; Liang, C.; Han, B. Identification of Breast Cancer through Spectroscopic Analysis of Cell-Membrane Sialic Acid Expression. *Anal. Chim. Acta* **2018**, *1033*, 148–155. [CrossRef]
147. Hernández-Arteaga, A.; de Jesús Zermeño Nava, J.; Kolosovas-Machuca, E.S.; Velázquez-Salazar, J.J.; Vinogradova, E.; José-Yacamán, M.; Navarro-Contreras, H.R. Diagnosis of Breast Cancer by Analysis of Sialic Acid Concentrations in Human Saliva by Surface-Enhanced Raman Spectroscopy of Silver Nanoparticles. *Nano Res.* **2017**, *10*, 3662–3670. [CrossRef]
148. Han, Y.; Qiang, L.; Gao, Y.; Gao, J.; He, Q.; Liu, H.; Han, L.; Zhang, Y. Large-Area Surface-Enhanced Raman Spectroscopy Substrate by Hybrid Porous GaN with Au/Ag for Breast Cancer MiRNA Detection. *Appl. Surf. Sci.* **2021**, *541*, 148456. [CrossRef]
149. Yarbakht, M.; Nikkhah, M.; Moshaii, A.; Weber, K.; Matthäus, C.; Cialla-May, D.; Popp, J. Simultaneous Isolation and Detection of Single Breast Cancer Cells Using Surface-Enhanced Raman Spectroscopy. *Talanta* **2018**, *186*, 44–52. [CrossRef]
150. Zheng, Z.; Wu, L.; Li, L.; Zong, S.; Wang, Z.; Cui, Y. Simultaneous and Highly Sensitive Detection of Multiple Breast Cancer Biomarkers in Real Samples Using a SERS Microfluidic Chip. *Talanta* **2018**, *188*, 507–515. [CrossRef] [PubMed]
151. Hameed, M.K.; Parambath, J.B.M.; Gul, M.T.; Khan, A.A.; Park, Y.; Han, C.; Mohamed, A.A. Arylated Gold Nanostars Aided SERS Study of Breast Cancer Cells. *Appl. Surf. Sci.* **2022**, *583*, 152504. [CrossRef]
152. Kapara, A.; Brunton, V.G.; Graham, D.; Faulds, K. Characterisation of Estrogen Receptor Alpha (ERα) Expression in Breast Cancer Cells and Effect of Drug Treatment Using Targeted Nanoparticles and SERS. *Analyst* **2020**, *145*, 7225–7233. [CrossRef] [PubMed]
153. Kapara, A.; Brunton, V.; Graham, D.; Faulds, K. Investigation of Cellular Uptake Mechanism of Functionalised Gold Nanoparticles into Breast Cancer Using SERS. *Chem. Sci.* **2020**, *11*, 5819–5829. [CrossRef]
154. Lee, S.; Chon, H.; Lee, J.; Ko, J.; Chung, B.H.; Lim, D.W.; Choo, J. Rapid and Sensitive Phenotypic Marker Detection on Breast Cancer Cells Using Surface-Enhanced Raman Scattering (SERS) Imaging. *Biosens. Bioelectron.* **2014**, *51*, 238–243. [CrossRef]
155. Choi, N.; Dang, H.; Das, A.; Sim, M.S.; Chung, I.Y.; Choo, J. SERS Biosensors for Ultrasensitive Detection of Multiple Biomarkers Expressed in Cancer Cells. *Biosens. Bioelectron.* **2020**, *164*, 112326. [CrossRef]
156. Meng, S.; Chen, R.; Xie, J.; Li, J.; Cheng, J.; Xu, Y.; Cao, H.; Wu, X.; Zhang, Q.; Wang, H. Surface-Enhanced Raman Scattering Holography Chip for Rapid, Sensitive and Multiplexed Detection of Human Breast Cancer-Associated MicroRNAs in Clinical Samples. *Biosens. Bioelectron.* **2021**, *190*, 113470. [CrossRef]
157. Weng, S.; Lin, D.; Lai, S.; Tao, H.; Chen, T.; Peng, M.; Qiu, S.; Feng, S. Highly Sensitive and Reliable Detection of MicroRNA for Clinically Disease Surveillance Using SERS Biosensor Integrated with Catalytic Hairpin Assembly Amplification Technology. *Biosens. Bioelectron.* **2022**, *208*, 114236. [CrossRef]

158. Li, Y.; Qi, X.; Lei, C.; Qifeng, Q.; Zhang, S. Simultaneous SERS Detection and Imaging of Two Biomarkers on the Cancer Cell Surface by Self-Assembly of Branched DNA-Gold Nanoaggregates. *Chem. Commun.* **2014**, *50*, 9907–9909. [CrossRef]
159. Lee, J.U.; Kim, W.H.; Lee, H.S.; Park, K.H.; Sim, S.J. Quantitative and Specific Detection of Exosomal MiRNAs for Accurate Diagnosis of Breast Cancer Using a Surface-Enhanced Raman Scattering Sensor Based on Plasmonic Head-Flocked Gold Nanopillars. *Small* **2019**, *15*, 1804968. [CrossRef]
160. Zhong, Q.; Zhang, K.; Huang, X.; Lu, Y.; Zhao, J.; He, Y.; Liu, B. In Situ Ratiometric SERS Imaging of Intracellular Protease Activity for Subtype Discrimination of Human Breast Cancer. *Biosens. Bioelectron.* **2022**, *207*, 114194. [CrossRef] [PubMed]
161. Li, L.; Liao, M.; Chen, Y.; Shan, B.; Li, M. Surface-Enhanced Raman Spectroscopy (SERS) Nanoprobes for Ratiometric Detection of Cancer Cells. *J. Mater. Chem. B* **2019**, *7*, 815–822. [CrossRef] [PubMed]
162. Wang, Y.; Kang, S.; Khan, A.; Ruttner, G.; Leigh, S.Y.; Murray, M.; Abeytunge, S.; Peterson, G.; Rajadhyaksha, M.; Dintzis, S.; et al. Quantitative Molecular Phenotyping with Topically Applied SERS Nanoparticles for Intraoperative Guidance of Breast Cancer Lumpectomy. *Sci. Rep.* **2016**, *6*, 21242. [CrossRef] [PubMed]
163. Zhang, Q.; Ma, R.; Zhang, Y.; Zhao, J.; Wang, Y.; Xu, Z. Dual-Aptamer-Assisted Ratiometric SERS Biosensor for Ultrasensitive and Precise Identification of Breast Cancer Exosomes. *ACS Sens.* **2023**. [CrossRef]
164. Shen, L.S.N.; Du, Y.; Wei, N.; Li, Q.; Li, S.M.; Sun, T.M.; Xu, S.; Wang, H.; Man, X.X.; Han, B. SERS Studies on Normal Epithelial and Cancer Cells Derived from Clinical Breast Cancer Specimens. *Spectrochim. Acta Part A Mol. Biomol. Spectrosc.* **2020**, *237*, 118364. [CrossRef]
165. Lin, Y.; Gao, S.; Zheng, M.; Tang, S.; Lin, K.; Xie, S.; Yu, Y.; Lin, J. A Microsphere Nanoparticle Based-Serum Albumin Targeted Adsorption Coupled with Surface-Enhanced Raman Scattering for Breast Cancer Detection. *Spectrochim. Acta Part A Mol. Biomol. Spectrosc.* **2021**, *261*, 120039. [CrossRef]
166. Lin, Y.; Gao, J.; Tang, S.; Zhao, X.; Zheng, M.; Gong, W.; Xie, S.; Gao, S.; Yu, Y.; Lin, J. Label-Free Diagnosis of Breast Cancer Based on Serum Protein Purification Assisted Surface-Enhanced Raman Spectroscopy. *Spectrochim. Acta Part A Mol. Biomol. Spectrosc.* **2021**, *263*, 120234. [CrossRef]
167. Moisoiu, V.; Socaciu, A.; Stefancu, A.; Iancu, S.D.; Boros, I.; Alecsa, C.D.; Rachieriu, C.; Chiorean, A.R.; Eniu, D.; Leopold, N.; et al. Breast Cancer Diagnosis by Surface-Enhanced Raman Scattering (SERS) of Urine. *Appl. Sci.* **2019**, *9*, 806. [CrossRef]
168. Akbar, S.; Majeed, M.I.; Nawaz, H.; Rashid, N.; Tariq, A.; Hameed, W.; Shakeel, S.; Dastgir, G.; Bari, R.Z.A.; Iqbal, M.; et al. Surface-Enhanced Raman Spectroscopic (SERS) Characterization of Low Molecular Weight Fraction of the Serum of Breast Cancer Patients with Principal Component Analysis (PCA) and Partial Least Square-Discriminant Analysis (PLS-DA). *Anal. Lett.* **2022**, *55*, 1588–1604. [CrossRef]
169. Feng, S.; Huang, S.; Lin, D.; Chen, G.; Xu, Y.; Li, Y.; Huang, Z.; Pan, J.; Chen, R.; Zeng, H. Surface-Enhanced Raman Spectroscopy of Saliva Proteins for the Noninvasive Differentiation of Benign and Malignant Breast Tumors. *Int. J. Nanomed.* **2015**, *10*, 537–547. [CrossRef]
170. Iancu, S.D.; Cozan, R.G.; Stefancu, A.; David, M.; Moisoiu, T.; Moroz-Dubenco, C.; Bajcsi, A.; Chira, C.; Andreica, A.; Leopold, L.F.; et al. SERS Liquid Biopsy in Breast Cancer. What Can We Learn from SERS on Serum and Urine? *Spectrochim. Acta Part A Mol. Biomol. Spectrosc.* **2022**, *273*, 120992. [CrossRef]
171. Lin, X.; Jia, X.; Lin, J.Y.; Wu, P.H.; Weng, Y.; Feng, S. A Comparative Study Based on Serum SERS Spectra in and on the Coffee Ring for High Precision Breast Cancer Detection. *J. Raman Spectrosc.* **2022**, *53*, 1371–1379. [CrossRef]
172. Ştiufiuc, G.F.; Toma, V.; Buse, M.; Mărginean, R.; Morar-Bolba, G.; Culic, B.; Tetean, R.; Leopold, N.; Pavel, I.; Lucaciu, C.M.; et al. Solid Plasmonic Substrates for Breast Cancer Detection by Means of SERS Analysis of Blood Plasma. *Nanomaterials* **2020**, *10*, 1212. [CrossRef] [PubMed]
173. Cervo, S.; Mansutti, E.; Del Mistro, G.; Spizzo, R.; Colombatti, A.; Steffan, A.; Sergo, V.; Bonifacio, A. SERS Analysis of Serum for Detection of Early and Locally Advanced Breast Cancer. *Anal. Bioanal. Chem.* **2015**, *407*, 7503–7509. [CrossRef] [PubMed]
174. Lin, D.; Wang, Y.; Wang, T.; Zhu, Y.; Lin, X.; Lin, Y.; Feng, S. Metabolite Profiling of Human Blood by Surface-Enhanced Raman Spectroscopy for Surgery Assessment and Tumor Screening in Breast Cancer. *Anal. Bioanal. Chem.* **2020**, *412*, 1611–1618. [CrossRef] [PubMed]
175. Gao, N.; Wang, Q.; Tang, J.; Yao, S.; Li, H.; Yue, X.; Fu, J.; Zhong, F.; Wang, T.; Wang, J. Non-Invasive SERS Serum Detection Technology Combined with Multivariate Statistical Algorithm for Simultaneous Screening of Cervical Cancer and Breast Cancer. *Anal. Bioanal. Chem.* **2021**, *413*, 4775–4784. [CrossRef]
176. Vargas-Obieta, E.; Martínez-Espinosa, J.C.; Martínez-Zerega, B.E.; Jave-Suárez, L.F.; Aguilar-Lemarroy, A.; González-Solís, J.L. Breast Cancer Detection Based on Serum Sample Surface Enhanced Raman Spectroscopy. *Lasers Med. Sci.* **2016**, *31*, 1317–1324. [CrossRef]
177. Ma, X.; Xiong, H.; Guo, J.; Liu, Z.; Han, Y.; Liu, M. Label-Free Breast Cancer Detection and Classification by Convolutional Neural Network-Based on Exosomes Surface-Enhanced Raman Scattering. *J. Innov. Opt. Health Sci.* **2022**, *2022*, 2244001. [CrossRef]
178. Zhang, Z.; Wang, J.; Shanmugasundaram, K.B.; Yeo, B.; Möller, A.; Wuethrich, A.; Lin, L.L.; Trau, M. Tracking Drug-Induced Epithelial–Mesenchymal Transition in Breast Cancer by a Microfluidic Surface-Enhanced Raman Spectroscopy Immunoassay. *Small* **2020**, *16*, 1905614. [CrossRef]
179. Zhu, J.; Zhou, J.; Guo, J.; Cai, W.; Liu, B.; Wang, Z.; Sun, Z. Surface-Enhanced Raman Spectroscopy Investigation on Human Breast Cancer Cells. *Chem. Cent. J.* **2013**, *7*, 37. [CrossRef]

180. Brozek-Pluska, B.; Kopec, M.; Surmacki, J. Surface-Enhanced Raman Spectroscopy Analysis of Human Breast Cancer via Silver Nanoparticles: An Examination of Fabrication Methods. *J. Spectrosc.* **2018**, *2018*, 4893274. [CrossRef]
181. Narayanan, N.; Kim, J.H.; Santhakumar, H.; Joseph, M.M.; Karunakaran, V.; Shamjith, S.; Saranya, G.; Sujai, P.T.; Jayasree, R.S.; Barman, I.; et al. Nanotheranostic Probe Built on Methylene Blue Loaded Cucurbituril [8] and Gold Nanorod: Targeted Phototherapy in Combination with SERS Imaging on Breast Cancer Cells. *J. Phys. Chem. B* **2021**, *125*, 13415–13424. [CrossRef] [PubMed]
182. Feng, J.; Chen, L.; Xia, Y.; Xing, J.; Li, Z.; Qian, Q.; Wang, Y.; Wu, A.; Zeng, L.; Zhou, Y. Bioconjugation of Gold Nanobipyramids for SERS Detection and Targeted Photothermal Therapy in Breast Cancer. *ACS Biomater. Sci. Eng.* **2017**, *3*, 608–618. [CrossRef] [PubMed]
183. Zeng, L.; Pan, Y.; Wang, S.; Wang, X.; Zhao, X.; Ren, W.; Lu, G.; Wu, A. Raman Reporter-Coupled Agcore@Aushell Nanostars for in Vivo Improved Surface Enhanced Raman Scattering Imaging and Near-Infrared-Triggered Photothermal Therapy in Breast Cancers. *ACS Appl. Mater. Interfaces* **2015**, *7*, 16781–16791. [CrossRef] [PubMed]
184. Xinyue, L.; Keshavarz, M.; Panagiotis, K.; Roddan, A.; Yeatman, E.; Thompson, A. SERS Detection of Breast Cancer-Derived Exosomes Using a Nanostructured Pt-Black Template. *Adv. Sens. Res.* **2023**, *2023*, 2200039. [CrossRef]
185. Pramanik, A.; Mayer, J.; Patibandla, S.; Gates, K.; Gao, Y.; Davis, D.; Seshadri, R.; Ray, P.C. Mixed-Dimensional Heterostructure Material-Based SERS for Trace Level Identification of Breast Cancer-Derived Exosomes. *ACS Omega* **2020**, *5*, 16602–16611. [CrossRef]
186. Li, G.; Zhu, N.; Zhou, J.; Kang, K.; Zhou, X.; Ying, B.; Yi, Q.; Wu, Y. A Magnetic Surface-Enhanced Raman Scattering Platform for Performing Successive Breast Cancer Exosome Isolation and Analysis. *J. Mater. Chem. B* **2021**, *9*, 2709–2716. [CrossRef]
187. Yang, Z.; Su, H.S.; You, E.M.; Liu, S.; Li, Z.; Zhang, Y. High Uniformity and Enhancement Au@AgNS 3D Substrates for the Diagnosis of Breast Cancer. *ACS Omega* **2022**, *7*, 15223–15230. [CrossRef]
188. Wang, X.P.; Walkenfort, B.; König, M.; König, L.; Kasimir-Bauer, S.; Schlücker, S. Fast and Reproducible ISERS Microscopy of Single HER2-Positive Breast Cancer Cells Using Gold Nanostars as SERS Nanotags. *Faraday Discuss.* **2017**, *205*, 377–386. [CrossRef]
189. Chen, Z.; Shen, X.; Chen, S.; Dai, K. *Gastric Cancer Prewarning and Early Diagnosis System*; Springer: Berlin/Heidelberg, Germany, 2017; ISBN 9789402409499.
190. Hunter, R.A.; Asare-Werehene, M.; Mandour, A.; Tsang, B.K.; Anis, H. Determination of Chemoresistance in Ovarian Cancer by Simultaneous Quantification of Exosomes and Exosomal Cisplatin with Surface Enhanced Raman Scattering. *Sens. Actuators B Chem.* **2022**, *354*, 131237. [CrossRef]
191. Moothanchery, M.; Perumal, J.; Mahyuddin, A.P.; Singh, G.; Choolani, M.; Olivo, M. Rapid and Sensitive Detection of Ovarian Cancer Biomarker Using a Portable Single Peak Raman Detection Method. *Sci. Rep.* **2022**, *12*, 12459. [CrossRef]
192. Sarkar, S.; Gogoi, M.; Mahato, M.; Joshi, A.B.; Baruah, A.J.; Kodgire, P.; Boruah, P. Biosensors for Detection of Prostate Cancer: A Review. *Biomed. Microdevices* **2022**, *24*, 32. [CrossRef]
193. Turan, E.; Zengin, A.; Suludere, Z.; Kalkan, N.Ö.; Tamer, U. Construction of a Sensitive and Selective Plasmonic Biosensor for Prostate Specific Antigen by Combining Magnetic Molecularly-Imprinted Polymer and Surface-Enhanced Raman Spectroscopy. *Talanta* **2022**, *237*, 122926. [CrossRef]
194. Haroon, M.; Tahir, M.; Nawaz, H.; Majeed, M.I.; Al-Saadi, A.A. Surface-Enhanced Raman Scattering (SERS) Spectroscopy for Prostate Cancer Diagnosis: A Review. *Photodiagn. Photodyn. Ther.* **2022**, *37*, 102690. [CrossRef] [PubMed]
195. Ashrafizadeh, M.; Aghamiri, S.; Tan, S.C.; Zarrabi, A.; Sharifi, E.; Rabiee, N.; Kadumudi, F.B.; Pirouz, A.D.; Delfi, M.; Byrappa, K.; et al. Nanotechnological Approaches in Prostate Cancer Therapy: Integration of Engineering and Biology. *Nano Today* **2022**, *45*, 101532. [CrossRef]
196. Gaba, F.; Tipping, W.J.; Salji, M.; Faulds, K.; Graham, D.; Leung, H.Y. Raman Spectroscopy in Prostate Cancer: Techniques, Applications and Advancements. *Cancers.* **2022**, *14*, 1535. [CrossRef]
197. Pandey, A.; Sarkar, S.; Pandey, S.K.; Srivastava, A. Silica Nanospheres Coated Silver Islands as an Effective Opto-Plasmonic SERS Active Platform for Rapid and Sensitive Detection of Prostate Cancer Biomarkers. *Molecules* **2022**, *27*, 7821. [CrossRef] [PubMed]
198. Wei, Y.; Zhu, Y.Y.; Wang, M.L. Surface-Enhanced Raman Spectroscopy of Gastric Cancer Serum with Gold Nanoparticles/Silicon Nanowire Arrays. *Optik* **2016**, *127*, 7902–7907. [CrossRef]
199. Ito, H.; Inoue, H.; Hasegawa, K.; Hasegawa, Y.; Shimizu, T.; Kimura, S.; Onimaru, M.; Ikeda, H.; Kudo, S. ei Use of Surface-Enhanced Raman Scattering for Detection of Cancer-Related Serum-Constituents in Gastrointestinal Cancer Patients. *Nanomed. Nanotechnol. Biol. Med.* **2014**, *10*, 599–608. [CrossRef]
200. Ge, S.; Ran, M.; Mao, Y.; Sun, Y.; Zhou, X.; Li, L.; Cao, X. A Novel DNA Biosensor for the Ultrasensitive Detection of DNA Methyltransferase Activity Based on a High-Density 'Hot Spot' SERS Substrate and Rolling Circle Amplification Strategy. *Analyst* **2021**, *146*, 5326–5336. [CrossRef]
201. Feng, S.; Chen, R.; Lin, J.; Pan, J.; Wu, Y.; Li, Y.; Chen, J.; Zeng, H. Gastric Cancer Detection Based on Blood Plasma Surface-Enhanced Raman Spectroscopy Excited by Polarized Laser Light. *Biosens. Bioelectron.* **2011**, *26*, 3167–3174. [CrossRef]
202. Pan, H.; Dong, Y.; Gong, L.; Zhai, J.; Song, C.; Ge, Z.; Su, Y.; Zhu, D.; Chao, J.; Su, S.; et al. Sensing Gastric Cancer Exosomes with MoS2-Based SERS Aptasensor. *Biosens. Bioelectron.* **2022**, *215*, 114553. [CrossRef]

203. Liu, Z.; Li, T.; Wang, Z.; Liu, J.; Huang, S.; Min, B.H.; An, J.Y.; Kim, K.M.; Kim, S.; Chen, Y.; et al. Gold Nanopyramid Arrays for Non-Invasive Surface-Enhanced Raman Spectroscopy-Based Gastric Cancer Detection via SEVs. *ACS Appl. Nano Mater.* **2022**, *5*, 12506–12517. [CrossRef]
204. Gayoung, E.; Hongki, K.; Ahreum, H.; Hye-Young, S.; Yuna, C.; Jeong, M.; Donghyeong, K.; Miyeon, L.; Eun-Kyung, L.; Jinyoung, J.; et al. Nanogap-Rich Au Nanowire SERS Sensor for Ultrasensitive Telomerase Activity Detection. *Adv. Funct. Mater.* **2017**, *27*, 1701832. [CrossRef]
205. Chen, Y.; Cheng, S.; Zhang, A.; Song, J.; Chang, J.; Wang, K.; Gao, G.; Zhang, Y.; Li, S.; Liu, H.; et al. Salivary Analysis Based on Surface Enhanced Raman Scattering Sensors Distinguishes Early and Advanced Gastric Cancer Patients from Healthy Persons. *J. Biomed. Nanotechnol.* **2018**, *14*, 1773–1784. [CrossRef]
206. Cao, X.; Ge, S.; Hua, W.; Zhou, X.; Lu, W.; Gu, Y.; Li, Z.; Qian, Y. A Pump-Free and High-Throughput Microfluidic Chip for Highly Sensitive SERS Assay of Gastric Cancer-Related Circulating Tumor DNA via a Cascade Signal Amplification Strategy. *J. Nanobiotechnol.* **2022**, *20*, 271. [CrossRef]
207. Chen, Y.; Zhang, Y.; Pan, F.; Liu, J.; Wang, K.; Zhang, C.; Cheng, S.; Lu, L.; Zhang, W.; Zhang, Z.; et al. Breath Analysis Based on Surface-Enhanced Raman Scattering Sensors Distinguishes Early and Advanced Gastric Cancer Patients from Healthy Persons. *ACS Nano* **2016**, *10*, 8169–8179. [CrossRef]
208. Huang, L.; Zhu, Y.; Xu, C.; Cai, Y.; Yi, Y.; Li, K.; Ren, X.; Jiang, D.; Ge, Y.; Liu, X.; et al. Noninvasive Diagnosis of Gastric Cancer Based on Breath Analysis with a Tubular Surface-Enhanced Raman Scattering Sensor. *ACS Sens.* **2022**, *7*, 1439–1450. [CrossRef]
209. Cao, D.; Lin, H.; Liu, Z.; Qiu, J.; Ge, S.; Hua, W.; Cao, X.; Qian, Y.; Xu, H.; Zhu, X. PCA-TLNN-Based SERS Analysis Platform for Label-Free Detection and Identification of Cisplatin-Treated Gastric Cancer. *Sens. Actuators B Chem.* **2023**, *375*, 132903. [CrossRef]
210. Guo, L.; Li, Y.; Huang, F.; Dong, J.; Li, F.; Yang, X.; Zhu, S.; Yang, M. Identification and Analysis of Serum Samples by Surface-Enhanced Raman Spectroscopy Combined with Characteristic Ratio Method and PCA for Gastric Cancer Detection. *J. Innov. Opt. Health Sci.* **2019**, *12*, 1950003. [CrossRef]
211. Ma, J.; Zhou, H.; Gong, L.; Liu, S.; Zhou, Z.; Mao, W.; Zheng, R. Distinction of Gastric Cancer Tissue Based on Surface-Enhanced Raman Spectroscopy. *Opt. Health Care Biomed. Opt. V* **2012**, *8553*, 855328. [CrossRef]
212. Chen, Y.; Chen, Y.; Chen, G.; Zheng, X.; He, C.; Feng, S.; Lin, X.; Chen, R.; Zeng, H. Discrimination of Gastric Cancer from Normal by Serum RNA Based on Surface-Enhanced Raman Spectroscopy (SERS) and Multivariate Analysis. *Med. Phys.* **2012**, *39*, 5664–5668. [CrossRef]
213. Feng, S.Y.; Pan, J.J.; Wu, Y.A.; Lin, D.; Chen, Y.P.; Xi, G.Q.; Lin, J.Q.; Chen, R. Study on Gastric Cancer Blood Plasma Based on Surface-Enhanced Raman Spectroscopy Combined with Multivariate Analysis. *Sci. China Life Sci.* **2011**, *54*, 828–834. [CrossRef]
214. Aslam, M.A.; Xue, C.; Wang, K.; Chen, Y.; Zhang, A.; Cai, W.; Ma, L.; Yang, Y.; Sun, X.; Liu, M.; et al. SVM Based Classification and Prediction System for Gastric Cancer Using Dominant Features of Saliva. *Nano Biomed. Eng.* **2020**, *12*, 1–13. [CrossRef]
215. Aslam, M.A.; Xue, C.; Liu, M.; Wang, K.; Cui, D. Classification and Prediction of Gastric Cancer from Saliva Diagnosis Using Artificial Neural Network. *Eng. Lett.* **2020**, *29*, 2.
216. Avram, L.; Iancu, S.D.; Stefancu, A.; Moisoiu, V.; Colnita, A.; Marconi, D.; Donca, V.; Buzdugan, E.; Craciun, R.; Leopold, N.; et al. SERS-Based Liquid Biopsy of Gastrointestinal Tumors Using a Portable Raman Device Operating in a Clinical Environment. *J. Clin. Med.* **2020**, *9*, 212. [CrossRef]
217. Li, X.; Yang, T.; Li, S.; Wang, D.; Song, Y.; Yu, K. Different Classification Algorithms and Serum Surface Enhanced Raman Spectroscopy for Noninvasive Discrimination of Gastric Diseases. *J. Raman Spectrosc.* **2016**, *47*, 917–925. [CrossRef]
218. Li, S.X.; Zhang, Y.J.; Zeng, Q.Y.; Li, L.F.; Guo, Z.Y.; Liu, Z.M.; Xiong, H.L.; Liu, S.H. Potential of Cancer Screening with Serum Surface-Enhanced Raman Spectroscopy and a Support Vector Machine. *Laser Phys. Lett.* **2014**, *11*, 065603. [CrossRef]
219. Moisoiu, T.; Dragomir, M.P.; Iancu, S.D.; Schallenberg, S.; Birolo, G.; Ferrero, G.; Burghelea, D.; Stefancu, A.; Cozan, R.G.; Licarete, E.; et al. Combined MiRNA and SERS Urine Liquid Biopsy for the Point-of-Care Diagnosis and Molecular Stratification of Bladder Cancer. *Mol. Med.* **2022**, *28*, 39. [CrossRef]
220. Gao, S.; Lin, Y.; Zhao, X.; Gao, J.; Xie, S.; Gong, W.; Yu, Y.; Lin, J. Label-Free Surface Enhanced Raman Spectroscopy Analysis of Blood Serum via Coffee Ring Effect for Accurate Diagnosis of Cancers. *Spectrochim. Acta Part A Mol. Biomol. Spectrosc.* **2022**, *267*, 120605. [CrossRef]
221. Meng, C.; Li, H.; Chen, C.; Wu, W.; Gao, J.; Lai, Y.; Ka, M.; Zhu, M.; Lv, X.; Chen, F.; et al. Serum Raman Spectroscopy Combined with Gaussian—Convolutional Neural Network Models to Quickly Detect Liver Cancer Patients. *Spectrosc. Lett.* **2022**, *55*, 79–90. [CrossRef]
222. Ni, J.-T.; Huang, M.-Y.; Ji, W.; Wang, L.; Sun, T.-D. Recent Advances in Surface-Enhanced Raman Scattering for Liver Cancer Detection. *Chin. J. Anal. Chem.* **2022**, *50*, 100180. [CrossRef]
223. Zhang, Q.; Hou, D.; Wen, X.; Xin, M.; Li, Z.; Wu, L.; Pathak, J.L. Gold Nanomaterials for Oral Cancer Diagnosis and Therapy: Advances, Challenges, and Prospects. *Mater. Today Bio* **2022**, *15*, 100333. [CrossRef] [PubMed]
224. Zadeh, F.A.; Shahhosseini, E.; Rasoolzadegan, S.; Özbolat, G.; Farahbod, F. Au Nanoparticles in the Diagnosis and Treatment of Ovarian Cancer: A New Horizon in the Personalized Medicine. *Nanomed. Res. J.* **2022**, *7*, 1–18. [CrossRef]
225. Wang, Y.; Zhang, Y.; Du, Q.; Cao, D.; Lu, X.; Meng, Z. Sensitive SERS Detection of Oral Squamous Cell Carcinoma-Related MiRNAs in Saliva via a Gold Nanohexagon Array Coupled with Hybridization Chain Reaction Amplification. *Anal. Methods* **2022**, *14*, 4563–4575. [CrossRef]

226. Fălămaș, A.; Rotaru, H.; Hedeșiu, M. Surface-Enhanced Raman Spectroscopy (SERS) Investigations of Saliva for Oral Cancer Diagnosis. *Lasers Med. Sci.* **2020**, *35*, 1393–1401. [CrossRef]
227. Wang, K.; Qiu, Y.; Wu, C.; Wen, Z.N.; Li, Y. Surface-Enhanced Raman Spectroscopy and Multivariate Analysis for the Diagnosis of Oral Squamous Cell Carcinoma. *J. Raman Spectrosc.* **2023**. [CrossRef]
228. Lin, Y.; Lin, J.; Zheng, M.; Gong, W.; Li, H.; Shu, Z.; Du, W.; Gao, S.; Yu, Y. Quantitative and Direct Serum Albumin Detection by Label-Free SERS Using Tunable Hydroxyapatite Nanostructure for Prostate Cancer Detection. *Anal. Chim. Acta* **2022**, *1221*, 340101. [CrossRef]
229. Zhao, J.; Wang, J.; Liu, Y.; Han, X.X.; Xu, B.; Ozaki, Y.; Zhao, B. Detection of Prostate Cancer Biomarkers via a SERS-Based Aptasensor. *Biosens. Bioelectron.* **2022**, *216*, 114660. [CrossRef]
230. Lu, Y.; Zhan, C.; Yu, L.; Yu, Y.; Jia, H.; Chen, X.; Zhang, D.; Gao, R. Multifunctional Nanocone Array as Solid Immunoassay Plate and SERS Substrate for the Early Diagnosis of Prostate Cancer on Microfluidic Chip. *Sens. Actuators B Chem.* **2023**, *376*, 133046. [CrossRef]
231. Munteanu, V.C.; Munteanu, R.A.; Gulei, D.; Mărginean, R.; Schițcu, V.H.; Onaciu, A.; Toma, V.; Știufiuc, G.F.; Coman, I.; Știufiuc, R.I. New Insights into the Multivariate Analysis of SER Spectra Collected on Blood Samples for Prostate Cancer Detection: Towards a Better Understanding of the Role Played by Different Biomolecules on Cancer Screening: A Preliminary Study. *Cancers* **2022**, *14*, 3227. [CrossRef] [PubMed]
232. Stefancu, A.; Moisoiu, V.; Couti, R.; Andras, I.; Rahota, R.; Crisan, D.; Pavel, I.E.; Socaciu, C.; Leopold, N.; Crisan, N. Combining SERS Analysis of Serum with PSA Levels for Improving the Detection of Prostate Cancer. *Nanomedicine* **2018**, *13*, 2455–2467. [CrossRef]
233. Liyanage, T.; Alharbi, B.; Quan, L.; Esquela-Kerscher, A.; Slaughter, G. Plasmonic-Based Biosensor for the Early Diagnosis of Prostate Cancer. *ACS Omega* **2022**, *7*, 2411–2418. [CrossRef] [PubMed]
234. Zhao, X.; Xu, Q.; Lin, Y.; Du, W.; Bai, X.; Gao, J.; Li, T.; Huang, Y.; Yu, Y.; Wu, X.; et al. Label-free surface-enhanced Raman spectroscopy detectionof prostate cancer combined with multivariate statistical algorithm. *J. Raman Spectrosc.* **2022**, *53*, 1861–1870. [CrossRef]
235. Sayyadi, N.; Justiniano, I.; Wang, Y.; Zheng, X.; Zhang, W.; Jiang, L.; Polikarpov, D.M.; Willows, R.D.; Gillatt, D.; Campbell, D.; et al. Detection of Rare Prostate Cancer Cells in Human Urine Offers Prospect of Non-Invasive Diagnosis. *Sci. Rep.* **2022**, *12*, 18452. [CrossRef] [PubMed]
236. Lu, S.; Lin, S.; Zhang, H.; Liang, L.; Shen, S. Methods of Respiratory Virus Detection: Advances towards Point-of-Care for Early Intervention. *Micromachines* **2021**, *12*, 697. [CrossRef] [PubMed]
237. Omidifar, N.; Lankarani, K.B.; Moghadami, M.; Shokripour, M.; Chashmpoosh, M.; Mousavi, S.M.; Hashemi, S.A.; Gholami, A. Different Laboratory Diagnosis Methods of COVID-19: A Systematic Review. *Arch. Clin. Infect. Dis.* **2021**, *16*, e110667. [CrossRef]
238. Mousavi, S.M.; Hashemi, S.A.; Rahmanian, V.; Kalashgrani, M.Y. Highly Sensitive Flexible SERS-Based Sensing Platform for Detection of Biosensors Highly Sensitive Flexible SERS-Based Sensing Platform for Detection of COVID-19. *Biosensors* **2022**, *12*, 466. [CrossRef]
239. Stöckel, S.; Kirchhoff, J.; Neugebauer, U.; Rösch, P.; Popp, J. The Application of Raman Spectroscopy for the Detection and Identification of Microorganisms. *J. Raman Spectrosc.* **2016**, *47*, 89–109. [CrossRef]
240. Soler, M.; Estevez, M.C.; Cardenosa-Rubio, M.; Astua, A.; Lechuga, L.M. How Nanophotonic Label-Free Biosensors Can Contribute to Rapid and Massive Diagnostics of Respiratory Virus Infections: COVID-19 Case. *ACS Sens.* **2020**, *5*, 2663–2678. [CrossRef]
241. Iravani, S. Nano- And Biosensors for the Detection of SARS-CoV-2: Challenges and Opportunities. *Mater. Adv.* **2020**, *1*, 3092–3103. [CrossRef]
242. Zhang, D.; Zhang, X.; Ma, R.; Deng, S.; Wang, X.; Wang, X.; Zhang, X.; Huang, X.; Liu, Y.; Li, G.; et al. Ultra-Fast and Onsite Interrogation of Severe Acute Respiratory Syndrome Coronavirus 2 (SARS-CoV-2) in Waters via Surface Enhanced Raman Scattering (SERS). *Water Res.* **2021**, *200*, 117243. [CrossRef] [PubMed]
243. Joung, Y.; Kim, K.; Lee, S.; Chun, B.-S.; Lee, S.; Hwang, J.; Choi, S.; Kang, T.; Lee, M.-K.; Chen, L.; et al. Rapid and Accurate On-Site Immunodiagnostics of Highly Contagious Severe Acute Respiratory Syndrome Coronavirus 2 Using Portable Surface-Enhanced Raman Scattering-Lateral Flow Assay Reader. *ACS Sens.* **2022**, *7*, 3470–3480. [CrossRef] [PubMed]
244. Saviñon-Flores, F.; Méndez, E.; López-Castaños, M.; Carabarin-Lima, A.; López-Castaños, K.A.; González-Fuentes, M.A.; Méndez-Albores, A. A Review on Sers-Based Detection of Human Virus Infections: Influenza and Coronavirus. *Biosensors* **2021**, *11*, 66. [CrossRef] [PubMed]
245. Lim, J.Y.; Nam, J.S.; Yang, S.E.; Shin, H.; Jang, Y.H.; Bae, G.U.; Kang, T.; Lim, K.I.; Choi, Y. Identification of Newly Emerging Influenza Viruses by Surface-Enhanced Raman Spectroscopy. *Anal. Chem.* **2015**, *87*, 11652–11659. [CrossRef] [PubMed]
246. Zhang, K.; Wang, Z.; Liu, H.; Perea-López, N.; Ranasinghe, J.C.; Bepete, G.; Minns, A.M.; Rossi, R.M.; Lindner, S.E.; Huang, S.X.; et al. Understanding the Excitation Wavelength Dependence and Thermal Stability of the SARS-CoV-2 Receptor-Binding Domain Using Surface-Enhanced Raman Scattering and Machine Learning. *ACS Photonics* **2022**, *9*, 2963–2972. [CrossRef]
247. Yang, Y.; Xu, B.; Murray, J.; Haverstick, J.; Chen, X.; Tripp, R.A.; Zhao, Y. Rapid and Quantitative Detection of Respiratory Viruses Using Surface-Enhanced Raman Spectroscopy and Machine Learning. *Biosens. Bioelectron.* **2022**, *217*, 114721. [CrossRef]
248. Ye, J.; Yeh, Y.; Xue, Y.; Wang, Z.; Zhang, N.; Liu, H.; Zhang, K.; Ricker, R.; Yu, Z.; Roder, A. Accurate Virus Identi Fi Cation with Interpretable Raman Signatures by Machine Learning. *Proc. Natl. Acad. Sci. USA* **2022**, *199*, e2118836119. [CrossRef] [PubMed]

249. Carlomagno, C.; Bertazioli, D.; Gualerzi, A.; Picciolini, S.; Banfi, P.I.; Lax, A.; Messina, E.; Navarro, J.; Bianchi, L.; Caronni, A.; et al. COVID-19 Salivary Raman Fingerprint: Innovative Approach for the Detection of Current and Past SARS-CoV-2 Infections. *Sci. Rep.* **2021**, *11*, 4943. [CrossRef]
250. Zavyalova, E.; Ambartsumyan, O.; Zhdanov, G.; Gribanyov, D.; Gushchin, V.; Tkachuk, A.; Rudakova, E.; Nikiforova, M.; Kuznetsova, N.; Popova, L.; et al. Sers-Based Aptasensor for Rapid Quantitative Detection of Sars-Cov-2. *Nanomaterials* **2021**, *11*, 1394. [CrossRef]
251. Hwang, C.S.H.; Lee, S.; Lee, S.; Kim, H.; Kang, T.; Lee, D.; Jeong, K.H. Highly Adsorptive Au-TiO2 Nanocomposites for the SERS Face Mask Allow the Machine-Learning-Based Quantitative Assay of SARS-CoV-2 in Artificial Breath Aerosols. *ACS Appl. Mater. Interfaces* **2022**, *14*, 54550–54557. [CrossRef] [PubMed]
252. Zhang, Z.; Jiang, S.; Wang, X.; Dong, T.; Wang, Y.; Li, D.; Gao, X.; Qu, Z.; Li, Y. A Novel Enhanced Substrate for Label-Free Detection of SARS-CoV-2 Based on Surface-Enhanced Raman Scattering. *Sens. Actuators B Chem.* **2022**, *359*, 131568. [CrossRef] [PubMed]
253. Wang, C.; Wang, C.; Wang, X.; Wang, K.; Zhu, Y.; Rong, Z.; Wang, W.; Xiao, R.; Wang, S. Magnetic SERS Strip for Sensitive and Simultaneous Detection of Respiratory Viruses. *ACS Appl. Mater. Interfaces* **2019**, *11*, 19495–19505. [CrossRef]
254. Zhang, D.; Huang, L.; Liu, B.; Ge, Q.; Dong, J.; Zhao, X. Rapid and Ultrasensitive Quantification of Multiplex Respiratory Tract Infection Pathogen via Lateral Flow Microarray Based on SERS Nanotags. *Theranostics* **2019**, *9*, 4849–4859. [CrossRef] [PubMed]
255. Yang, Y.; Peng, Y.; Lin, C.; Long, L.; Hu, J.; He, J.; Zeng, H.; Huang, Z.; Li, Z.Y.; Tanemura, M.; et al. Human ACE2-Functionalized Gold "Virus-Trap" Nanostructures for Accurate Capture of SARS-CoV-2 and Single-Virus SERS Detection. *Nano-Micro Lett.* **2021**, *13*, 109. [CrossRef]
256. Peng, Y.; Lin, C.; Long, L.; Masaki, T.; Tang, M.; Yang, L.; Liu, J.; Huang, Z.; Li, Z.; Luo, X.; et al. Charge-Transfer Resonance and Electromagnetic Enhancement Synergistically Enabling MXenes with Excellent SERS Sensitivity for SARS-CoV-2 S Protein Detection. *Nano-Micro Lett.* **2021**, *13*, 52. [CrossRef]
257. Gu, M.M.; Guan, P.C.; Xu, S.S.; Li, H.M.; Kou, Y.C.; Lin, X.D.; Kathiresan, M.; Song, Y.; Zhang, Y.J.; Jin, S.Z.; et al. Ultrasensitive Detection of SARS-CoV-2 S Protein with Aptamers Biosensor Based on Surface-Enhanced Raman Scattering. *J. Chem. Phys.* **2023**, *158*, 024203. [CrossRef]
258. Lim, J.Y.; Nam, J.S.; Shin, H.; Park, J.; Song, H.I.; Kang, M.; Lim, K.I.; Choi, Y. Identification of Newly Emerging Influenza Viruses by Detecting the Virally Infected Cells Based on Surface Enhanced Raman Spectroscopy and Principal Component Analysis. *Anal. Chem.* **2019**, *91*, 5677–5684. [CrossRef]
259. Eom, G.; Hwang, A.; Kim, H.; Yang, S.; Lee, D.K.; Song, S.; Ha, K.; Jeong, J.; Jung, J.; Lim, E.K.; et al. Diagnosis of Tamiflu-Resistant Influenza Virus in Human Nasal Fluid and Saliva Using Surface-Enhanced Raman Scattering. *ACS Sens.* **2019**, *4*, 2282–2287. [CrossRef]
260. Zhang, M.; Li, X.; Pan, J.; Zhang, Y.; Zhang, L.; Wang, C.; Yan, X.; Liu, X.; Lu, G. Ultrasensitive Detection of SARS-CoV-2 Spike Protein in Untreated Saliva Using SERS-Based Biosensor. *Biosens. Bioelectron.* **2021**, *190*, 113421. [CrossRef]
261. Yadav, S.; Sadique, M.A.; Ranjan, P.; Kumar, N.; Singhal, A.; Srivastava, A.K.; Khan, R. Sers Based Lateral Flow Immunoassay for Point-of-Care Detection of Sars-Cov-2 in Clinical Samples. *ACS Appl. Bio Mater.* **2021**, *4*, 2974–2995. [CrossRef]
262. Chen, S.; Meng, L.; Wang, L.; Huang, X.; Ali, S.; Chen, X.; Yu, M.; Yi, M.; Li, L.; Chen, X.; et al. SERS-Based Lateral Flow Immunoassay for Sensitive and Simultaneous Detection of Anti-SARS-CoV-2 IgM and IgG Antibodies by Using Gap-Enhanced Raman Nanotags. *Sens. Actuators B Chem.* **2021**, *348*, 130706. [CrossRef]
263. Liu, H.; Dai, E.; Xiao, R.; Zhou, Z.; Zhang, M.; Bai, Z.; Shao, Y.; Qi, K.; Tu, J.; Wang, C.; et al. Development of a SERS-Based Lateral Flow Immunoassay for Rapid and Ultra-Sensitive Detection of Anti-SARS-CoV-2 IgM/IgG in Clinical Samples. *Sens. Actuators B Chem.* **2021**, *329*, 129196. [CrossRef] [PubMed]
264. Antoine, D.; Mohammadi, M.; Vitt, M.; Dickie, J.M.; Jyoti, S.S.; Tilbury, M.A.; Johnson, P.A.; Wawrousek, K.E.; Wall, J.G. Rapid, Point-of-Care ScFv-SERS Assay for Femtogram Level Detection of SARS-CoV-2. *ACS Sens.* **2022**, *7*, 866–873. [CrossRef] [PubMed]
265. Li, Z.; Luo, Y.; Song, Y.; Zhu, Q.; Xu, T.; Zhang, X. One-Click Investigation of Shape Influence of Silver Nanostructures on SERS Performance for Sensitive Detection of COVID-19. *Anal. Chim. Acta* **2022**, *1234*, 340523. [CrossRef]
266. Kim, W.; Kim, S.; Han, J.; Kim, T.G.; Bang, A.; Choi, H.W.; Min, G.E.; Shin, J.H.; Moon, S.W.; Choi, S. An Excitation Wavelength-Optimized, Stable SERS Biosensing Nanoplatform for Analyzing Adenoviral and AstraZeneca COVID-19 Vaccination Efficacy Status Using Tear Samples of Vaccinated Individuals. *Biosens. Bioelectron.* **2022**, *204*, 114079. [CrossRef] [PubMed]
267. Karunakaran, V.; Joseph, M.M.; Yadev, I.; Sharma, H.; Shamna, K.; Saurav, S.; Sreejith, R.P.; Anand, V.; Beegum, R.; Regi David, S.; et al. A Non-Invasive Ultrasensitive Diagnostic Approach for COVID-19 Infection Using Salivary Label-Free SERS Fingerprinting and Artificial Intelligence. *J. Photochem. Photobiol. B Biol.* **2022**, *234*, 112545. [CrossRef]
268. Shanmukh, S.; Jones, L.; Driskell, J.; Zhao, Y.; Dluhy, R.; Tripp, R.A. Rapid and Sensitive Detection of Respiratory Virus Molecular Signatures Using a Silver Nanorod Array SERS Substrate. *Nano Lett.* **2006**, *6*, 2630–2636. [CrossRef] [PubMed]
269. Dluhy, R.A.; Shanmukh, S.; Jones, L.; Zhao, Y.P.; Driskell, J.D.; Tripp, R.A. Identification and Classification of Respiratory Syncytial Virus (RSV) Strains by Surface-Enhanced Raman Spectroscopy and Multivariate Statistical Techniques. *Anal. Bioanal. Chem.* **2008**, *390*, 1551–1555. [CrossRef]
270. Huang, J.; Wen, J.; Zhou, M.; Ni, S.; Le, W.; Chen, G.; Wei, L.; Zeng, Y.; Qi, D.; Pan, M.; et al. On-Site Detection of SARS-CoV-2 Antigen by Deep Learning-Based Surface-Enhanced Raman Spectroscopy and Its Biochemical Foundations. *Anal. Chem.* **2021**, *93*, 9174–9182. [CrossRef] [PubMed]

271. Yeh, Y.J.; Le, T.N.; Hsiao, W.W.W.; Tung, K.L.; Ostrikov, K.; Chiang, W.H. Plasmonic Nanostructure-Enhanced Raman Scattering for Detection of SARS-CoV-2 Nucleocapsid Protein and Spike Protein Variants. *Anal. Chim. Acta* **2023**, *1239*, 340651. [CrossRef] [PubMed]
272. Liu, Z.; Wang, C.; Zheng, S.; Yang, X.; Han, H.; Dai, Y.; Xiao, R. Simultaneously Ultrasensitive and Quantitative Detection of Influenza A Virus, SARS-CoV-2, and Respiratory Syncytial Virus via Multichannel Magnetic SERS-Based Lateral Flow Immunoassay. *Nanomed. Nanotechnol. Biol. Med.* **2023**, *47*, 102624. [CrossRef] [PubMed]
273. Chisanga, M.; Williams, H.; Boudreau, D.; Pelletier, J.N.; Trottier, S. Label-Free SERS for Rapid Differentiation of SARS-CoV-2-Induced Serum Metabolic Profiles in Non-Hospitalized Adults. *Anal. Chem.* **2023**, *95*, 3638–3646. [CrossRef] [PubMed]
274. Bacteria. Available online: https://microbiologysociety.org/why-microbiology-matters/what-is-microbiology/bacteria.html#:~:text=Bacteria%20are%20classified%20into%20five,)%20or%20corkscrew%20(spirochaetes) (accessed on 30 January 2023).
275. Bacteria as Pathogens. Available online: https://sphweb.bumc.bu.edu/otlt/mph-modules/ph/ph709_infectiousagents/PH709_InfectiousAgents4.html#:~:text=While%20only%20about%205%25%20of,of%20human%20disease%20and%20death (accessed on 30 January 2023).
276. Nanda, M.; Kumar, V.; Sharma, D.K. Multimetal Tolerance Mechanisms in Bacteria: The Resistance Strategies Acquired by Bacteria That Can Be Exploited to 'Clean-up' Heavy Metal Contaminants from Water. *Aquat. Toxicol.* **2019**, *212*, 1–10. [CrossRef] [PubMed]
277. Glick, B.R. Phytoremediation: Synergistic Use of Plants and Bacteria to Clean up the Environment. *Biotechnol. Adv.* **2003**, *21*, 383–393. [CrossRef]
278. Kulshreshtha, A.; Agrawal, R.; Barar, M.; Saxena, S. A Review on Bioremediation of Heavy Metals in Contaminated Water. *IOSR J. Environ. Sci. Toxicol. Food Technol.* **2014**, *8*, 44–50. [CrossRef]
279. Zhou, X.; Hu, Z.; Yang, D.; Xie, S.; Jiang, Z.; Niessner, R.; Haisch, C.; Zhou, H.; Sun, P. Bacteria Detection: From Powerful SERS to Its Advanced Compatible Techniques. *Adv. Sci.* **2020**, *7*, 2001739. [CrossRef] [PubMed]
280. Wang, R.; Kim, K.; Choi, N.; Wang, X.; Lee, J.; Jeon, J.H.; Rhie, G.E.; Choo, J. Highly Sensitive Detection of High-Risk Bacterial Pathogens Using SERS-Based Lateral Flow Assay Strips. *Sens. Actuators B Chem.* **2018**, *270*, 72–79. [CrossRef]
281. Zhu, T.; Hu, Y.; Yang, K.; Dong, N.; Yu, M.; Jiang, N. A Novel SERS Nanoprobe Based on the Use of Core-Shell Nanoparticles with Embedded Reporter Molecule to Detect *E. coli* O157:H7 with High Sensitivity. *Microchim. Acta* **2018**, *185*, 30. [CrossRef]
282. Chisanga, M.; Muhamadali, H.; Ellis, D.I.; Goodacre, R. Surface-Enhanced Raman Scattering (SERS) in Microbiology: Illumination and Enhancement of the Microbial World. *Appl. Spectrosc.* **2018**, *72*, 987–1000. [CrossRef]
283. Kim, J.A.; Wales, D.J.; Thompson, A.J.; Yang, G.Z. Fiber-optic SERS probes fabricated using two-photon polymerization for rapid detection of bacteria. *Adv. Opt. Mater.* **2020**, *8*, 1901934. [CrossRef]
284. Zhou, H.; Yang, D.; Ivleva, N.P.; Mircescu, N.E.; Niessner, R.; Haisch, C. SERS Detection of Bacteria in Water by in Situ Coating with Ag Nanoparticles. *Anal. Chem.* **2014**, *86*, 1525–1533. [CrossRef]
285. Yang, Y.; Zeng, C.; Huang, J.; Wang, M.; Qi, W.; Wang, H.; He, Z. Specific and Quantitative Detection of Bacteria Based on Surface Cell Imprinted SERS Mapping Platform. *Biosens. Bioelectron.* **2022**, *215*, 114524. [CrossRef]
286. Pearson, B.; Wang, P.; Mills, A.; Pang, S.; McLandsborough, L.; He, L. Innovative Sandwich Assay with Dual Optical and SERS Sensing Mechanisms for Bacterial Detection. *Anal. Methods* **2017**, *9*, 4732–4739. [CrossRef]
287. Hudson, S.D.; Chumanov, G. Bioanalytical Applications of SERS (Surface-Enhanced Raman Spectroscopy). *Anal. Bioanal. Chem.* **2009**, *394*, 679–686. [CrossRef] [PubMed]
288. Mosier-Boss, P.A. Review on SERS of Bacteria. *Biosensors* **2017**, *7*, 51. [CrossRef] [PubMed]
289. Jarvis, R.M.; Goodacre, R. Characterisation and Identification of Bacteria Using SERS. *Chem. Soc. Rev.* **2008**, *37*, 931–936. [CrossRef]
290. Xia, J.; Li, W.; Sun, M.; Wang, H. Application of SERS in the Detection of Fungi, Bacteria and Viruses. *Nanomaterials* **2022**, *12*, 3572. [CrossRef]
291. Wang, C.; Meloni, M.M.; Wu, X.; Zhuo, M.; He, T.; Wang, J.; Wang, C.; Dong, P. Magnetic Plasmonic Particles for SERS-Based Bacteria Sensing: A Review. *AIP Adv.* **2019**, *9*, 010701. [CrossRef]
292. Efrima, S.; Zeiri, L. Understanding SERS of Bacteria. *J. Raman Spectrosc.* **2009**, *40*, 277–288. [CrossRef]
293. Liu, H.B.; Du, X.J.; Zang, Y.X.; Li, P.; Wang, S. SERS-Based Lateral Flow Strip Biosensor for Simultaneous Detection of Listeria Monocytogenes and Salmonella Enterica Serotype Enteritidis. *J. Agric. Food Chem.* **2017**, *65*, 10290–10299. [CrossRef]
294. Mungroo, N.A.; Oliveira, G.; Neethirajan, S. SERS Based Point-of-Care Detection of Food-Borne Pathogens. *Microchim. Acta* **2016**, *183*, 697–707. [CrossRef]
295. Lin, H.Y.; Huang, C.H.; Hsieh, W.H.; Liu, L.H.; Lin, Y.C.; Chu, C.C.; Wang, S.T.; Kuo, I.T.; Chau, L.K.; Yang, C.Y. On-Line SERS Detection of Single Bacterium Using Novel SERS Nanoprobes and a Microfl Uidic Dielectrophoresis Device. *Small* **2014**, *10*, 4700–4710. [CrossRef]
296. Wang, Y.; Lee, K.; Irudayaraj, J. Silver Nanosphere SERS Probes for Sensitive Identification of Pathogens. *J. Phys. Chem. C* **2010**, *114*, 16122–16128. [CrossRef]
297. Witkowska, E.; Korsak, D.; Kowalska, A.; Janeczek, A.; Kamińska, A. Strain-Level Typing and Identification of Bacteria—A Novel Approach for SERS Active Plasmonic Nanostructures. *Anal. Bioanal. Chem.* **2018**, *410*, 5019–5031. [CrossRef]
298. Pang, Y.; Wan, N.; Shi, L.; Wang, C.; Sun, Z.; Xiao, R.; Wang, S. Dual-Recognition Surface-Enhanced Raman Scattering (SERS)Biosensor for Pathogenic Bacteria Detection by Using Vancomycin-SERS Tags and Aptamer-Fe3O4@Au. *Anal. Chim. Acta* **2019**, *1077*, 288–296. [CrossRef] [PubMed]

299. Gao, X.; Yin, Y.; Wu, H.; Hao, Z.; Li, J.; Wang, S.; Liu, Y. Integrated SERS Platform for Reliable Detection and Photothermal Elimination of Bacteria in Whole Blood Samples. *Anal. Chem.* **2021**, *93*, 1569–1577. [CrossRef]
300. Zhou, Z.; Xiao, R.; Cheng, S.; Wang, S.; Shi, L.; Wang, C.; Qi, K.; Wang, S. A Universal SERS-Label Immunoassay for Pathogen Bacteria Detection Based on Fe3O4@Au-Aptamer Separation and Antibody-Protein A Orientation Recognition. *Anal. Chim. Acta* **2021**, *1160*, 338421. [CrossRef]
301. Wang, J.; Wu, X.; Wang, C.; Rong, Z.; Ding, H.; Li, H.; Li, S.; Shao, N.; Dong, P.; Xiao, R.; et al. Facile Synthesis of Au-Coated Magnetic Nanoparticles and Their Application in Bacteria Detection via a SERS Method. *ACS Appl. Mater. Interfaces* **2016**, *8*, 19958–19967. [CrossRef]
302. Wang, C.; Wang, J.; Li, M.; Qu, X.; Zhang, K.; Rong, Z.; Xiao, R.; Wang, S. A Rapid SERS Method for Label-Free Bacteria Detection Using Polyethylenimine-Modified Au-Coated Magnetic Microspheres and Au@Ag Nanoparticles. *Analyst* **2016**, *141*, 6226–6238. [CrossRef]
303. Huang, L.; Sun, D.W.; Wu, Z.; Pu, H.; Wei, Q. Reproducible, Shelf-Stable, and Bioaffinity SERS Nanotags Inspired by Multivariate Polyphenolic Chemistry for Bacterial Identification. *Anal. Chim. Acta* **2021**, *1167*, 338570. [CrossRef] [PubMed]
304. Bi, L.; Wang, X.; Cao, X.; Liu, L.; Bai, C.; Zheng, Q.; Choo, J.; Chen, L. SERS-Active Au@Ag Core-Shell Nanorod (Au@AgNR) Tags for Ultrasensitive Bacteria Detection and Antibiotic-Susceptibility Testing. *Talanta* **2020**, *220*, 121397. [CrossRef] [PubMed]
305. Hunter, R.; Sohi, A.N.; Khatoon, Z.; Berthiaume, V.R.; Alarcon, E.I.; Godin, M.; Anis, H. Optofluidic Label-Free SERS Platform for Rapid Bacteria Detection in Serum. *Sens. Actuators B Chem.* **2019**, *300*, 126907. [CrossRef]
306. Sivanesan, A.; Witkowska, E.; Adamkiewicz, W.; Dziewit, Ł.; Kamińska, A.; Waluk, J. Nanostructured Silver-Gold Bimetallic SERS Substrates for Selective Identification of Bacteria in Human Blood. *Analyst* **2014**, *139*, 1037–1043. [CrossRef]
307. Witkowska, E.; Szymborski, T.; Kamińska, A.; Waluk, J. Polymer Mat Prepared via Forcespinning™ as a SERS Platform for Immobilization and Detection of Bacteria from Blood Plasma. *Mater. Sci. Eng. C* **2017**, *71*, 345–350. [CrossRef]
308. Zhang, L.; Xu, J.; Mi, L.; Gong, H.; Jiang, S.; Yu, Q. Multifunctional Magnetic-Plasmonic Nanoparticles for Fast Concentration and Sensitive Detection of Bacteria Using SERS. *Biosens. Bioelectron.* **2012**, *31*, 130–136. [CrossRef]
309. Krafft, B.; Tycova, A.; Urban, R.D.; Dusny, C.; Belder, D. Microfluidic device for concentration and SERS-based detection of bacteria in drinking water. *Electrophoresis* **2021**, *42*, 86–94. [CrossRef]
310. Yang, E.; Li, D.; Yin, P.; Xie, Q.; Li, Y.; Lin, Q.; Duan, Y. A Novel Surface-Enhanced Raman Scattering (SERS) Strategy for Ultrasensitive Detection of Bacteria Based on Three-Dimensional (3D) DNA Walker. *Biosens. Bioelectron.* **2021**, *172*, 112758. [CrossRef]
311. Wang, C.; Wang, C.; Li, J.; Tu, Z.; Gu, B.; Wang, S. Ultrasensitive and Multiplex Detection of Four Pathogenic Bacteria on a Bi-Channel Lateral Flow Immunoassay Strip with Three-Dimensional Membrane-like SERS Nanostickers. *Biosens. Bioelectron.* **2022**, *214*, 114525. [CrossRef]
312. Kearns, H.; Goodacre, R.; Jamieson, L.E.; Graham, D.; Faulds, K. SERS Detection of Multiple Antimicrobial-Resistant Pathogens Using Nanosensors. *Anal. Chem.* **2017**, *89*, 12666–12673. [CrossRef]
313. Gracie, K.; Correa, E.; Mabbott, S.; Dougan, J.A.; Graham, D.; Goodacre, R.; Faulds, K. Simultaneous Detection and Quantification of Three Bacterial Meningitis Pathogens by SERS. *Chem. Sci.* **2014**, *5*, 1030–1040. [CrossRef]
314. Mosier-Boss, P.A.; Sorensen, K.C.; George, R.D.; Obraztsova, A. SERS Substrates Fabricated Using Ceramic Filters for the Detection of Bacteria. *Spectrochim. Acta Part A Mol. Biomol. Spectrosc.* **2016**, *153*, 591–598. [CrossRef] [PubMed]
315. Walter, A.; März, A.; Schumacher, W.; Rösch, P.; Popp, J. Towards a Fast, High Specific and Reliable Discrimination of Bacteria on Strain Level by Means of SERS in a Microfluidic Device. *Lab Chip* **2011**, *11*, 1013–1021. [CrossRef]
316. Wu, X.; Xu, C.; Tripp, R.A.; Huang, Y.W.; Zhao, Y. Detection and Differentiation of Foodborne Pathogenic Bacteria in Mung Bean Sprouts Using Field Deployable Label-Free SERS Devices. *Analyst* **2013**, *138*, 3005–3012. [CrossRef] [PubMed]
317. Ankamwar, B.; Sur, U.K.; Das, P. SERS Study of Bacteria Using Biosynthesized Silver Nanoparticles as the SERS Substrate. *Anal. Methods* **2016**, *8*, 2335–2340. [CrossRef]
318. Ciloglu, F.U.; Caliskan, A.; Saridag, A.M.; Kilic, I.H.; Tokmakci, M.; Kahraman, M.; Aydin, O. Drug-Resistant Staphylococcus Aureus Bacteria Detection by Combining Surface-Enhanced Raman Spectroscopy (SERS) and Deep Learning Techniques. *Sci. Rep.* **2021**, *11*, 18444. [CrossRef]
319. Nerve Agents, BBC Report. Available online: https://www.bbc.com/news/uk-43431537 (accessed on 30 January 2023).
320. Lister, A.P.; Sellors, W.J.; Howle, C.R.; Mahajan, S. Raman Scattering Techniques for Defense and Security Applications. *Anal. Chem.* **2021**, *93*, 417–429. [CrossRef]
321. Sadayoshi, O.H.B.U.; Yamashina, A.; Takasu, N.; Yamaguchi, T.; Murai, T.; Nakano, K.; Hinohara, S. Sarin poisoning on Tokyo subway. *South. Med. J.* **1997**, *90*, 587–593.
322. Inscore, F.E.; Gift, A.D.; Maksymiuk, P.; Farquharson, S. Characterization of Chemical Warfare G-Agent Hydrolysis Products by Surface-Enhanced Raman Spectroscopy. *Chem. Biol. Point Sens. Homel. Def. II* **2004**, *5585*, 46. [CrossRef]
323. Mukherjee, S.; Gupta, R.D. Organophosphorus Nerve Agents: Types, Toxicity, and Treatments. *J. Toxicol.* **2020**, *2020*, 3007984. [CrossRef]
324. Saylan, Y.; Akgönüllü, S.; Denizli, A. Plasmonic Sensors for Monitoring Biological and Chemical Threat Agents. *Biosensors* **2020**, *10*, 142. [CrossRef] [PubMed]
325. Primera-Pedrozo, O.M.; Jerez-Rozo, J.I.; De La Cruz-Montoya, E.; Luna-Pineda, T.; Pacheco-Londoño, L.C.; Hernández-Rivera, S.P. Nanotechnology-Based Detection of Explosives and Biological Agents Simulants. *IEEE Sens. J.* **2008**, *8*, 963–973. [CrossRef]

326. Yan, F.; Vo-Dinh, T. Surface-Enhanced Raman Scattering Detection of Chemical and Biological Agents Using a Portable Raman Integrated Tunable Sensor. *Sens. Actuators B Chem.* **2007**, *121*, 61–66. [CrossRef]
327. Pearman, W.F.; Fountain, A.W. Classification of Chemical and Biological Warfare Agent Simulants by Surface-Enhanced Raman Spectroscopy and Multivariate Statistical Techniques. *Appl. Spectrosc.* **2006**, *60*, 356–365. [CrossRef] [PubMed]
328. Hakonen, A.; Rindzevicius, T.; Schmidt, M.S.; Andersson, P.O.; Juhlin, L.; Svedendahl, M.; Boisen, A.; Käll, M. Detection of Nerve Gases Using Surface-Enhanced Raman Scattering Substrates with High Droplet Adhesion. *Nanoscale* **2016**, *8*, 1305–1308. [CrossRef] [PubMed]
329. Juhlin, L.; Mikaelsson, T.; Hakonen, A.; Stenbæk, M. Selective Surface-Enhanced Raman Scattering Detection of Tabun, VX and Cyclosarin Nerve Agents Using 4-Pyridine Amide Oxime Functionalized Gold Nanopillars Talanta Selective Surface-Enhanced Raman Scattering Detection of Tabun, VX and Cyclosarin Nerve A. *Talanta* **2020**, *211*, 120721. [CrossRef]
330. Farquharson, S.; Gift, A.; Maksymiuk, P.; Inscore, F. Surface-Enhanced Raman Spectra of VX and Its Hydrolysis Products. *Appl. Spectrosc.* **2005**, *59*, 654–660. [CrossRef]
331. Heleg-Shabtai, V.; Sharabi, H.; Zaltsman, A.; Ron, I.; Pevzner, A. Surface-Enhanced Raman Spectroscopy (SERS) for Detection of VX and HD in the Gas Phase Using a Hand-Held Raman Spectrometer. *Analyst* **2020**, *145*, 6334–6341. [CrossRef]
332. Zhao, Q.; Liu, G.; Zhang, H.; Zhou, F.; Li, Y.; Cai, W. SERS-Based Ultrasensitive Detection of Organophosphorus Nerve Agents via Substrate's Surface Modification. *J. Hazard. Mater.* **2017**, *324*, 194–202. [CrossRef]
333. Spencer, K.M.; Sylvia, J.M.; Clauson, S.L.; Janni, J.A. Surface-Enhanced Raman as a Water Monitor for Warfare Agents. *Vib. Spectrosc. Sens. Syst.* **2002**, *4577*, 158. [CrossRef]
334. Bertone, J.F.; Cordeiro, K.L.; Sylvia, J.M.; Spencer, K.M. A Nanoengineered Sensor to Detect Vibrational Modes of Warfare Agents/Explosives Using Surface-Enhanced Raman Scattering. *Sens. Command. Control. Commun. Intell. Technol. Homel. Secur. Homel. Def. III* **2004**, *5403*, 387. [CrossRef]
335. Kim, Y.T.; Kim, D.; Park, S.; Zhexembekova, A.; Byeon, M.; Hong, T.E.; Lee, J.; Lee, C.Y. Aqueous Microlenses for Localized Collection and Enhanced Raman Spectroscopy of Gaseous Molecules. *Adv. Opt. Mater.* **2021**, *9*, 2101209. [CrossRef]
336. Wu, J.; Zhu, Y.; Gao, J.; Chen, J.; Feng, J.; Guo, L.; Xie, J. A Simple and Sensitive Surface-Enhanced Raman Spectroscopic Discriminative Detection of Organophosphorous Nerve Agents. *Anal. Bioanal. Chem.* **2017**, *409*, 5091–5099. [CrossRef] [PubMed]
337. Lafuente, M.; Berenschot, E.J.W.; Tiggelaar, R.M.; Mallada, R.; Tas, N.R.; Pina, M.P. 3D Fractals as SERS Active Platforms: Preparation and Evaluation for Gas Phase Detection of G-Nerve Agents. *Micromachines* **2018**, *9*, 60. [CrossRef] [PubMed]
338. Goel, A.K. Anthrax: A Disease of Biowarfare and Public Health Importance. *World J. Clin. Cases* **2015**, *3*, 20. [CrossRef] [PubMed]
339. Farrell, M.E.; Pellegrino, P.M. Army Relevant Biological Hazards Detection with Commercial SERS Substrates. *Biosen. Nanomed. V* **2012**, *8460*, 84600J. [CrossRef]
340. Sajanlal, P.R.; Pradeep, T. Functional Hybrid Nickel Nanostructures as Recyclable SERS Substrates: Detection of Explosives and Biowarfare Agents. *Nanoscale* **2012**, *4*, 3427–3437. [CrossRef]
341. Wang, T.; Dong, P.; Zhu, C.; Sha, P.; Gao, W.; Wu, Y.; Wu, X. Trace Detection of Anthrax Protective Antigens via a Competitive Method Based on Surface-Enhanced Raman Scattering. *Sens. Actuators B Chem.* **2021**, *346*, 130467. [CrossRef]
342. Gao, R.; Ko, J.; Cha, K.; Ho Jeon, J.; Rhie, G.E.; Choi, J.; de Mello, A.J.; Choo, J. Fast and Sensitive Detection of an Anthrax Biomarker Using SERS-Based Solenoid Microfluidic Sensor. *Biosens. Bioelectron.* **2015**, *72*, 230–236. [CrossRef]
343. Naqvi, T.K.; Bajpai, A.; Bharati, M.S.S.; Kulkarni, M.M.; Siddiqui, A.M.; Soma, V.R.; Dwivedi, P.K. Ultra-Sensitive Reusable SERS Sensor for Multiple Hazardous Materials Detection on Single Platform. *J. Hazard. Mater.* **2021**, *407*, 124353. [CrossRef]
344. Sengupta, A.; Shende, C.; Farquharson, S.; Inscore, F. Detection of Bacillus Anthracis Spores Using Peptide Functionalized SERS-Active Substrates. *Int. J. Spectrosc.* **2012**, *2012*, 176851. [CrossRef]
345. Yilmaz, M.; Senlik, E.; Biskin, E.; Yavuz, M.S.; Tamer, U.; Demirel, G. Combining 3-D Plasmonic Gold Nanorod Arrays with Colloidal Nanoparticles as a Versatile Concept for Reliable, Sensitive, and Selective Molecular Detection by SERS. *Phys. Chem. Chem. Phys.* **2014**, *16*, 5563–5570. [CrossRef]
346. Li, B.; Wang, T.; Bai, W.; Su, Q.; Wu, X.; Dong, P. Label-Free and Rapid Detection of Anthrax Protective Antigen by Surface-Enhanced Raman Scattering on Au Nanorods. *IEEE Sens. J.* **2021**, *21*, 18425–18434. [CrossRef]
347. Cheung, M.; Lee, W.W.Y.; Cowcher, D.P.; Goodacre, R.; Bell, S.E.J. SERS of Meso-Droplets Supported on Superhydrophobic Wires Allows Exquisitely Sensitive Detection of Dipicolinic Acid, an Anthrax Biomarker, Considerably below the Infective Dose. *Chem. Commun.* **2016**, *52*, 9925–9928. [CrossRef] [PubMed]
348. Félix-Rivera, H.; González, R.; Rodríguez, G.D.M.; Primera-Pedrozo, O.M.; Ríos-Velázquez, C.; Hernández-Rivera, S.P. Improving SERS Detection of Bacillus Thuringiensis Using Silver Nanoparticles Reduced with Hydroxylamine and with Citrate Capped Borohydride. *Int. J. Spectrosc.* **2011**, *2011*, 989504. [CrossRef]
349. FountainIII, A.W.; Pearman, W.F. Multivariate Statistical Classification of Surface Enhanced Raman Spectra of Chemical and Biological Warfare Agent Simulants. *Chem. Biol. Sens. Ind. Environ. Secur.* **2005**, *5994*, 180–193. [CrossRef]
350. Arano-Martinez, J.A.; Martínez-González, C.L.; Salazar, M.I.; Torres-Torres, C. A Framework for Biosensors Assisted by Multiphoton Effects and Machine Learning. *Biosensors* **2022**, *12*, 710. [CrossRef]
351. Luo, R.; Popp, J.; Bocklitz, T. Deep Learning for Raman Spectroscopy: A Review. *Analytica* **2022**, *3*, 287–301. [CrossRef]
352. Ralbovsky, N.M.; Lednev, I.K. Towards Development of a Novel Universal Medical Diagnostic Method: Raman Spectroscopy and Machine Learning. *Chem. Soc. Rev.* **2020**, *49*, 7428–7453. [CrossRef]

353. Ding, Y.; Sun, Y.; Liu, C.; Jiang, Q.Y.; Chen, F.; Cao, Y. SERS-Based Biosensors Combined with Machine Learning for Medical Application. *ChemistryOpen* **2023**, *12*, e202200192. [CrossRef]
354. Cui, F.; Yue, Y.; Zhang, Y.; Zhang, Z.; Zhou, H.S. Advancing Biosensors with Machine Learning. *ACS Sens.* **2020**, *5*, 3346–3364. [CrossRef]
355. Lv, R.; Wang, Z.; Ma, Y.; Li, W.; Tian, J. Machine Learning Enhanced Optical Spectroscopy for Disease Detection. *J. Phys. Chem. Lett.* **2022**, *13*, 9238–9249. [CrossRef]
356. Schackart, K.E.; Yoon, J.Y. Machine Learning Enhances the Performance of Bioreceptor-Free Biosensors. *Sensors* **2021**, *21*, 5519. [CrossRef]
357. Lussier, F.; Missirlis, D.; Spatz, J.P.; Masson, J.F. Machine-Learning-Driven Surface-Enhanced Raman Scattering Optophysiology Reveals Multiplexed Metabolite Gradients Near Cells. *ACS Nano* **2019**, *13*, 1403–1411. [CrossRef] [PubMed]
358. Nguyen, P.H.L.; Hong, B.; Rubin, S.; Fainman, Y. Machine Learning for Composition Analysis of SsDNA Using Chemical Enhancement in SERS. *Biomed. Opt. Express* **2020**, *11*, 5092. [CrossRef] [PubMed]
359. Narla, L.M.; Rao, S.V. Identification of Metals and Alloys Using Color CCD Images of Laser-Induced Breakdown Emissions Coupled with Machine Learning. *Appl. Phys. B Lasers Opt.* **2020**, *126*, 113. [CrossRef]
360. Beeram, R.; Banerjee, D.; Narlagiri, L.M.; Soma, V.R. Machine Learning for Rapid Quantification of Trace Analyte Molecules Using SERS and Flexible Plasmonic Paper Substrates. *Anal. Methods* **2022**, *14*, 1788–1796. [CrossRef]
361. Murthy, N.L.; Abdul Salam, S.; Rao, S.V. Stand-off Femtosecond Laser Induced Breakdown Spectroscopy of Metals, Soil, Plastics and Classification Studies. In Proceedings of the 2019 Workshop on Recent Advances in Photonics (WRAP), Guwahati, India, 13–14 December 2019; pp. 1–3. [CrossRef]
362. Boehmke, B.; Greenwell, B. *Hands-On Machine Learning with R*; Chapman and Hall/CRC: Boca Raton, FL, USA, 2019; ISBN 9781492032649.
363. Li, D.; Zhang, Q.; Deng, B.; Chen, Y.; Ye, L. Rapid, Sensitive Detection of Ganciclovir, Penciclovir and Valacyclovir-Hydrochloride by Artificial Neural Network and Partial Least Squares Combined with Surface Enhanced Raman Spectroscopy. *Appl. Surf. Sci.* **2021**, *539*, 148224. [CrossRef]
364. Boulesteix, A.L.; Strimmer, K. Partial Least Squares: A Versatile Tool for the Analysis of High-Dimensional Genomic Data. *Brief. Bioinform.* **2007**, *8*, 32–44. [CrossRef]
365. Deng, W.; Huang, Z.; Zhang, J.; Xu, J. A Data Mining Based System for Transaction Fraud Detection. In Proceedings of the 2021 IEEE International Conference on Consumer Electronics and Computer Engineering (ICCECE), Guangzhou, China, 15–17 January 2021; pp. 542–545. [CrossRef]
366. Fan, X.; Ming, W.; Zeng, H.; Zhang, Z.; Lu, H. Deep Learning-Based Component Identification for the Raman Spectra of Mixtures. *Analyst* **2019**, *144*, 1789–1798. [CrossRef]
367. Zhou, H.; Xu, L.; Ren, Z.; Zhu, J.; Lee, C. Machine Learning-Augmented Surface-Enhanced Spectroscopy toward next-Generation Molecular Diagnostics. *Nanoscale Adv.* **2023**, *5*, 538–570. [CrossRef]
368. Malinick, A.S.; Stuart, D.D.; Lambert, A.S.; Cheng, Q. Surface Plasmon Resonance Imaging (SPRi) in Combination with Machine Learning for Microarray Analysis of Multiple Sclerosis Biomarkers in Whole Serum. *Biosens. Bioelectron. X* **2022**, *10*, 100127. [CrossRef]
369. Pradhan, P.; Guo, S.; Ryabchykov, O.; Popp, J.; Bocklitz, T.W. Deep Learning a Boon for Biophotonics? *J. Biophotonics* **2020**, *13*, e201960186. [CrossRef] [PubMed]
370. Moon, G.; Lee, J.; Lee, H.; Yoo, H.; Ko, K.; Im, S.; Kim, D. Machine Learning and Its Applications for Plasmonics in Biology. *Cell Rep. Phys. Sci.* **2022**, *3*, 101042. [CrossRef]
371. Sun, Y.; Shi, L.; Mi, L.; Guo, R.; Li, T. Recent Progress of SERS Optical Nanosensors for MiRNA Analysis. *J. Mater. Chem. B* **2020**, *8*, 5178–5183. [CrossRef] [PubMed]
372. Raji, H.; Tayyab, M.; Sui, J.; Mahmoodi, S.R.; Javanmard, M. Biosensors and Machine Learning for Enhanced Detection, Stratification, and Classification of Cells: A Review. *Biomed. Microdevices* **2022**, *24*, 26. [CrossRef] [PubMed]
373. Banerjee, A.; Maity, S.; Mastrangelo, C.H. Nanostructures for biosensing, with a brief overview on Cancer Detection, IoT, and the Role of Machine Learning In Smart Biosensors. *Sensors* **2021**, *21*, 1253. [CrossRef]
374. Beeram, R.; Vendamani, V.S.; Soma, V.R. Spectrochimica Acta Part A: Molecular and Biomolecular Spectroscopy Deep Learning Approach to Overcome Signal Fluctuations in SERS for Efficient On-Site Trace Explosives Detection. *Spectrochim. Acta Part A Mol. Biomol. Spectrosc.* **2023**, *289*, 122218. [CrossRef] [PubMed]
375. Moon, G.; Son, T.; Lee, H.; Kim, D. Deep Learning Approach for Enhanced Detection of Surface Plasmon Scattering. *Anal. Chem.* **2019**, *91*, 9538–9545. [CrossRef]
376. Gupta, A.K.; Hsu, C.H.; Lai, C.S. Enhancement of the Au/ZnO-NA Plasmonic SERS Signal Using Principal Component Analysis as a Machine Learning Approach. *IEEE Photonics J.* **2020**, *12*, 1–11. [CrossRef]
377. Vendamani, V.S.; Beeram, R.; Neethish, M.M.; Rao, S.V.S.N.; Rao, S.V. Wafer-Scale Silver Nanodendrites with Homogeneous Distribution of Gold Nanoparticles for Biomolecules Detection. *iScience* **2022**, *25*, 104849. [CrossRef]
378. Erzina, M.; Trelin, A.; Guselnikova, O.; Dvorankova, B.; Strnadova, K.; Perminova, A.; Ulbrich, P.; Mares, D.; Jerabek, V.; Elashnikov, R.; et al. Precise Cancer Detection via the Combination of Functionalized SERS Surfaces and Convolutional Neural Network with Independent Inputs. *Sens. Actuators B Chem.* **2020**, *308*, 127660. [CrossRef]

379. Wang, S.; Dong, H.; Shen, W.; Yang, Y.; Li, Z.; Liu, Y.; Wang, C.; Gu, B.; Zhang, L. Rapid SERS Identification of Methicillin-Susceptible and Methicillin-Resistant: Staphylococcus Aureus via Aptamer Recognition and Deep Learning. *RSC Adv.* **2021**, *11*, 34425–34431. [CrossRef]
380. Kazemzadeh, M.; Hisey, C.L.; Dauros-Singorenko, P.; Swift, S.; Zargar-Shoshtari, K.; Xu, W.; Broderick, N.G.R. Label-Free Classification of Bacterial Extracellular Vesicles by Combining Nanoplasmonic Sensors with Machine Learning. *IEEE Sens. J.* **2022**, *22*, 1128–1137. [CrossRef]
381. Dong, R.; Weng, S.; Yang, L.; Liu, J. Detection and Direct Readout of Drugs in Human Urine Using Dynamic Surface-Enhanced Raman Spectroscopy and Support Vector Machines. *Anal. Chem.* **2015**, *87*, 2937–2944. [CrossRef] [PubMed]
382. Lin, C.; Liang, S.; Li, Y.; Peng, Y.; Huang, Z.; Li, Z.; Yang, Y.; Luo, X. Localized Plasmonic Sensor for Direct Identifying Lung and Colon Cancer from the Blood. *Biosens. Bioelectron.* **2022**, *211*, 114372. [CrossRef] [PubMed]
383. Peng, S.; Lu, D.; Zhang, B.; You, R.; Chen, J.; Xu, H.; Lu, Y. Machine Learning—Assisted Internal Standard Calibration Label—Free SERS Strategy for Colon Cancer Detection. *Anal. Bioanal. Chem.* **2023**. [CrossRef] [PubMed]
384. Seifert, S.; Merk, V.; Kneipp, J. Identification of Aqueous Pollen Extracts Using Surface Enhanced Raman Scattering (SERS) and Pattern Recognition Methods. *J. Biophotonics* **2016**, *9*, 181–189. [CrossRef] [PubMed]
385. Hassoun, M.; Rüger, J.; Kirchberger-Tolstik, T.; Schie, I.W.; Henkel, T.; Weber, K.; Cialla-May, D.; Krafft, C.; Popp, J. A Droplet-Based Microfluidic Chip as a Platform for Leukemia Cell Lysate Identification Using Surface-Enhanced Raman Scattering. *Anal. Bioanal. Chem.* **2018**, *410*, 999–1006. [CrossRef] [PubMed]
386. Mühlig, A.; Bocklitz, T.; Labugger, I.; Dees, S.; Henk, S.; Richter, E.; Andres, S.; Merker, M.; Stöckel, S.; Weber, K.; et al. LOC-SERS: A Promising Closed System for the Identification of Mycobacteria. *Anal. Chem.* **2016**, *88*, 7998–8004. [CrossRef] [PubMed]
387. Bratchenko, L.A.; Al-Sammarraie, S.Z.; Tupikova, E.N.; Konovalova, D.Y.; Lebedev, P.A.; Zakharov, V.P.; Bratchenko, I.A. Analyzing the Serum of Hemodialysis Patients with End-Stage Chronic Kidney Disease by Means of the Combination of SERS and Machine Learning. *Biomed. Opt. Express* **2022**, *13*, 4926. [CrossRef] [PubMed]
388. Gao, K.; Zhu, H.; Charron, B.; Mochizuki, T.; Dong, C.; Ding, H.; Cui, Y.; Lu, M.; Peng, W.; Zhu, S.; et al. Combining Dense Au Nanoparticle Layers and 2D Surface-Enhanced Raman Scattering Arrays for the Identification of Mutant Cyanobacteria Using Machine Learning. *J. Phys. Chem. C* **2022**, *126*, 9446–9455. [CrossRef]
389. Ikponmwoba, E.; Ukorigho, O.; Moitra, P.; Pan, D.; Gartia, M.R.; Owoyele, O. A Machine Learning Framework for Detecting COVID-19 Infection Using Surface-Enhanced Raman Scattering. *Biosensors* **2022**, *12*, 589. [CrossRef] [PubMed]
390. Pérez-Jiménez, A.I.; Lyu, D.; Lu, Z.; Liu, G.; Ren, B. Surface-Enhanced Raman Spectroscopy: Benefits, Trade-Offs and Future Developments. *Chem. Sci.* **2020**, *11*, 4563–4577. [CrossRef] [PubMed]
391. Hou, M.; Huang, Y.; Ma, L.; Zhang, Z. Quantitative Analysis of Single and Mix Food Antiseptics Basing on SERS Spectra with PLSR Method. *Nanoscale Res. Lett.* **2016**, *11*, 296. [CrossRef]
392. Yan, S.; Liu, C.; Fang, S.; Ma, J.; Qiu, J.; Xu, D.; Li, L.; Yu, J.; Li, D.; Liu, Q. SERS-Based Lateral Flow Assay Combined with Machine Learning for Highly Sensitive Quantitative Analysis of *Escherichia coli* O157:H7. *Anal. Bioanal. Chem.* **2020**, *412*, 7881–7890. [CrossRef] [PubMed]
393. Nguyen, C.Q.; Thrift, W.J.; Bhattacharjee, A.; Ranjbar, S.; Gallagher, T.; Darvishzadeh-Varcheie, M.; Sanderson, R.N.; Capolino, F.; Whiteson, K.; Baldi, P.; et al. Longitudinal Monitoring of Biofilm Formation via Robust Surface-Enhanced Raman Scattering Quantification of Pseudomonas Aeruginosa -Produced Metabolites. *ACS Appl. Mater. Interfaces* **2018**, *10*, 12364–12373. [CrossRef] [PubMed]
394. Lee, K.M.; Herrman, T.J. Determination and Prediction of Fumonisin Contamination in Maize by Surface–Enhanced Raman Spectroscopy (SERS). *Food Bioprocess Technol.* **2016**, *9*, 588–603. [CrossRef]
395. Kuligowski, J.; El-Zahry, M.R.; Sánchez-Illana, Á.; Quintás, G.; Vento, M.; Lendl, B. Surface Enhanced Raman Spectroscopic Direct Determination of Low Molecular Weight Biothiols in Umbilical Cord Whole Blood. *Analyst* **2016**, *141*, 2165–2174. [CrossRef]
396. Tan, A.; Zhao, Y.; Sivashanmugan, K.; Squire, K.; Wang, A.X. Quantitative TLC-SERS Detection of Histamine in Seafood with Support Vector Machine Analysis. *Food Control* **2019**, *103*, 111–118. [CrossRef]
397. Rahman, A.; Kang, S.; Wang, W.; Huang, Q.; Kim, I.; Vikesland, P.J. Lectin-Modified Bacterial Cellulose Nanocrystals Decorated with Au Nanoparticles for Selective Detection of Bacteria Using Surface-Enhanced Raman Scattering Coupled with Machine Learning. *ACS Appl. Nano Mater.* **2022**, *5*, 259–268. [CrossRef]
398. Banaei, N.; Moshfegh, J.; Mohseni-Kabir, A.; Houghton, J.M.; Sun, Y.; Kim, B. Machine Learning Algorithms Enhance the Specificity of Cancer Biomarker Detection Using SERS-Based Immunoassays in Microfluidic Chips. *RSC Adv.* **2019**, *9*, 1859–1868. [CrossRef]
399. Cheng, N.; Chen, D.; Lou, B.; Fu, J.; Wang, H. A Biosensing Method for the Direct Serological Detection of Liver Diseases by Integrating a SERS-Based Sensor and a CNN Classifier. *Biosens. Bioelectron.* **2021**, *186*, 113246. [CrossRef] [PubMed]
400. Zhang, X.; Liang, B.; Zhang, J.; Hao, X.; Xu, X.; Chang, H.M.; Leung, P.C.K.; Tan, J. Raman Spectroscopy of Follicular Fluid and Plasma with Machine-Learning Algorithms for Polycystic Ovary Syndrome Screening. *Mol. Cell. Endocrinol.* **2021**, *523*, 111139. [CrossRef] [PubMed]
401. Barucci, A.; D'Andrea, C.; Farnesi, E.; Banchelli, M.; Amicucci, C.; De Angelis, M.; Hwang, B.; Matteini, P. Label-Free SERS Detection of Proteins Based on Machine Learning Classification of Chemo-Structural Determinants. *Analyst* **2021**, *146*, 674–682. [CrossRef]

402. Kazemzadeh, M.; Hisey, C.L.; Zargar-Shoshtari, K.; Xu, W.; Broderick, N.G.R. Deep Convolutional Neural Networks as a Unified Solution for Raman Spectroscopy-Based Classification in Biomedical Applications. *Opt. Commun.* **2022**, *510*, 127977. [CrossRef]
403. Othman, N.H.; Yoot Lee, K.; Mohd Radzol, A.R.; Mansor, W.; Amanina Yusoff, N. PCA-Polynomial-ELM Model Optimal for Detection of NS1 Adulterated Salivary SERS Spectra. *J. Phys. Conf. Ser.* **2019**, *1372*, 012064. [CrossRef]
404. Yang, Y.; Xu, B.; Haverstick, J.; Ibtehaz, N.; Muszyński, A.; Chen, X.; Chowdhury, M.E.H.; Zughaier, S.; Zhao, Y. Differentiation and Classification of Bacterial Endotoxins Based on Surface Enhanced Raman Scattering and Advanced Machine Learning. *Nanoscale* **2022**, *14*, 8806–8817. [CrossRef]
405. Lin, D.; Hsieh, C.L.; Hsu, K.C.; Liao, P.H.; Qiu, S.; Gong, T.; Yong, K.T.; Feng, S.; Kong, K.V. Geometrically Encoded SERS Nanobarcodes for the Logical Detection of Nasopharyngeal Carcinoma-Related Progression Biomarkers. *Nat. Commun.* **2021**, *12*, 3430. [CrossRef]
406. Wang, G.; Lipert, R.J.; Jain, M.; Kaur, S.; Chakraboty, S.; Torres, M.P.; Batra, S.K.; Brand, R.E.; Porter, M.D. Detection of the Potential Pancreatic Cancer Marker MUC4 in Serum Using Surface-Enhanced Raman Scattering. *Anal. Chem.* **2011**, *83*, 2554–2561. [CrossRef]
407. Lu, L.; Guan, S.; Guan, Y.; Hong, M. Dual-Modal Fluorescence-SERS Detection of Blood Glucose Engineered by Hierarchical Laser-Induced Micro/Nano Structures for Diabetes Screening. *Adv. Mater. Interfaces* **2022**, *9*, 2102532. [CrossRef]
408. Sun, J.; Gong, L.; Wang, W.; Gong, Z.; Wang, D.; Fan, M. Surface-Enhanced Raman Spectroscopy for on-Site Analysis: A Review of Recent Developments. *Luminescence* **2020**, *35*, 808–820. [CrossRef] [PubMed]
409. Kho, K.W.; Fu, C.Y.; Dinish, U.S.; Olivo, M. Clinical SERS: Are We There Yet? *J. Biophotonics* **2011**, *4*, 667–684. [CrossRef] [PubMed]

biosensors

MDPI

Article

An Immunochromatographic Assay for the Rapid and Qualitative Detection of Mercury in Rice

Shuai Lv [1], Xinxin Xu [1], Shanshan Song [1], Liguang Xu [1], Liqiang Liu [1], Chuanlai Xu [1] and Hua Kuang [1,2,*]

1 International Joint Research Laboratory for Biointerface and Biodetection, School of Food Science and Technology, Jiangnan University, Wuxi 214122, China
2 State Key Laboratory of Food Science and Technology, Jiangnan University, Wuxi 214122, China
* Correspondence: kuangh@jiangnan.edu.cn; Tel./Fax: +86-510-85329077

Abstract: Mercury is a major pollutant in food crops. In this study, we synthesized an anti-mercury monoclonal antibody (mAb; IC_{50} was 0.606 ng mL^{-1}) with high sensitivity and specificity and different immunogens and coating antigens and developed an immuno-chromatographic assay (ICA) for the detection of mercury in rice. The ICA strip had a visible detection limit of 20 ng g^{-1} and a cut-off value of 500 ng g^{-1} in rice. The performance of the ICA strip was consistent with that of ICP-MS and ic-ELISA. The recoveries of mercury in rice ranged from 94.5% to 113.7% with ic-ELISA and from 93.6% to 116.45% with ICP-MS. Qualitative analysis by ICA can be obtained with the naked eye. The ICA strip is an effective and practical method for the rapid and high-throughput determination of mercury in rice.

Keywords: rice sample; mercury; immunochromatographic assay; qualitatively; strip

Citation: Lv, S.; Xu, X.; Song, S.; Xu, L.; Liu, L.; Xu, C.; Kuang, H. An Immunochromatographic Assay for the Rapid and Qualitative Detection of Mercury in Rice. *Biosensors* **2022**, *12*, 694. https://doi.org/10.3390/bios12090694

Received: 9 August 2022
Accepted: 22 August 2022
Published: 28 August 2022

1. Introduction

Mercury, one of nature's most toxic heavy metals, causes serious harm to the environment and humans [1,2]. Mercury is the only metallic element in liquid form at room temperature, and its vapor is highly toxic [3]. Studies have reported that mercury in grains results from environmental contamination [4,5]. Once the soil and water become contaminated, the food produced in the region will have excessive levels of mercury. In excess, mercury causes damage to the kidneys, stomach, and intestines and, in severe cases, damage to the central nervous system, reproductive mutations, and DNA. The World Health Organization reported that the per capita intake of mercury should not exceed 27 μg kg^{-1}. The GB 2762-2017 standard [6] issued by the China Food and Drug Administration (CFDA) in 2017 stipulates that the limit for total mercury in rice, barley, and wheat should not exceed 0.02 mg kg^{-1}.

Current detection methods for heavy metals include atomic fluorescence spectrometry [7], inductively coupled plasma mass spectrometry (ICP-MS), inductively coupled plasma atomic emission spectrometry, atomic absorption spectrometry, cold atomic absorption, and heavy metal detection techniques based on antibody recognition [8].

These traditional instrumental methods, especially ICP-MS, have significant advantages, including high accuracy, strong anti-interference, and trace and ultra-trace multi-element analysis [9]. However, these methods require expensive instruments and highly trained staff and are not portable. Immunoassays can overcome the shortcomings of instrumental detection and are sensitive and have high throughput [10].

The preparation of anti-mercury monoclonal antibody (mAb) for mercury detection has been reported [11,12]. However, studies reported low sensitivity and cross-reactivity with copper [13]. Additionally, there are no reports on using immunoassay test strips in grains [14].

This study aimed to prepare a mAb with high sensitivity and low cross-reactivity by using different chelating agents for synthetic immunogens and heterologous coating antigens [15]. While developing the immuno-chromatographic assay (ICA) strip, we optimized the surface activity and antigen-antibody concentration to detect mercury in rice.

2. Material and Methods

2.1. Chemicals and Instrumentation

We obtained mercury (II), mercury (I), MeHg, cadmium (II), lead (II), copper (II), magnesium (II), arsenic (III), nickel (II), calcium (II), manganese (II), and chromium (III) at 1 mg mL^{-1} in 5% HNO_3 from the National Analysis Center for Iron and Steel of China (Beijing, China) and isothiocyanobenzyl-EDTA (ITCBE) from Dojindo Molecular Technologics, Tnc (Osaka, Japan). J&K Scientific (Beijing, China) provided ethylene diamine tetraacetic acid (EDTA) and 6-mercaptonicotinic-acid (MNA). Keyhole limpet hemocyanin (KLH), bovine serum albumin (BSA), goat anti-mouse immunoglobulin, Freund's complete adjuvant, Freund's incomplete adjuvant, and gold chloride trihydrate were acquired from Sigma-Aldrich (St. Louis, MO, USA). From Gibco BRL (Paisley, UK), we obtained RPMI-1640 cell culture medium and polyethyleneglycol 1500 (PEG), and from Every Green Co. Ltd. (Hangzhou, China), we acquired fetal calf serum. All chemicals were of analytical grade or higher. We soaked the glassware in acid for more than an hour.

Thermo Fisher (Thermo Fisher Scientific, Massachusetts, USA) supplied the microplate reader, centrifuge, and ICP-MS instrument. Other equipment included an electrophoresis apparatus (HBIO, MA, USA), magnetic stirrer, and vortex (Wiggens, Beijing, China). A UV/Vis scanner (Bokin Instruments, Tsushima, Japan) was used to characterize the immunogen and antigen.

2.2. Synthesis of Immunogen and Coating Antigen

2.2.1. Synthesis of Immunogen

Mercury, which has a molecular weight of 201 Da and no surface-active sites, cannot be directly coupled to a protein. Therefore, with some modifications, we used a bifunctional chelating agent-ITCBE based on Zhang's [16] method. Figure 1a shows the synthesis of the immunogen. First, we dissolved 6 mg KLH (1.3 μM) in HBS, added 1.17 mg ITCBE (2.6 μM, dissolved in DMSO) dropwise, and adjusted the pH to 12 with 0.5 mol mL^{-1} NaOH. After stirring at 4 °C for 12 h, we added the mixture to 0.643 mg Hg^{2+} (3.2 μM) while ensuring that the pH was maintained at 8 with 0.1 mol mL^{-1} NaOH. To remove the excess free Hg^{2+}, we filtered Hg-ITCBE-KLH three times using an ultrafiltration tube. All conjugates were adjusted to 2 mol mL^{-1} with 0.01 M HBS.

2.2.2. Synthesis of Coating Conjugate

Figure 1b shows the synthesis of the coating antigens by reference to the method of Wang [17]. An amount of 0.41 mg MNA (1.76 μmol) was dissolved in 200 μL DMSO and 0.923 mg (5.29 μmol) NHS, then 1.523 mg (5.29 μmol) EDC were added. After stirring for 6 h, the activation solution was added dropwise to BSA dissolved in CB (coating buffer), and the reaction was carried out overnight. The dialysate was then dialyzed for three days to remove the excess MNA. The dialysate was solution A, and the Hg^{2+} solution was solution B. An amount of 354 μL of the B solution was added dropwise to the A solution, and the pH was controlled at around 7.4 with 0.5 M sodium hydroxide. The excess Hg^{2+} was removed by dialysis for three more days to obtain the coating antigen (Hg-MNA-BSA).

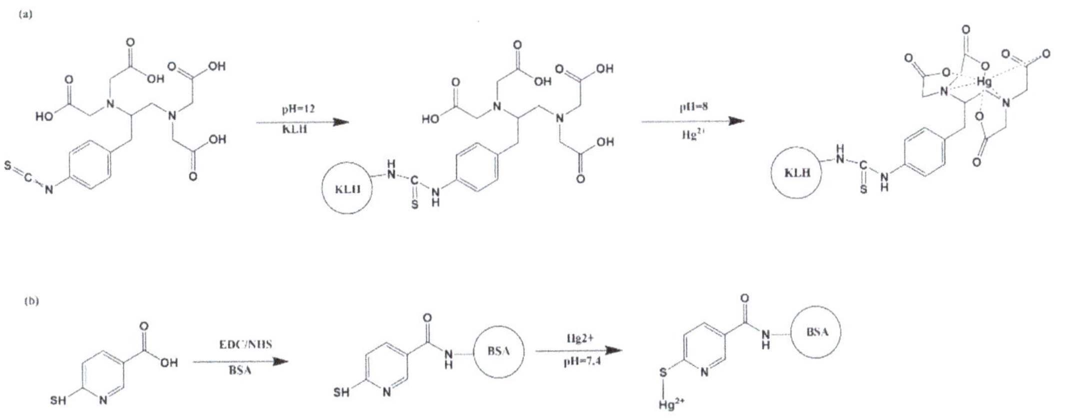

Figure 1. (**a**) The scheme of synthesizing immunogen (Hg-ITCBE-KLH). (**b**) The synthesis of coating antigen (Hg-MNA-BSA).

2.3. Preparation of Anti-Mercury mAb

All animal experiments were in strict accordance with Chinese laws and guidelines, which were approved by the Animal Ethics Committee of Jiangnan University. Female BALB/c mice (20, aged 6–8 weeks old) were assigned into two groups of 20. Referring to the previous monoclonal antibody preparation method [18,19], 20 BALB/c female mice were selected and divided into two groups. For the first immunization, a mixture of Freund's adjuvant and immune antigen was emulsified and injected subcutaneously into BALB/c mice at three-week intervals. An indirect competitive enzyme-linked immunosorbent assay (ic-ELISA) was used to evaluate the titer and specificity of the mouse serum to mercury following each immunization. The mouse with the highest titer and specificity to mercury was chosen for cell fusion. Cell fusion was performed according to the previous literature [20]. A hybridoma fusion method was used to fuse myeloma cells, and spleen cells in a ratio of 1:10. Cells with strong signals and high specificity were selected by ic-ELISA. After three subclonations, stable antibody-producing clones were expanded and stored in liquid nitrogen. The stable subclones were injected intra-peritoneally into paraffin-primed BALB/c mice for fluid production. After a week, the fluid was extracted from the mice and purified by the caprylic acid-ammonium sulfate method to obtain mAb against mercury. After three days of dialysis in PBS, the purified mAb was stored at −20 °C [21].

2.4. Characterization of mAb

Sensitivity and specificity are important factors in assessing antibodies. In this research, affinity constants and maximum half inhibition were measured by indirect ELISA. We assessed the sensitivity and stability of the anti-mercury mAb by establishing a standard curve. The specificity of mAb against mercury was assessed based on the cross-reactivity (CR) [22] of mAb with Cu^{2+}, MeHg, Pb^{2+}, Cd^{2+}, Cr^{3+}, Ni^{2+}, Mg^{2+}, Ca^{2+}, Mn^{2+}, Fe^{3+}.

$$\text{CR}\ (\%) = (\text{IC}_{50} \text{ of } \text{Hg}^{2+}/\text{IC}_{50} \text{ of other metals}) \times 100\%.$$

2.5. Preparation of Colloidal Gold-mAb

The preparation of colloidal gold-mAb refers to the Liu's methods [23]. Colloidal gold with a diameter of 20 nm was prepared using the sodium citrate reduction method. Briefly, 4 mL of 1% trisodium citrate solution was added to 100 mL of a 0.01% (*w*/*v*) boiling solution of chloroauric acid in a flask with vigorous stirring. The mixture was stirred continuously until the colour of the solution changed from pale yellow to burgundy. The solution was then cooled to room temperature and stored at 4 ± 1 °C to label the mAb. The colloidal gold was characterised by transmission electron microscopy (TEM). After the preparation of gold standard antibodies, the colloidal gold-mAb was stored at 4 °C for later use.

2.6. Preparation of the ICA Strip

As shown in Figure 2, the ICA strip contains five components: an absorbent pad, a nitrocellulose (NC) membrane, a conjugate pad, a polyvinyl chloride (PVC) plate and a sample pad [24]. The test and control lines (T- and C-lines) were generated by spraying the encapsulated antigen (0.5 mg/mL) and goat anti-mouse IgG (0.2 mg/mL) onto the NC membrane at 0.9 µL/cm, respectively, using a dispenser. The NC membranes were then dried at 37 °C for 12 h. The GNP-labelled mAb was diluted five-fold with a suspension buffer (0.02 M Tris-HCl) containing 0.1% Tween-20, 0.1% PEG, 5% sucrose, 5% alginate and 0.2% BSA. The coupling pad was then sprayed at a concentration of 3.5 µL/cm. After drying at 37 °C for 12 h, the NC membrane, binding pad, sample pad and absorbent pad were laminated and adhered to a PVC backing pad.

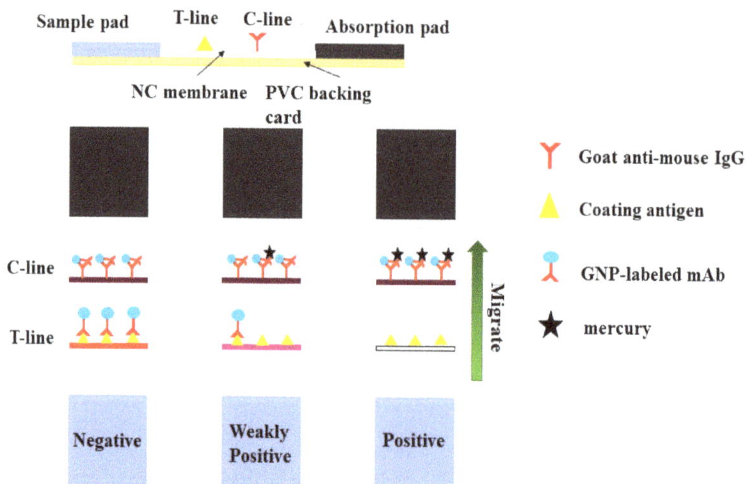

Figure 2. Composition of the test strip and the schematic for sample detection.

2.7. Evaluation of the ICA Strip

The optimal loading buffer solution and the concentration of test capture reagents (Hg-MNA-BSA) and mAb are first determined in the test strip evaluation and optimization process. Then, we established the standard concentration profile for mercury (0, 10, 20, 50, 100 ng mL^{-1}), which was diluted with HBS [25]. For the evaluation, 80 µL of Hg standard solution and 40 µL of the colloidal gold-mAb were added dropwise to the ICA strip. After 15 min, the T- and C-lines will show a color reaction, and the sensitivity of the ICA will be evaluated according to the difference in color between the T- and C-lines [26]. Finally, the qualitative analysis could be performed with the naked eyes to assess the sensitivity of the ICA.

2.8. True Samples Characterized by ICP-MS

Four true samples (1*, 2*, 3*, 4*) were purchased from the National Research Centre for Reference Materials (NRCCRM), Chinese Academy of Metrology. The four samples were analysed simultaneously by ICP-MS and ICA strips to compare their stability and authenticity. The pre-treatment steps for ICP-MS are described in the SI (supporting information). As shown in Figure S1, the ICP-MS standard curve for mercury was y = 192.468 + 1022.76x with a value of R^2 of 0.998. As shown in Table S1, the mercury content in the four samples was 2287.5, 507, 58.5, and 5.4 ng g^{-1}, respectively, where sample number four was negative, and the rest were positive samples.

2.9. Sample Pretreatment

The ICA strips were used to assess the authenticity and accuracy of the actual sample [27]. The rice was powdered in a grinder, and the sample was submerged in 0.75 M nitric acid (1 g of sample in 5 mL of nitric acid solution) and vortexed continuously for 5 min. After vortexing, the sample was centrifuged at 6500 r/min for 15 min, and the supernatant was removed. The pH of the supernatant was adjusted to 7.4 with 0.75 M sodium hydroxide [28]. Five instances of dilution with HBS were used to obtain the final sample extract. Finally, the Hg standards were added to the extracts to prepare samples of different concentrations, verified by ICA test strips.

3. Results and Discussion

3.1. Antigen Characterization

We characterized the immunogen (Hg-ITCBE-KLH) and coating antigen (Hg-MNA-BSA) using UV/Vis. Figure 3a shows that ITCBE had UV absorption peaks at 265 nm, KLH had absorption peaks at 280 nm and 350 nm, and Hg-ITCBE-KLH had UV absorption in both ranges, indicating that ITCBE had successfully coupled to the carrier protein. Figure 3b shows absorption peaks at 296 nm and 342 nm for MNA, 280 nm for BSA, and 280 nm and 350 nm for MNA-BSA, indicating a successful conjugation of MNA and BSA. When bound to mercury, the UV absorption of metal ions such as Hg^{2+} is weak; the absorption peak of Hg-MNA-BSA shifted, further demonstrating the successful synthesis of the antigen.

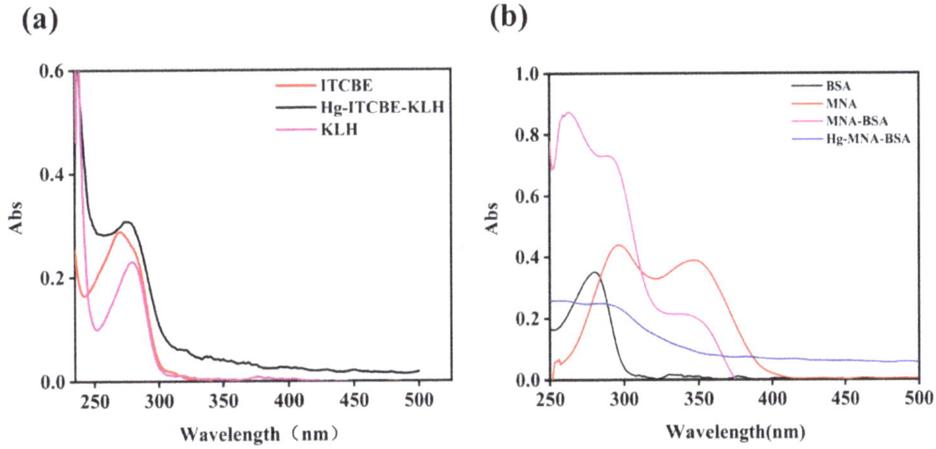

Figure 3. The UV–Vis spectroscopy of antigen. (**a**) Confirmation of immunogen (Hg–ITCBE–KLH). (**b**) Confirmation of coating antigen (Hg–MNA-BSA).

3.2. Characterization of Anti-Mercury mAb

The mAb against mercury was classified by analyzing the isotype using a mouse mAb isotyping kit. Figure 4a revealed that the anti-mercury mAb belonged to the isotype IgG2b and the light chain type kappa. Following purification, we identified mAb against mercury using ic-ELISA. Figure 4b shows that the equation was $y = 0.094 + (1.629-0.094)/(x/0.04)^{1.539}]$ with a correlation coefficient (R^2) of 0.997. The half-maximal inhibitory concentration (IC_{50}) and limit of detection (LOD) of anti-mercury mAb were 0.606 ng mL^{-1} and 0.07 ng mL^{-1}, respectively, with a linear range of detection of 0.302–20 ng mL^{-1}. Therefore, anti-mercury mAb had high sensitivity and specificity [29].

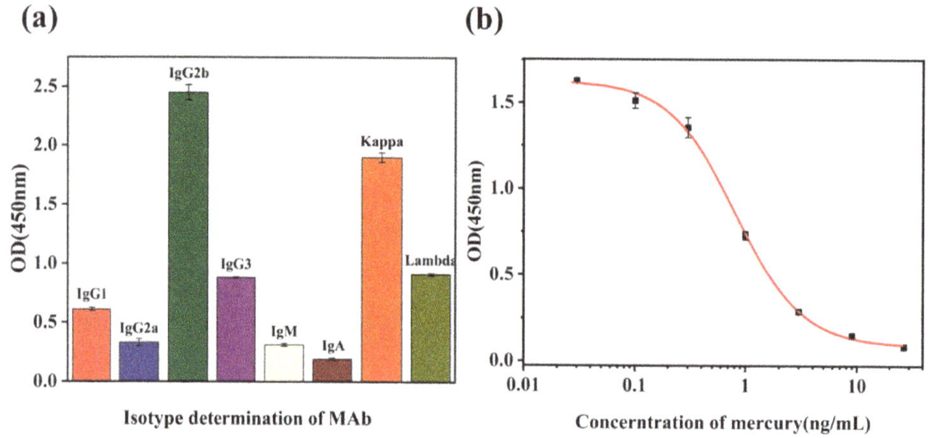

Figure 4. Characterization of mAb against mercury. (**a**) Isotype determination; (**b**) standard curves for mercury detection.

The cross-reactivity (CR) of anti-mercury mAb to other metals reflects the specificity of the antibody and its tolerance to matrix interference [30]. In this study, we used two different metal chelators, ITCBE and MNA, to synthesize immunogen and coating conjugate, respectively. Antibodies against the chelators could be eliminated during the screening of the mouse sera and hybridoma cells. MNA binds to mercury much more strongly than other metal ions, which prevents cross-reactivity and increases the specificity of anti-mercury mAb. Table 1 shows that anti-mercury mAb had no cross-reactivity with other metal ions (Cu^{2+}, MeHg, Pb^{2+}, Cd^{2+}, Cr^{3+}, Ni^{2+}, Mg^{2+}, Ca^{2+}, Mn^{2+}, and Fe^{3+}).

Table 1. The cross-reactivity of mAb against mercury with Hg^{2+}, Cu^{2+}, MeHg, Pb^{2+}, Cd^{2+}, and other metals.

Metals	IC_{50}(ng·mL^{-1})	CR (%)
Hg^{2+} (Mercury II)	0.606	100
MeHg (Methyl mercury)	8.9	9.05
Hg^{+} (Mercury I)	25	3.22
Pb^{2+} (Lead)	>10,000	<0.001
Cd^{2+} (Cadmium)	>10,000	<0.001
Cr^{3+} (chromium)	>10,000	<0.001
Ni^{2+} (Nickel)	>10,000	<0.001
Mg^{2+} (Magnesium)	>10,000	<0.001
Ca^{2+} (Calcium)	>10,000	<0.001
Cu^{2+} (Copper)	>10,000	<0.001
Mn^{2+} (Manganese)	>10,000	<0.001

3.3. *Performance of the ICA Strip*

When tuning the performance of ICA strips, choose the appropriate concentration of colloidal gold-mAb, coating antigen, resuspension buffer and other physical parameters (ionic strength, pH). Referring to the previous laboratory optimisation, we chose 0.47% HEPES, pH = 7.4 HBS buffer in the ICA test strips to dilute the specimens. If the running buffer is not suitable, a blockage can occur, leading to color development failure, and affecting the results[27]. For optimisation, we chose basic buffer and four surfactants (5% PEG, 5% BSA, 5% Polyvinylpyrrolidone and 5% On-870) for commissioning. As shown in Figure 5a, they were evaluated by testing with 0 and 20 ng mL^{-1} mercury. The results showed that the basic buffer showed darker T and C color development, significant inhibition and no blockage. This indicates that the basic buffer can help the antibody and antigen react quickly to produce results. As shown in Figure 5b, the basic buffer is the optimal surfactant for the test strips. The best colour development and sensitivity when 0.5 μg mL^{-1} of antigen and 10 μg mL^{-1} of mAb were selected. Therefore, the concentrations of 0.5 mg mL^{-1} of antigen and 10 μg mL^{-1} of mAb were combined to assess the sensitivity and matrix effect of the ICA test strips.

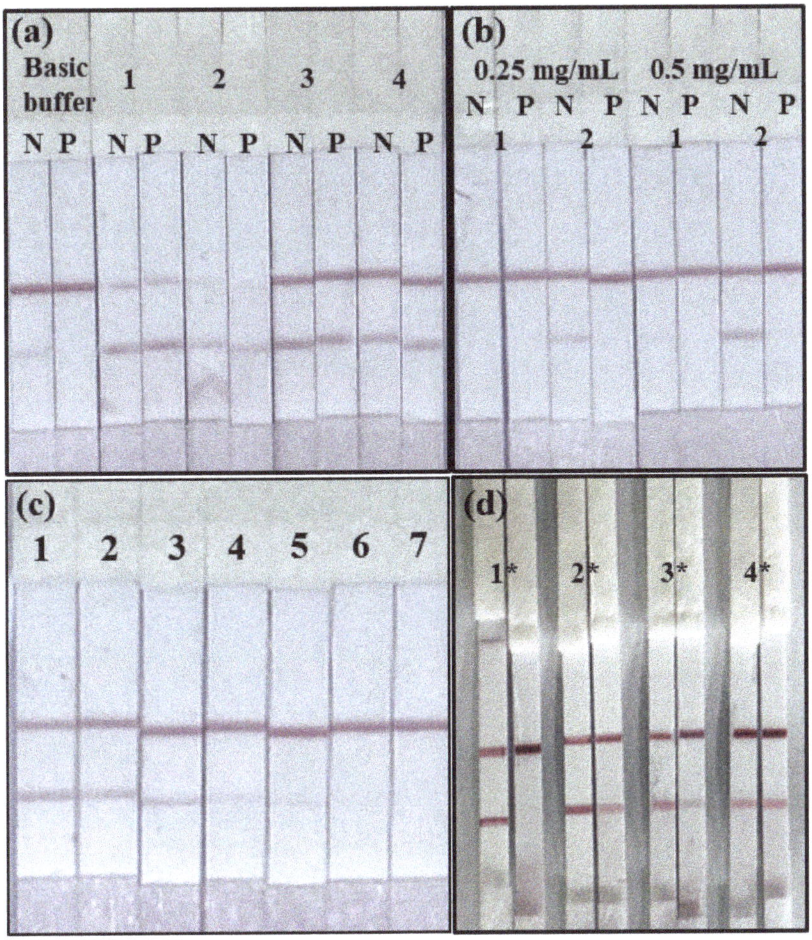

Figure 5. ICA strip performance and applied in the spike and true samples. (**a**) ICA strip with different running basic buffer and 1, 2, 3, 4, which represent 5% PVP, 5% On-870, 5% BSA, 5% PEG respectively (N, 0 ng mL^{-1}; P, 20 ng mL^{-1} of mercury). (**b**) The optimization of the antigen and the colloidal gold-mAb concentration (1, 5 μg mL^{-1} mAb; 2, 10 μg mL^{-1}; N, 0 ng mL^{-1}; P, 20 ng mL^{-1} of mercury). (**c**) The numbers (1, 2, 3, 4, 5, 6, 7 represent 0, 10, 20, 50, 100, 200, 500 ng mL^{-1}, respectively. (**d**) Sample analysis with ICA strips (n = 4). (1*, 2*, 3*, 4*) are four real samples of rice purchased from the National Reference Materials Centre.

3.4. Matrix Evaluation of the ICA Strip

Rice contains copper, magnesium, and macronutrients [31,32]. The presence of macronutrients interferes with the ICA strip results. To overcome this interference, a highly sensitive and specific mAb is required. Additionally, the test strip conditions need to be optimized. We chose rice as a substrate to assess the performance of the ICA strip (rice was mercury-negative by ICP-MS). According to the national standard, GB 5009.17-2014, the mercury content in grain should not exceed 20 μg kg^{-1}. We spiked rice samples with 0, 10, 20, 50, 100, 200, and 500 ng mL^{-1} mercury and analyzed them using the ICA strip. The cut-off value and visual limit of detection (vLOD) were used to assess the sensitivity of the ICA strip. The vLOD refers to the minimum concentration of mercury that results in a weak T-line. The cut-off value refers to the minimum concentration of mercury that results in a colorless T-line. Figure 5c shows that at 20 ng g^{-1} mercury, the T-line became lighter than at 0 ng g^{-1}; therefore, vLOD was 20 ng g^{-1}. At 200 ng g^{-1} mercury, the color of the T-line disappeared completely. Therefore, the cut-off value was 200 ng g^{-1} for mercury in rice.

Therefore, despite the presence of macronutrients, the ICA strip could detect mercury in the sample with good accuracy. The ICA strip meets the national standard (0.02 mg kg^{-1}) of detection for mercury in rice. The results can be visualized with the naked eye, which is convenient for the rapid determination of mercury in rice.

3.5. Analysis of Mercury in True Samples

Rice samples from the National Research Centre for Reference Materials (NRCCRM) of the Chinese Academy of Metrology included mercury-negative and -positive samples. The confirm the accuracy and stability of the ICA strip, standards of different concentrations of mercury (0, 10, 20, 50, 100, 200, 500 ng mL^{-1}) were added to the negative samples, as shown in Figure 5c (vLOD of 20 ng mL^{-1} and a cut-off value of 500 ng mL^{-1}). Additionally, we spiked mercury-negative samples with mercury and analyzed them using ic-ELISA, ICP-MS, and ICA strips. Table 2 shows the recoveries were 93.6–116.45% for ICP-MS and 94.5–113.7% for ic-ELISA.

Table 2. Analysis of mercury in rice samples by the ICA strip assay (n = 5).

Samples	Spiked Level (ng/mL)	ic-ELISA			ICP-MS			ICA Strip				
		Mean ± SD (ng/mL)	Recovery ± SD (%)	CV (%)	Mean ± SD (ng/mL)	Recovery ± SD (%)	CV (%)					
Rice	0	ND [a]	NC [b]	NC [b]	ND [a]	NC [b]	NC [b]	−	−	−	−	+
	10	9.86 ± 1.23	98.6 ± 4.4	4.4	9.93 ± 1.34	99.3 ± 4.66	4.6	−	−	±	+	+
	20	18.89 ± 4.23	94.5 ± 10.16	10.7	22.91 ± 4.13	114.5 ± 6.33	5.5	±	±	+	+	−
	100	112.7 ± 9.81	112.7 ± 13.59	12	93.6 ± 8.12	93.6 ± 13.12	14	+	+	+	+	±
	200	226.8 ± 13.44	113.4 ± 8.93	7.8	232.9 ± 9.87	116.45 ± 10.9	9.3	+	+	+	+	+

Notes: [a], ND, not detectable; [b], NC, not calculated. −, negative: the concentration of Hg^{2+} was <5 ng mL^{-1}; ±, weakly positive: the concentration of Hg^{2+} was 5–20 ng mL^{-1} in rice samples; +, positive: the concentration of Hg^{2+} was ≥20 ng mL^{-1} in rice sample.

The ICA strip qualitatively detected the concentration of mercury in rice, and the results were consistent with those obtained from ICP-MS and ic-ELISA. The ICA strip is often more reliable and less time-consuming than instrumental methods. The four rice samples were tested using mercury ICA strips. Figure 5d and Table 3 show that there were significant color differences between the T-line color of samples 1* through 3* and the blank, indicating that samples 1* through 3* were positive. Sample 3* was a weak positive sample. In contrast, there was no visible difference in the T-line color of sample 4*. The ICA strips results were consistent with the ICP-MS results.

Previously reported test strips for the detection of mercury were limited to simple matrices such as pure and river water [33,34]. In contrast, the ICA strip developed in this study can not only detect mercury in a complex matrix like rice but also has low detection costs, simple preparation methods, and low operator requirements. Additionally, it is suitable for most detection scenarios [35]. Based on this method, the identification of metal content in other grains can be further developed.

Table 3. Analysis of mercury in true samples by the ICA strip assay (n = 4).

Sample	ICP-MS	Visual			
1* Rice	2287.5 ± 50.23	+ [a]	+ [a]	+ [a]	+ [a]
2* Rice	507 ± 17.66	± [b]	+ [a]	+ [a]	+ [a]
3* Rice	58.5 ± 8.17	± [b]	± [b]	± [b]	+ [a]
4* Rice	5.4 ± 0.83	− [c]	− [c]	− [c]	± [b]

Notes: [a], positive; [b], weak positive; [c], not detectable.

4. Conclusions

We synthesized different immunogens and coating antigen and screened a highly sensitive and specific mAb against mercury using ic-ELISA and hybridoma technology. We developed a GNP-based lateral-flow ICA strip assay for the rapid detection of mercury. We optimized the running buffer, colloidal gold mAb, and antigen concentration of the ICA strip of vLOD was 20 ng g^{-1}, and the cut-off value was 500 ng g^{-1}. The mercury recovery in rice ranged from 93.6% to 116.45% (ICP-MS) and from 94.5% to 113.7% (ic-ELISA). Therefore, the ICA strip can be used to qualitatively determine mercury in the true rice, and the results are consistent with ICP-MS. Furthermore, our method can be used for large-scale mercury screening in rice and developed for other food crops.

Supplementary Materials: The following supporting information can be downloaded at: https://www.mdpi.com/article/10.3390/bios12090694/s1, Figure S1: The standard curve of mercury for ICP-MS (y = 192.468 + 1022.726x); Table S1: ICP-MS determi-nation of mercury in real samples (n = 4).

Author Contributions: Conceptualization, H.K. and C.X.; methodology, S.L. and L.X.; software, L.L. and X.X.; validation, S.L. and X.X.; formal analysis, S.L. and L.X.; investigation, S.L.; resources, S.L.; data curation, S.L.; writing—original draft preparation, S.L.; writing—review and editing, H.K. and C.X.; visualization, X.X.; supervision, L.L., H.K. and C.X; project administration, H.K. and C.X; funding acquisition, S.S., H.K. and C.X. All authors have read and agreed to the published version of the manuscript.

Funding: This project was supported by the National Key Research and Development Program of China (2019YFC1604604).

Institutional Review Board Statement: Not applicable.

Informed Consent Statement: Not applicable.

Data Availability Statement: Not applicable.

Conflicts of Interest: The authors declare no potential conflict of interest.

References

1. Jan, A.T.; Azam, M.; Siddiqui, K.; Ali, A.; Choi, I.; Haq, Q.M.R. Heavy Metals and Human Health: Mechanistic Insight into Toxicity and Counter Defense System of Antioxidants. *Int. J. Mol. Sci.* **2015**, *16*, 29592–29630. [PubMed]
2. Clemens, S.; Ma, J.F. Toxic Heavy Metal and Metalloid Accumulation in Crop Plants and Foods. In *Annual Review of Plant Biology*; Merchant, S.S., Ed.; Annual Reviews: Santa Clara, CA, USA; Palo Alto: Santa Clara, CA, USA, 2016; Volume 67, pp. 489–512.
3. Gworek, B.; Dmuchowski, W.; Baczewska-Dabrowska, A.H. Mercury in the terrestrial environment: A review. *Environ. Sci. Eur.* **2020**, *32*, 19.
4. Azimi, A.; Azari, A.; Rezakazemi, M.; Ansarpour, M. Removal of Heavy Metals from Industrial Wastewaters: A Review. *ChemBioEng Rev.* **2017**, *4*, 37–59.
5. Sommar, J.; Zhu, W.; Shang, L.H.; Lin, C.J.; Feng, X.B. Seasonal variations in metallic mercury (Hg-0) vapor exchange over biannual wheat-corn rotation cropland in the North China Plain. *Biogeosciences* **2016**, *13*, 2029–2049.
6. Chen, X.; Lyu, H.Y.; Zhang, J.; Bai, L.; Wang, J. National Food Safety Standards Related to Microbiological Contaminants in China: Recent Progress and Challenges. *Foodborne Pathog. Dis.* **2021**, *18*, 528–537.
7. da Silva, D.L.F.; da Costa, M.A.P.; Silva, L.O.B.; dos Santos, W.N.L. Simultaneous determination of mercury and selenium in fish by CVG AFS. *Food Chem.* **2019**, *273*, 24–30.

8. Vacchina, V.; Epova, E.N.; Berail, S.; Medina, B.; Donard, O.F.X.; Seby, F. Tin and mercury and their speciation (organotin compounds and methylmercury) in worldwide red wine samples determined by ICP-MS and GC-ICP-MS. *Food Addit. Contam. Part B-Surveill.* **2020**, *13*, 88–98.
9. Saleh, T.A.; Fadillah, G.; Ciptawati, E.; Khaled, M. Analytical methods for mercury speciation, detection, and measurement in water, oil, and gas. *Trac-Trends Anal. Chem.* **2020**, *132*, 116016.
10. Wang, Y.H.; Ding, M.Y.; Ma, H.Q.; Wu, J.; Zhao, H.W.; Wan, Y.L. Development of a specific monoclonal antibody-based icELISA for detection of arecoline in traditional Chinese medicines and fresh areca nuts. *Food Agric. Immunol.* **2022**, *33*, 113–126. [CrossRef]
11. Wang, X.; Liu, F.; Shao, Q.; Yin, Z.; Wang, L.; Fu, Z. A novel chemiluminescent immunochromatographic assay strip for rapid detection of mercury ions. *Anal. Methods* **2017**, *9*, 2401–2406.
12. Wu, Z.; Shen, H.; Hu, J.; Fu, Q.; Yao, C.; Yu, S.; Xiao, W.; Tang, Y. Aptamer-based fluorescence-quenching lateral flow strip for rapid detection of mercury (II) ion in water samples. *Anal. Bioanal. Chem.* **2017**, *409*, 5209–5216. [PubMed]
13. Zhang, K.Z.; Cai, H.D.; Lu, M.; Wei, D.X.; Yin, J.Q.; Ding, N.S.; Lai, W.H.; Peng, J. Quantum dot nanobead immunochromatographic assay based on bispecific monoclonal antibody for the simultaneous detection of aflatoxin B-1 and amantadine. *Food Agric. Immunol.* **2022**, *33*, 403–418.
14. She, P.; Chu, Y.; Liu, C.; Guo, X.; Zhao, K.; Li, J.; Du, H.; Zhang, X.; Wang, H.; Deng, A. A competitive immunoassay for ultrasensitive detection of Hg(2+) in water, human serum and urine samples using immunochromatographic test based on surface-enhanced Raman scattering. *Anal. Chim. Acta* **2016**, *906*, 139–147. [PubMed]
15. Song, S.S.; Chen, Z.W.; Liu, J.; Xu, L.G.; Kuang, H.; Zhu, J.P.; Liu, L.Q. Development of an immunocolloidal strip for rapid detection of picoxystrobin. *Food Agric. Immunol.* **2020**, *31*, 711–722.
16. Zhang, Y.; Li, X.; Liu, G.; Wang, Z.; Kong, T.; Tang, J.; Zhag, P.; Yang, W.; Li, D.; Liu, L.; et al. Development of ELISA for detection of mercury based on specific monoclonal antibodies against mercury-chelate. *Biol. Trace Elem Res.* **2011**, *144*, 854–864.
17. Wang, Y.; Yang, H.; Pschenitza, M.; Niessner, R.; Li, Y.; Knopp, D.; Deng, A. Highly sensitive and specific determination of mercury(II) ion in water, food and cosmetic samples with an ELISA based on a novel monoclonal antibody. *Anal. Bioanal. Chem.* **2012**, *403*, 2519–2528.
18. Guo, L.L.; Wu, X.L.; Liu, L.Q.; Kuang, H.; Xu, C.L. Gold Immunochromatographic Assay for Rapid On-Site Detection of Lincosamide Residues in Milk, Egg, Beef, and Honey Samples. *Biotechnol. J.* **2020**, *15*, 1900174.
19. Xu, X.; Liu, L.; Wu, X.; Kuang, H.; Xu, C. Ultrasensitive immunochromatographic strips for fast screening of the nicarbazin marker in chicken breast and liver samples based on monoclonal antibodies. *Anal. Methods* **2020**, *12*, 2143–2151.
20. Zeng, L.; Xu, X.; Song, S.; Xu, L.; Liu, L.; Xiao, J.; Xu, C.; Kuang, H. Synthesis of haptens and gold-based immunochromatographic paper sensor for vitamin B6 in energy drinks and dietary supplements. *Nano Res.* **2021**, *15*, 2479–2488.
21. Zeng, L.; Xu, X.; Ding, H.; Song, S.; Xu, L.; Xu, C.; Kuang, H. A gold nanoparticle based colorimetric sensor for the rapid detection of Yersinia enterocolitica serotype O:8 in food samples. *J. Mater. Chem. B* **2022**, *10*, 909–914.
22. Zou, S.; Cui, G.; Liu, L.; Song, S.; Kuang, H. Development of ic-ELISA and an immunochromatographic strip assay for the detection of methylmercury. *Food Agric. Immunol.* **2017**, *28*, 699–710. [CrossRef]
23. Liu, J.; Xu, X.; Wu, A.; Song, S.; Kuang, H.; Liu, L.; Wang, Z.; Xu, L.; Sun, M.; Xu, C. An immunochromatographic assay for the rapid detection of oxadixyl in cucumber, tomato and wine samples. *Food Chem.* **2022**, *379*, 132131. [PubMed]
24. Liu, J.; Xu, X.X.; Wu, A.H.; Wang, Z.Y.; Song, S.S.; Kuang, H.; Liu, L.Q.; Xu, C.L. Preparing monoclonal antibodies and developing immunochromatographic assay strips for the determination of propamocarb levels. *Food Chem.* **2022**, *370*, 131284. [PubMed]
25. Zeng, L.; Guo, L.; Wang, Z.; Xu, X.; Ding, H.; Song, S.; Xu, L.; Kuang, H.; Xu, C. Gold nanoparticle-based immunochromatographic assay for detection Pseudomonas aeruginosa in water and food samples. *Food Chem. X* **2021**, *9*, 100117.
26. Jiang, H.T.; Xu, X.X.; Song, S.S.; Wu, A.H.; Liu, L.Q.; Kuang, H.; Xu, C.L. A monoclonal antibody-based colloidal gold immunochromatographic strip for the analysis of novobiocin in beef and chicken. *Food Addit. Contam. Part A* **2022**, *39*, 1053–1064.
27. Zheng, S.; Wu, T.; Li, J.X.; Jin, Q.; Xiao, R.; Wang, S.Q.; Wang, C.W. Difunctional immunochromatographic assay based on magnetic quantum dot for ultrasensitive and simultaneous detection of multiple mycotoxins in foods. *Sens. Actuators B Chem.* **2022**, *359*, 131528.
28. Lin, L.; Song, S.S.; Wu, X.L.; Liu, L.Q.; Kuang, H. A colloidal gold immunochromatography test strip based on a monoclonal antibody for the rapid detection of triadimefon and triadimenol in foods. *Food Agric. Immunol.* **2020**, *31*, 475–488.
29. Safaei, S.; Kazemian, H.; Junk, P.C. Dual functional MOF as a selective fluorescent naked-eye detector and effective sorbent for mercury ion. *J. Solid State Chem.* **2021**, *300*, 122267.
30. Wu, X.Y.; Cobbina, S.J.; Mao, G.H.; Xu, H.; Zhang, Z.; Yang, L.Q. A review of toxicity and mechanisms of individual and mixtures of heavy metals in the environment. *Environ. Sci. Pollut. Res.* **2016**, *23*, 8244–8259.
31. Li, P.; Feng, X.B.; Qiu, G.L.; Shang, L.H.; Li, Z.G. Mercury pollution in Asia: A review of the contaminated sites. *J. Hazard. Mater.* **2009**, *168*, 591–601.
32. Basu, N.; Abass, K.; Dietz, R.; Krummel, E.; Rautio, A.; Weihe, P. The impact of mercury contamination on human health in the Arctic: A state of the science review. *Sci. Total Environ.* **2022**, *831*, 154793. [PubMed]
33. Kannappan, S.; Ramisetty, B.C.M. Engineered Whole-Cell-Based Biosensors: Sensing Environmental Heavy Metal Pollutants in Water-a Review. *Appl. Biochem. Biotechnol.* **2022**, *194*, 1814–1840. [PubMed]

34. Xing, Y.P.; Xue, B.Y.; Qi, P.S.; Chen, G.Y.; Zhou, X.H. A rapid and sensitive fluorescence biosensor for Hg2+ detection in environmental samples. *Sens. Actuator Rep.* **2022**, *4*, 100101. [CrossRef]
35. Cheng, Z.H.; Wei, J.C.; Gu, L.Q.; Zou, L.; Wang, T.; Chen, L.; Li, Y.Q.; Yang, Y.; Li, P. DNAzyme-based biosensors for mercury (II) detection: Rational construction, advances and perspectives. *J. Hazard. Mater.* **2022**, *431*, 128606.

biosensors

Article

Quantitative SERS Detection of TBBPA in Electronic Plastic Based on Hydrophobic Cu-Ag Chips

Pei Dai [1,2,*], Xianzhi Huang [1], Yaqian Cui [1] and Lihua Zhu [1,*]

[1] School of Chemistry and Chemical Engineering, Huazhong University of Science and Technology, Wuhan 430074, China
[2] Yellow Crane Tower Science and Technology Park (Group) Co., Ltd., Wuhan 430074, China
* Correspondence: peidai94@hust.edu.cn (P.D.); lhzhu63@hust.edu.cn (L.Z.); Tel./Fax: +86-27-8755-6723 (L.Z.)

Abstract: Tetrabromobisphenol A (TBBPA) was one of the most widely used brominated flame retardants. However, it easily contaminates nature and harms the environment and human health during its production and use. Therefore, it is necessary to strictly control the content of TBBPA in electronics. Surface-enhanced Raman spectroscopy has the advantages of being fast and sensitive, but it is difficult to obtain the SERS spectra of TBBPA because the hydrophobic TBBPA molecule is difficult to approach with the hydrophilic surface of common noble metal SERS substrates. In the present work, a hydrophobic Cu-Ag chip was developed for the SERS detection of TBBPA. The integration of the hydrophobic interaction and the Ag-Br bonding promoted the adsorption of TBBPA on the Cu-Ag chip, allowing for SERS detection. It was observed that both the hydrophobicity and bimetallic composition of the Cu-Ag chip played important roles in the SERS detection of TBBPA. Under the optimized conditions, the low limit of detection of the established SERS method for TBBPA was 0.01 mg L^{-1}, within a linear range of 0.1–10 mg L^{-1}. Combined with ultrasonic-assisted extraction, the substrate could be used for the quantitative determination of TBBPA in electronic products. Compared with the HPLC-UV method used as a national standard, the relative error of the SERS method for quantifying the TBBPA content in a mouse cable and shell was ±3% and ±7.7%, respectively. According to the SERS results, the recovery of TBBPA in the spiked mouse shell was 95.6%.

Keywords: tetrabromobisphenol A; surface-enhanced Raman scattering; hydrophobic interaction; bi-metal chips; synergistic coupling

Citation: Dai, P.; Huang, X.; Cui, Y.; Zhu, L. Quantitative SERS Detection of TBBPA in Electronic Plastic Based on Hydrophobic Cu-Ag Chips. *Biosensors* **2022**, *12*, 881. https://doi.org/10.3390/bios12100881

Received: 6 September 2022
Accepted: 12 October 2022
Published: 17 October 2022

1. Introduction

Tetrabromobisphenol A (TBBPA) is currently the most widely used brominated flame retardant (BFRs) worldwide [1,2], and is often used as an additive flame retardant in the electronic plastic resins ABS (Acrylonitrile-Butadiene-Styrene) and HIPS (High Impact Polystyrene). It accounts for more than 60% of the total output of BFRs, with an annual demand of nearly 200,000 tons [3]. With the continuous development of the manufacturing industry, the production and demand of flame retardants are increasing. However, TBBPA is highly polluting and persistent, and it interferes in the development of brain and bone, harms the endocrine and hormonal systems with long-time exposure [4]. Due to its volatilization during production and use, TBBPA can penetrate into the atmosphere, water and soil. With the circulation of matter and energy in the biosphere, it may be enriched and eventually endanger the life and health of various organisms [5]. According to Annex II of the RoHS Directive, the permitted limit of TBBPA as an additive flame retardant is 0.1 wt% in homogeneous materials, such as shell and packaging, and the reactive TBBPA in the circuit board may be exempted. Therefore, it is necessary to monitor and control the amount of TBBPA in these products.

Liquid chromatography [6–9] and gas chromatography [10,11] are the main detection methods of TBBPA at the present time. They can realize the accurate qualitative and quantitative detection of TBBPA, with the advantages of high sensitivity and reproducibility. However, the analysis methods based on chromatography still have problems, such as the high cost of equipment, the complex and tedious sample pretreatment and the long detection cycle. Moreover, derivatization of TBBPA is always needed for its detection by gas chromatography, which may lead to low recovery of TBBPA. Xie et al. [12] developed a GC-MS method for quantitative detection of TBBPA, and the recovery was only $79 \pm 1\%$. Alternatively, the electrochemical method can provide a more effective strategy for TBBPA detection, having the advantages of low cost, high sensitivity, and being simple and convenient [13–15]. Zhao et al. [16] successfully detected TBBPA on a g-C_3N_4-modified glassy carbon electrode (GCE) with a limit of detection (LOD) of 5 nM. Wang [17] modified gold nanorob/polycysteine composite on a GCE and decreased the LOD to 3.2 nM. However, the electrochemical detection of TBBPA may have poor selectivity. This can be improved by combining it with molecularly imprinted technology [18] and electrochemical immunosensors [19].

Surface-enhanced Raman spectroscopy (SERS) can reflect the fingerprint information of target molecules [20,21]. It has been used in the rapid detection of organic molecules due to its high sensitivity, convenience and efficiency [22–25]. TBBPA has a large oil-water distribution coefficient (log K_{ow} = 4.5), showing strong hydrophobicity with a water solubility of only 4.15 mg L^{-1} (pH = 7). Therefore, TBBPA molecules are difficult to adsorb on the surface of noble metal hydrosols and other hydrophilic substrates to realize the SERS detection. Therefore, there are few reports on the SERS detection of TBBPA. Kadasala used magnetic Au nanoclusters modified with 4-dimethylaminopyridine (DMAP) as an enhanced substrate to realize the SERS detection of TBBPA with a LOD of 1 nM [26]. However, modification with the organic ligand DMAP introduced a large number of strong background peaks, seriously interfering with the discrimination of the characteristic peaks of TBBPA. We anticipated that the establishment of surface-clean hydrophobic SERS substrates can overcome the difficulties in the SERS detection of hydrophobic molecules.

In recent years, hydrophobic materials have attracted more and more attention. They have broad application prospects in metal corrosion prevention [27,28], self-cleaning [29,30], oil-water separation [31,32], surface-enhanced Raman spectroscopy [33,34] and other aspects. In general, the surface hydrophobicity of solid materials is mainly affected by two factors, the rough structure and chemical composition of the surface. Therefore, in addition to modifying low surface energy molecules on the material surface, hydrophobicity could also be realized by constructing a rough surface to form an air cushion which cannot be wet by water. There are many methods to prepare the hydrophobic materials, including the template method [35], the chemical etching method [36], the hydrothermal method [37], the electrodeposition method [38] and so on. Most of them have the problems of involving complex preparation processes and large instruments, and they cannot be applied in large scale production. To simplify the preparation process and develop an environmentally friendly preparation method, in the early work of our group, the superhydrophobic copper coating was prepared by a chemical reduction method and applied to the oil-water separation [39]. With the aid of hydrazine reduction without adding any modified agents, Cu NPs were deposited on the fabrics to form the superhydrophobic surface.

Therefore, aiming to achieve SERS detection of the hydrophobic TBBPA, we developed a hydrophobic Cu-Ag SERS chip to improve the SERS enhancement of substrate and its binding ability with TBBPA. Firstly, the hydrophobic copper-coated fabric with ordered micro-nano structure was prepared. Then, a hydrophobic Cu-Ag chip was constructed by introducing Ag onto the Cu surface through a replacement reaction. The synergistic coupling of Cu-Ag bimetal promised the SERS chips a good SERS enhancement. The surface of the Cu-Ag chips was hydrophobic and had abundant Ag. It made it so that TBBPA could approach the surface of the Cu-Ag chips based on the hydrophobic interaction, and finally bonded with the substrate through Ag-Br. Therefore, the SERS detection of TBBPA was realized with low interference and high sensitivity. Finally, the SERS method was combined with ultrasound-assisted extraction technology to achieve rapid quantitative determination of the TBBPA content in electronics.

2. Experimental Section

2.1. Reagents and Materials

$AgNO_3$, $N_2H_4 \cdot H_2O$ (85%), citrate sodium, crystal violet (CV), tetrahydrofuran (THF), toluene, methanol and ethanol were purchased from Sinopharm Chemical Reagent Co., Ltd. (Shanghai, China). $Cu(CH_3COO)_2 \cdot H_2O$ was purchased from Shanghai Lingfeng Chemical Reagents Co., Ltd. (Shanghai, China). Sudan I and Sudan III were purchased from Macklin (Shanghai, China). TBBPA (98%) was purchased from Aladdin Chemistry Co., Ltd. (Shanghai, China). Nylon66 microporous filtration membranes were purchased from Tianjin Jinteng Experimental Equipment Co., Ltd. (Tianjin, China). Fabrics were made of ordinary pure cotton. All the chemicals were analytical grade and used without further purification. Deionized water was used throughout the experiments.

2.2. Preparation of Hydrophobic Cu-Ag Chips

The hydrophobic Cu-Ag chip was prepared with a modified chemical reduction method. Fabrics were firstly cut into pieces with a size of 2 × 2 cm, followed by washing with water and ethanol. After drying, it was immersed in a solution of $Cu(CH_3COO)_2 \cdot H_2O$ (20 mL, 10 g L^{-1}) for several minutes. The Cu NPs were reduced and deposited on the fabric by adding hydrazine hydrate (800 μL, 80%) drop by drop until quiescence at room temperature for 24 h. Then, the Cu-coated fabric was taken out to wash with water and ethanol, and immersed in the ethanol solution of $AgNO_3$ (0.1 M) for 5 min. After being washed again and vacuum dried at 60 °C for 1 h, the hydrophobic Cu-Ag chip was finally obtained for subsequent SERS detection.

The morphology of the obtained Cu-Ag chip was characterized by a GeminiSEM 300 scanning electron microscope (ZEISS, Oberkochen, Germany). Contact angles (CAs) were measured with an OCA20 contact angle instrument (Dataphysics, Filderstadt, Germany). X-ray diffraction (XRD) patterns were obtained with a SmartLab-SE diffractometer (Rigaku, Tokyo, Japan).

2.3. Preparation of Hydrophilic Ag Chips

To show the advantages of the above Cu-Ag chip in the SERS detection of hydrophobic targets, a hydrophilic Ag chip was prepared for comparison. Firstly, Ag NPs were prepared by a one-step reduction method with using sodium citrate as the reducing agent. The aqueous solution of $AgNO_3$ (100 mL, 1.0 mmol L^{-1}) was heated to boiling, and then sodium citrate (2.0 mL, 1 wt%) was quickly added to it. After the reaction proceeded for 30 min and the solution cooled to room temperature, the Ag nanoparticles were obtained. To prepare the Ag chip, the organic filter membrane (0.22 μm) was placed on the vacuum filtration device, then 5 mL Ag sol and 5 mL toluene were added into the container, on the filter membrane, to form the immiscible two-phase fluid. Then, 2 mL ethanol was injected into the interface of the two phases with a syringe, so that the Ag NPs self-assembled in the interface to form the Ag array. The vacuum pump was started to filter the solution, and the assembled Ag array was transferred to the filter membrane. After drying, the hydrophilic Ag chip was finally obtained for subsequent SERS detection.

2.4. Analysis of TBBPA

The SERS detection of TBBPA was achieved with the hydrophobic Cu-Ag chip. The Cu-Ag chip was cut into small pieces each with a size of 0.5 × 0.5 cm and immersed into the ethanol solution of target molecules for 10 min for further detection. SERS detection was performed with the ATR8100 microscopic Raman spectrometer (Optosky, Xiamen, China) with a 785 nm laser as the excitation source. The laser power was set to 50 mW, and the exposure time was 5 s. During the test, ethanol was dropped onto the SERS substrate to keep it wet. If not specified elsewhere, 5 different positions were randomly selected on each SERS chip for parallel tests, and the average spectrum was taken for further analysis.

To verify the accuracy of the SERS method, the content of TBBPA was also quantitatively detected by high-performance liquid chromatography with an HPLC-1260 (Agilent, Santa Clara, CA, USA). A ZORBAX SB-C18 reversed-phase column was used to separate the targets. The injection volume was 20 μL and the column temperature was 30 °C. The mobile phase was a mixture of methanol/water = 3/7 (v/v), with a flow rate of 1.0 mL min^{-1}. The detection wavelength was set to 230 nm for the quantitative analysis.

2.5. Sample Pretreatment of the Electronics

To quantitatively detect the content of TBBPA in practical electronics, ultrasonic assisted extraction (UAE) was used to extract TBBPA from solid samples. Firstly, 0.1 g of electronic products were cut into small particles with a size less than 1 mm. UAE was performed with a KQ-600KDE ultrasonic cleaner (Shumei, Kunshan, China) at 240 W, using 5 mL tetrahydrofuran (THF) as the extractant. After 20 min of extraction, 15 mL methanol was added to the extraction solution to precipitate the plastic. Then, the organic filter was used to remove the precipitates, and the obtained filter liquor was retained for subsequent detection. The analytical methods for TBBPA detection were described in Section 2.3.

3. Results and Discussion

3.1. Characterization of the Cu-Ag Chip and Its SERS Detection of TBBPA

In the present work, Cu NPs were firstly deposited on the fabrics through a chemical reduction method. By using hydrazine hydrate as the reducing agent, Cu NPs were deposited uniformly and closely on the fabric. Figure 1a shows the SEM image of the surface morphology of the Cu coating with a deposition time of 24 h (this Cu coating was referred to as the 24 h Cu chip). It was observed that the spherical Cu NPs with diameters of 300–500 nm were packed tightly together. The structure of the Cu coating made the fabric hydrophobic, and the contact angle of the water droplet on its surface was about 149° (Figure 1e). As shown in Figure 1g, the XRD pattern of the Cu coating showed two peaks at 43.3° and 50.5°, which matched well with the diffraction peaks of (1 1 1) and (2 0 0) of Cu (JCPDS 04-0836), indicating that the state was that of metallic copper. Therefore, Ag could be deposited on the surface of the Cu metal through the replacement reaction.

Compared with the Cu chip, the XRD pattern of the Cu-Ag chip (Figure 1g) showed three new peaks appearing at 38.1°, 44.3° and 64.4°, corresponding to the (1 1 1), (2 0 0), and (2 2 0) crystal planes of silver (JCPDS 04-0783), respectively. Along with the prolonging of replacement reaction time, the size of the Ag nanoparticles on the 24 h Cu coating increased. As shown in Figure 1b–d, with increases in the replacement time of Ag from 3 min to 10 min, the small-size Ag NPs (Figure 1b) gradually became too large to completely coat the Cu NPs (Figure 1c), and finally formed the Ag sheet structure at the micrometer scale (Figure 1d). For the 24 h Cu-5 min Ag chip, Ag particles completely covered the underlayer Cu NPs and presented a multilateral shape of 500~800 nm. Although its surface roughness decreased, the tight packing of Cu-Ag NPs made the Cu-Ag chip still able to retain good hydrophobic properties, with a contact angle of 140.1°. Figure 1h shows the Vis-NIR diffuse reflectance spectra of the above Cu and Cu-Ag chips. The surface plasmon resonance (SPR) peak of the Cu chip appeared at 557 nm, and the Cu-Ag chip showed stronger SPR absorption with a red shift to 716 nm.

The SERS response of the TBBPA solution (10 mg L^{-1}) was detected with the 24 h Cu-5 min Ag chip, as shown in Figure 1i. The background interference of the Cu-Ag chip was little, and the four peaks at 1607 cm^{-1}, 1405 cm^{-1}, 1129 cm^{-1} and 856 cm^{-1} might be attributed to some organic compounds of the fabric in the chip. The introduction of TBBPA solution introduced three new peaks at 1446 cm^{-1}, 1283 cm^{-1} and 710 cm^{-1} in the SERS spectrum, consistent with the normal Raman peak of TBBPA solid. This indicated that the Cu-Ag chip without modification could directly adsorb TBBPA molecules and realize their SERS detection.

Figure 1. Characterization of Cu-Ag chips and their SERS detection of TBBPA. SEM images of the (**a**) Cu coating with 24 h deposition and (**b–d**) Cu-Ag coating with replacement times of (**b**) 3 min, (**c**) 5 min and (**d**) 10 min. (**e**,**f**) Contact angles of the coatings of (**e**) 24 h Cu and (**f**) 24 h Cu-5 min Ag on the fabric. (**g**) XRD patterns and (**h**) Vis-NIR diffuse reflectance spectra of the Cu and Cu-Ag chips. (**i**) Comparison between SERS spectra of (1) a blank background, (2) TBBPA solution (10 mg L^{-1}) obtained from the hydrophobic Cu-Ag chips and (3) Raman spectrum of TBBPA powders. (**j**) The molecular structure of TBBPA.

3.2. Effects of the Preparation Conditions of the Cu-Ag Chip on Its Performance

The introduction of Ag in the Cu-Ag SERS chip increases not only the electromagnetic enhancement of the Cu-Ag chip but also the Ag-Br interaction between the SERS substrate and TBBPA. The deposition amounts of Cu and Ag were controlled by changing the deposition time of Cu and the replacement time of Ag, respectively. As shown in Figure 2a, when the replacement time for depositing Ag was fixed at 5 min, the SERS intensity of TBBPA on the Cu-Ag chip gradually increased with increases in the Cu deposition time. When the Cu deposition time was as short as two hours, all the deposited copper would be completely dissolved by the replacement reaction during the Ag deposition, so that the diffraction peaks of Cu were hardly observed on the XRD pattern of the 2 h Cu-5 min Ag chip (Figure 2b). When the Cu deposition time was extended to 24 h, the SERS intensity of TBBPA was the strongest. With further extensions to the Cu deposition time, the SERS intensity of TBBPA decreased. This may be due to the decrease in the relative content of Ag in the SERS substrate, which led to the weakening of both the SERS enhancement and the interaction of Cu-Ag chip with TBBPA.

Then, the influence of Ag replacement time on the SERS intensity was investigated by fixing the Cu deposition time at 24 h. As shown in Figure 2c, by prolonging the Ag replacement time, the peak intensity of TBBPA at 1283 cm^{-1} first increased and then decreased, and reached the maximum value at 5 min. Figure 2d shows the XRD pattern of Cu-Ag chips with different replacement times. As the replacement time increased from 1 min to 5 min, the Ag content on the SERS substrate increased continuously, and the characteristic peaks of metallic Cu could still be observed. This proved that the obtained SERS substrate was composed of Cu and Ag composite materials. When the replacement time was further extended to more than 5 min, Cu NPs were completely dissolved, and the SERS substrate became a monometallic Ag material. Therefore, the surface state had basically remained unchanged (Figure 2d). The contact angles of the Cu-Ag chips with replacement times of 8 min and 10 min were 143.2° and 140.0° respectively, which seemed to be nearly consistent with those at 5 min.

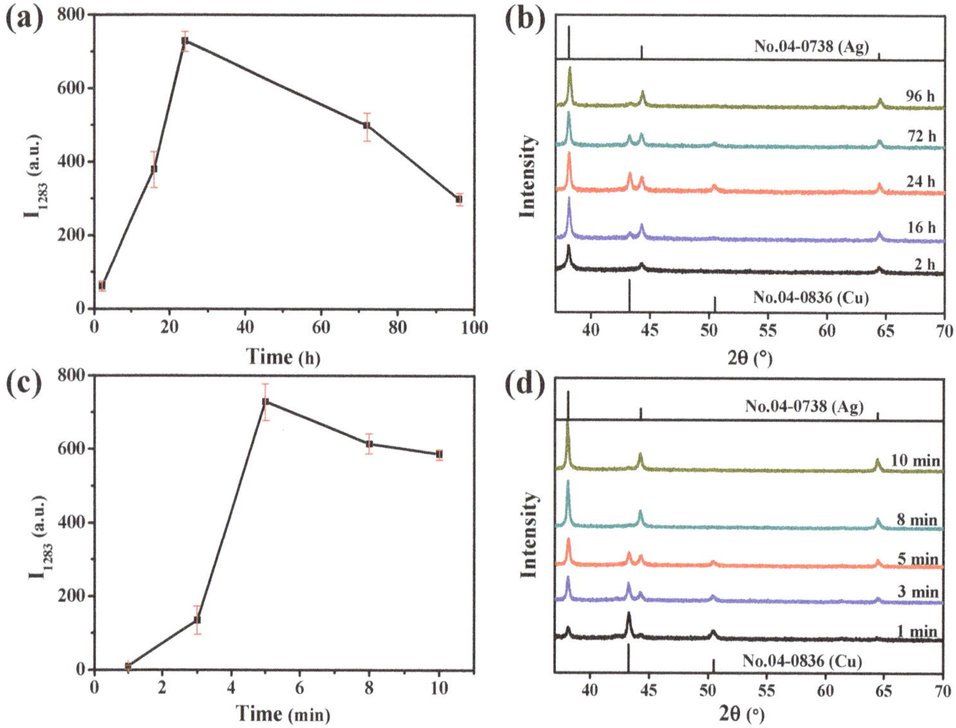

Figure 2. (**a**,**b**) Influence of the Cu deposition time on (**a**) the peak intensity of TBBPA at 1283 cm^{-1} and (**b**) the XRD pattern of the Cu-Ag chips. The replacement time of Ag was set as 5 min. (**c**,**d**) Influence of the Ag replacement time on (**c**) the peak intensity of TBBPA at 1283 cm^{-1} and (**d**) the XRD pattern of the Ag-Cu chips. The deposition time of Cu was set as 24 h.

Among these SERS substrates prepared under different conditions, the 24 h Cu-5 min Ag chip had the strongest SERS enhancement for TBBPA. This could be explained by the above SEM and diffuse reflectance spectroscopy characterizations (Figure 1). Ag of the 24 h Cu-5 min Ag chip formed a polygonized structure with corner angle on the surface of the Cu, its tip provided great SERS performance. The core-shell composite structure also provided the synergistic coupling of Cu and Ag, resulting in the red shift of its SPR peak, which matched well with the 785 nm excitation laser. In addition, the Cu-Ag chip still maintained hydrophobicity under these preparation conditions, and the exposure surface of Ag increased. Both of them enhanced the adsorption capacity of the 24 h Cu-5 min Ag chip to TBBPA. Based on the above discussions, the increase in Ag content improved the Ag-Br interaction and the electromagnetic enhancement of the Cu-Ag substrate for TBBPA, so as to realize the SERS detection of TBBPA. The existence of a small amount of Cu could further enhance the SERS intensity of TBBPA on this Cu-Ag chip. Therefore, in the present work, the Cu deposition time and Ag replacement time were selected to be 24 h and 5 min, respectively.

3.3. *Effect of the Hydrophobicity of the Cu-Ag Chip on the Detection of Hydrophobic Targets*

In order to evaluate the influence of the hydrophobicity of the Cu-Ag chip on the SERS detection of TBBPA, a completely hydrophilic Ag chip with a contact angle of 0° was prepared by interface assembly and transfer for comparison with the above hydrophobic Cu-Ag chip. Figure 3a,b gives the SERS responses of two hydrophilic targets, crystal violet (CV) and rhodamine 6G (R6G). Except for the four peaks of the blank background

($1607\ cm^{-1}$, $1405\ cm^{-1}$, $1129\ cm^{-1}$ and $856\ cm^{-1}$), there were abundant characteristic peaks of target molecules in their SERS spectra obtained from the hydrophobic Cu-Ag chips. The SERS peaks marked in the pink area were the fingerprint peaks of CV and R6G. It could be found that the fingerprint peaks obtained from the Cu-Ag chips were consistent with the results obtained from the hydrophilic Ag chips, and the signal intensity was slightly lower than that of the hydrophilic substrate. This indicated that the hydrophobic Cu-Ag chip was applicable to the hydrophilic targets with good dispersion in solvent, and its SERS detection of target molecules was universal. The enhancement factor (EF) of CV on the Cu-Ag chips was 4.8×10^6. The hydrophilic Ag chip had a good affinity for the hydrophilic target; its SERS performance was similar to that of the hydrophobic Cu-Ag chip.

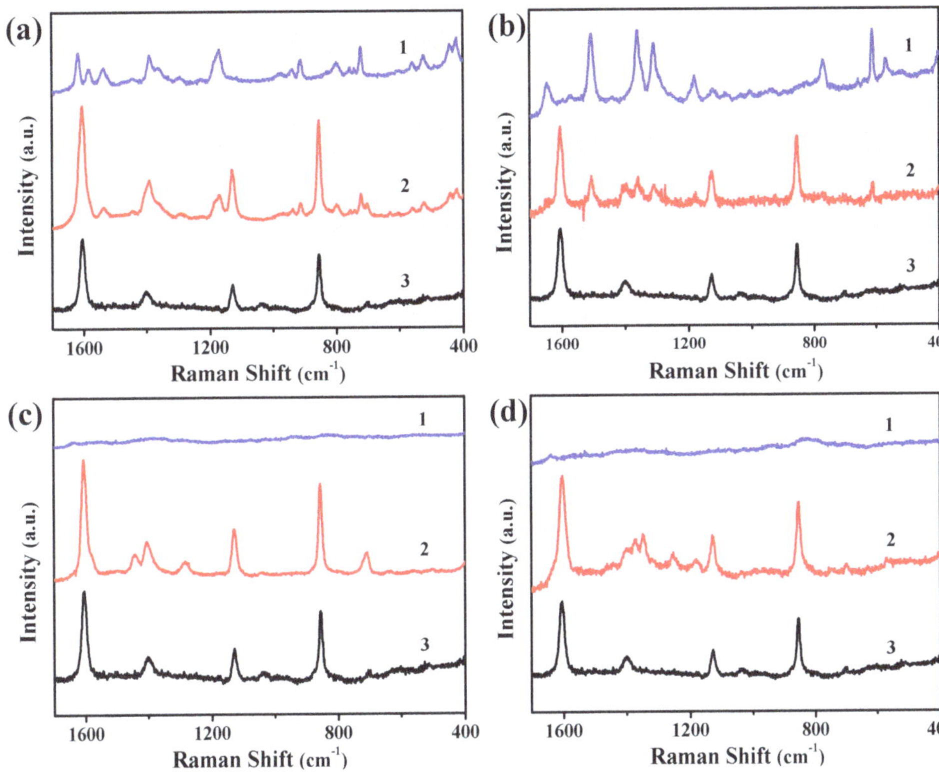

Figure 3. SERS spectra of the hydrophilic targets of (**a**) CV, (**b**) R6G, (**c**) TBBPA, and (**d**) Sudan I obtained from the hydrophilic Ag chips (1) and the hydrophobic Cu-Ag chips (2), with the blank background of hydrophobic Cu-Ag chips (3) for comparison.

In contrast, when the two substrates were applied for the SERS detection of the hydrophobic targets (as shown in Figure 3c,d), TBBPA and Sudan I dye, there was no characteristic peak in their SERS spectra obtained from the hydrophilic Ag chip. However, all their characteristic peaks could be observed in their SERS spectra obtained from the hydrophobic Cu-Ag chips. The EF of TBBPA on the Cu-Ag chips was 5.3×10^7. These results showed that the hydrophobicity of the Cu-Ag chips could improve their adsorption of hydrophobic targets, so as to realize the SERS detection of the hydrophobic targets with weak binding ability. This was probably because the hydrophobic targets existed as polymolecular aggregates in the ethanol solution, which had weak binding force with the noble metal. Therefore, they were hardly adsorbed on the hydrophilic Ag substrate. These aggregate droplets could approach the surface of the Cu-Ag chip based on the hydrophobic interaction, and then adsorbed on the substrate through the Ag-Br bond; the SERS detection of TBBPA was finally realized on this hydrophobic Cu-Ag chip.

3.4. Quantitative SERS Detection of TBBPA in Electronic Products

The Cu-Ag chip was used for quantitative detection of TBBPA, and the SERS spectra of TBBPA at various concentrations are shown in Figure 4a. It could be observed that the SERS intensity of the characteristic peaks of TBBPA increased gradually with the increase in its concentration. By taking the characteristic peak of TBBPA at 1283 cm^{-1} as the representative, the SERS intensity was plotted against the concentration of TBBPA (Figure 4b). It was found that there was a good linear relationship between the SERS intensity and TBBPA concentration on a logarithmic scale within the range of 0.1 to 10 mg L^{-1}. The wide linear range and low LOD (0.01 mg L^{-1}) indicated that the use of Cu-Ag chips provided a good quantitative analysis method for TBBPA.

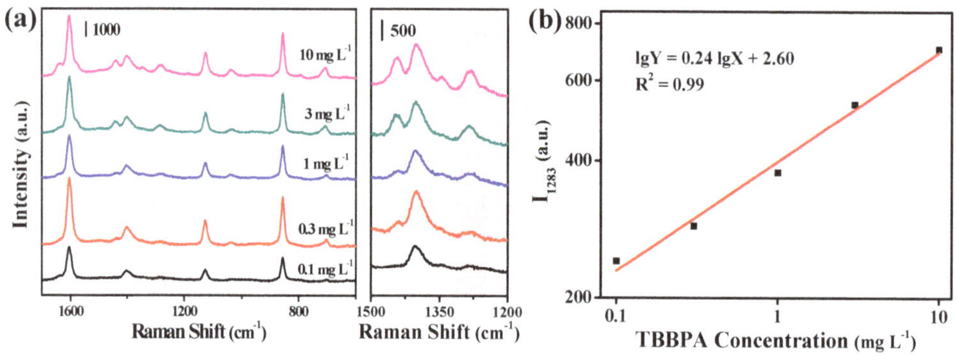

Figure 4. (**a**) SERS spectra of TBBPA in the different concentrations obtained from the hydrophobic Ag-Cu chips. (**b**) Standard curve for quantitative SERS detection of TBBPA from the hydrophobic Ag-Cu chips.

Our previous work reported a method that could completely extract brominated flame retardants in electronics through ultrasonic assisted extraction with THF as the extraction agent [40]. By combining that extraction method and the presently developed SERS method based on the hydrophobic Ag-Cu chip, we realized the rapid quantitative detection of TBBPA in electronics. In order to verify the accuracy of the SERS method, we also used the HPLC-UV method as the national standard detection method of TBBPA to quantitatively detect the TBBPA content in the same mouse samples. As shown in Table 1, according to the SERS method, the content of TBBPA in the cable was 1980.1 mg kg^{-1}, showing a relative error of −3.0% in comparison with the detection result (2040.3 mg kg^{-1}) of the HPLC-UV method. As for the mouse shell sample, the detected content of TBBPA by the SERS and HPLC-UV methods was 316.9 mg kg^{-1} and 343.5 mg kg^{-1}, respectively, with a relative error of −7.7%. All of these detected TBBPA contents were much lower than the permitted limit of TBBPA (0.1 wt%, or 1 g kg^{-1}) in electronics [41].

Table 1. TBBPA contents in the mouse and its spiked sample determined by SERS and HPLC-UV methods.

Sample (mg kg^{-1})	SERS	HPLC-UV
Cable	1980.1	2040.3
Shell	316.9	343.5
Spiked shell	1750.9	1899.0

The feasibility and accuracy of the UAE extraction-SERS detection method for quantitative detection of TBBPA in actual electronics were also verified through the spiked samples. Firstly, the spiked shell (spiked amount of 1500 mg kg^{-1}) was prepared by soaking the mouse shell in the standard solution of TBBPA followed by drying the solvent. After the same extraction operation, the TBBPA content of the spiked shell was detected to be 1750.9 mg kg^{-1} and 1899.0 mg kg^{-1} by the SERS and HPLC-UV methods, respectively. Compared to the original content of TBBPA in shell, the spiked amount was calculated as 1434.0 mg kg^{-1} and 1555.5 mg kg^{-1}, with recoveries of 95.6% and 103.7%, respectively. This proved that the ultrasonic-assisted extraction method had high extraction efficiency for the spiked TBBPA in the mouse shell, and the detection result of SERS method was accurate to use for the quantitative detection of TBBPA in electronic products.

4. Conclusions

A hydrophobic Cu-Ag chip was prepared for the SERS detection of TBBPA as a representative of hydrophobic substances. The hydrophobicity of the Cu-Ag chip enhanced the affinity of the substrate for the hydrophobic targets (such as TBBPA), which let TBBPA close to the surface of SERS substrate and then bonded with it through Ag-Br interactions. In addition, the synergistic coupling between the Cu and Ag bimetals improved the SERS enhancement of the substrate. These made it possible for the SERS detection of TBBPA on the Cu-Ag chip. By combining it with-ultrasonic assisted extraction (UAE), the quantitative detection of TBBPA in the electronics was realized. After UAE, the TBBPA content in electronics could be completely extracted and accurately determined, and the LOD of TBBPA was 2.0 mg kg^{-1} (0.01 mg L^{-1} for the TBBPA solution). Compared with the HPLC-UV, the SERS method was quick, convenient and sensitive; the quantitative detection results could be used to distinguish if the TBBPA content in electronics was appropriate.

Author Contributions: Methodology, P.D.; Conceptualization, L.Z.; Data curation, X.H.; Validation, Y.C. All authors have read and agreed to the published version of the manuscript.

Funding: This research was funded by the National Natural Science Foundation of China grant number 22076052 and 21976063.

Institutional Review Board Statement: Not applicable.

Informed Consent Statement: Not applicable.

Data Availability Statement: Not applicable.

Acknowledgments: The authors acknowledge the financial support and the contributions from all authors.

Conflicts of Interest: The authors declare no conflict of interest.

References

1. Okeke, E.S.; Huang, B.; Mao, G.; Chen, Y.; Zeng, Z.; Qian, X.; Wu, X.; Feng, W. Review of the environmental occurrence, analytical techniques, degradation and toxicity of TBBPA and its derivatives. *Environ. Res.* **2022**, *206*, 112594. [CrossRef]
2. Wang, R.; Xu, Z. Recycling of non-metallic fractions from waste electrical and electronic equipment (WEEE): A review. *Waste Manag.* **2014**, *34*, 1455–1469. [CrossRef] [PubMed]
3. Law, R.J.; Allchin, C.R.; de Boer, J.; Covaci, A.; Herzke, D.; Lepom, P.; Morris, S.; Tronczynski, J.; de Wit, C.A. Levels and trends of brominated flame retardants in the European environment. *Chemosphere* **2006**, *64*, 187–208. [CrossRef] [PubMed]
4. Zhang, H.; Liu, W.; Chen, B.; He, J.; Chen, F.; Shan, X.; Du, Q.; Li, N.; Jia, X.; Tang, J. Differences in reproductive toxicity of TBBPA and TCBPA exposure in male Rana nigromaculata. *Environ. Pollut.* **2018**, *243*, 394–403. [CrossRef] [PubMed]
5. Barghi, M.; Shin, E.S.; Choi, S.D.; Behrooz, R.D.; Chang, Y.S. HBCD and TBBPA in human scalp hair: Evidence of internal exposure. *Chemosphere* **2018**, *207*, 70–77. [CrossRef]
6. Tan, F.; Lu, B.; Liu, Z.; Chen, G.; Liu, Y.; Cheng, F.; Zhou, Y. Identification and quantification of TBBPA and its metabolites in adult zebrafish by high resolution liquid chromatography tandem mass spectrometry. *Microchem. J.* **2020**, *154*, 104566. [CrossRef]
7. Liu, L.; Liu, A.; Zhang, Q.; Shi, J.; He, B.; Yun, Z.; Jiang, G. Determination of tetrabromobisphenol-A/S and their main derivatives in water samples by high performance liquid chromatography coupled with inductively coupled plasma tandem mass spectrometry. *J. Chromatogr. A* **2017**, *1497*, 81–86. [CrossRef] [PubMed]

8. Song, X.L.; Lv, H.; Wang, D.D.; Liao, K.C.; Wu, Y.Y.; Li, G.M.; Chen, Y. Graphene oxide composite microspheres as a novel dispersive solid-phase extraction adsorbent of bisphenols prior to their quantitation by HPLC–mass spectrometry. *Microchem. J.* **2022**, *172*, 106920. [CrossRef]
9. Sun, X.; Peng, J.; Wang, M.; Wang, J.; Tang, C.; Yang, L.; Lei, H.; Li, F.; Wang, X.; Chen, J. Determination of nine bisphenols in sewage and sludge using dummy molecularly imprinted solid-phase extraction coupled with liquid chromatography tandem mass spectrometry. *J. Chromatogr. A* **2018**, *1552*, 10–16. [CrossRef]
10. Barrett, C.A.; Orban, D.A.; Seebeck, S.E.; Lowe, L.E.; Owens, E.J. Development of a low-density-solventdispersive liquid-liquid microextractionwith gas chromatography and massspectrometry method for the quantitationof tetrabromobisphenol-A from dust. *J. Sep. Sci.* **2015**, *38*, 2503–2509. [CrossRef] [PubMed]
11. Chokwe, T.B.; Okonkwo, J.O.; Sibali, L.L.; Ncube, E.J. An integrated method for the simultaneous determination of alkylphenol ethoxylates and brominated flame retardants in sewage sludge samples by ultrasonic-assisted extraction, solid phase clean-up, and GC-MS analysis. *Microchem. J.* **2015**, *123*, 230–236. [CrossRef]
12. Xie, Z.; Ebinghaus, R.; Lohmann, R.; Heemken, O.; Caba, A.; Püttmann, W. Trace determination of the flame retardant tetrabromobisphenol A in the atmosphere by gas chromatography-mass spectrometry. *Anal. Chim. Acta* **2007**, *584*, 333–342. [CrossRef] [PubMed]
13. Luo, S.; Yang, M.; Wu, Y.; Li, J.; Qin, J.; Feng, F. A Low Cost Fe_3O_4-sctivated biochar electrode sensor by resource utilization of excess sludge for detecting tetrabromobisphenol A. *Micromachines* **2022**, *13*, 115. [CrossRef] [PubMed]
14. Zhao, Q.; Zhou, H.; Wu, W.; Wei, X.; Jiang, S.; Zhou, T.; Liu, D.; Lu, Q. Sensitive electrochemical detection of tetrabromobisphenol A based on poly(diallyldimethylammonium chloride) modified graphitic carbon nitride-ionic liquid doped carbon paste electrode. *Electrochim. Acta* **2017**, *254*, 214–222. [CrossRef]
15. Zhou, F.; Wang, Y.; Wu, W.; Jing, T.; Mei, S.; Zhou, Y. Synergetic signal amplification of multi-walled carbon nanotubes-Fe_3O_4 hybrid and trimethyloctadecylammonium bromide as a highly sensitive detection platform for tetrabromobisphenol A. *Sci. Rep.* **2016**, *6*, 38000. [CrossRef]
16. Zhao, Q.; Wu, W.; Wei, X.; Jiang, S.; Zhou, T.; Li, Q.; Lu, Q. Graphitic carbon nitride as electrode sensing material for tetrabromobisphenol—A determination. *Sens. Actuators B Chem.* **2017**, *248*, 673–681. [CrossRef]
17. Wang, Y.; Liu, G.; Hou, X.; Huang, Y.; Li, C.; Wu, K. Assembling gold nanorods on a poly-cysteine modified glassy carbon electrode strongly enhance the electrochemical reponse to tetrabromobisphenol A. *Mikrochim. Acta* **2016**, *183*, 689–696. [CrossRef]
18. Zhang, K.; Kwabena, A.S.; Wang, N.; Lu, Y.; Cao, Y.; Luan, Y.; Liu, T.; Peng, H.; Gu, X.; Xu, W. Electrochemical assays for the detection of TBBPA in plastic products based on rGO/AgNDs nanocomposites and molecularly imprinted polymers. *J. Electroanal. Chem.* **2020**, *862*, 114022. [CrossRef]
19. Yakubu, S.; Xiao, J.; Gu, J.; Cheng, J.; Wang, J.; Li, X.; Zhang, Z. A competitive electrochemical immunosensor based on bimetallic nanoparticle decorated nanoflower-like MnO_2 for enhanced peroxidase-like activity and sensitive detection of Tetrabromobisphenol A. *Sens. Actuators B Chem.* **2020**, *325*, 128909. [CrossRef]
20. Albrecht, M.G.; Creighton, J.A. Anomalously intense Raman spectra of pyridine at a silver electrode. *J. Am. Chem. Soc.* **1977**, *99*, 5215–5217. [CrossRef]
21. Jeanmaire, D.L.; Van Duyne, R.P. Surface Raman spectroelectrochemistry Part I. Heterocyclic, aromatic, and aliphatic amines adsorbed on the anodized silver electrode. *J. Electroanal. Chem. Interfacial Electrochem.* **1977**, *84*, 1–20. [CrossRef]
22. Perumal, J.; Wang, Y.; Attia, A.B.E.; Dinish, U.S.; Olivo, M. Towards a point-of-care SERS sensor for biomedical and agri-food analysis applications: A review of recent advancements. *Nanoscale* **2021**, *13*, 553–580. [CrossRef]
23. Pu, H.; Huang, Z.; Xu, F.; Sun, D.W. Two-dimensional self-assembled Au-Ag core-shell nanorods nanoarray for sensitive detection of thiram in apple using surface-enhanced Raman spectroscopy. *Food Chem.* **2021**, *343*, 128548. [CrossRef] [PubMed]
24. Dai, P.; Zhang, Z.; Hou, X.; Ouyang, L.; Zhu, L. Rapid SERS inspection of carcinogenic aromatic amines in textiles by using liquid interfacial assembled Au array. *Talanta* **2021**, *234*, 122651. [CrossRef]
25. Dai, P.; Li, H.; Huang, X.; Zhu, L. Highly Sensitive and Stable Copper-Based SERS Chips Prepared by a Chemical Reduction Method. *Nanomaterials* **2021**, *11*, 2770. [CrossRef] [PubMed]
26. Kadasala, N.R.; Wei, A. Trace detection of tetrabromobisphenol A by SERS with DMAP-modified magnetic gold nanoclusters. *Nanoscale* **2015**, *7*, 10931–10935. [CrossRef] [PubMed]
27. Xiang, T.; Han, Y.; Guo, Z.; Wang, R.; Zheng, S.; Li, S.; Li, C.; Dai, X. Fabrication of Inherent Anticorrosion Superhydrophobic Surfaces on Metals. *ACS Sustain. Chem. Eng.* **2018**, *6*, 5598–5606. [CrossRef]
28. Zhang, X.; Zhou, T.; Liu, J.; Wang, R.; Hu, W.; Liu, L. Volcano-like hierarchical superhydrophobic surface synthesized via facile one-step secondary anodic oxidation. *Appl. Surf. Sci.* **2021**, *540*, 148337. [CrossRef]
29. Latthe, S.S.; Sutar, R.S.; Kodag, V.S.; Bhosale, A.K.; Kumar, A.M.; Sadasivuni, K.K.; Xing, R.; Liu, S. Self-cleaning superhydrophobic coatings: Potential industrial applications. *Prog. Org. Coat.* **2019**, *128*, 52–58. [CrossRef]
30. Sutar, R.S.; Kalel, P.J.; Latthe, S.S.; Kumbhar, D.A.; Mahajan, S.S.; Chikode, P.P.; Patil, S.S.; Kadam, S.S.; Gaikwad, V.H.; Bhosale, A.K.; et al. Superhydrophobic PVC/SiO_2 coating for self-cleaning application. *Macromol. Symp.* **2020**, *393*, 2000034. [CrossRef]
31. Qu, M.; Liu, Q.; He, J.; Li, J.; Liu, L.; Yang, C.; Yang, X.; Peng, L.; Li, K. A multifunctional superwettable material with excellent pH-responsive for controllable in situ separation multiphase oil/water mixture and efficient separation organics system. *Appl. Surf. Sci.* **2020**, *515*, 145991. [CrossRef]

32. Li, Z.; Zhang, T.; Wang, M.; Qiu, F.; Yue, X.; Yang, D. Hierarchical structurized waste brick with opposite wettability for on-demand oil/water separation. *Chemosphere* **2020**, *251*, 126348. [CrossRef]
33. Chen, F.; Zhao, Y.; Zhang, S.; Wei, S.; Ming, A.; Mao, C. Hydrophobic Wafer-Scale High-Reproducibility SERS Sensor Based on Silicon Nanorods Arrays Decorated with Au Nanoparticles for Pesticide Residue Detection. *Biosensors* **2022**, *12*, 273. [CrossRef] [PubMed]
34. Wu, J.; Cai, J.; Fan, Y.; Zhang, Y.; Fang, H.; Yan, S. Effective Enrichment of Plasmonic Hotspots for SERS by Spinning Droplets on a Slippery Concave Dome Array. *Biosensors* **2022**, *12*, 270. [CrossRef]
35. Fang, Z.; Luo, W.; Hu, H.; Xia, M.; Sun, Z.; Zhang, Y.; He, P. Ice-template triggered roughness: A facile method to prepare robust superhydrophobic surface with versatile performance. *Prog. Org. Coat.* **2019**, *135*, 345–351. [CrossRef]
36. Kumar, A.; Gogoi, B. Development of durable self-cleaning superhydrophobic coatings for aluminium surfaces via chemical etching method. *Tribol. Int.* **2018**, *122*, 114–118. [CrossRef]
37. Tuo, Y.; Chen, W.; Zhang, H.; Li, P.; Liu, X. One-step hydrothermal method to fabricate drag reduction superhydrophobic surface on aluminum foil. *Appl. Surf. Sci.* **2018**, *446*, 230–235. [CrossRef]
38. Liu, J.; Fang, X.; Zhu, C.; Xing, X.; Cui, G.; Li, Z. Fabrication of superhydrophobic coatings for corrosion protection by electrodeposition: A comprehensive review. *Colloids Surf. A Physicochem. Eng. Asp.* **2020**, *607*, 125498. [CrossRef]
39. Yang, W.; Li, J.; Zhou, P.; Zhu, L.; Tang, H. Superhydrophobic copper coating: Switchable wettability, on-demand oil-water separation, and antifouling. *Chem. Eng. J.* **2017**, *327*, 849–854. [CrossRef]
40. Dai, P.; Hou, D.; Guo, S.; Zhu, L.; Lei, M.; Tang, H. Ion chromatographic determination of total bromine in electronic devices in conjunction with a catalytic reduction debromination pretreatment. *Anal. Chim. Acta* **2019**, *1082*, 49–55. [CrossRef] [PubMed]
41. EU Pack15. ROHS Annex II Dossier for TBBP-A Restriction Proposal for Substances in Electrical and Electronic Equipment under RoHS. Available online: https://rohs.exemptions.oeko.info/fileadmin/user_upload/RoHS_Pack_15 (accessed on 12 October 2022).

biosensors

MDPI

Article

Screening and Evaluation of Thiamethoxam Aptamer Based on Pressurized GO-SELEX and Its Sensor Application

Yaqi Yue [1], De Zhang [1], Kangfei Tian [1], Dejiang Ni [1], Fei Guo [1], Zhi Yu [1], Pu Wang [1,*] and Pei Liang [2,*]

[1] Key Laboratory of Horticultural Plant Biology, Ministry of Education, College of Horticulture & Forestry Sciences, Huazhong Agricultural University, Wuhan 430070, China
[2] College of Optical and Electronic Technology, China Jiliang University, Hangzhou 310018, China
* Correspondence: pwang@mail.hzau.edu.cn (P.W.); plianghust@cjlu.edu.cn (P.L.); Tel./Fax: +86-571-8687-5622 (P.L.)

Abstract: Thiamethoxam, a nicotinic pesticide used worldwide, can cause great harm to the environment and even to human health, and aptamers, known as chemical antibodies, have high affinity and specificity for the target, as well as great potential in detecting small molecules such as pesticides. In this paper, we report a highly sensitive biosensor system for thiamethoxam residue detection based on aptamer technology. After 15 rounds of screening with the pressurized GO-SELEX technology, we found that the aptamer libraries of the 5th and 9th rounds showed high affinity by the capture method. Four candidate aptamers were obtained by high-throughput sequencing and secondary structure prediction. Among them, the aptamer named Thi-5R-18 from the 5th round was demonstrated to possess the highest affinity by isothermal titration calorimetry, with a dissociation constant (Kd) of 4.935×10^{-5} M. The results of molecular docking showed that thiamethoxam and Thi-5R-18 were combined with bases G-15, A-19, and T-71 through hydrogen bonding and π–π interaction.Thi-5R-18 was used as a recognition element to construct a AuNPs colorimetric aptasensor, achieving an ultralow detection limit of 0.37 nM. More importantly, this colorimetric aptasensor can be used for quantitative detection of thiamethoxam on tea leaves, with a recovery of 96.94%~105.86%. This study provides a highly sensitive biosensor for detection of thiamethoxam residue.

Keywords: GO-SELEX; ITC; colorimetric sensor; pesticide; aptamer

Citation: Yue, Y.; Zhang, D.; Tian, K.; Ni, D.; Guo, F.; Yu, Z.; Wang, P.; Liang, P. Screening and Evaluation of Thiamethoxam Aptamer Based on Pressurized GO-SELEX and Its Sensor Application. *Biosensors* **2023**, *13*, 155. https://doi.org/10.3390/bios13020155

Received: 13 November 2022
Revised: 9 January 2023
Accepted: 17 January 2023
Published: 18 January 2023

1. Introduction

Thiamethoxam, the second generation of new nicotine toxic insecticide widely used all over the world [1,2], can destroy the normal conduction of insect central nerves and cause insect death [3] by binding with nicotinic acetylcholine receptor, leading to its popular application to control aphids, planthoppers, leafhoppers, whiteflies, and other pests. However, studies have shown that thiamethoxam has adverse effects on bees and fish in the ecosystem, such as impairing the flying ability of bees, reducing their return rate and survival rate [4,5], and damaging fish liver and affecting their reproductive capacity [6]. In the past 30 years, the use of thiamethoxam has caused great damage to the environment [7], as well as potential harm to mammals and even human beings [8]. Many countries and regions have issued laws and regulations to control pesticide use and set maximum residue limits (MRLs). Therefore, it is necessary to establish a rapid and accurate detection method for thiamethoxam. At present, the detection methods of thiamethoxam include high performance liquid chromatography (HPLC) [9], Liquid Chromatograph-Mass Spectrometer (LC-MS) [10,11], enzyme linked immunosorbent assay (ELISA) [12], etc. However, most of these methods require large and high-precision instruments, experienced operators, and time-consuming and cumbersome data processing, suggesting that it is imperative to develop a simple and convenient method for thiamethoxam detection.

The concept of nucleic acid aptamer was first proposed by Ellington and Tuerk in 1990 and since then, aptamers have been gradually and extensively developed [13,14].

Nucleic acid aptamers, known as chemical antibodies, are single stranded DNA or RNA. Aptamers that specifically bind to targets can be screened from synthetic libraries in vitro by Systematic Evolution of Ligands by Exponential Enrichment (SELEX) technology [15]. The targets of aptamers include proteins, cells, toxins, antibiotics, molecular markers, drugs, heavy metals, etc. [16]. Compared with antibodies, aptamers are easier to synthesize, more convenient to label, lower in immunogenicity, and higher in thermal stability [17]. There are mainly two types of SELEX technologies for pesticides: fixed and non-fixed SELEX, with fixed SELEX technology including microplate SELEX, magnetic bead SELEX, and capture SELEX, while non-fixed SELEX technology including GO-SELEX and capillary electrophoresis SELEX. Microwell plate SELEX couples pesticides with macromolecular proteins such as bovine serum albumin (BSA), or fixes small molecules to the microplate through a protein carrier, which can be more easily bound to ssDNA than pesticides. Magnetic bead SELEX connects the target molecules to the magnetic bead surface through a chemical coupling reaction and can quickly obtain the target binding sequence through magnetic separation. The magnetic bead (MB) surface can be modified with amino, carboxyl, streptavidin, etc. Targets with functional groups such as carboxyl, amino, and biotin can be coupled with MB, but there are few functional groups on the pesticide's surface that can be coupled or modified. Therefore, the MB SELEX method to fix the target may change the target conformation and affect the binding site of ssDNA and the target. In practical applications, the target was in a free state, and the aptamers screened by MB SELEX may have a low affinity with the target. Capture SELEX fixes the library on the solid-phase carrier by using short strand ssDNA as the bridging sequence. In the presence of the target, the ssDNA binds to the target and allows its conformation to change, enabling ssDNA to detach from the solid carrier. Capture SELEX fixed libraries could effectively solve the impact of a fixed target. However, the fixation of the libraries will affect the diversity of oligonucleotide sequences, leading to unstable fixation efficiency, and thus the loss of sequences with high affinity and specificity to the target in the initial library. Capillary electrophoresis SELEX separates target-bound ssDNA from unbound ssDNA through mobility difference, which is efficient and highly selective, but is not applicable to small molecules with high mobility. Graphene oxide (GO) is obtained by a chemical method after strong oxidation of graphite [18]. The strong van der Waals force of GO makes it hydrophobic and easy to gather, and the oxygen-containing groups on the surface make it hydrophilic and dispersible, allowing GO to adsorb molecules containing aromatic rings through π-π force and hydrogen bonds [19]. The free ssDNA is separated from the target ssDNA complex through the hydrophobic interaction and unique adsorption on GO surface [20], and candidate sequences of aptamers are obtained by centrifugation. GO-SELEX can solve the problem that the targets do not have functional groups and have low molecular weight, making it suitable for screening of nucleic acid aptamers for small molecular targets.

Generally, GO-SELEX has the significant advantages of not requiring fixed targets or libraries, high screening efficiency, and simple operation. GO addition is a key factor affecting the adsorption of ssDNA, and during the screening process of GO-SELEX, the screening pressure usually increases with the increase of rounds, such as shortening the incubation time, increasing the proportion of GO/ssDNA or GO/target, etc. In this paper, we aimed to screen a thiamethoxam aptamer based on pressurized GO-SELEX and test the applicability of its sensor in thiamethoxam detection. Specifically, based on previous studies [21,22], aptamers of small pesticide molecular target of thiamethoxam were obtained by raising the additional amount of GO by an order of magnitude, and analyzed the impact of increased screening pressure on screening results. After 15 rounds of screening, an innovative capture method was set up to evaluate the affinity of the aptamer libraries of each round. High-throughput sequencing was carried out according to the rounds of affinity, and 4 candidate aptamers were found. The affinity of 4 aptamers was determined by isothermal titration calorimetry (ITC), and Thi-5R-18 was selected as the best aptamer. For thiamethoxam detection, the aptamer Thi-5R-18 was utilized as a component to identify targets and establish the AuNPs colorimetric sensor, which can be applied to thiamethoxam detection in actual tea leaves.

2. Materials and Methods

2.1. Materials

Thiamethoxam was obtained from MedChemExpress (Monmouth Junction, NJ, USA); GO from Shenzhen Suiheng Graphene Technology Co., Ltd. (Shenzhen, China); LA Taq from TaKaRa (Beijing, China); 50 bp DNA Ladder from Beijing Tsingke Biological Technology Co., Ltd. (Beijing, China); Agarose, 10 × TBE, 2 × Urea DNA Loading Buffer, DNA Urea-PAGE electrophoresis kit from Beijing Coolaber Technology Co., Ltd. (Beijing, China); UNIQ-10 Spin Column DNA Gel Extraction Kit for PAGE from Sangon Biotech (Shanghai, China) Co., Ltd. (Shanghai, China); and Tris-HCl, KCl, NaCl, and $MgCl_2$, 2-[(2-Hydroxy-1,1-bis(hydroxymethyl)ethyl)amino]ethanesulfonic acid (TES) (99% purity) from China Pharmaceutical Group Shanghai Chemical Reagent Co., Ltd. (Shanghai, China). All reagents were of analytical grade and ultrapure water was used throughout. Primers for screening aptamers and verifying effects, including FAM-labeled forward primers and Spacer18-labeled reverse primers (5′-ATAGGAGTCACGACGACCAGAA-3′; 5′-dA20-Spacer18-ATTAGTCAAGAGGTAGACGCACATA-3′), as well as unlabeled primers and library (5′-ATAGGAGTCACGACGACCAGAA-N_{40}-TATGTGCGTCTACCTCTTGACTAAT-3′), were all obtained from Optimus Beijing Tsingke Biological Technology Co., Ltd. (Beijing, China) and purified by HPLC.

2.2. SELEX

ssDNA library (10μM) was dissolved in 1 × Binding buffer (50 mM Tris-HCl, 5 mM KCl, 100 mM NaCl, 1 mM $MgCl_2$, pH 7.4), followed by boiling at 95 °C for 8 min and quenching immediately for 8 min. The adsorption capacity of GO to the target was verified as follows. Briefly, the GO/target mass ratios were set at 100:1, 200:1, 500:1, 1000:1, 2000:1, and 5000:1, respectively, followed by incubating GO and target together for 1 h and then centrifugation at 12000 rpm for 10 min. The supernatant was collected three times to measure the light absorbance value. Next, the GO/ssDNA mass ratio was optimized by setting, the GO/ssDNA mass ratios at 100:1, 200:1, 300:1, 400:1, 500:1, 1000:1, 2000:1, 3000:1, 4000:1, and 5000:1, respectively, followed by incubating GO and ssDNA together for 1 h, centrifugation at 12000 rpm for 10 min, and collecting the supernatant three times to measure the concentration of ssDNA. In the first round of screening, 5000:1 and 500:1 were selected as the mass ratios of GO and ssDNA. Meanwhile, the library was mixed with thiamethoxam solution and incubated at room temperature for 1 h, followed by adding 15 mg/mL GO, incubation for 1 h, and centrifugation at 12000 rpm for 10 min to collect the supernatant as the PCR template. The 50 μL PCR system mixture included 2.5 μL forward primer (10μM), 2.5 μL reverse primer (10 μM), 0.5 μL Takara La Taq, 5 μL La Taq buffer, 8 μL dNTP Mix and 31.5 μL PCR template. The PCR was performed under the conditions of 94 °C pre-denaturation for 5 min, 30 cycles of 98 °C denaturation for 45 s, 61 °C annealing for 15 s, 72 °C extension for 15 s, and 72 °C final extension for 1 min, performed successively. The same amount of 2 × urea loading buffer was added into the PCR product and was boiled at 95 °C for 3 min. After purification and recovery by 8% Urea-PAGE electrophoresis, the FAM-labeled ssDNA was used as the aptamer library for the next round of aptamer screening. A total of 15 rounds of aptamer screening were conducted.

2.3. Round Affinity Analysis

For round affinity analysis, each round of the aptamer library was used as a template, and biotin-labeled forward primer and dA20-Spacer18-labeled reverse primer were used for PCR. PCR products were purified and recovered by 8% Urea-PAGE to obtain a biotin-labeled aptamer library. After denaturation and renaturation, the aptamer library was attached to streptavidin beads, followed by adding an equal concentration of thiamethoxam to the complex solution and incubated for 1 h. Next, the supernatant was separated by magnetic absorption, followed by cleaning the magnetic beads twice with TES, adding ultrapure water in a 60 °C metal bath for 5 min, and then measuring absorption value of the supernatant at 255 nm.

2.4. HTS and Secondary Structure Prediction

The aptamers obtained in the 5th and 9th rounds were tested by PCR using unlabeled common primers. The purified product was sent to Novogene Co. Ltd. (Beijing, China), and the Illumina Platform was used for library construction and HTS. The results of HTS were analyzed by self-compiled programs. The secondary structure and free energy of the aptamers were predicted through online websites (www.unafold.org, accessed on 11 October 2022).

2.5. Measurement Dissociation Constant (Kd) by ITC

The Kd of aptamer sequence was measured by ITC. Briefly, the sample was degassed for 10 min at a vacuum degree above 400 mmHg. Both the aptamers and pesticide were dissolved in 1 × binding buffer, followed by adding 350 μL of 25 μM aptamers to the titrator and 50 μL of 1mM thiamethoxam to the sample cell. ITC was performed under the following conditions: total titration number, 20; volume of each titration, 2.5 μL; time, 200 s; reaction temperature, 25 °C; and stirring speed, 250 rpm. After running, the titration results were analyzed by Launch NanoAnalyze software to obtain the Kd of each aptamer.

2.6. Analysis and Verification of the Binding of Aptamer and Thiamethoxam

The secondary structure file of the aptamer was saved in Vienna output format, which was used to build the 3D structure of the aptamer (http://biophy.hust.edu.cn/new/, accessed on 6 January 2023). The 3D structure file of thiamethoxam molecule was download from the website (https://pubchem.ncbi.nlm.nih.gov/, accessed on 6 January 2023). AutoDock 4.2.6 software was used to predict the docking of aptamer and thiamethoxam molecule, and OpenBabel 2.4.1 software was used to convert the docking result from pdb format to pdbqt format. The possible interaction sites between the aptamer and thiamethoxam molecules were analyzed, and PyMoL 1.7.6 software was used to visualize the molecular docking results. After the docking sites of thiamethoxam and the aptamer were known, bases were truncated at both ends of the sequence (Figure S1A). The Kd of truncated sequence was verified by ITC.

2.7. Thiamethoxam Detection with AuNPs Colorimetric Sensor

The AuNPs colorimetric sensor was constructed with Thi-5R-18 to verify the practicability of the aptamer. The AuNPs were prepared as reported by Zheng et al. [23]. Briefly, 10 μM aptamers (5 μL) was added to each EP tube and incubated with 250 μL of different concentrations of thiamethoxam for 10 min, followed by adding 135 μL of nano particles and incubating for 5 min. After adding 70 μL of 0.25 M NaCl rapidly to the solution and incubating for 5 min, 200 μL of the obtained solution was collected and placed in an enzyme standard plate for 200 nm~800 nm spectral scanning. The linear regression equation was established between the concentration of thiamethoxam and the absorbance value of the mixed system. The specificity of the aptamer was tested with 10^{-8} M of thiamethoxam and 10^{-7} M of dinotefuran, admire, clothianidin, and glyphosate. Meanwhile, the AuNPs colorimetric sensor was used to detect thiamethoxam in real samples by spraying 300 μL of different concentrations of thiamethoxam (20nM, 40nM, 80nM, and 100nM) on tea leaves, followed by drying at room temperature, washing with 300 μL ultrapure water to obtain thiamethoxam solution, and then using the aptasensor to obtain the thiamethoxam concentration by measuring the $A_{650/520}$.

3. Results and Discussion

3.1. Screening of Thiamethoxam Aptamers

In this study, GO-SELEX technology was used to screen thiamethoxam aptamers. During screening, the ssDNA library was first incubated with thiamethoxam to facilitate the combination of thiamethoxam and ssDNA to form a complex. After adding GO, a large amount of ssDNA with no affinity to thiamethoxam was captured on the GO surface through π-π stacking, and only a small amount of target-bound ssDNA remained in the supernatant. After PCR amplification, product recovery, and purification, ssDNA was obtained as a secondary library to enter the next round of screening (Figure 1).

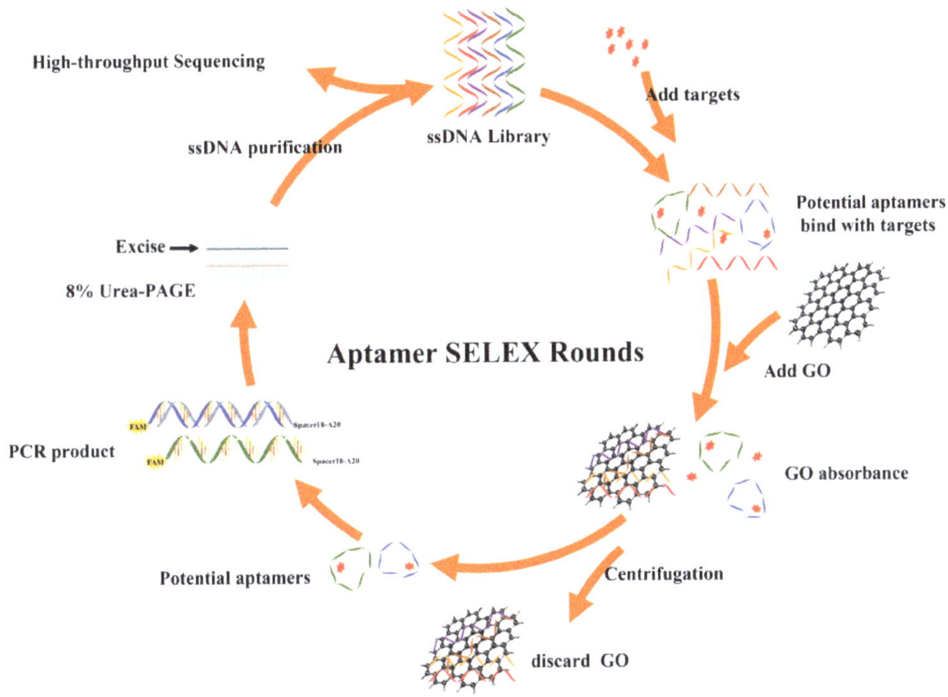

Figure 1. Schematic diagram for screening thiamethoxam aptamers by GO-SELEX.

GO-SELEX was suitable for screening the aptamers of targets which cannot be absorbed by GO. ΔA255 was the difference value between the absorbance value at 255 nm of the supernatant incubated with GO/target and GO/1 × Binding buffer (control). As shown in Figure 2A, after adding GO and the target in the incubation system at the ratio of 5000:1, free target still existed in the solution. Based on the previous report [21] and cost, 500:1 was used as the mass ratio of GO/target.

The key to aptamer screening is to obtain and isolate high affinity binding sequences from random ssDNA libraries. Since different lengths of ssDNA show different binding kinetics on the GO surface [24], the GO/ssDNA mass ratio was optimized before GO-SELEX experiments, enabling GO to fully adsorb ssDNA. Commonly, the GO/ssDNA quality ratio was between 500:1 and 1000:1 [21,25], allowing the ssDNA concentration to reach the baseline level. In Figure 2B, the nucleic acid concentration was seen to decrease with the increase of GO/ssDNA, and at the mass ratio of 500:1, the DNA content in the solution reached the first equilibrium point, indicating that this concentration reached the baseline level, consistent with previous studies [21]. At GO/ssDNA mass ratio was between 500:1~4000:1, although the nucleic acid concentration could reach the baseline level, there was still free ssDNA not bound to the target, which would become non-specific enrichment sequences and existed in each round of the library, thus affecting the screening efficiency. When the mass ratio further increased to 5000:1, the concentration of ssDNA decreased significantly and the content of free ssDNA in the solution decreased to the minimum at the same time. This experiment aimed to compare the difference between pressurized SELEX and ordinary SELEX, so the high pressure GO/ssDNA ratio was selected for the SELEX process. In the subsequent SELEX process, 5000:1 was used as the mass ratio of GO/ssDNA. In this study, a FAM-labeled forward primer and a dA20-spacer18-labeled reverse primer were used to separate the aptamer sequence by 8% Urea-PAGE electrophoresis, with FAM-labeled forward primer PCR product being used in the next round of the aptamer library. In this experiment, 15 rounds of screening were conducted to explore the effect of pressurized GO-SELEX on the round affinity.

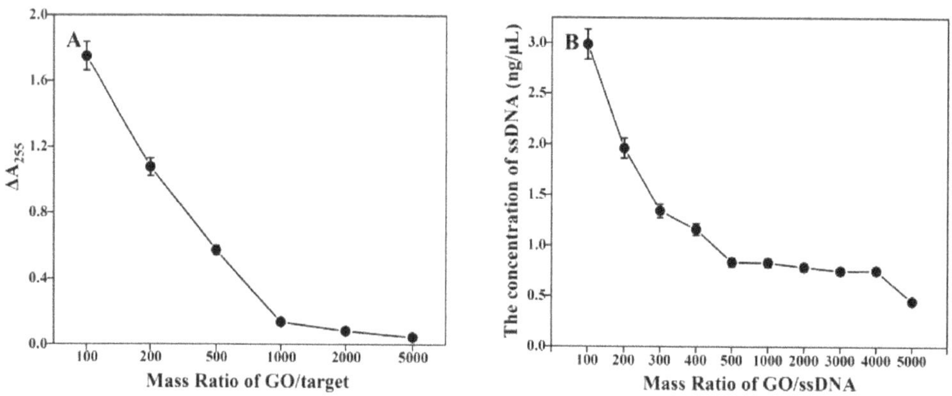

Figure 2. (**A**) Optimization of the mass ratio of GO and target. ΔA255 indicated the difference value between the absorbance value at 255nm of the supernatant (GO/target) and the control (GO/1× Binding buffer). (**B**) Optimization of the mass ratio of GO and ssDNA. The concentration of ssDNA indicated the content of free ssDNA in the supernatant.

3.2. Round Affinity Analysis

The round affinity between aptamer library and target is usually determined by the recovery rate [26], which was expressed by the ratio of ssDNA concentration before incubation and after incubation. Thiamethoxam has a maximum absorption wavelength at 255 nm, close to the maximum absorption wavelength of ssDNA. As the measured ssDNA concentration was inaccurate, this study used an innovative capture method to measure the affinity of each round. Specifically, each round of biotin-labeled aptamer library was fixed onto the magnetic beads with streptavidin magnetic beads, allowing the aptamers to capture the free targets in the solution. The more affinity sequences in a round, the more targets could be captured, and the higher the absorbance at 255 nm. As shown in Figure 3A, the proportion of ssDNA with affinity to thiamethoxam in the library was low at initial screening, and after several rounds of screening, the affinity of ssDNA was gradually enriched. The aptamers obtained from the 5th and 9th rounds had the highest affinity. With the further increase of rounds, the affinity showed no further improvement, and in Figure 3B, the result of 8% Urea-PAGE gel also showed that the brightness of gel bands was unstable from the 10th round. The brightness of gel bands indirectly indicated the DNA content. The quality ratio of GO/ssDNA in the screening process was 5000:1, which was one order of magnitude higher than that in general GO-SELEX experiments, hence the higher screening pressure inferring that the adsorption of GO on ssDNA surface was unstable, which would lead to the loss of affinity sequences.

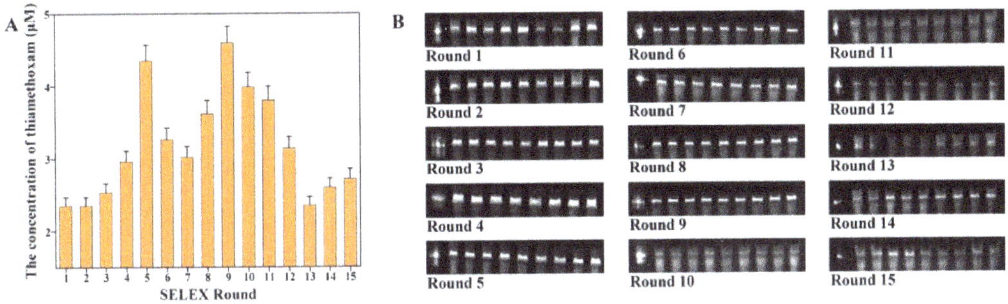

Figure 3. (**A**) The concentration of eluted thiamethoxam in each round. (**B**) Result for each round of 8% urea-PAGE. The corresponding electrophoretic bands from left to right are self-made 87nt marker, FAM-labeled 1~15 round candidate aptamer bands.

3.3. HTS and Secondary Structure Analysis of Sequences

The number of aptamer sequences screened in this study was more than one million and the first 30 sequences with the highest enrichment degree were selected as candidate aptamers in each round, which were named Thi-5R-1~Thi-5R-30 and Thi-9R-1~Thi-9R-30, according to the enrichment degree. Candidate aptamers were selected based on the representative secondary structure (Figure 4) and low free energy sequences (Table 1). Low free energy represented the stability of the secondary structure formed by the sequences. Each candidate sequence of aptamer contained the stem ring structure formed by hydrogen bond connection within the molecule, which was the structural basis for combination with thiamethoxam.

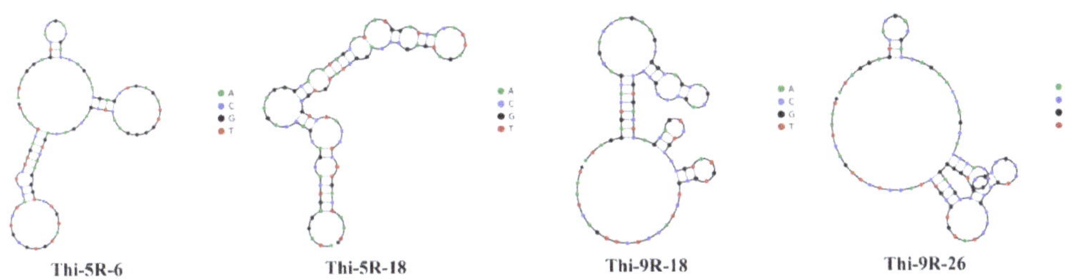

Figure 4. The secondary structures of aptamer candidates.

Table 1. Sequences of aptamer candidates.

Aptamer	40nt Sequences (5′-3′)	Free Energy (kcal/mol)
Thi-5R-6	CGAGCTGAGATTGGGGAACTCGACGACAGTCAAGGGTCTG	−9.78
Thi-5R-18	GGGCAAATAGCATAATGGATCACATTAGATGAGCCCAGGC	−7.4
Thi-9R-18	GCGGCAGCAGCAGCCCGCCCGTGACTCAGCAGTCTGCCCG	−11.4
Thi-9R-26	CCAGCACCCACCGGTGGGGACGCCGCGCCGCCTGCCGCCC	−8.24

3.4. Measurement of Sequence Affinity by ITC

In this study, the affinity of candidate aptamers was evaluated by Kd values calculated through the ITC. ITC could directly measure the heat released in the process of biomolecular binding, and the experimental data were expressed in the form of thermogram, which could accurately obtain the complete thermodynamic information of biomolecular interaction. The potential molecular interaction mechanism could be clarified by analyzing the ITC data. According to the ITC results (Figure 5), the Kd value was 4.935×10^{-5} M for Thi-5R-18 combined with the target to release heat, and the Kd values of two published thiamethoxam aptamers FAM-Thi13 [27] and seq.20 [28] were 3.661×10^{-4} M and 2.115×10^{-4} M, respectively, indicating that the aptamer Thi-5R-18 in this paper had the lowest Kd value and the highest affinity among the three aptamers. Therefore, Thi-5R-18 was used as the aptamer for thiamethoxam detection in the subsequent sensor performance test.

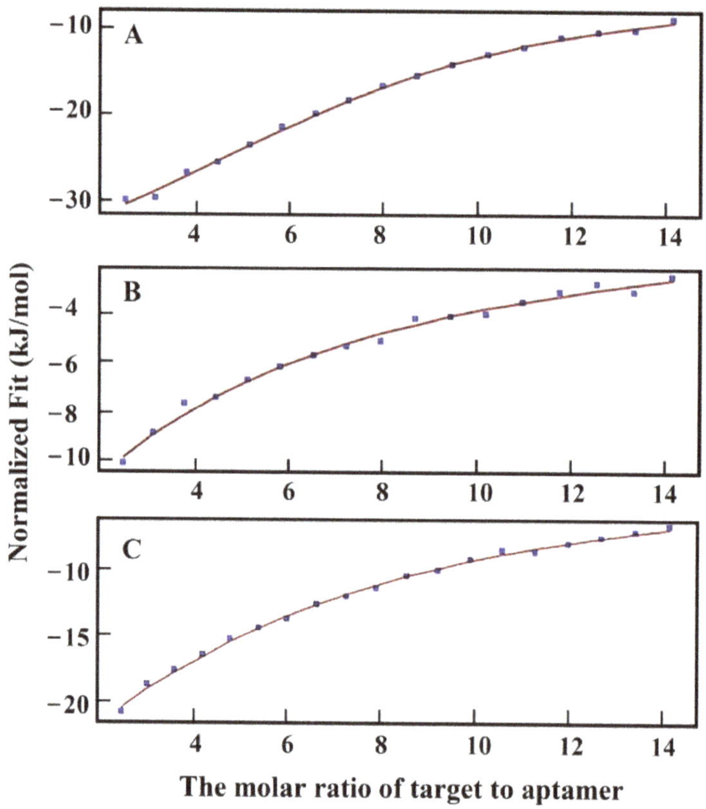

Figure 5. ITC analysis of (**A**) Thi-5R-18 (**B**) FAM-Thi13 (**C**) seq.20 with thiamethoxam.

3.5. The Binding of Aptamer and Thiamethoxam

The binding model of Thi-5R-18 and thiamethoxam was further analyzed by molecular docking. The results of molecular docking (Figure S1B) showed that the small molecule of thiamethoxam and Thi-5R-18 were combined with hydrogen bond and π–π interaction. Thiamethoxam was combined with bases G-15, A-19, and T-71. Further analysis showed that the oxygen atom at the end of G-15 of Thi-5R-18 formed a hydrogen bond with the 1st oxygen atom of thiamethoxam (The bond length was 2.8 Å). The oxygen atom at the end of T-71 of Thi-5R-18 formed a hydrogen bond with the 7th oxygen atom of thiamethoxam (The bond length was 3.7 Å). The benzene ring of thiamethoxam formed π–π bond with the benzene ring of T-71 (The bond length was 4.7 Å). These hydrogen bonds and π–π bond further stabilized the binding of thiamethoxam to Thi-5R-18. Previous studies have shown that binding may also be related to van der Waals force and electrostatic force [27]. A total of 19 bases that included G-15 and A-19 and 17 bases that included T-71 at both ends of the sequence were cut off. Figure S2 showed that the Kd value of truncated aptamer sequence measured by ITC could not be fitted. It was confirmed that after deletion of the binding site, the affinity of the aptamer disappeared.

3.6. Detection of Thiamethoxam by Aunps Colorimetric Sensor

To verify the applicability of the aptamer Thi-5R-18 in thiamethoxam detection, a colorimetric sensor based on AuNPs was constructed with Thi-5R-18 as the recognition element. As displayed in Figure S3, the size of nanoparticles was relatively uniform with an average particle size of 15.90 nm. As previously reported, there was a large amount of positive charge on the surface of AuNPs [29], and when NaCl was added, the Cl^- could destroy the charge balance on the surface of AuNPs, making the nanoparticles aggregate and the solution blue. However, the aptamers adsorbed on the surface of AuNPs through electrostatic adsorption would keep the nanoparticles in a dispersed state, enabling the solution to stay red and the nanoparticles not to aggregate even after adding NaCl (Figure 6A). After adding the target, the targets were combined with aptamers and left on the surface of the AuNPs, allowing the nanoparticles to aggregate and the solution to turn blue after adding NaCl. As shown in Figure 7A, 520 nm was the maximum absorption wavelength of AuNPs, and the aggregation of particles reduced the light absorption value here. After adding the target, the absorbance value declined at 520 nm but increased at 650 nm. In Figure 7B, a good linear correlation was shown between the intensity ratio of light absorption values at 650 nm and 520 nm ($A_{650/520}$) and the concentrations of thiamethoxam. The linear response range of the colorimetric sensor was 5 to 120nM. The linear regression equation was calculated as y = 0.0071x + 0.31533 and the limit of detection (LOD) of the sensor was 0.37 nM (with calculation details shown in supporting information), indicating the high sensitivity of the aptsensor system. Figure 7C showed the specificity test result of the thiamethoxam aptamer based on the colorimetry method. Four pesticides with structures similar to thiamethoxam were detected by the aptsensor system and the binding ability of the aptamer to the four targets was less than one tenth of that of thiamethoxam, although their concentration was ten times higher than that of thiamethoxam, indicating the high specificity of the aptasensor system for thiamethoxam.

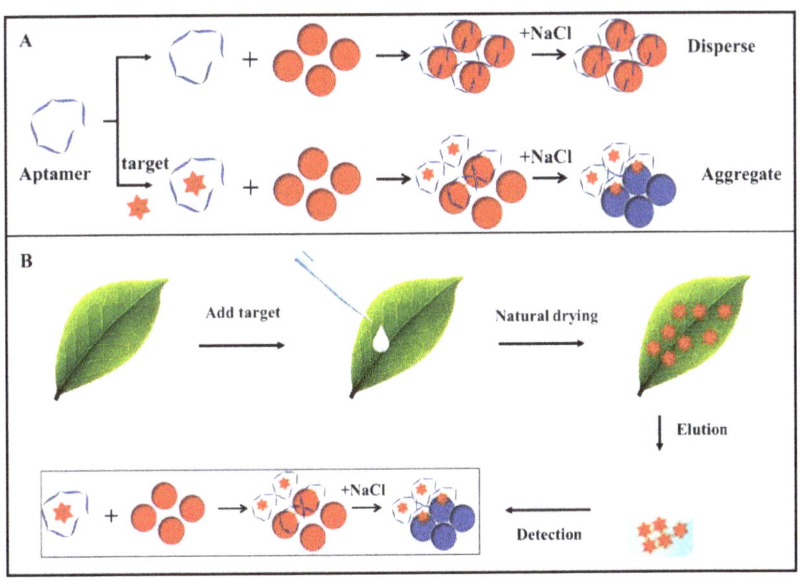

Figure 6. Schematic diagram of (**A**) color development principle of AuNPs and (**B**) surface detection of actual tea leaf samples.

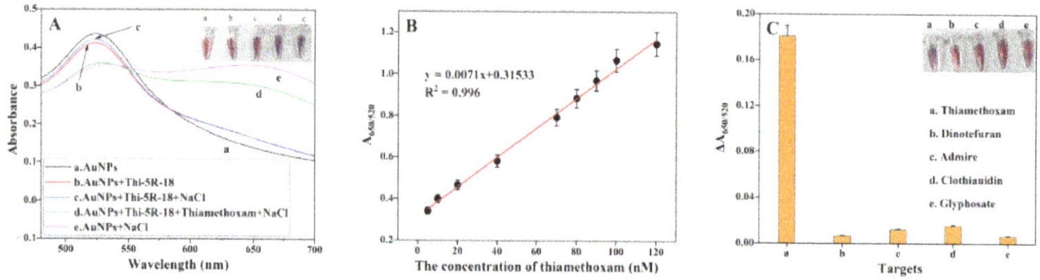

Figure 7. (**A**) UV-vis spectroscopy of AuNPs under different conditions. (**B**)The linear relationship between the value of $A_{650/520}$ and the concentration of thiamethoxam. (**C**) The value of $A_{650/520}$ for different targets.

In order to test the applicability of this aptamer colorimetric sensor in actual samples, the thiamethoxam pesticide was sprayed on the leaf surface and eluted for colorimetric detection after natural drying. The concentration of thiamethoxam after elution was 20, 40, 80, and 100 nM (Figure 6B), respectively. The results (Table 2) showed that the recovery was 96.94%~105.86% and RSD (Relative Standard Deviation) was 0.41%~3.76%, implying that the constructed aptamer colorimetric sensor had good repeatability and reliability.

Table 2. Determination of thiamethoxam pesticide in spiked real samples.

Sample	Spiked Concentration (nM)	Measured Concentration [a] (nM)	Recovery (%)	RSD (%)
1	100	105.40	105.40	1.88
2	80	84.69	105.86	0.41
3	40	38.78	96.94	3.76
4	20	19.61	98.03	0.94

[a] Mean values were calculated from three determinations.

4. Conclusions

In this study, non-immobilized GO-SELEX was used to screen ssDNA aptamers that specifically bind to thiamethoxam. In the screening process, the screening pressure was improved by increasing the additional amount of GO. The 5th and 9th rounds of libraries showed the highest affinity using a capture method. This method was innovatively used to determine the affinity of round libraries, and effectively solved the problem that the maximum absorption wavelength was close between target and ssDNA. Aptamer Thi-5R-18 obtained from the 5th round can specifically bind to the target and its affinity was proved to be the best by ITC, demonstrating that high screening pressure was conducive to shortening screening rounds and reducing experimental cost. The results of molecular docking showed that thiamethoxam and Thi-5R-18 were combined with bases G-15, A-19, and T-71 through hydrogen bonding and π–π interaction, which was also verified by ITC. Moreover, a colorimetric aptasensor based on AuNPs was constructed by the aptamer Thi-5R-18, with the linear regression equation $y = 0.0071x + 0.31533$ and an LOD of 0.37 nM. Furthermore, the aptasensor system was applied to actual sample detection, achieving the recovery of 96.94%~105.86% and the RSD of 0.41%~3.76%, suggesting that the aptasensor could be used for rapid and sensitive detection of thiamethoxam residues.

Supplementary Materials: The following supporting information can be downloaded at: https://www.mdpi.com/article/10.3390/bios13020155/s1, Figure S1: (A) Secondary structures of Thi-5R-18. The base cutting sites were labeled by scissors. G-15, A-19, T-71 were the binding sites. (B) Molecular docking of thiamethoxam and Thi-5R-18. Figure S2: ITC analysis of truncated Thi-5R-18 with thiamethoxam. Figure S3: (A) TEM images of AuNPs. (B) Size distribution of AuNPs.

Author Contributions: Conceptualization, P.W. and Y.Y.; methodology, Y.Y. and D.Z.; software, Y.Y. and P.W.; validation, D.Z. and K.T.; date analysis, Y.Y. and D.Z.; investigation, Y.Y., D.Z. and K.T.; resources, Y.Y., P.W. and P.L.; writing, Y.Y.; proofread, D.Z.; funding acquisition, Z.Y.; supervision, D.N. and F.G.; project administration, P.W. and P.L. All authors have read and agreed to the published version of the manuscript.

Funding: This research was funded by the Sub topic (Program No. 2021YFD1000401) of the National Key Research and Development Program of China (Program No. 2021YFD1000400), national key R&D project "Construction of County Demonstration Area of Tea Industry Chain Integration" (2021YFD1000401), Hubei Provincial Key Water Conservancy Promotion and Application Project "Research and Demonstration of Key Technologies for Tea Processing in Danjiangkou City, Danjiangkou Reservoir Area" (HBSLKJTG2020001), national nature science foundation of China (NO. 32202540) and China Postdoctoral Science Foundation (No. 2022M711277).

Institutional Review Board Statement: Not applicable.

Informed Consent Statement: Not applicable.

Data Availability Statement: Not applicable.

Conflicts of Interest: The authors declare no conflict of interest.

References

1. Maienfisch, P.; Angst, M.; Brandl, F.; Fischer, W.; Hofer, D.; Kayser, H.; Kobel, W.; Rindlisbacher, A.; Senn, R.; Steinemann, A.; et al. Chemistry and biology of thiamethoxam: A second generation neonicotinoid. *Pest Manag. Sci.* **2001**, *57*, 906–913. [CrossRef] [PubMed]
2. Maienfisch, P.; Huerlimann, H.; Rindlisbacher, A.; Gsell, L.; Dettwiler, H.; Haettenschwiler, J.; Sieger, E.; Walti, M. The discovery of thiamethoxam: A second-generation neonicotinoid. *Pest Manag. Sci.* **2001**, *57*, 165–176. [CrossRef]
3. Yamamoto, I.; Tomizawa, M.; Saito, T.; Miyamoto, T.; Walcott, E.C.; Sumikawa, K. Structural factors contributing to insecticidal and selective actions of neonicotinoids. *Arch. Insect Biochem. Physiol.* **1998**, *37*, 24–32. [CrossRef]
4. Henry, M.; Beguin, M.; Requier, F.; Rollin, O.; Odoux, J.F.; Aupinel, P.; Aptel, J.; Tchamitchian, S.; Decourtye, A. A Common Pesticide Decreases Foraging Success and Survival in Honey Bees. *Science* **2012**, *336*, 348–350. [CrossRef] [PubMed]
5. Tosi, S.; Burgio, G.; Nieh, J.C. A common neonicotinoid pesticide, thiamethoxam, impairs honey bee flight ability. *Sci. Rep.* **2017**, *7*, 1201. [CrossRef] [PubMed]
6. Zhu, L.; Li, W.; Zha, J.; Li, N.; Wang, Z. Chronic thiamethoxam exposure impairs the HPG and HPT axes in adult Chinese rare minnow (*Gobiocypris rarus*): Docking study, hormone levels, histology, and transcriptional responses. *Ecotoxicol. Environ. Saf.* **2019**, *185*, 109683. [CrossRef]
7. Goulson, D.; Kleijn, D. REVIEW: An overview of the environmental risks posed by neonicotinoid insecticides. *J. Appl. Ecol.* **2013**, *50*, 977–987. [CrossRef]
8. Han, W.; Tian, Y.; Shen, X. Human exposure to neonicotinoid insecticides and the evaluation of their potential toxicity: An overview. *Chemosphere* **2018**, *192*, 59–65. [CrossRef]
9. Jyot, G.; Singh, B. Development and Validation of an HPLC Method for Determination of Thiamethoxam and Its Metabolites in Cotton Leaves and Soil. *J. Aoac Int.* **2017**, *100*, 796–803. [CrossRef]
10. Suganthi, A.; Nikita, S.A.; Kousika, J.; Bhuvaneswari, K.; Sridharan, S. Determination of thiamethoxam residues in banana stem and fruit through LC-MS/MS. *Environ. Monit. Assess.* **2018**, *190*, 293. [CrossRef]
11. Ting, G.; Gui, F.J.; Jin, X.; Shi, J.G.; Xiao, F.L.; Yu, P.Z.; De, Y.H. Determination of the residue dynamics and dietary risk of thiamethoxam and its metabolite clothianidin in citrus and soil by LC-MS/MS. *J. Environ. Sci. Health B (Pestic. Food Contam. Agric. Wastes)* **2019**, *54*, 326–335. [CrossRef]
12. Ma, H.X.; Xu, Y.J.; Li, Q.X.; Xu, T.; Wang, X.T.; Li, J. Application of enzyme-linked immunosorbent assay for quantification of the insecticides imidacloprid and thiamethoxam in honey samples. *Food Addit. Contam. Part A-Chem. Anal. Control. Exp. Risk Assess.* **2009**, *26*, 713–718. [CrossRef] [PubMed]
13. Ellington, A.D.; Szostak, J.W. Invitro Selection of Rna Molecules That Bind Specific Ligands. *Nature* **1990**, *346*, 818–822. [CrossRef]
14. Tuerk, C.; Gold, L. Systematic Evolution of Ligands by Exponential Enrichment-Rna Ligands to Bacteriophage-T4 Dna-Polymerase. *Science* **1990**, *249*, 505–510. [CrossRef] [PubMed]
15. Saito, S. SELEX-based DNA Aptamer Selection: A Perspective from the Advancement of Separation Techniques. *Anal. Sci.* **2021**, *37*, 17–26. [CrossRef]
16. Ruscito, A.; DeRosa, M.C. Small-Molecule Binding Aptamers: Selection Strategies, Characterization, and Applications. *Front. Chem.* **2016**, *4*, 14. [CrossRef]
17. Chen, A.L.; Yang, S.M. Replacing antibodies with aptamers in lateral flow immunoassay. *Biosens. Bioelectron.* **2015**, *71*, 230–242. [CrossRef]

18. Xu, S.T.; Liu, J.K.; Xue, Y.; Wu, T.Y.; Zhang, Z.F. Appropriate conditions for preparing few-layered graphene oxide and reduced graphene oxide. *Fuller. Nanotub. Carbon Nanostruct.* **2017**, *25*, 40–46. [CrossRef]
19. Park, S.; Ruoff, R.S. Chemical methods for the production of graphenes. *Nat. Nanotechnol.* **2009**, *4*, 217–224. [CrossRef]
20. Cui, L.; Chen, Z.R.; Zhu, Z.; Lin, X.Y.; Chen, X.; Yang, C.J. Stabilization of ssRNA on Graphene Oxide Surface: An Effective Way to Design Highly Robust RNA Probes. *Anal. Chem.* **2013**, *85*, 2269–2275. [CrossRef]
21. Balaban, S.; Man, E.; Durmus, C.; Bor, G.; Ceylan, A.E.; Gumus, Z.P.; Evran, S.; Coskunol, H.; Timur, S. Sensor Platform with a Custom-Tailored Aptamer for Diagnosis of Synthetic Cannabinoids. *Electroanalysis* **2020**, *32*, 656–665. [CrossRef]
22. Bor, G.; Man, E.; Ugurlu, O.; Ceylan, A.E.; Balaban, S.; Durmus, C.; Pinar Gumus, Z.; Evran, S.; Timur, S. in vitro Selection of Aptamer for Imidacloprid Recognition as Model Analyte and Construction of a Water Analysis Platform. *Electroanalysis* **2020**, *32*, 1922–1929. [CrossRef]
23. Zheng, Y.; Wang, Y.; Yang, X.R. Aptamer-based colorimetric biosensing of dopamine using unmodified gold nanoparticles. *Sens. Actuators B-Chem.* **2011**, *156*, 95–99. [CrossRef]
24. Titelman, G.I.; Gelman, V.; Bron, S.; Khalfin, R.L.; Cohen, Y.; Bianco-Peled, H. Characteristics and microstructure of aqueous colloidal dispersions of graphite oxide. *Carbon* **2005**, *43*, 641–649. [CrossRef]
25. Guan, J.; He, K.; Gunasekaran, S. Selection of ssDNA aptamer using GO-SELEX and development of DNA nanostructure-based electrochemical aptasensor for penicillin. *Biosens. Bioelectron. X* **2022**, *12*, 100220. [CrossRef]
26. Lee, B.H.; Kim, S.H.; Ko, Y.; Park, J.C.; Ji, S.; Gu, M.B. The sensitive detection of ODAM by using sandwich-type biosensors with a cognate pair of aptamers for the early diagnosis of periodontal disease. *Biosens. Bioelectron.* **2019**, *126*, 122–128. [CrossRef]
27. Luo, Y.; Jin, Z.; Wang, J.; Ding, P.; Pei, R. The isolation of a DNA aptamer to develop a fluorescent aptasensor for the thiamethoxam pesticide. *Analyst* **2021**, *146*, 1986–1995. [CrossRef]
28. Kong, Q.; Yue, F.; Liu, M.; Huang, J.; Yang, F.; Liu, J.; Li, J.; Li, F.; Sun, X.; Guo, Y.; et al. Non-immobilized GO-SELEX of aptamers for label-free detection of thiamethoxam in vegetables. *Anal. Chim. Acta* **2022**, *1202*, 339677. [CrossRef]
29. Maye, M.M.; Han, L.; Kariuki, N.N.; Ly, N.K.; Chan, W.B.; Luo, J.; Zhong, C.J. Gold and alloy nanoparticles in solution and thin film assembly: Spectrophotometric determination of molar absorptivity. *Anal. Chim. Acta* **2003**, *496*, 17–27. [CrossRef]

biosensors

Perspective

Highly Sensitive Flexible SERS-Based Sensing Platform for Detection of COVID-19

Seyyed Mojtaba Mousavi [1], Seyyed Alireza Hashemi [2], Vahid Rahmanian [3], Masoomeh Yari Kalashgrani [4], Ahmad Gholami [4,*], Navid Omidifar [5] and Wei-Hung Chiang [1,*]

1 Department of Chemical Engineering, National Taiwan University of Science and Technology, Taipei City 106335, Taiwan; kempo.smm@gmail.com
2 Nanomaterials and Polymer Nanocomposites Laboratory, School of Engineering, University of British Columbia, Kelowna, BC V1V 1V7, Canada; s.a.hashemi0@gmail.com
3 Centre of Molecular and Macromolecular Studies, Polish Academy of Sciences, Sienkiewicza 112, 90-363 Lodz, Poland; vahid.ra@cbmm.lodz.pl
4 Biotechnology Research Center, Shiraz University of Medical Science, Shiraz 71468-64685, Iran; masoomeh.yari.72@gmail.com
5 Department of Pathology, School of Medicine, Shiraz University of Medical Sciences, Shiraz 71468-64685, Iran; omidifarn@sums.ac.ir
* Correspondence: gholami@sums.ac.ir (A.G.); whchiang@mail.ntust.edu.tw (W.-H.C.)

Abstract: COVID-19 continues to spread and has been declared a global emergency. Individuals with current or past infection should be identified as soon as possible to prevent the spread of disease. Surface-enhanced Raman spectroscopy (SERS) is an analytical technique that has the potential to be used to detect viruses at the site of therapy. In this context, SERS is an exciting technique because it provides a fingerprint for any material. It has been used with many COVID-19 virus subtypes, including Deltacron and Omicron, a novel coronavirus. Moreover, flexible SERS substrates, due to their unique advantages of sensitivity and flexibility, have recently attracted growing research interest in real-world applications such as medicine. Reviewing the latest flexible SERS-substrate developments is crucial for the further development of quality detection platforms. This article discusses the ultra-responsive detection methods used by flexible SERS substrate. Multiplex assays that combine ultra-responsive detection methods with their unique biomarkers and/or biomarkers for secondary diseases triggered by the development of infection are critical, according to this study. In addition, we discuss how flexible SERS-substrate-based ultrasensitive detection methods could transform disease diagnosis, control, and surveillance in the future. This study is believed to help researchers design and manufacture flexible SERS substrates with higher performance and lower cost, and ultimately better understand practical applications.

Keywords: flexible SERS substrates; coronavirus; ultrasensitive detection; biomarker

Citation: Mousavi, S.M.; Hashemi, S.A.; Rahmanian, V.; Kalashgrani, M.Y.; Gholami, A.; Omidifar, N.; Chiang, W.-H. Highly Sensitive Flexible SERS-Based Sensing Platform for Detection of COVID-19. *Biosensors* **2022**, *12*, 466. https://doi.org/10.3390/bios12070466

Received: 17 May 2022
Accepted: 21 June 2022
Published: 28 June 2022

1. Introduction

Infectious diseases such as coronaviruses (CoV), especially coronavirus disease 2019 (COVID-19), infect and kill millions of people worldwide [1,2]. These viruses are transmitted by individual contact with a contaminated surface and subsequent contact with the mouth or nose, as well as by inhalation of tiny droplets exhaled by infected individuals when they cough or sneeze [3]. Individuals who have influenza present with clinical features of symptomatic COVID-19 infection, including dyspnea, cough, and fever, which can lead to medical complications such as kidney damage and pneumonia. Therefore, a complete diagnosis with sensitive and accurate analytical methods or rapid diagnosis is required for optimal patient therapy [4–6]. Significant advances in surface-enhanced Raman spectroscopy (SERS) as a suitable point-of-care testing (POCT), since its introduction in 1973, have been demonstrated for a number of analytes, including viruses [7,8]. The complete molecular composition of microorganisms, bacteria, and viruses is reflected in the Raman spectrum, which includes many different vibrational states of reliable classifiers (proteins, carbohydrates, DNA/RNA, and lipids) as well as nonspecific constituents of species (such as carotenoids). Thus, Raman spectra contain both genetic and phenotypic signatures of the studied microorganisms, bacteria, and viruses, since all cellular components are based on the expression of different parts of the genome [9,10]. One of the features of surface-enhanced Raman spectroscopy (SERS) is the identification of characteristic peaks of the outer-membrane proteins of molecules in the sensitive and rapid detection of various molecules, including viruses such as coronavirus [11]. Other features of SERS for coronavirus detection include the use of electromagnetic-field enhancement, by exciting local-surface plasmon resonances in nanostructured-metal surfaces such as gold or silver [12]. The detection of analytes at very low concentrations, and the performance of assays without pretreatment are among the capabilities of SERS. When used in conjunction with an immunoassay, the technology provides an extremely high level of specificity [13,14]. In addition, multiplex SERS immunoassays can detect a variety of analytes, which significantly increases the versatility of the technique compared to the polymerase chain reaction (PCR), which can only analyze materials containing genetic material [15,16]. Therefore, the most promising option for faster detection of COVID-19 than PCR is the use of surface-enhanced Raman spectroscopy (SERS). The PCR protocol amplifies single-stranded DNA to 100 billion copies after 40 cycles of doubling, to achieve a sufficient fluorescent-signal strength for virus detection, which takes hours. Identification of unique gene sequences and single nucleotide polymorphisms by SERS-amplified signals provides more accurate and cost-effective diagnosis, while DNA amplification from COVID-19 does not require a long time [17].

Among the advantages of SERS over other COVID-19-detection methods is the identification of gene sequences through the commercial availability of multiple Raman dyes, leading to the development of nonoverlapping gene-detection probes. SERS tags in ultrasensitive detection also include antibodies, aptamers, and DNA; the detection of proteins, DNA, and other components is based on SERS tags [18,19]. Other advantages of SERS include the speed of analysis, emergence of robust and commercially available Raman spectrometers, simplicity of sample manipulation, and on-site detection of analytes [20]. Limitations of the SERS method include the need for close contact between the analyte and the amplifier surface, degradation of substrates over time that reduces the signal, limited reuse of substrates, problems with homogeneity and reproducibility of the SERS signal in a substrate, and limited substrate selection for a given analyte [21]. In recent years, the SERS-based ultrasensitive sensing platform has emerged as the most promising option for the detection of selective analytes at the nanostructured level, due to its specificity, high sensitivity, nondestructive sensing, and narrow linewidth. The presence of a target analyte near a metal surface leads to SERS events. Adsorbed targets, incident light, and metal nanoparticles are three important factors in optimizing the SERS measurement model for Raman signals from adsorbed targets. resulting in several great reviews and monographs [22]. Moreover, the composition, size, and morphology of the plasmonic metal nanostructures allow for the configuration of a wide range of optimization parameters. Currently, the design of the SERS platform is guided by the visual application of some commonly accepted rules. Gold and silver metals are widely used to make nanoparticles because the surface transitions cause localized surface-plasmon resonance in the visible region of the electromagnetic spectrum. Gold is used because of its chemical stability, oxidation resistance, and ability to be functionalized with more organic materials. Plasmon resonance of silver nanoparticles occurs at lower wavelengths and with greater intensity than gold nanoparticles. Silver nanoparticles have stronger peaks than gold and are, therefore, more sensitive to the refractive index of the environment [2,23]. This means that silver is the most advanced of the plasmonic materials, although gold is more versatile in practice due to the greater variety of available nanoparticle shapes and higher chemical stability. A simple and effective parameter to optimize SERS is the particle density in SERS substrates [24,25].

The aim of this review study was to review recent advances in the fabrication of highly sensitive flexible-SERS substrates for COVID-19-detection work in this field. That is in addition to techniques for resonance amplification and SERS spectroscopy. Numerous researchers have thoroughly investigated the processes behind the benefits of SERS, high-quality nanostructures for SERS-based detection, the principles of SERS and amplification mechanisms, analyte detection, multiplex analysis for coronavirus detection, and SERS COVID-19 detection, all of which were evaluated.

2. Techniques for Resonance Enhancement

When the excitation laser resonates with the electronic transmission of the analyte, the surface-enhanced resonance Raman-scattering (SERRS) effect occurs, as noted by Yuan et al. [26]. Molecular-resonance excitation can increase the effective cross-sectional area of the target analytes by many orders of magnitude. Numerous articles do not distinguish between SERS and SERRS, although both have different properties. Compared with SERS, the SERRS mechanism exhibits a significant improvement. Moreover, the intensity of SERRS is wavelength-dependent. SERRS spectroscopy has evolved to the point where it can identify single resonant molecules, as both the molecular-resonance effect and the surface-plasmon-resonance (SPR) enhancement contribute to the generation of exceptionally strong Raman-scattering signals [27]. Excellent candidates for SERRS research are fluorescent protein molecules (such as proteins, rhodamine 6G (R6G), malachite green, antibodies, purple crystals, etc.) or luminescent dyes due to large Raman-scattering cross-sections and intense optical absorption in the visible region. The single molecule can be uniquely identified by SERRS spectroscopy through initial observations in the visible window. Matching the broad adsorption range of the surface-plasmon resonance (SPR) of aggregated silver colloids, with stimulation by Raman-scattering resonance, is due to the most commonly used substrate, as shown in Table 1. Non-invasive photothermal imaging has gained considerable attention in recent years, due to the cross-sectional area of massive absorption of nanomaterials in the near-infrared region. Despite the minimal energy loss during penetration, NIR light using heat-generating nanomaterials is a good choice for targeting cells, because it does not damage adjacent normal tissue. For example, the fabrication of unique gold nanorods with SERS coding (SPR band, 790 nm) shows high optical absorption at 810 nm for photothermal heating, although it does not fully match the NIR-excitation source at 785 nm [28].

Table 1. Examples of SERS substrates based on their characteristics.

SERS Substrate	Enhancement Factor (EF)	Analyte	Limit of Detection (LOD)	Ref.
Ag nanoparticles/microporous silicon	-	rhodamine 6G	10^{-9} M	[29]
Au nanothorns/macroporous silicon	10^8	crystal violet	10^{-12}–10^{-15} M	[30]
Ag nanoparticles/silicon nanopillars	10^{11}	acetone	0.0037 ng	[31]
Au nanoparticles/mesoporous silicon	-	benzenethio	10^{-6} M	[32]
Ag nanoparticles/mesoporous silicon	2.8×10^8	p-thiocresol	5.2×10^{-9} M	[33]

2.1. SERS Spectroscopy

Surface-enhanced Raman spectroscopy is an extended form of Raman spectroscopy, in which metal nanostructures are used to enhance Raman scattering. In this case, the efficiency of Raman scattering for molecules adsorbed on metal nanostructures increases dramatically, from 10^{12} to 10^{14}. A small number of scattered photons are accessible for detection due to the intrinsic weakness of Raman signals, particularly when excited by visible light. Utilizing surface-enhanced Raman scattering is one way to increase weak Raman signals (SERS). SERS employs nanoscale, roughened metal surfaces consisting of gold (Au) or silver (Ag), The schematic representation of SERS can be found in Figure 1. Therefore, surface-enhanced Raman spectroscopy (SERS) is currently a widely used optical tool for the analysis of the molecular components of chemicals and biological samples, with the potential to detect single molecules [34]. The overlap of the excitation wavelength of the Raman laser with the resonance wavelength of the plasmon structure has made it possible to use SERS spectroscopy in sensors and various laboratory analyses. This is because the resonance of the substituted surface plasmons plays an important role in electromagnetic amplification [35,36]. Therefore, the preparation of colloids and metal substrates with the ability to generate and maintain plasmonic effects is of great interest [37,38]. Therefore, the results of SERS spectroscopy depend on the performance of the substrate in signal amplification and its reproducibility [39]. Among the various methods to fabricate a suitable substrate, electrostatic self-assembly of plasmonic nanoparticles, especially gold and silver, on a functionalized substrate is an easy way to obtain a uniform, reproducible, and cost-effective substrate. Different plasmonic nanoparticles can be used depending on laboratory conditions. However, reports show that silver nanoparticles provide the highest Raman-signal enhancement in the visible-wavelength range [16,40]. Considering the influence of nanoparticle size on Raman-signal amplification [41], it is important to choose an appropriate method to prepare nanoparticles of the desired size. Therefore, to prepare silver nanoparticles, a better chemical method was used to obtain nanoparticles of desired size for the SERS substrate. Various probes are used for SERS, among which pigment molecules are the best and most commonly used [42]. The reason for this is the structure of the pigments, which allows them to bind to the surface of the metal nanoparticles. All pigments are water-soluble, so they are suitable for use in the colloids of nanoparticles synthesized in water, as a better chemical process. Moreover, these structures can be used in solid substrates such as self-assembling substrates. On the other hand, the pigments produce very intense background fluorescence at the visible wavelength of the laser, which interferes with the Raman signals. Numerous factors, such as the type of pigment, the distance between the nanoparticles and the pigment molecules, and the dimensions of the substrate nanoparticles used, affect the Raman spectrum obtained [43,44]. By using a suitable substrate, the Raman signal can be enhanced, and the fluorescence effects of the pigments can be eliminated [45].

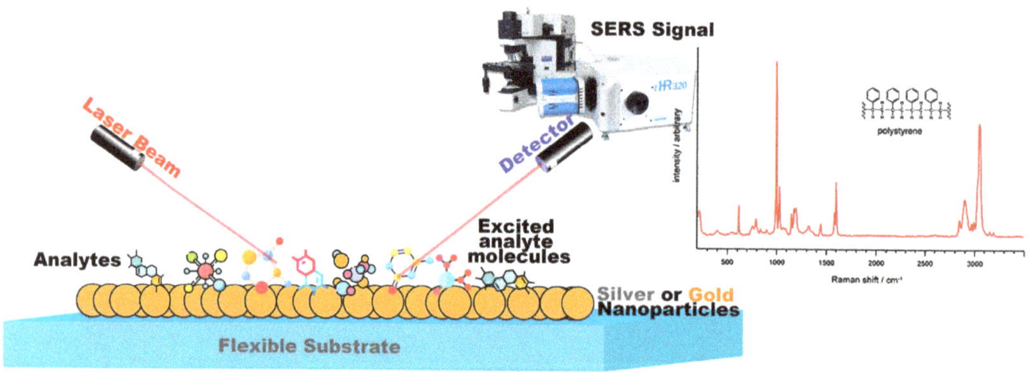

Figure 1. The Raman signal can be amplified further when the roughened metal surface is used in combination with laser light that is matched to the absorption maxima of the molecule. This effect is known as surface-enhanced resonance Raman scattering (SERS), which is shown in the figure.

2.2. High-Quality Nanostructures for SERS-Based Detection

The SPR-enhancement property, which is very important for ultrasensitive diagnosis, depends strongly on the type of nanoparticles used. Variations in the properties of the nanostructures, including composition, stability, shape, dielectric parameters, size, and surface modification, contribute significantly to the generation of SPR-induced EM enhancement. In recent decades, various types of prominent nanostructures have been designed and developed for SERS-based ultrasensitive diagnostic studies, including aggregated nanostructures, self-assembled nanostructures, uniform-plasmonic nanostructures, alloyed nanostructures, magnetic nanostructures, carbon-based materials, three-dimensional X-ray structures, silicon-based nanomaterials, and metal–organic frameworks [46]. Nanostructured materials often exhibit new physical and chemical properties relative to the bulk state. Due to the properties of nanomaterials and their special applications, various types of metal nanostructures, including nanoparticles, single nanoparticles, composite nanoparticles, and nanoparticle clusters, are used in SERS. Usually metal nanoparticles of gold, silver, and copper are used as active ingredients in SERS to detect COVID-19, due to their plasmonic peak in the visible region. These metal nanoparticles provide localized surface-plasmon resonance, which creates a large localized electromagnetic field for the detection of COVID-19. Metal nanoparticles with different sizes and shapes create distinct LSPR properties and dramatically increase the effect of SERS [47].

To get a better idea of the reinforcement properties, the SERS-reinforcement properties of the different nanostructures are discussed in more detail in the following sections. Aggregated nanostructures serve as initial sensing platforms for ultrasensitive detection based on SERS. Usually, most substrates used for simple silver aggregates are prepared by boiling silver nitrate with sodium-citric acid (Lee-Meisel technique) [48]. Ideally, the broad absorption range of nanoaggregates under laser irradiation is in the size range of 10–150 nm, to obtain and enhance significant metal SPR for single-molecule detection. The SPR-adsorption band of silver nanoparticles has a significant impact on the generation of SERS effects, when used to efficiently activate metallic SPR. Activation of silver colloids by the analytes themselves occurs through the modulation of colloidal aggregation states, when the target analytes contain active atoms such as nitrogen or sulphur. Salt-induced aggregation (sodium chloride or potassium chloride) is able to solve the problem, when the target molecules are unable to produce colloidal aggregation due to the altered surface charge [49].

SERS-based detection of single molecules showed that the sensitivity of detection by addition of activation solutions strongly depends on factors such as random adsorption sites (hot spots) of the nanoaggregates. Recently, it was reported that an ultrasensitive method for the detection of ozone was developed using accumulated gold nanoparticles as SERS substrate [50,51].

2.3. Principles of SERS and Enhancement Mechanisms

Theoretically, when a molecule is engraved on the surface of gold, silver, or other precious metals, the spectrum of the adsorbed molecule is reduced, resulting in a relaxation of Raman's selection rules; as a result, more frequencies are seen than in conventional Raman spectroscopy, as shown in Figure 2. The classical theory of optical scattering provides a qualitative concept for the SERS process. Since the most important feature of metal nanoparticles is their optical property, which is not reflected in the size of their mass., metallic nanoparticles typically absorb and scatter light at different frequencies depending on their size, shape, and material. Surface plasmons are excited in small metal particles, but on smooth metal surfaces it is not possible to excite surface plasmons directly. This phenomenon is the best effect in observing the light adsorption of SERS [52]. Among the auxiliary factors in SERS' optical absorption during COVID-19 diagnosis, we can mainly mention five factors: size, shape, constituents of nanoparticles, interparticle distance, and refractive index of the nanoparticle environment. From the optical point of view, the results of increasing polarization are very important. One of them is the simultaneous increase in scattering efficiency and light absorption by the metal nanoparticles [53]. The incident light beam induces an oscillating dipole (μ) in the particle. The particle scatters the light with oscillating bipolar frequencies. The bipolar moment is generally composed of different coordinated-frequency components. Due to the intensity of the Raman spectrum induced, which is proportional to the quadratic moment, increasing the polarization of the molecule (molecular effect) and increasing the external field acting on the molecule (field effect) are two ways to increase the Raman spectrum [54]. Theoretical models also provide for two main types of excitation: (1) irradiating matter molecules with light, i.e., applying the electric field of light (external field) to the molecule, a larger alternative field is generated near the metal surface due to electromagnetic resonance, which is called the electromagnetic effect. (2) Another method to enhance the Raman spectrum is the chemical effect (molecular effect), in which the polarizability of the molecule is enhanced by the interaction between the molecular surface and the metal surface [55]. The effect of synthesis methods on the ultra-responsive detector ensures that the molecules desired for the detector can be attached to the surface of the metal substrate or at least placed in its vicinity. Therefore, Raman-signal amplification is provided by pulsed resonances in the metal substrate [56].

In the electromagnetic effect, the irradiation of the metal surface with incident light excites the embedded surface plasmons. By tuning the frequency of the incident light with the plasmon frequency (ωp) of the metal, the electric field is enhanced and maximized. Thus, the increase in intensity of the Raman spectrum for the species adsorbed on the metal surfaces is due to the enhancement of an electric field near the metal surface. For scattering to occur, the plasmon oscillations must be perpendicular to the surface. If the oscillations are parallel to the surface, scattering will not occur. For this reason, uneven surfaces or nanoparticle arrays are often used in SERS experiments, because these surfaces create a region where alternating group oscillations can occur [57]. Incident light on the surface can cause a variety of phenomena on the plane. If the surface is such that its properties, such as dimensions and smoothness, are smaller than the wavelength of light, then only the contribution of dipole radiation is considered, and the dipole set contributes to the plasmon oscillations, which enhances the Raman spectrum. The complexity of the SERS effect is due to the fact that the field amplification occurs twice. First, the field enhancement increases the intensity of the incident light, which excites the Raman modes of the molecule under study. This increases the Raman-scattering spectrum. The Raman spectrum itself is enhanced by a process similar to the process of increasing the intensity of the incident light, further increasing the Raman output spectrum. In both phases, the electric field is enhanced by E^2, so that the final amplitude of the spectrum is proportional to E^4 [58,59]. The amplitude of the Raman spectrum is not the same for all frequencies. At frequencies where the Raman spectrum has a small shift with respect to the incident light, both the incident laser light and the Raman spectrum may be close to the plasmonic frequency resonance, enhancing E^4. However, when the shift is large, the incident light and Raman spectrum cannot fluctuate with ωp, so the gain in both phases cannot be maximal [60]. The material from which the metal surface is selected is determined by the resonant frequency of the plasmon. Visible and near-infrared light is used to evoke Raman modes, and is usually of gold and silver for the desired metal surface used in the SERS test. Since the plasmon-resonance frequencies of gold and silver are in this wavelength range and can amplify the electric field of visible and near-infrared light, the absorption spectrum of copper is also in the range that can be used in SERS experiments [61]. Platinum and palladium nanostructures also have plasmonic resonance in the near-visible and infrared regions [62]. Although the electromagnetic model or the electromagnetic effect can explain several properties associated with SERS, it cannot express some specific properties of SERS. When using the electromagnetic model, the chemical structure of the analytical species is not investigated, so the model or chemical effect is also considered. One of the important processes that increases molecular polarization is the mechanism of charge transfer or the chemical effect between the metal and the material (analyte) adsorbed on the metal. The chemical effect occurs simultaneously with the electromagnetic effect, but is used only in special cases [63].

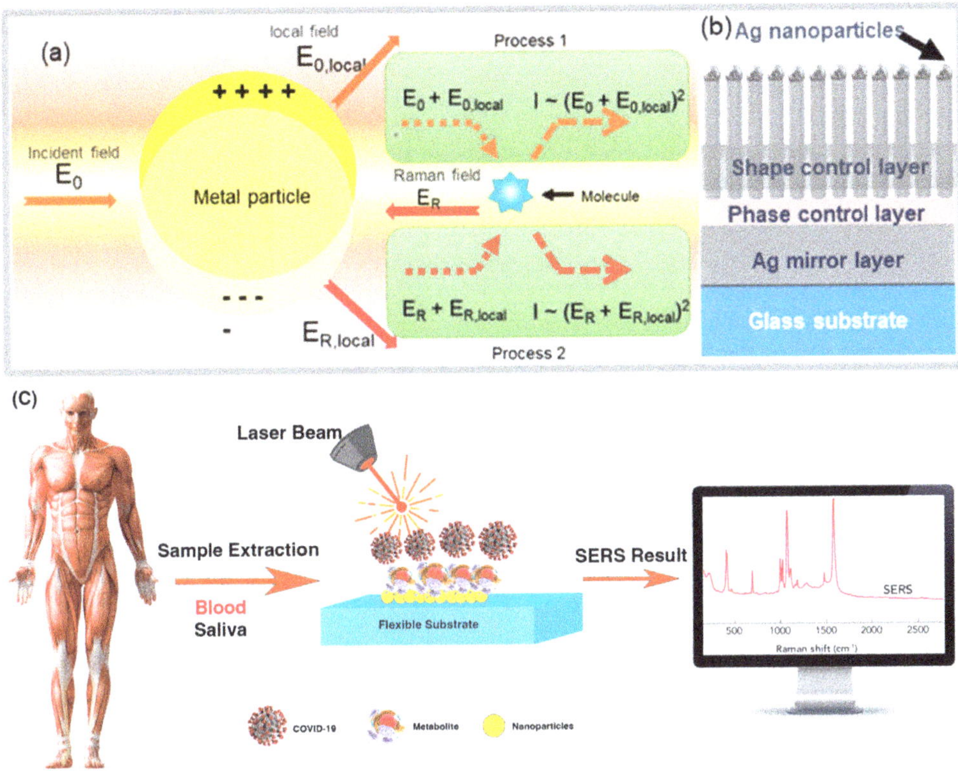

Figure 2. Schematic shows (**a**) mechanism of the surface−enhanced Raman scattering (SERS) electromagnetic (EM) effect, electromagnetic SERS enhancement, and (**b**) multilayer thin−film "local plasmon resonators". Reprinted with permission [64]. Copyright © 2020 American Chemical Society. (**c**) Principles of COVID−19 detection from biopsies with SERS.

2.4. SERS Measurements

Dry or wet measurements of SERS on solid substrates are possible. The dry method is the most common and is perform after the sample has evaporated from the substrate. By immersing the substrate in the sample for a certain period of time (the incubation process), the evaporation process is initiate. Methods of wet experiments include placing a drop of the liquid sample on the substrate SERS or using complex microfluidic systems lab-on-a-chip surface-enhanced Raman spectroscopy (LoC-SERS) on a glass substrate and covering it with a thin coverslip. They can also be perform using the "droplet approach", where the signal is detected at the edge of a sample droplet on the SERS substrate. Therefore, these approaches affect the final results of SERS due to their drawbacks. When the sample evaporates, the absorption process is abruptly stope, and there is heterogeneity in the generated layer or the "coffee ring" effect. In addition, the metallic nanoparticles may oxidize or dissolve during incubation, resulting in the loss of the perpendicular arrangement of molecules in Raman observations. Inefficient Raman signals arise from the use of different interfaces or the effect of different refractive indices, when a coating glass is used in a liquid environment [65]. The "droplet approach" has the advantage that a suitable change in the hydrophobicity of the substrate can lead to a pre-concentration of the analyte, bypassing diffusion limitations and enhancing the SERS signal. However, for recording measurements by a camera integrated into a Raman spectrometer, it is important to create an ideal distance from the droplet boundary to its center. This distance must be sufficient, to prevent the sample from evaporating at the intended location throughout the experiment, but not so short that the substrates do not respond well [66].

3. Flexible SERS Substrate

The best flexible SERS substrates are simply those that support the strongest plasmon resonances; in other words, those that provide the strongest amplification or gain. Metallic nanostructures can amplify the optical field (hotspots) and act like an antenna. When molecules or nanoparticles are in the hotspot, Raman scattering increases dramatically [67]. Most flexible SERS substrates are designed for excitation in the visible infrared range (about 400–1000 nm), the usual range for molecular Raman-scattering experiments. Structures with dimensions below the wavelength range and typically less than 100 nm are the most suitable flexible substrates for SERS. Generally, a good upgrade rate is obtained from structures made of gold or silver. These two metals, gold and silver, are the most commonly used metals in SERS and plasmons [68]. This is due to their good optical properties for the realization of plasmonic resonances in the visible infrared range (400–1000 nm), which is the range of interest for SERS. In principle, there are no limits to the small size of the metal components that make up the flexible SERS substrate. For example, a simple, rough metal surface can also be used as a flexible SERS substrate. However, much less enhancement is achieved than with conventional metal nanostructures. SERS can be measured on structures made of a variety of metals, such as copper or platinum, but even in this case, less enhancement is achieved than with the commonly used metals, gold and silver. Enhancement is not the only important property of a flexible SERS substrate. Among the various aspects, the surface area and surface roughness are of particular importance. SERS is a surface-spectroscopy method, and it is obvious that the surface properties of the flexible substrate play an important role. Surface roughness is an important and crucial factor in SERS experiments and increases the surface sensitivity, resulting in an inelastic scattering of light [69], because at low concentrations it is limited by the inherent strength of the signals from the molecules SERS. However, in situations where the molecules are bound to the flexible substrate by direct contact with the metal, the maximum Raman signal available is limited by the maximum number of molecules in that layer. If the molecule in question is a weak Raman scatterer and the maximum achievable SERS signal is too low, the SERS signal cannot be measured. There are several ways to avoid this problem: using a flexible substrate with higher average gain (which increases the average SERS signal of each molecule), using a flexible substrate with larger surface area (which increases the number of signaling molecules), and increasing the laser power. In addition to these basic features, the simplicity and cost of sample fabrication and preparation should also be considered. The interactions between the flexible substrate and probe play an important role in SERS, so it is useful to use special flexible substrates (or even substrates designed specifically for this molecule) for a certain type of molecules [70]. Flexible SERS substrates can be experimentally divided into three main categories: (1) metal particles (usually nanoparticles) in solution, such as colloidal solutions [71]; (2) flat metal structures; e.g., arrays of metal nanoparticles lying on a flat surface (e.g., glass, silicon, or metal) [72]; (3) metal electrodes [73]. With these interpretations, it can be said that a flexible SERS substrate is a kind of naming for any structure that provides plasmon resonance, which can lead to the development of suitable amplifications in the Raman range and can include metal particles in colloidal solutions; two-dimensional metal structures, such as an array of metal nanoparticles lying on a two-dimensional substrate of glass, silicon, or metal; or metal electrodes [74]. Table 2 shows a comparison of the SERS enhancement of a normal substrate with a flexible substrate.

Table 2. Comparing the SERS enhancement of a normal substrate with a flexible substrate.

Types of Flexible SERS Platforms	Compositions	Unique Features including Flexibility	Detection Limit	Fabrication Methods	Applications	Ref.
In situ detection for SERS	Ag NPs	Floating metal film	10^{-11} m (4-ATP)	One-step electronic reduction	Liquid-phase detection	[75]
	Ag/Au nanowires	3D cross-point nanostructures	10^{-11} m (R6G)	Nanotransfer printing	Glucose detection	[76]
Actively tunable SERS	Au NPs	An open-to-closed system	-	Cast method	Bio-macromolecules' detection	[77]
	Wrinkled graphene/ Au NPs	50% tensile strain without performance degradation	10^{-9} m (R6G)	Graphene transfer/ physical deposition	Multiple analytes' detection	[78]
Swab-sampling approach	Au NPs	"Sticky" feature	0.24 ng cm^{-2} (Thiram)	Drop-dry method	Pesticide residues' detection	[79]
	Ag NWs	High capture capability of pesticides	40.2 ng cm^{-2} (PQ)	Mixing and vacuum filtration	Onsite residual-pesticide detection	[80]

4. Detection of Analytes

SERS can be used for direct or indirect detection of analytes. Adsorption of elements, such as antibodies, aptamers, or similar molecules stabilized on nanostructured surfaces, the use of molecular linkers held close enough to the substrate, or adsorption of analytes on the substrate, are direct measurement requirements. This approach is suitable for analytes with a large cross section for Raman scattering, as shown in Figure 3. It benefits from the thorough control and precision of the quantification procedure, as well as the ability to identify and chemically characterize the analyte by examining its binding properties [81,82]. By correlating changes in the spectrum of the SERS metabolites, reaction products, or reporter molecules (RM), indirect-detection measurements of the concentration of the target analyte are made. This approach allows for the detection of analytes with few or no Raman-vibrational modes as well as for multiplex detection. The most common approach for indirect detection in biological samples is the use of reporter molecules. In this approach, substrates are functionalized with one or more molecules (monoplex or multiplex detection) that undergo a change in Raman cross section, upon contact with the target analyte. RMs are usually tiny molecules with large Raman cross sections. Not only are they photochemically stable, but only a small number of their bands overlap with the matrix or analyte, due to limitations in the Raman spectrum [83].

4.1. Determination of Viruses by SERS

The combination of reporter molecules (indirect detection) with a sandwich immunoassay is a more sensitive and specific method for virus detection by SERS. In this way, the decision technique is usually composed of the following elements: (1) a SERS tag consisting of a Raman-reporter molecule and an identification element consisting of a specific antibody (detection antibody) entrapped on the active SERS nanoparticles, and (2) a template called the adsorption substrate (which does not have to be a metal surface) functionalized with a binding antibody (an adsorption antibody) to bind the antigen SERS-tag complex. Using the SERS tag, quantification is performed by monitoring the Raman signal from the reporter molecule, before and after the interaction with the absorbing element. The limited specificity and inherent limitations of the direct detection of biological analytes, such as the low affinity for common noble-metal SERS substrates, small scattering cross sections, and large molecule sizes, can be overcome with this technique, as shown in Figure 4 [84].

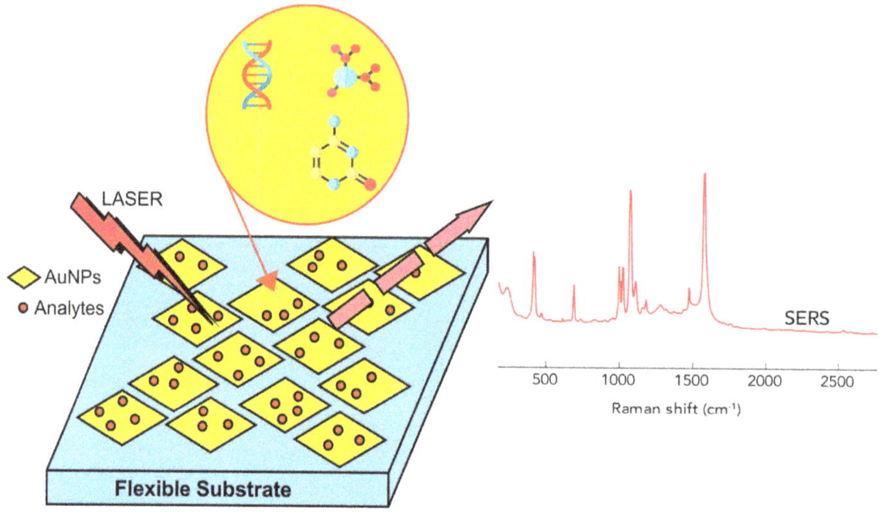

Figure 3. Surface-enhanced Raman spectroscopy (SERS) is a specialized kind of Raman spectroscopy, in which the analyte of interest interacts with gold or silver nanostructures to greatly boost the Raman signal. It allows Raman spectroscopy, a method typically used for identifying analytes, to be used for the detection of trace levels of potentially hazardous or physiologically significant substances.

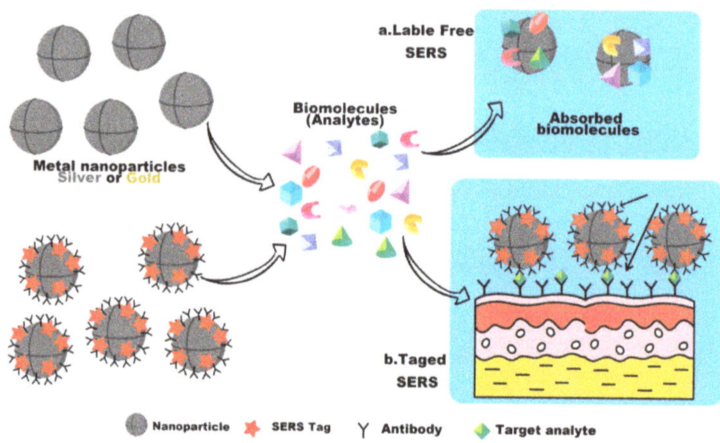

Figure 4. The figure illustrates the differences between a label-free SERS technique and a label-based SERS approach. In label-free SERS, the spectroscopic signal arises from all analytes that adsorb on the SERS substrate (direct detection), while in SERS, the spectroscopic signal results from the labels on a SERS tag that selectively attach to a target analyte (indirect detection).

4.2. Multiplex Analysis for Coronavirus Detection

Multiplex analysis is considered the gold standard for COVID-19 identification, because of its ability to replicate small amounts of the coronavirus' genetic material. Currently, multiplex testing for SARS-CoV-2 is usually performed on the specimens collected with a swab from the upper-respiratory tract. In addition, several studies have been performed using serum, feces, or ocular secretions [85]. Recently, the Rutgers Clinical Genomics Laboratory developed the (TaqPath COVID-19 Combo Kit) RT-PCR using saliva samples collected from the patient. This method is faster and less painful than other methods of sample collection, reduces potential risks to treatment personnel, and increases sample size [86]. As shown in Figure 5, multiplex analysis begins with the conversion of genomic viral RNA to DNA by DNA-dependent DNA polymerase (RNA-reverse transcriptase). This reaction is designed based on small DNA-primer sequences that specifically identify complementary sequences in the viral RNA genome for reverse transcriptase, to synthesize a short complementary DNA copy (cDNA) of the viral RNA. This is done using a fluorescent dye or a probe labeled with a fluorescent molecule and a quenching molecule, as in the TaqMan method. This is an automated system then repeats the amplification process for approximately 47 cycles, until the viral cDNA is normally detected by a fluorescent or electrical signal [87]. Multiplex analysis is usually performed in a one- or two-step procedure. In the one-step multiplex analysis of the entire reaction, from cDNA synthesis to propagation, the multiplex is performed in a single microtube. In the two-step method, cDNA synthesis and proliferation are performed in separate tubes. Although the two-step method is more flexible and sensitive than the one-step method and also requires fewer raw materials to initiate the reaction, the one-step method is the preferred method for detecting COVID-19. Since it can be started quickly and requires little sample management, the risks of pipetting and contamination during the reverse transcription and proliferation phases are lower [88]. To date, most molecular diagnostic experiments have used multiplex technology targeting different regions of the coronavirus genome, including the ORF1b or ORF8 regions and the nucleocapsid (N) genes, spike protein (S), DNA-dependent RNA polymerase (RdRp), and envelope protein (E). Assays targeting the E envelope protein have been shown to be similar to other coronavirus strains. On the other hand, the low similarity of the N, RdRp, and S genes of coronavirus to those of other bat viruses has made these genes specific targets for identification. Several methods have been used in laboratories to increase diagnostic sensitivity, by examining multiple genes simultaneously or identifying different regions in the same target gene. In general, multiplex methods with high sensitivity and specificity, as well as the ability to process a large number of samples, are the most common methods for detecting COVID-19. Therefore, considering that COVID-19 is likely to remain in the population like influenza viruses, a multiplex-testing method for multiple diseases should be considered as a routine test in the future [89].

4.3. SERS COVID-19 Detection

Currently, the analytical approach for diagnosis of the SARS-CoV-2 virus is specific detection by RT-PCR on respiratory specimens, such as nasopharyngeal swabs, oropharyngeal swabs, tracheal aspirates, or alveolar bronchial lavage, as shown in Figure 6. Samples of nasal secretions from both nostrils are collected with a large swab inserted into the nasopharynx by passing it around in a circle at least four times for a total of 15 s. In this approach, two different amplification sections of primers or RNA-dependent RNA polymerase (RpRd) genes, nucleoprotein (NP protein), and envelope protein (E protein) are used. However, the nucleic-acid test has the disadvantage of requiring a high risk of viral RNA degradation during collection, transport, and storage, as well as large amounts of good quality viral RNA (which varies greatly from person to person). Moreover, despite the high sensitivity of the PCR technique in patients with high clinical suspicion, a negative nucleic-acid test does not rule out SARS-CoV-2 infection [90]. When a negative nucleic-acid test result is detected once or twice, other diagnostic options such as serological ELISA tests for IgG and IgM antibodies, formed with bat coronavirus nucleoprotein, must be explored. One of the new alternative techniques used to quantify SARS-CoV-2 is SERS. Zhang et al. investigated the diagnosis of SARS-CoV-2 using SERS in combination with multivariate-statistical analysis. A functional receptor for the human coronavirus spike glycoprotein SARS-CoV-2 in its S1 component is named ACE-2. ACE-2, thus, performed a dual role as a reporter molecule and molecular-recognition element, as its Raman signal was suppressed at a stimulation wavelength of 780 nm after the recognition and binding of the receptor-binding domain (RBD) of the SARS-CoV-2 spike protein. The ACE-2@SN-SERS assay was used to analyze multiple real-water samples from hospitals and pipelines before and during various biological-wastewater-treatment methods, in situ and without pretreatment. A real-time polymerase-chain reaction (PCR) was previously used to detect the presence or absence of SARS-CoV-2 viruses in the samples [91].

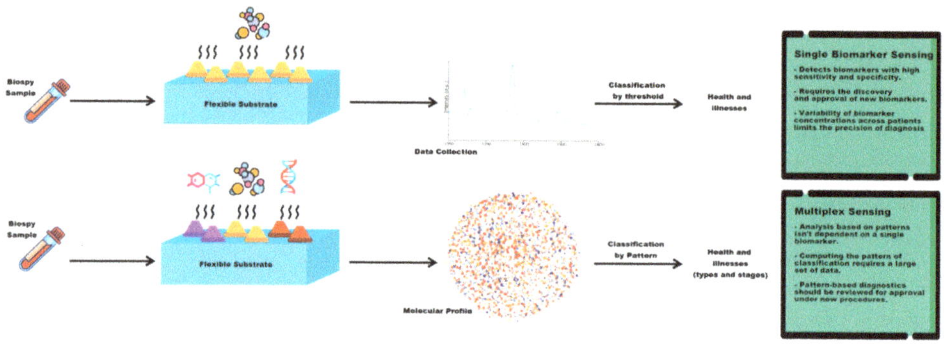

(a)

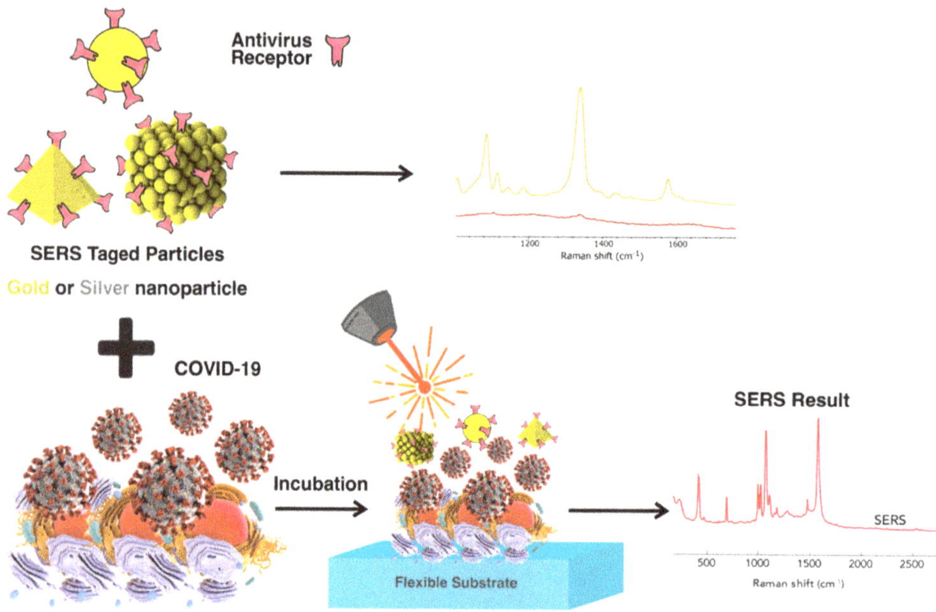

(b)

Figure 5. (**a**) Advancing from single−biomarker sensing to multiplex sensing. Diagnostic screening of patient−derived biopsies with multiplex sensors demonstrates high sensitivity and specificity in comparison of single sensors. (**b**) Principles of multiplexed detection using the surface−enhanced Raman scattering (SERS) technique, with metal nanoparticles that are different in size and shape and have unique Raman signals with narrow peaks, thus, they could be used as SERS tags. In addition, schematic illustration of simultaneous detection of COVID−19-associated antigen expressed by SERS imaging.

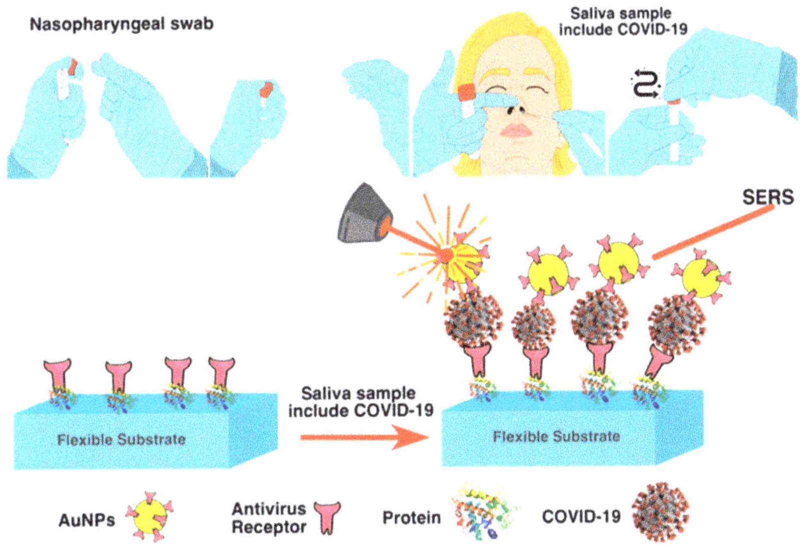

Figure 6. A big swab put into the nasopharynx is passed around in a circle at least four times for a total of 15 s to collect samples of nasal secretions from both nostrils, and mechanism of SERS for COVID-19 detection with saliva sample, which contains COVID-19 virus.

Based on the attenuation of the induced SERS signal in the presence of the virus, the detection of the SARS-CoV-2 spike protein in water samples is performed using two markers to classify positive and negative samples. The first indicator is based on the behavior of the band at 1182 cm^{-1}, which is associated with the amide II band, for the N-H bending and C-N stretching of the ACE-2 enzyme. In the presence of the SARS-CoV-2 protein, the band changes to 1189 cm^{-1}, which is due to a change in the structure of ACE-2 upon binding. Therefore, the ratio of Raman intensity at 1182 cm^{-1} to 1189 cm^{-1} (ratio 1189/1182) was used as a biomarker. The second indicator, which was determined by doing a principal-component analysis with linear-discriminant analysis (PCA-LDA) on the total spectral changes of the data, utilized the first LDA score (LD1 score) to distinguish the positive and negative groups. The results of the two indicators indicated that the detection of the SARS-CoV-2 spike protein was equivalent to that of real-time PCR, with the exception of the samples handled biologically where PCR failed to identify the virus, in contrast to the SERS results of the 1182/1189 ratios and the LD1 score. This was explained by the SARS-CoV-2 spike protein having greater resistance to disinfection compared to RNA. This approach resulted in either a positive or a negative finding; therefore, no LOD value was reported. On the other hand, the lack of a study investigating the selectivity against other ACE-2 target viruses or interference, as well as the fact that the assay reacts to viral coatings or free-spike proteins, which may lead to overestimation of the presence of SARS-CoV-2, are some limitations of this method [92,93]. SERS based on COVID-19 detection and its limitations are shown in Table 3.

Table 3. SERS based on COVID-19 detection and its limitations.

Target of Virus	Technique	Material Coating	Diagnosis of COVID-19 in Clinical	on Surfaces	Limit of Detection	Limitations	Ref.
COVID-19	SERS microfluid	Au/Ag	✓	✗	NA	• Weak signal relative to background. • Low sensitivity with low protein concentration. • Consumes time to collect sample. • Laser wavelength is unstable. • Mass transport challenge. • Heterogeneity of surface.	[12]
MERS	SERS-LSPR	Silver nanodot	✓	✗	1–10^6 nM		[94]
COVID-19	SERS	Gold nanoparticles	✓	✗	17.7 pM		[95]
COVID-19	SERS-LSPR	Silver nanodot	✓	✗	153.53, 230.37 pM		[96]
COVID-19/ spike protein	LSPR	AuNIs	✓	✗	0.22 ± 0.08 pM		[97]
COVID-19 RNA	Fluorescence	Gold	✓	✗	1000 TU mL^{-1}	• Collection of sample consumes time. • Low sample size.	[98]
Coronavirus/ N-protein	SPR	NA	✓	✗	2.17 nM	• Low selectivity. • A small perception depth. • Mass transport challenge. • Heterogeneity of surface. • Misinterpretation of data.	[99]

✓: Yes; ✗: No.

5. Future and Perspectives

Specifically, for the detection of COVID-19 virus, the preparation of a flexible SERS substrate is emphasized, and its substrate design is a critical factor because no SERS substrate is capable of detecting viruses. Therefore, there is no need to perform SERS testing in the BSL-3 laboratory using the appropriate SARS-CoV-2 pseudovirus. Reconciliation of positive and negative identification with subsequent IVD applications and EUA regulations is possible through the SERS system. To assess the high demand during the disease outbreak, flexible SERS substrates for virus detection should be manufactured on a large scale in the near future, due to the advancement of modular-design technology and high-performance production.

6. Conclusions

COVID-19 is a pandemic disease, and because progression of coronavirus infection can lead to severe respiratory problems and possibly death, there is an urgent need for various diagnostic strategies for early detection of the disease. The technique of culturing microorganisms and using a microscope are essential for a deep understanding of the new coronavirus and its relationship to host cells, which is considered one way to detect the virus. However, in the case of the COVID-19 pandemic, this method is very time-consuming and costly. Therefore, to identify different subtypes of the new SARS-CoV-2 coronavirus, the flexible SERS-substrates strategies used in this study were analyzed. According to the new SARS-CoV-2 virus, its quantification was performed using flexible SERS substrates along with multivariate-statistical analysis, by detecting the binding domain of the spike-protein receptor. In this regard, it can be concluded that indirect SERS procedures, especially those based on SIA, have good potential for gaining a foothold as a point-of-care technology for the detection of viral infectious diseases such as coronavirus. Likewise, new analytical strategies for the future are being understood, such as multiplex assays that combine the diagnosis of RVsZO with specific biomarkers or biomarkers of secondary disease due to the progression of infection. The above configuration can be used as a test to screen a patient's health status or as a dual-confirmatory test.

Author Contributions: S.M.M. and A.G. developed the idea and structure of the review article; V.R., S.A.H., and M.Y.K. wrote the manuscript collecting the materials from databases; N.O. and A.G.

revised and improved the manuscript; A.G. and W.-H.C. supervised the manuscript. All authors have read and agreed to the published version of the manuscript.

Funding: This work is sponsored by the Ministry of Science and Technology, Taiwan (grant numbers: MOST 110-2628-E-011-003, MOST 109-2923-E-011-003-MY and MOST 111-NU-E-011-001-NU).

Institutional Review Board Statement: Not applicable.

Informed Consent Statement: Not applicable.

Data Availability Statement: All data generated or analyzed during this study are included in this published article.

Conflicts of Interest: The authors declare no conflict of interest.

References

1. Kalashgarani, M.Y.; Babapoor, A. Application of nano-antibiotics in the diagnosis and treatment of infectious diseases. *Adv. Appl. NanoBio-Technol.* **2022**, *3*, 22–35.
2. Mousavi, S.M.; Hashemi, S.A.; Parvin, N.; Gholami, A.; Ramakrishna, S.; Omidifar, N.; Moghadami, M.; Chiang, W.-H.; Mazraedoost, S. Recent biotechnological approaches for treatment of novel COVID-19: From bench to clinical trial. *Drug Metab. Rev.* **2021**, *53*, 141–170. [CrossRef] [PubMed]
3. Negahdaripour, M.; Rahbar, M.R.; Mosalanejad, Z.; Gholami, A. Theta-Defensins to Counter COVID-19 as Furin Inhibitors: In Silico Efficiency Prediction and Novel Compound Design. *Comput. Math. Methods Med.* **2022**, *2022*, 9735626. [CrossRef] [PubMed]
4. Mousavi, S.M.; Hashemi, S.A.; Kalashgrani, M.Y.; Gholami, A.; Omidifar, N.; Babapoor, A.; Vijayakameswara Rao, N.; Chiang, W.-H. Recent Advances in Plasma-Engineered Polymers for Biomarker-Based Viral Detection and Highly Multiplexed Analysis. *Biosensors* **2022**, *12*, 286. [CrossRef]
5. Omidifar, N.; Bagheri Lankarani, K.; Moghadami, M.; Shokripour, M.; Chashmpoosh, M.; Mousavi, S.M.; Hashemi, S.A.; Gholami, A. Different laboratory diagnosis methods of COVID-19: A systematic review. *Arch. Clin. Infect. Dis.* **2021**, *16*, e110667. [CrossRef]
6. Asadi, K.; Gholami, A. Virosome-based nanovaccines; a promising bioinspiration and biomimetic approach for preventing viral diseases: A review. *Int. J. Biol. Macromol.* **2021**, *182*, 648–658. [CrossRef]
7. Wu, N.N. (Sensor Division Outstanding Achievement Award Address) Paper-Based Lateral Flow Strips as Point-of-Care Testing Tools for Detection of Biomarkers and Viruses. *ECS Meet. Abstr.* **2020**, *66*, 3328. [CrossRef]
8. Mousavi, S.M.; Hashemi, S.A.; Kalashgrani, M.Y.; Omidifar, N.; Bahrani, S.; Vijayakameswara Rao, N.; Babapoor, A.; Gholami, A.; Chiang, W.-H. Bioactive Graphene Quantum Dots Based Polymer Composite for Biomedical Applications. *Polymers* **2022**, *14*, 617. [CrossRef]
9. Stöckel, S.; Kirchhoff, J.; Neugebauer, U.; Rösch, P.; Popp, J. The application of Raman spectroscopy for the detection and identification of microorganisms. *J. Raman Spectrosc.* **2016**, *47*, 89–109. [CrossRef]
10. Mousavi, S.-M.; Nejad, Z.M.; Hashemi, S.A.; Salari, M.; Gholami, A.; Ramakrishna, S.; Chiang, W.-H.; Lai, C.W. Bioactive agent-loaded electrospun nanofiber membranes for accelerating healing process: A review. *Membranes* **2021**, *11*, 702. [CrossRef]
11. Sitjar, J.; Xu, H.-Z.; Liu, C.-Y.; Wang, J.-R.; Liao, J.-D.; Tsai, H.-P.; Lee, H.; Liu, B.H.; Chang, C.-W. Synergistic surface-enhanced Raman scattering effect to distinguish live SARS-CoV-2 S pseudovirus. *Anal. Chim. Acta* **2022**, *1193*, 339406. [CrossRef] [PubMed]
12. Jadhav, S.A.; Biji, P.; Panthalingal, M.K.; Krishna, C.M.; Rajkumar, S.; Joshi, D.S.; Sundaram, N. Development of integrated microfluidic platform coupled with Surface-enhanced Raman Spectroscopy for diagnosis of COVID-19. *Med. Hypotheses* **2021**, *146*, 110356. [CrossRef] [PubMed]
13. Sitjar, J.; Liao, J.-D.; Lee, H.; Tsai, H.-P.; Wang, J.-R.; Liu, P.-Y. Challenges of SERS technology as a non-nucleic acid or-antigen detection method for SARS-CoV-2 virus and its variants. *Biosens. Bioelectron.* **2021**, *181*, 113153. [CrossRef] [PubMed]
14. Mousavi, S.M.; Zarei, M.; Hashemi, S.A.; Ramakrishna, S.; Chiang, W.-H.; Lai, C.W.; Gholami, A.; Omidifar, N.; Shokripour, M. Asymmetric membranes: A potential scaffold for wound healing applications. *Symmetry* **2020**, *12*, 1100. [CrossRef]
15. Guo, J.; Chen, S.; Guo, J.; Ma, X. Nanomaterial labels in lateral flow immunoassays for point-of-care-testing. *J. Mater. Sci. Technol.* **2021**, *60*, 90–104. [CrossRef]
16. Mousavi, S.M.; Hashemi, S.A.; Zarei, M.; Bahrani, S.; Savardashtaki, A.; Esmaeili, H.; Lai, C.W.; Mazraedoost, S.; Abassi, M.; Ramavandi, B. Data on cytotoxic and antibacterial activity of synthesized Fe_3O_4 nanoparticles using Malva sylvestris. *Data Brief* **2020**, *28*, 104929. [CrossRef]
17. Taha, B.A.; Al Mashhadany, Y.; Bachok, N.N.; Bakar, A.A.; Hafiz Mokhtar, M.H.; Dzulkefly Bin Zan, M.S.; Arsad, N. Detection of COVID-19 virus on surfaces using photonics: Challenges and perspectives. *Diagnostics* **2021**, *11*, 1119. [CrossRef]
18. Larson, T.; Culbreath, K.; Chavez, D.; Larson, R.; Crossey, M.; Grenache, D.G. Modeling SARS-CoV-2 positivity using laboratory data: Timing is everything. *Clin. Chem.* **2020**, *66*, 981–983. [CrossRef]
19. Mousavi, S.M.; Low, F.W.; Hashemi, S.A.; Lai, C.W.; Ghasemi, Y.; Soroshnia, S.; Savardashtaki, A.; Babapoor, A.; Pynadathu Rumjit, N.; Goh, S.M. Development of graphene based nanocomposites towards medical and biological applications. *Artif. Cells Nanomed. Biotechnol.* **2020**, *48*, 1189–1205. [CrossRef]

20. McNay, G.; Eustace, D.; Smith, W.E.; Faulds, K.; Graham, D. Surface-enhanced Raman scattering (SERS) and surface-enhanced resonance Raman scattering (SERRS): A review of applications. *Appl. Spectrosc.* **2011**, *65*, 825–837. [CrossRef]
21. Mosier-Boss, P.A. Review of SERS substrates for chemical sensing. *Nanomaterials* **2017**, *7*, 142. [CrossRef] [PubMed]
22. Ahmadi, S.; Fazilati, M.; Nazem, H.; Mousavi, S.M. Green synthesis of magnetic nanoparticles using Satureja hortensis essential oil toward superior antibacterial/fungal and anticancer performance. *BioMed Res. Int.* **2021**, *2021*, 8822645. [CrossRef] [PubMed]
23. Lu, X.; Rycenga, M.; Skrabalak, S.E.; Wiley, B.; Xia, Y. Chemical synthesis of novel plasmonic nanoparticles. *Annu. Rev. Phys. Chem.* **2009**, *60*, 167–192. [CrossRef] [PubMed]
24. Solís, D.M.; Taboada, J.M.; Obelleiro, F.; Liz-Marzán, L.M.; García de Abajo, F.J. Optimization of nanoparticle-based SERS substrates through large-scale realistic simulations. *ACS Photonics* **2017**, *4*, 329–337. [CrossRef] [PubMed]
25. Abootalebi, S.N.; Mousavi, S.M.; Hashemi, S.A.; Shorafa, E.; Omidifar, N.; Gholami, A. Antibacterial effects of green-synthesized silver nanoparticles using *Ferula asafoetida* against *Acinetobacter baumannii* isolated from the hospital environment and assessment of their cytotoxicity on the human cell lines. *J. Nanomater.* **2021**, *2021*, 6676555. [CrossRef]
26. Yuan, Y.; Panwar, N.; Yap, S.H.K.; Wu, Q.; Zeng, S.; Xu, J.; Tjin, S.C.; Song, J.; Qu, J.; Yong, K.-T. SERS-based ultrasensitive sensing platform: An insight into design and practical applications. *Coord. Chem. Rev.* **2017**, *337*, 1–33. [CrossRef]
27. Abdollahifar, A.; Hashemi, S.A.; Mousavi, S.M.; Rahsepar, M.; Amani, A.M. Fabrication of graphene oxide-lead oxide epoxy based composite with enhanced chemical resistance, hydrophobicity and thermo-mechanical properties. *Adv. Polym. Technol.* **2018**, *37*, 3792–3803. [CrossRef]
28. Taranova, N.; Berlina, A.; Zherdev, A.; Dzantiev, B. 'Traffic light' immunochromatographic test based on multicolor quantum dots for the simultaneous detection of several antibiotics in milk. *Biosens. Bioelectron.* **2015**, *63*, 255–261. [CrossRef]
29. Kosović, M.; Balarin, M.; Ivanda, M.; Đerek, V.; Marciuš, M.; Ristić, M.; Gamulin, O. Porous silicon covered with silver nanoparticles as surface-enhanced Raman scattering (SERS) substrate for ultra-low concentration detection. *Appl. Spectrosc.* **2015**, *69*, 1417–1424. [CrossRef]
30. Khajehpour, K.J.; Williams, T.; Bourgeois, L.; Adeloju, S. Gold nanothorns–macroporous silicon hybrid structure: A simple and ultrasensitive platform for SERS. *Chem. Commun.* **2012**, *48*, 5349–5351. [CrossRef]
31. Wong, C.L.; Dinish, U.; Schmidt, M.S.; Olivo, M. Non-labeling multiplex surface enhanced Raman scattering (SERS) detection of volatile organic compounds (VOCs). *Anal. Chim. Acta* **2014**, *844*, 54–60. [CrossRef] [PubMed]
32. Jiao, Y.; Koktysh, D.S.; Phambu, N.; Weiss, S.M. Dual-mode sensing platform based on colloidal gold functionalized porous silicon. *Appl. Phys. Lett.* **2010**, *97*, 153125. [CrossRef]
33. Lai, Y.; Wang, J.; He, T.; Sun, S. Improved surface enhanced Raman scattering for nanostructured silver on porous silicon for ultrasensitive determination of 2,4,6-trinitrotoluene. *Anal. Lett.* **2014**, *47*, 833–842. [CrossRef]
34. Saviñon-Flores, F.; Méndez, E.; López-Castaños, M.; Carabarin-Lima, A.; López-Castaños, K.A.; González-Fuentes, M.A.; Méndez-Albores, A. A review on SERS-based detection of human virus infections: Influenza and coronavirus. *Biosensors* **2021**, *11*, 66. [CrossRef] [PubMed]
35. Wang, S.; Jiang, J.; He, X.; Yang, S.; Wu, H.; Qin, Z.; Chu, M.; Zhang, Z.; Liao, J.; Wang, X. Research progress of SERS on uranyl ions and uranyl compounds: A review. *J. Mater. Chem. C* **2022**, *10*, 4006–4018. [CrossRef]
36. Mousavi, S.M.; Hashemi, S.A.; Gholami, A.; Kalashgrani, M.Y.; Vijayakameswara Rao, N.; Omidifar, N.; Hsiao, W.W.-W.; Lai, C.W.; Chiang, W.-H. Plasma-Enabled Smart Nanoexosome Platform as Emerging Immunopathogenesis for Clinical Viral Infection. *Pharmaceutics* **2022**, *14*, 1054. [CrossRef]
37. Mosier-Boss, P.A. Review on SERS of Bacteria. *Biosensors* **2017**, *7*, 51. [CrossRef]
38. Kazemi, K.; Ghahramani, Y.; Kalashgrani, M.Y. Nano biofilms: An emerging biotechnology applications. *Adv. Appl. NanoBio-Technol.* **2022**, *3*, 8–15.
39. Natan, M.J. Concluding remarks surface enhanced Raman scattering. *Faraday Discuss.* **2006**, *132*, 321–328. [CrossRef]
40. Tian, L.; Morrissey, J.J.; Kattumenu, R.; Gandra, N.; Kharasch, E.D.; Singamaneni, S. Bioplasmonic paper as a platform for detection of kidney cancer biomarkers. *Anal. Chem.* **2012**, *84*, 9928–9934. [CrossRef]
41. Stamplecoskie, K.G.; Scaiano, J.C.; Tiwari, V.S.; Anis, H. Optimal size of silver nanoparticles for surface-enhanced Raman spectroscopy. *J. Phys. Chem. C* **2011**, *115*, 1403–1409. [CrossRef]
42. Joseph, V.; Matschulat, A.; Polte, J.; Rolf, S.; Emmerling, F.; Kneipp, J. SERS enhancement of gold nanospheres of defined size. *J. Raman Spectrosc.* **2011**, *42*, 1736–1742. [CrossRef]
43. Sen, T.; Patra, A. Resonance energy transfer from rhodamine 6G to gold nanoparticles by steady-state and time-resolved spectroscopy. *J. Phys. Chem. C* **2008**, *112*, 3216–3222. [CrossRef]
44. Mousavi, S.; Arjmand, O.; Hashemi, S.; Banaei, N. Modification of the epoxy resin mechanical and thermal properties with silicon acrylate and montmorillonite nanoparticles. *Polym. Renew. Resour.* **2016**, *7*, 101–113. [CrossRef]
45. Walters, C.M.; Pao, C.; Gagnon, B.P.; Zamecnik, C.R.; Walker, G.C. Bright Surface-Enhanced Raman Scattering with Fluorescence Quenching from Silica Encapsulated J-Aggregate Coated Gold Nanoparticles. *Adv. Mater.* **2018**, *30*, 1705381. [CrossRef] [PubMed]
46. Hashemi, S.A.; Mousavi, S.M.; Bahrani, S.; Gholami, A.; Chiang, W.-H.; Yousefi, K.; Omidifar, N.; Rao, N.V.; Ramakrishna, S.; Babapoor, A. Bio-enhanced polyrhodanine/graphene Oxide/Fe_3O_4 nanocomposite with kombucha solvent supernatant as ultra-sensitive biosensor for detection of doxorubicin hydrochloride in biological fluids. *Mater. Chem. Phys.* **2022**, *279*, 125743. [CrossRef]

47. Chen, X.; Jiang, C.; Yu, S. Nanostructured materials for applications in surface-enhanced Raman scattering. *CrystEngComm* **2014**, *16*, 9959–9973. [CrossRef]
48. Lou, B.; Li, T.-D.; Zheng, S.-F.; Su, Y.-Y.; Li, Z.-Y.; Liu, W.; Yu, F.; Ge, S.-X.; Zou, Q.-D.; Yuan, Q. Serology characteristics of SARS-CoV-2 infection after exposure and post-symptom onset. *Eur. Respir. J.* **2020**, *56*, 2000763. [CrossRef]
49. Xu, K.; Zhou, R.; Takei, K.; Hong, M. Toward flexible surface-enhanced Raman scattering (SERS) sensors for point-of-care diagnostics. *Adv. Sci.* **2019**, *6*, 1900925. [CrossRef]
50. Taubenberger, J.K.; Morens, D.M. 1918 Influenza: The mother of all pandemics. *Rev. Biomed.* **2006**, *17*, 69–79. [CrossRef]
51. Gholami, A.; Hashemi, S.A.; Yousefi, K.; Mousavi, S.M.; Chiang, W.-H.; Ramakrishna, S.; Mazraedoost, S.; Alizadeh, A.; Omidifar, N.; Behbudi, G. 3D nanostructures for tissue engineering, cancer therapy, and gene delivery. *J. Nanomater.* **2020**, *2020*, 1852946. [CrossRef]
52. Sarkar, P.; Bhui, D.K.; Bar, H.; Sahoo, G.P.; De, S.P.; Misra, A. Synthesis and photophysical study of silver nanoparticles stabilized by unsaturated dicarboxylates. *J. Lumin.* **2009**, *129*, 704–709. [CrossRef]
53. Israelsen, N. *Surface-Enhanced Raman Spectroscopy-Based Biomarker Detection for B-Cell Malignancies*; Utah State University: Logan, UT, USA, 2015.
54. Li, Y.-S.; Church, J.S. Raman spectroscopy in the analysis of food and pharmaceutical nanomaterials. *J. Food Drug Anal.* **2014**, *22*, 29–48. [CrossRef] [PubMed]
55. Berlina, A.N.; Taranova, N.A.; Zherdev, A.V.; Vengerov, Y.Y.; Dzantiev, B.B. Quantum dot-based lateral flow immunoassay for detection of chloramphenicol in milk. *Anal. Bioanal. Chem.* **2013**, *405*, 4997–5000. [CrossRef] [PubMed]
56. Aroca, R.; Alvarez-Puebla, R.; Pieczonka, N.; Sanchez-Cortez, S.; Garcia-Ramos, J. Surface-enhanced Raman scattering on colloidal nanostructures. *Adv. Colloid Interface Sci.* **2005**, *116*, 45–61. [CrossRef]
57. Do, Y.S.; Park, J.H.; Hwang, B.Y.; Lee, S.M.; Ju, B.K.; Choi, K.C. Plasmonic color filter and its fabrication for large-area applications. *Adv. Opt. Mater.* **2013**, *1*, 133–138. [CrossRef]
58. Moskovits, M. Surface-enhanced Raman spectroscopy: A brief perspective. In *Surface-Enhanced Raman Scattering*; Springer: Berlin/Heidelberg, Germany, 2006; pp. 1–17.
59. Mousavi, S.M.; Hashemi, S.A.; Zarei, M.; Gholami, A.; Lai, C.W.; Chiang, W.H.; Omidifar, N.; Bahrani, S.; Mazraedoost, S. Recent progress in chemical composition, production, and pharmaceutical effects of kombucha beverage: A complementary and alternative medicine. *Evid.-Based Complement. Altern. Med.* **2020**, *2020*, 4397543. [CrossRef]
60. Campion, A.; Kambhampati, P. Surface-enhanced Raman scattering. *Chem. Soc. Rev.* **1998**, *27*, 241–250. [CrossRef]
61. Creighton, J.A.; Eadon, D.G. Ultraviolet–visible absorption spectra of the colloidal metallic elements. *J. Chem. Soc. Faraday Trans.* **1991**, *87*, 3881–3891. [CrossRef]
62. Langhammer, C.; Yuan, Z.; Zorić, I.; Kasemo, B. Plasmonic properties of supported Pt and Pd nanostructures. *Nano Lett.* **2006**, *6*, 833–838. [CrossRef]
63. Lombardi, J.R.; Birke, R.L. A unified approach to surface-enhanced Raman spectroscopy. *J. Phys. Chem. C* **2008**, *112*, 5605–5617. [CrossRef]
64. Kumar, S.; Tokunaga, K.; Namura, K.; Fukuoka, T.; Suzuki, M. Experimental evidence of a twofold electromagnetic enhancement mechanism of surface-enhanced Raman scattering. *J. Phys. Chem. C* **2020**, *124*, 21215–21222. [CrossRef]
65. Bobosha, K.; Tjon Kon Fat, E.M.; van den Eeden, S.J.; Bekele, Y.; van der Ploeg-van Schip, J.J.; de Dood, C.J.; Dijkman, K.; Franken, K.L.; Wilson, L.; Aseffa, A. Field-evaluation of a new lateral flow assay for detection of cellular and humoral immunity against *Mycobacterium leprae*. *PLoS Negl. Trop. Dis.* **2014**, *8*, e2845. [CrossRef]
66. Schwartzberg, A.M.; Oshiro, T.Y.; Zhang, J.Z.; Huser, T.; Talley, C.E. Improving nanoprobes using surface-enhanced Raman scattering from 30-nm hollow gold particles. *Anal. Chem.* **2006**, *78*, 4732–4736. [CrossRef]
67. Kalachyova, Y.; Erzina, M.; Postnikov, P.; Svorcik, V.; Lyutakov, O. Flexible SERS substrate for portable Raman analysis of biosamples. *Appl. Surf. Sci.* **2018**, *458*, 95–99. [CrossRef]
68. Li, Z.; Huang, X.; Lu, G. Recent developments of flexible and transparent SERS substrates. *J. Mater. Chem. C* **2020**, *8*, 3956–3969. [CrossRef]
69. Bharati, M.S.S.; Soma, V.R. Flexible SERS substrates for hazardous materials detection: Recent advances. *Opto-Electron. Adv.* **2021**, *4*, 210048. [CrossRef]
70. Ma, P.; Liang, F.; Diao, Q.; Wang, D.; Yang, Q.; Gao, D.; Song, D.; Wang, X. Selective and sensitive SERS sensor for detection of Hg^{2+} in environmental water base on rhodamine-bonded and amino group functionalized SiO_2-coated Au–Ag core–shell nanorods. *RSC Adv.* **2015**, *5*, 32168–32174. [CrossRef]
71. Kneipp, K.; Wang, Y.; Kneipp, H.; Perelman, L.T.; Itzkan, I.; Dasari, R.R.; Feld, M.S. Single molecule detection using surface-enhanced Raman scattering (SERS). *Phys. Rev. Lett.* **1997**, *78*, 1667. [CrossRef]
72. Toderas, F.; Baia, M.; Baia, L.; Astilean, S. Controlling gold nanoparticle assemblies for efficient surface-enhanced Raman scattering and localized surface plasmon resonance sensors. *Nanotechnology* **2007**, *18*, 255702. [CrossRef]
73. Sauer, G.; Brehm, G.; Schneider, S. Preparation of SERS-active gold film electrodes via electrocrystallization: Their characterization and application with NIR excitation. *J. Raman Spectrosc.* **2004**, *35*, 568–576. [CrossRef]
74. Suresh, V.; Ding, L.; Chew, A.B.; Yap, F.L. Fabrication of large-area flexible SERS substrates by nanoimprint lithography. *ACS Appl. Nano Mater.* **2018**, *1*, 886–893. [CrossRef]

75. Wang, Z.; Li, M.; Wang, W.; Fang, M.; Sun, Q.; Liu, C. Floating silver film: A flexible surface-enhanced Raman spectroscopy substrate for direct liquid phase detection at gas–liquid interfaces. *Nano Res.* **2016**, *9*, 1148–1158. [CrossRef]
76. Jeong, J.W.; Arnob, M.M.P.; Baek, K.M.; Lee, S.Y.; Shih, W.C.; Jung, Y.S. 3D cross-point plasmonic nanoarchitectures containing dense and regular hot spots for surface-enhanced Raman spectroscopy analysis. *Adv. Mater.* **2016**, *28*, 8695–8704. [CrossRef]
77. Mitomo, H.; Horie, K.; Matsuo, Y.; Niikura, K.; Tani, T.; Naya, M.; Ijiro, K. Active gap SERS for the sensitive detection of biomacromolecules with plasmonic nanostructures on hydrogels. *Adv. Opt. Mater.* **2016**, *4*, 259–263. [CrossRef]
78. Chen, W.; Gui, X.; Zheng, Y.; Liang, B.; Lin, Z.; Zhao, C.; Chen, H.; Chen, Z.; Li, X.; Tang, Z. Synergistic effects of wrinkled graphene and plasmonics in stretchable hybrid platform for surface-enhanced Raman spectroscopy. *Adv. Opt. Mater.* **2017**, *5*, 1600715. [CrossRef]
79. Chen, J.; Huang, Y.; Kannan, P.; Zhang, L.; Lin, Z.; Zhang, J.; Chen, T.; Guo, L. Flexible and adhesive surface enhance Raman scattering active tape for rapid detection of pesticide residues in fruits and vegetables. *Anal. Chem.* **2016**, *88*, 2149–2155. [CrossRef]
80. Koh, E.H.; Mun, C.; Kim, C.; Park, S.-G.; Choi, E.J.; Kim, S.H.; Dang, J.; Choo, J.; Oh, J.-W.; Kim, D.-H. M13 bacteriophage/silver nanowire surface-enhanced raman scattering sensor for sensitive and selective pesticide detection. *ACS Appl. Mater. Interfaces* **2018**, *10*, 10388–10397. [CrossRef]
81. Chen, H.; Park, S.-G.; Choi, N.; Moon, J.-I.; Dang, H.; Das, A.; Lee, S.; Kim, D.-G.; Chen, L.; Choo, J. SERS imaging-based aptasensor for ultrasensitive and reproducible detection of influenza virus A. *Biosens. Bioelectron.* **2020**, *167*, 112496. [CrossRef]
82. Alipour, A.; Kalashgarani, M.Y. Nano Protein and Peptides for Drug Delivery and Anticancer Agents. *Adv. Appl. NanoBio-Technol.* **2022**, *3*, 60–64.
83. Wang, Y.; Zong, S.; Li, N.; Wang, Z.; Chen, B.; Cui, Y. SERS-based dynamic monitoring of minimal residual disease markers with high sensitivity for clinical applications. *Nanoscale* **2019**, *11*, 2460–2467. [CrossRef] [PubMed]
84. Zhou, P.; Yang, X.-L.; Wang, X.-G.; Hu, B.; Zhang, L.; Zhang, W.; Si, H.-R.; Zhu, Y.; Li, B.; Huang, C.-L. A pneumonia outbreak associated with a new coronavirus of probable bat origin. *Nature* **2020**, *579*, 270–273. [CrossRef] [PubMed]
85. Xia, J.; Tong, J.; Liu, M.; Shen, Y.; Guo, D. Evaluation of coronavirus in tears and conjunctival secretions of patients with SARS-CoV-2 infection. *J. Med. Virol.* **2020**, *92*, 589–594. [CrossRef] [PubMed]
86. Takeuchi, Y.; Furuchi, M.; Kamimoto, A.; Honda, K.; Matsumura, H.; Kobayashi, R. Saliva-based PCR tests for SARS-CoV-2 detection. *J. Oral Sci.* **2020**, *62*, 350–351. [CrossRef]
87. VanGuilder, H.D.; Vrana, K.E.; Freeman, W.M. Twenty-five years of quantitative PCR for gene expression analysis. *Biotechniques* **2008**, *44*, 619–626. [CrossRef]
88. Wong, M.L.; Medrano, J.F. Real-time PCR for mRNA quantitation. *Biotechniques* **2005**, *39*, 75–85. [CrossRef]
89. Carter, L.J.; Garner, L.V.; Smoot, J.W.; Li, Y.; Zhou, Q.; Saveson, C.J.; Sasso, J.M.; Gregg, A.C.; Soares, D.J.; Beskid, T.R. Assay techniques and test development for COVID-19 diagnosis. *ACS Cent. Sci.* **2020**, *6*, 591–605. [CrossRef]
90. Kobayashi, T.; Jung, S.-M.; Linton, N.M.; Kinoshita, R.; Hayashi, K.; Miyama, T.; Anzai, A.; Yang, Y.; Yuan, B.; Akhmetzhanov, A.R. Communicating the risk of death from novel coronavirus disease (COVID-19). *J. Clin. Med.* **2020**, *9*, 580. [CrossRef]
91. Li, Y.; Lin, C.; Peng, Y.; He, J.; Yang, Y. High-sensitivity and point-of-care detection of SARS-CoV-2 from throat and nasal swabs by magnetic SERS biosensor. *Sens. Actuators B Chem.* **2022**, *365*, 131974. [CrossRef]
92. Espejo, A.P.; Akgun, Y.; Al Mana, A.F.; Tjendra, Y.; Millan, N.C.; Gomez-Fernandez, C.; Cray, C. Review of current advances in serologic testing for COVID-19. *Am. J. Clin. Pathol.* **2020**, *154*, 293–304. [CrossRef]
93. Hwang, Y.; Ferhan, A.R.; Yoon, B.K.; Sut, T.N.; Jeon, W.-Y.; Koo, D.J.; Jackman, J.A.; Cho, N.-J. Surface engineering of plasmonic gold nanoisland platforms for high-sensitivity refractometric biosensing applications. *Appl. Mater. Today* **2022**, *26*, 101280. [CrossRef]
94. Kim, H.; Hwang, J.; Kim, J.H.; Lee, S.; Kang, M. Sensitive detection of multiple fluoresence probes based on surface-enhanced raman scattering (sers) for mers-cov. In Proceedings of the 2019 IEEE 14th International Conference on Nano/Micro Engineered and Molecular Systems (NEMS), Bangkok, Thailand, 11–14 April 2019; pp. 498–501.
95. Yang, Y.; Peng, Y.; Lin, C.; Long, L.; Hu, J.; He, J.; Zeng, H.; Huang, Z.; Li, Z.-Y.; Tanemura, M. Human ACE2-functionalized gold "virus-trap" nanostructures for accurate capture of SARS-CoV-2 and single-virus SERS detection. *Nano-Micro Lett.* **2021**, *13*, 109. [CrossRef] [PubMed]
96. Kim, E.J.; Kim, H.; Park, E.; Kim, T.; Chung, D.R.; Choi, Y.-M.; Kang, M. Based Multiplex Surface-Enhanced Raman Scattering Detection Using Polymerase Chain Reaction Probe Codification. *Anal. Chem.* **2021**, *93*, 3677–3685. [CrossRef] [PubMed]
97. Qiu, G.; Gai, Z.; Tao, Y.; Schmitt, J.; Kullak-Ublick, G.A.; Wang, J. Dual-functional plasmonic photothermal biosensors for highly accurate severe acute respiratory syndrome coronavirus 2 detection. *ACS Nano* **2020**, *14*, 5268–5277. [CrossRef] [PubMed]
98. Wang, D.; He, S.; Wang, X.; Yan, Y.; Liu, J.; Wu, S.; Liu, S.; Lei, Y.; Chen, M.; Li, L. Rapid lateral flow immunoassay for the fluorescence detection of SARS-CoV-2 RNA. *Nat. Biomed. Eng.* **2020**, *4*, 1150–1158. [CrossRef]
99. Chen, H.; Gill, A.; Dove, B.K.; Emmett, S.R.; Kemp, C.F.; Ritchie, M.A.; Dee, M.; Hiscox, J.A. Mass spectroscopic characterization of the coronavirus infectious bronchitis virus nucleoprotein and elucidation of the role of phosphorylation in RNA binding by using surface plasmon resonance. *J. Virol.* **2005**, *79*, 1164–1179. [CrossRef]

MDPI
St. Alban-Anlage 66
4052 Basel
Switzerland
Tel. +41 61 683 77 34
Fax +41 61 302 89 18
www.mdpi.com

Biosensors Editorial Office
E-mail: biosensors@mdpi.com
www.mdpi.com/journal/biosensors

www.ingramcontent.com/pod-product-compliance
Lightning Source LLC
Chambersburg PA
CBHW040808250926
51850CB00007BA/1108
9783036572840